£10.95

COMPLEXES AND TRANSITION EL

A Macmillan Chemistry Text

Consulting Editor: Dr Peter Sykes, University of Cambridge

Other Titles of Related Interest

THE HEAVY TRANSITION ELEMENTS: *S. A. Cotton and F. A. Hart*

COMPLEXES AND FIRST-ROW TRANSITION ELEMENTS

DAVID NICHOLLS

Senior Lecturer in the Department of Inorganic, Physical and Industrial Chemistry, University of Liverpool

MACMILLAN

First published 1974
Reprinted 1979, 1986, 1987, 1988, 1989

Published by
MACMILLAN EDUCATION LTD
Houndmills, Basingstoke, Hampshire RG21 2XS
and London
Companies and representatives
throughout the world

Printed in Hong Kong

ISBN 0-333-17088-1

To Wendy

Preface

Transition-metal chemistry is very largely concerned with the chemistry of complexes and so it is fitting therefore that both subjects should be dealt with in the same textbook. We deal here with the general aspects of complexes first and then with the specific chemistry of the first-row transition elements. For several reasons this book deals only with the chemistry of the first-row transition elements at a level suitable for first- or second-year undergraduates in British universities. The chief of these is that the first member of each group of transition elements differs significantly in properties from the subsequent members, and thus vertical comparisons within the groups are less valuable than horizontal ones. Further, the explanations of the magnetic and spectral properties are more simply dealt with in the first-row transition elements and the aqueous chemistry can be systematised more easily. The second- and third-row transition elements are covered in a companion volume (S. A. Cotton and F. A. Hart: *The Heavy Transition Elements*), which also gives a more detailed treatment of organometallic compounds and the bonding therein (as well as other topics such as metal carbonyls and metal–metal bonding) at a level more appropriate to final-year honours students in British universities.

In the present text a little basic physical chemistry is assumed, in particular the concepts of atomic orbitals, elementary kinetics and thermodynamics and electrode potentials. Throughout the text emphasis is placed on the properties and reactions of metal aquo-ions and the donor–acceptor properties of transition-metal compounds. In the author's opinion a knowledge of these concepts can give the student a general understanding of the more elementary reactions of transition-metal compounds.

Liverpool   David Nicholls
1974

Contents

1 The Development of Co-ordination Chemistry

1.1 Introduction to Complexes

The chemistry of the transition elements is very largely concerned with the chemistry of co-ordination compounds. Co-ordination compounds or, as they are perhaps more usually called, *complexes* play a very important part in our lives today; the study of them has contributed greatly to our understanding of the chemical bond and of inorganic chemistry as a whole. The number of possible co-ordination compounds in chemistry is almost infinite. Present-day research workers prepare many new complexes every week; most of these will be of academic interest only, but just a few will be of considerable economic importance as well. For many years complexes were regarded as of interest to the theoretical and inorganic chemist only, but now they are playing vital roles in analytical chemistry, in the synthesis of organic chemicals, in polymerisation processes, and in our understanding of biological processes. Perhaps the most important are the naturally occurring complexes such as chlorophyll, the magnesium complex that is important in plant photosynthesis, and haemoglobin, the iron complex that is a carrier of oxygen in blood.

What do we understand by the term 'complexes'? It is difficult to define the term in a formal way. However, as a guide we can say that a complex is formed when a number of ions or molecules combine with a central atom to form an entity in which the number of atoms directly attached to the central atom exceeds the normal covalency (oxidation state) of this atom. Complexes may be neutral (no charge), cationic (positively charged), or anionic (negatively charged). Thus silver chloride (not a complex) dissolves in ammonia solution to form the ion $[Ag(NH_3)_2]^+$ which is a complex because the number of bonded groups exceeds the +1 oxidation state of silver. The reader will almost certainly have come across this and many other reactions in which complex ions are formed. The pretty blue solution of copper(II) sulphate, which we meet very early in our experience of chemistry, is a solution of complex ions formed by the reaction

$$CuSO_4 + H_2O \text{ (excess)} \rightarrow [Cu(H_2O)_6]^{2+} + SO_4{}^{2-} \text{ (aq.)}$$
white   blue

This complex ion is responsible for the blue colour of the solution; indeed most of the familiar blue copper(II) salts are complexes containing co-ordinated water molecules. So we begin to see that complex formation has a marked effect on the properties (for example solubility and colour) of simple salts. Some other well-known complex compounds are listed in Table 1.1.

The groups bonded to the central atom are called *ligands* or donor molecules; in the formation of a complex, the ligands are often said to donate a pair of electrons to the metal or other central atom. Complexes are thus the products of co-ordinate bond formation. It is because water is a ligand that it is impossible to place a transition-metal ion in aqueous solution without complex formation

TABLE 1.1   SOME COMPLEX COMPOUNDS

'Simple' compound or ion	Ligand	Complex
BF_3	NMe_3	$Me_3N.BF_3$
$AgCl$	Cl^-	$[AgCl_2]^-$
Cu^{2+}	NH_3	$[Cu(NH_3)_6]^{2+}$
Al^{3+}	F^-	$[AlF_6]^{3-}$
Ni^{2+}	H_2O	$[Ni(H_2O)_6]^{2+}$
Co^{3+}	NO_2^-	$[Co(NO_2)_6]^{3-}$

occurring; if no other co-ordinating groups are present, this complex will usually be the hexa-aquo-ion $[M(H_2O)_6]^{n+}$. In this connexion the reader should notice that the solvents have not been specified for the 'reactions' cited in Table 1.1. Reaction of the Cu^{2+} ion with ammonia only (that is in liquid ammonia) yields the hexammine-copper(II) cation, while in concentrated aqueous solutions of ammonia there is competition between the ammonia and water molecules for the Cu^{2+} ion; the composition $[Cu(NH_3)_4(H_2O)_2]^{2+}$ (sometimes abbreviated to $[Cu(NH_3)_4]^{2+}$) of the so-called cupritetrammine ion is largely reached.

We can further qualify our description of complex compounds to distinguish them from double salts. In a complex ion, the anion or cation concerned is actually present as a discrete entity either in solution or in the solid state or in both. In double salts, the individuality of the constituent salts is maintained. For example, potash alum $KAl(SO_4)_2.12H_2O$ reacts chemically as a mixture of potassium sulphate and aluminium sulphate and does not contain or show reactions of the $[Al(SO_4)_2]^-$ ion. However, there is a complex ion within this double salt. Potash alum contains the ion $[Al(H_2O)_6]^{3+}$ and like aluminium sulphate it gives this ion in aqueous solution. A more instructive way to write the formula for potash alum is thus $K^+[Al(H_2O)_6]^{3+}(SO_4^{2-})_2.6H_2O$.

1.2   The Historical Development of Co-ordination Chemistry

The first complexes to be the subject of any great study were the cobaltammines. As early as 1798 Tassaert observed that a solution of a cobalt(II) salt in aqueous ammonia becomes brown on exposure to air, the colour changing to wine red on boiling. Some decades later, Frémy showed that the cobalt had been oxidised to cobalt(III) and that the new salt is associated with up to six ammonia molecules, for example $CoCl_3.6NH_3$. This compound stimulated interest because it was difficult to understand how the two compounds $CoCl_3$ and NH_3, each with its valency satisfied, could combine together to form a stable compound. Many more complexes of ammonia were prepared subsequently. The reactions of nickel chloride and copper sulphate solutions with an excess of ammonia were found to give deeply coloured solutions from which the purple $NiCl_2.6NH_3$ and the deep blue $CuSO_4.4NH_3.H_2O$ could be isolated. However, it was with cobalt(III) and platinum(II) that a wide variety of ammonia complexes (ammines) could be prepared, and it was the study of the properties of these series of ammines that was to shed much light on the structures of co-ordination compounds.

While their structures were unknown, these new compounds were often named according to their colour. Thus the orange-yellow $CoCl_3.6NH_3$ was called *luteocobaltic chloride,* the purple $CoCl_3.5NH_3$ *purpureocobaltic chloride,* and the green isomer of $CoCl_3.4NH_3$ *praseocobaltic chloride.* Other complexes were named after their discoverers, for example Magnus' green salt $PtCl_2.2NH_3$ (now formulated as $[Pt(NH_3)_4[PtCl_4])$ and Reinecke's salt $Cr(SCN)_3.NH_4SCN.2NH_3$ (now formulated as $NH_4[Cr(NH_3)_2(NCS)_4])$. Around 1830 Zeise found that ethylene could form compounds with platinum salts, and he was able to isolate complexes of formulae $PtCl_2.C_2H_4$ and $PtCl_2.KCl.C_2H_4$. These compounds were the first organometallic compounds of the transition elements to be prepared; their structures have been elucidated only fairly recently and the potassium compound (now formulated as $K[PtCl_3(C_2H_4)])$ is still often called Zeise's salt. Today complexes are named systematically; rules for naming them are given at the end of this chapter.

1.2.1   The Werner Theory

Our understanding of co-ordination compounds dates from the time of Alfred Werner. Werner presented his first paper on this subject in 1891 at the age of twenty-five; in 1913 he received the Nobel prize for chemistry. Prior to Werner, other theories had been put forward to explain the structures of complex compounds. One such theory was the *chain theory* in which, for example, $CoCl_3.6NH_3$ was formulated as

$$Co \overset{\diagup NH_3-Cl}{\underset{\diagdown NH_3-Cl}{-\!\!-NH_3-\!\!-NH_3-\!\!-NH_3-\!\!-NH_3-\!\!-Cl}}$$

This theory followed naturally from the well-known catenation of carbon in aliphatic chains, but such structures were soon shown to be incorrect. It will be appropriate to discuss here some of the experimental evidence that was available at the time of Werner.

1.2.2   Ionisable Chloride

When the cobaltammine chlorides are treated with silver nitrate solution, they do not all behave similarly; all the chlorine present in the complexes is not always precipitated instantly and quantitatively at room temperature. This statement is elaborated in Table 1.2. In $CoCl_3.5NH_3$, for example, two-thirds of the total chlorine present is precipitated by silver nitrate solution at room temperature; this is known as the ionisable chloride. The other one-third can be liberated only by breaking down the complex, for example by boiling with sodium hydroxide solution; this chlorine is known as the nonionisable chloride. From these facts Werner concluded that the nonionisable chlorines are bonded covalently to the cobalt. The ionisable chlorides are just the number of free chloride ions required to balance the charge on the cobalt cation as a whole; hence the formulation of these complexes as in Table 1.2. The cobaltammine $CoCl_3.3NH_3$ would then be expected to have

TABLE 1.2   CHLORIDE ION PRECIPITATION FROM COBALTAMMINES

Complex	No. of Cl⁻ ions precipitated	Present formulation
$CoCl_3.6NH_3$	3	$[Co(NH_3)_6]Cl_3$
$CoCl_3.5NH_3$	2	$[Co(NH_3)_5Cl]Cl_2$
$CoCl_3.4NH_3$	1	$[Co(NH_3)_4Cl_2]Cl$

the structure $[CoCl_3(NH_3)_3]$ with no ionisable chlorides. Unfortunately this compound cannot be prepared but the iridium analogue $IrCl_3.3NH_3$ is found to have no ionisable chlorides. This was an important result in showing the weakness of the chain theory, since it would predict the same number of ionisable chlorides (two) for $CoCl_3.4NH_3$ and $IrCl_3.3NH_3$, namely

$$Co \diagup_{\diagdown Cl}^{Cl} \!\!\!\!-\!\!-\!\!- NH_3NH_3NH_3NH_3Cl \qquad\qquad Ir \diagup_{\diagdown Cl}^{Cl} \!\!\!\!-\!\!-\!\!- NH_3NH_3NH_3Cl$$

1.2.3   Molar Conductivity

A further method of testing the degree of ionisation of a complex is the measurement of its electrical conductivity in solution. The more ions that a complex liberates in solution, the greater the conductivity; a complex that does not ionise will have negligible conductivity. For comparison purposes, the molar conductivities (Λ_M) are most useful. By measuring the molar conductivities of some simple salts, the relationship between the number of ions present per molecule and the conductivity can be established. Thus Λ_M values in water (at 10^{-3} M) are about 120 ohm^{-1} cm^2 for salts giving two ions (for example NaCl), about 260 ohm^{-1} cm^2 for three ions (for example $BaCl_2$), and about 400 ohm^{-1} cm^2 for four ions (for example $CeCl_3$). Table 1.3 shows the molar conductivities and the consequent formulations of some platinum ammines.

TABLE 1.3   CONDUCTIVITIES OF SOME PLATINUM AMMINES

Complex	Λ_M (ohm^{-1} cm^2)	No. of ions	Present formulation
$PtCl_4.6NH_3$	523	5	$[Pt(NH_3)_6]^{4+}\,4Cl^-$
$PtCl_4.5NH_3$	404	4	$[Pt(NH_3)_5Cl]^{3+}\,3Cl^-$
$PtCl_4.4NH_3$	229	3	$[Pt(NH_3)_4Cl_2]^{2+}\,2Cl^-$
$PtCl_4.3NH_3$	97	2	$[Pt(NH_3)_3Cl_3]^+\,Cl^-$
$PtCl_4.2NH_3$	0	0	$[Pt(NH_3)_2Cl_4]$

1.2.4   Werner's Postulates

Not only did Werner suggest the formulations that are considered correct today for the cobaltammines, as for example in Table 1.2, but he went on to develop a theory that is still applicable without essential alteration. We may summarise Werner's postulates as follows.

(1)   Metals possess two types of valency, the primary (ionisable) valency and the secondary (nonionisable) valency. We now call these *oxidation state* and *co-ordination number* respectively.

(2)   Primary valencies are satisfied by negative ions, and secondary valencies by negative ions or neutral molecules, both valencies being satisfied.

(3)   The secondary valencies are directed to fixed positions in space about the central metal ion.

The first two postulates give a straightforward rationalisation of the structural formulae of complex compounds. Thus in the cobaltammines (Table 1.2) the primary valency of cobalt is in each case three and the secondary valency is six. The second postulate is exemplified by the formulations for the ammines in Tables 1.2 and 1.3. In every case the primary and secondary valencies are satisfied; for the platinum ammines the primary valency is four but the secondary valency is still six. We shall see that many metals have a preferred co-ordination number of six, that is they prefer to have six groups (ligands) directly attached to the metal atom; these six ligands are said to be in the co-ordination sphere of the metal. The ligands in the co-ordination sphere are bonded to the metal by co-ordinate bonds whereas the ions outside the co-ordination sphere are bonded ionically to the co-ordination sphere as a whole. Thus, when $[Co(NH_3)_5 Cl] Cl_2$ is dissolved in water, the ions $[Co(NH_3)_5 Cl]^{2+}$ and Cl^- are present in the solution. The addition of silver nitrate precipitates the free Cl^- ions but the chlorine bonded co-ordinately in the $[Co(NH_3)_5 Cl]^{2+}$ cation is not readily available as chloride ion.

The third postulate was perhaps the most significant but it was the most difficult to prove. It predicted the occurrence of various types of isomers (compounds having the same empirical formula but different structures), and it took Werner nearly twenty years to prove this by the resolution of an inorganic complex into its predicted optical isomers. The significance of the third postulate lies in its contribution to the development of inorganic stereochemistry. The fact that so many of the then known complexes had a co-ordination number of six raised the important question of how the six ligands were distributed spatially around the metal ion. Before the advent of direct methods for structural determination, for example by use of X-rays, the choice of structure was made on the basis of the number of stereoisomers obtainable.

Let us consider the ways in which six groups might be arranged spatially around a metal. The most symmetrical structures are those illustrated in Figure 1.1. Now let us consider the number of isomers that each of these would give for a compound of formula $MA_4 B_2$. The hexagonal planar structure (Figure 1.1a) would give three stereoisomers (equivalent to the benzene *ortho-, meta-,* and *para-*isomers) with the B groups in, for example, positions 1,2, 1,3, or 1,4. Similarly the trigonal prismatic structure (Figure 1.1b) would give rise to three isomers with B groups in, for example, 1,2, 1,4, and 1,5 positions. However, the octahedral structure (Figure 1.1c) can give rise to only two isomers, with B groups in, for example, 1,2 *(cis)* or 1,6 *(trans)* positions. For numerous complexes of the formula $MA_4 B_2$, only two isomers were in fact found; in no case could three isomers be isolated, so the evidence indicated that the octahedral structure was correct. Prior to confirmation

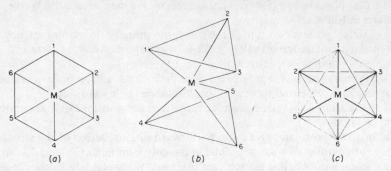

Figure 1.1   Arrangements of six groups around a metal: (*a*) hexagonal planar; (*b*) trigonal prismatic; (*c*) octahedral

of this by X-ray analysis, Werner set about proving the octahedral structure in a more positive way than that so far described. This involved the resolution into optical isomers of complexes of the type M(AA)₃ where AA is a chelating ligand (see chapter 4). Neither the hexagonal planar nor the trigonal prismatic structure would give optical isomers with these complexes, so the octahedral structure was proved to be correct.

1.3   Nomenclature of Co-ordination Compounds

Because of the enormous number and wide variety of co-ordination compounds it is important to have a systematic way of naming compounds unambiguously. Fortunately, definitive rules for the nomenclature of inorganic compounds were laid down in 1958 by a commission of the International Union of Pure and Applied Chemistry; these *IUPAC rules* (with very minor modifications) are now in widespread use and are the basis of the nomenclature presented here.

Perhaps the biggest advance in recent years has been the general acceptance of the Stock notation. In this, the oxidation state of a metal is designated by a Roman numeral in parentheses immediately after its name; the endings -ous, -ic, -yl, etc., are thus rendered obsolete. A few examples of the names of simple compounds will suffice to show the advantages of this method. The old problem to the student beginning chemistry of whether ferrous ion is di- or tri-valent now no longer arises; iron(II) chloride is clearly $FeCl_2$. Other simple examples are given in Table 1.4.

TABLE 1.4   STOCK NOTATION COMPARED WITH OLDER NOMENCLATURE

Formula	Stock notation	Older name
$CuSO_4$	copper(II) sulphate	cupric sulphate
Hg_2Cl_2	mercury(I) chloride	mercurous chloride
WO_3	tungsten(VI) oxide	tungstic oxide
$VOCl_2$	vanadium(IV) oxide dichloride	vanadyl chloride
$VOCl_3$	vanadium(V) oxide trichloride	vanadyl trichloride

There is no need to use the notation in unambiguous cases such as sodium chloride or calcium sulphate, and, particularly for other than transition elements, the prefixes mono-, di-, tri-, tetra-, etc. are frequently used to denote the stoichiometry instead of using oxidation states, for example carbon monoxide and sulphur trioxide.

For complex compounds, as with simple salts, the cation is named first followed by the anion. The ligands are arranged in front of the central metal atom, negative ones first then neutral ones; the oxidation state is written immediately after the metal as in naming the simple salts. The number of each type of ligand is specified by the Greek prefixes mono-, di-, tri-, tetra-, penta-, hexa-, hepta-, and octa-. It will be noticed that negative ligands end in -o, for example chloro (Cl^-) and cyano (CN^-), while neutral ligands keep their usual names except for water which is *aquo* and ammonia which is *ammine*.

Anionic complexes. If the complex is in an anionic form, the ending -ate is attached to the name of the metal and this is followed by the oxidation state according to the Stock notation, for example

K_2TiCl_6	potassium hexachlorotitanate(IV)
$Na_3Co(NO_2)_6$	sodium hexanitrocobaltate(III)
$K_4Fe(CN)_6$	potassium hexacyanoferrate(II)
$NaMn(CO)_5$	sodium pentacarbonylmanganate(−I)

The names of five metals change when they are in an anionic form; we have seen the example above of iron which becomes ferrate. The others are copper, which becomes cuprate, gold which becomes aurate, lead which becomes plumbate, and tin which becomes stannate. It is not necessary to specify the number of cations in the complexes named above; for example, we do not say 'dipotassium hexachlorotitanate(IV)' because, if the complex has $TiCl_6{}^{2-}$ anions, it must have two K^+ ions to balance the charges.

Cationic complexes. These have the cation, named according to the aforementioned rules, followed by a space and then the name of the anion, for example

$[Co(NH_3)_6]Cl_3$	hexa-amminecobalt(III) chloride
$[CoCl(NH_3)_5]Cl_2$	chloropenta-amminecobalt(III) chloride
$[CrCl_2(H_2O)_4]Cl$	dichlorotetra-aquochromium(III) chloride

Neutral complexes. These are named in one word, for example

$[PtCl_2(NH_3)_2]$	dichlorodiammineplatinum(II)
$[Co(NO_2)_3(NH_3)_3]$	trinitrotriamminecobalt(III)

Complicated ligands. When the ligand has the prefix di- or tri- in its name, for example diethyl ether or triphenylphosphine, the number of ligand molecules present in the complex is denoted by the prefixes bis-, tris-, tetrakis-, pentakis-, hexakis-, etc., and the name of the ligand is placed in parentheses, for example

$TiCl_4(Et_2O)_2$	tetrachlorobis(diethyl ether)titanium(IV)
$Ni(CO)_2(Ph_3P)_2$	dicarbonylbis(triphenylphosphine)nickel
$[Fe(en)_3][Fe(CO)_4]$	tris(ethylenediamine)iron(II) tetracarbonylferrate(−II)

These prefixes are also used when any ambiguity might arise; for example, the compound $SiCl_4(NH_2Me)_2$ might be wrongly named tetrachloro-dimethylaminesilicon(IV) whereas the name tetrachlorobis(methylamine)silicon(IV) is unambiguous.

Organic ligands that lose a proton when they react with a metal ion are treated as anionic and give the ending -ato, for example

$Ni(C_4H_7O_2N_2)_2$ bis(dimethylglyoximato)nickel(II)

$CoCl_2(C_4H_8O_2N_2)_2$ dichlorobis(dimethylglyoxime)cobalt(II)

$$\left(\text{dimethylglyoxime} = \begin{array}{c} CH_3C=NOH \\ | \\ CH_3C=NOH \end{array} ; \text{dimethylglyoximato} = \begin{array}{c} CH_3C=N-O^- \\ | \\ CH_3C=NOH \end{array} \right)$$

$Cr(acac)_3$ tris(acetylacetonato)chromium(III)

(acetylacetone = $CH_3COCH_2COCH_3$; acetylacetonato = $CH_3CO\bar{C}HCOCH_3$)

Bridging groups.   A bridging group is indicated by placing the Greek letter μ immediately before its name, for example

$[(NH_3)_5Cr-OH-Cr(NH_3)_5]Cl_5$ μ-hydroxo-bis[penta-amminechromium(III)] chloride.

$(CO)_3Fe(CO)_3Fe(CO)_3$ tri-μ-carbonyl-bis(tricarbonyliron).

2 Lewis Acids and Bases

2.1 The Simple Lewis Approach

We saw in chapter 1 that complex compounds are often formed by direct reaction between a 'simple' molecule or ion and a ligand. How do we know which 'simple' substances will react with ligands? One way of dealing with this problem is to use the theory of acids and bases put forward by G.N. Lewis in 1938. Lewis called those substances capable of acting as ligands *bases* and the substances with which these react *acids*. The formal definitions may be summarised as follows

Lewis acid − an acceptor of an electron pair
Lewis base − a donor of an electron pair

The reaction between a Lewis base and a Lewis acid is thus a donor−acceptor reaction involving the formation of a co-ordinate bond and, more often than not, a complex compound. The importance of the word 'pair' in the definitions must be stressed. Lewis acids are not necessarily oxidising agents; these become reduced by accepting any number of electrons but Lewis acids can accept only pairs of electrons, and these pairs are normally not transferred completely to the acid but are used in the formation of covalent bonds. The terms *electrophile* and *nucleophile*, as used in organic chemistry, are synonymous with Lewis acid and Lewis base respectively. The more important types of species that are capable of acting as Lewis acids and bases are listed in Tables 2.1 and 2.2.

TABLE 2.1   LEWIS ACIDS (ELECTROPHILES)

Type	Examples
1. Positive ions	H^+, Ag^+, Cu^{2+}, Al^{3+}
2. Molecules, especially halides, formed by elements in the first row of the periodic table, that have incomplete octets	$BeCl_2$, BF_3
3. Compounds, especially halides, in which the central atom may exceed its octet (for example by by use of d orbitals)	$SiCl_4$, $TiCl_4$, VCl_4, $TaCl_5$

2.1.1   Types of Lewis Acid

Let us now consider these various types of acid and base, and give examples of the reactions that can occur. The types of acids and bases in Tables 2.1 and 2.2 (with the exception of the type 3 bases, which react only with certain acids) will normally react together; we are thus provided immediately with a rough guide as to whether or not two substances will react together, given favourable circumstances.

TABLE 2.2   LEWIS BASES (LIGANDS)

Type	Examples
1. Negative ions	OH^-, F^-, CN^-
2. Molecules with one or two lone pairs of electrons	NH_3, NEt_3, $P(C_6H_5)_3$, H_2O, Me_2S
3. Molecules having carbon–carbon multiple bonds	$CH_2=CH_2$, $HC\equiv CH$, C_6H_6

Obviously positive ions will react with negative ions to form salts, but a complex compound may be produced if there is an excess of the ligand, for example

$$Ag^+ + 2Cl^- \rightarrow [AgCl_2]^-$$
$$Co^{2+} + 4Cl^- \rightarrow [CoCl_4]^{2-}$$
$$Fe^{3+} + 6CN^- \rightarrow [Fe(CN)_6]^{3-}$$

Positive ions may also react with type 2 (Table 2.2) bases; complex cations then result, for example

$$Ag^+ + 2NH_3 \rightarrow [Ag(NH_3)_2]^+$$
$$Cu^{2+} + 6H_2O \rightarrow [Cu(H_2O)_6]^{2+}$$
$$Ni^{2+} + 6NH_3 \rightarrow [Ni(NH_3)_6]^{2+}$$

Other things being equal, the acid strength of the cations increases with increasing charge and with decreasing radius. While all cations attract strong bases such as water molecules to some extent, the interactions of cations with unsaturated organic molecules are largely confined to the later transition elements, for example

$$[PtCl_4]^{2-} + C_2H_4 \rightarrow [PtCl_3(C_2H_4)]^- + Cl^-$$

Apart from cations, nearly all the other Lewis acids are elemental halides (types 2 and 3 in Table 2.1). Other neutral molecules, such as $AlMe_3$ and $B(OR)_3$, are Lewis acids but not usually such strong acids as the corresponding compounds containing the electron-withdrawing halogen atoms. In the first row of the periodic table, beryllium and boron halides react with two molecules and one molecule of a unidentate base respectively, thus completing the octet around the central atom in forming a tetrahedral complex, for example

$$BeCl_2 + 2Cl^- \rightarrow [BeCl_4]^{2-}$$
$$BF_3 + F^- \rightarrow [BF_4]^-$$
$$BeCl_2 + 2Et_2O \rightarrow BeCl_2(OEt_2)_2$$
$$BBr_3 + Et_3N \rightarrow BBr_3(NEt_3)$$

For these halides the central atoms use 2s and 2p atomic orbitals for bonding; the 3d orbitals are too high in energy to be used, so no more than eight electrons can be accommodated. In carbon tetrachloride the carbon atom already has a complete octet, so it does not show acceptor properties; unlike the halides of beryllium and boron, the carbon halides do not react under normal conditions with, for example, water, alcohols, ethers, and chloride ions.

With the halides of the second and subsequent rows of the periodic table, the octet rule no longer applies, and twelve or more electrons form stable electronic arrangements about the central atoms. Instead of the formation of tetrahedral complexes, six-co-ordination is more common, especially with the halides of the transition elements, for example

$$SiF_4 \quad + \ 2F^- \quad \rightarrow \quad SiF_6{}^{2-}$$
$$VCl_4 \quad + \ 2py \quad \rightarrow \quad VCl_4(py)_2$$
$$TiBr_4 \quad + \ 2OEt_2 \quad \rightarrow \quad TiBr_4(OEt_2)_2$$

The ease of hydrolysis of the silicon tetrahalides compared with the relative inertness of the carbon compounds is attributed to the failure of the carbon halides to form the intermediate adduct. The hydrolysis reaction is not thermodynamically unfavourable but the activation-energy barrier in forming the intermediate is too high to be reached under ordinary conditions. For silicon (and other covalent halides of the type MX_4) we can write

$$SiCl_4 \quad + \ 2H_2O \quad \rightarrow \quad SiCl_4(H_2O)_2$$

This intermediate undergoes further reactions in the presence of an excess of water; for simplicity we may write these as gas-phase equilibria

$$SiCl_4(H_2O)_2 \qquad + \ H_2O \quad \rightleftharpoons \quad SiCl_3(OH)(H_2O)_2 \ + \ HCl$$
$$SiCl_3(OH)(H_2O)_2 \ + \ H_2O \quad \rightleftharpoons \quad SiCl_2(OH)_2(H_2O)_2 \ + \ HCl$$
$$SiCl_2(OH)(H_2O)_2 \ + \ H_2O \quad \rightleftharpoons \quad SiCl(OH)_3(H_2O)_2 \ + \ HCl$$
$$SiCl(OH)_3(H_2O)_2 \ + \ H_2O \quad \rightleftharpoons \quad Si(OH)_4(H_2O)_2 \ + \ HCl$$

In aqueous solution they are more appropriately treated as ionic equilibria of the type

$$(H^+)_2 \, [SiCl_4(OH)_2 \,]^{2-} \ + \ H_2O \quad \rightleftharpoons \quad (H^+)_2 \, [SiCl_3(OH)_3 \,]^{2-} \ + \ HCl$$

in which the nature of the predominant hydroxochloro-species present depends on the concentration of the hydrochloric acid. Thus, with many halides of this type, addition to concentrated hydrochloric acid gives the net result of chloride ion acceptance rather than hydrolysis, for example

$$TiCl_4 \quad + \ conc. \, HCl \quad \rightarrow \quad (H^+)_2 \, [TiCl_6]^{2-}$$
$$SnCl_4 \quad + \ conc. \, HCl \quad \rightarrow \quad (H^+)_2 \, [SnCl_6]^{2-}$$

2.2 Types of Lewis Base (Ligands)

We have already seen many examples of anions acting as ligands, and since all simple anions are capable of acting in this way we need not further classify them at this stage. However, with neutral molecules it is important to be able to distinguish between molecules that can act as ligands and those that cannot. The simple compounds formed by elements in the first row of the periodic table can be represented by the six structures shown, where X is any other atom or group and two dots denote a lone pair of electrons.

X——Be——X   B   C   N   O   F

acidic   acidic   neutral   basic   basic   neutral or acidic

We can now see that the molecules possessing one or two pairs of electrons, over and above those used in bonding, are those containing trivalent nitrogen and divalent oxygen atoms. The base strength will depend on the nature of X. If X is an electron-withdrawing atom or group such as chlorine, the lone pair of electrons is not so readily available for co-ordinate bond formation as when X is an electron-repelling group such as CH_3. The generally expected trend for base strengths of nitrogen donors is thus

$$NF_3 \ < \ NH_3 \ < \ NMe_3$$

The other elements in group VB, for example phosphorus and arsenic, will be similarly basic in their trivalent compounds, and in group VIB sulphur, like oxygen, has two lone pairs of electrons in compounds in which it is divalent. Some of the more important ligands formed by these elements are given in Table 2.3. Note that virtually any organic molecule containing trivalent nitrogen or phosphorus, or divalent oxygen or sulphur, is potentially capable of acting as a ligand. Thus all amines, phosphines, alcohols, ethers, thioalcohols, thioethers, aldehydes, and ketones act in this way, as well as many other types of compound. The number of co-ordination compounds that can be synthesised is thus enormous.

TABLE 2.3   SOME IMPORTANT LIGANDS

Formula	Name	Abbreviation
NH_3	ammonia	–
NH_2NH_2	hydrazine	–
C_5H_5N	pyridine	py
$NH_2CH_2CH_2NH_2$	ethylenediamine	en
	2,2'-bipyridyl	bipy
	1,10-phenanthroline	phen
$(C_6H_5)_3P$	triphenylphosphine	Ph_3P
	o-phenylenebisdimethylarsine	diars
	tetrahydrofuran	THF
$CH_3CO\bar{C}HCOCH_3$	acetylacetonato anion	acac
	ethylenediaminetetraacetate anion	EDTA
$(C_2H_5)_2S$	diethyl sulphide	Et_2S

Ligands having only one atom possessing lone-pair electrons are called *unidentate ligands* since they can bond via one atom only; examples are NH_3, H_2O, and pyridine. When two atoms possessing lone-pair electrons are present in a molecule it may (but not necessarily) bond via both of these atoms; it is then called a *bidentate ligand,* examples being hydrazine, ethylenediamine, and 2,2-bipyridyl. Similarly, ligands with three, four, five, and six donor sites are known. These multidentate ligands may act in either of two different ways. They may bond with all their donor atoms, linked to the same metal atom as in the tris(ethylenediamine)cobalt(III) cation

In this example, the ethylenediamine is acting as a *chelate* ligand; chelating ligands form stable rings with metal ions and are the most important type of multidentate ligand. The alternative mode of bonding is that in which the multidentate ligand is bonded to two metal atoms with the formation of a bridge, for example in the polymeric $NiCl_2(N_2H_4)_2$

Anionic ligands are also capable of bridging and chelation. Unidentate anions can form bridges; for example, $PtCl_2(PPh_3)$, which is dimeric, has the structure

In this structure the chlorines in the bridges are still unidentate in the sense that only one atom is acting as the donor. Hydroxide ions commonly form bridges (*ol* bridges) in the deprotonation of hexa-aquo-cations. Chelating anions include oxalato and acetylacetonato, and the analytically important EDTA anion, the magnesium salt of which is illustrated in Figure 2.1.

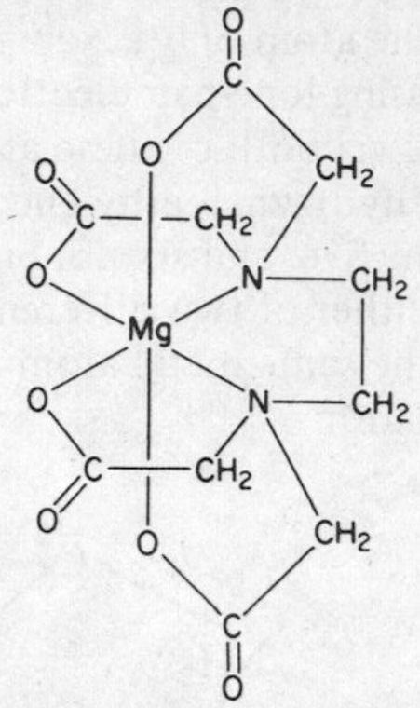

Figure 2.1   Magnesium complex of the anion from ethylenediaminetetraacetic acid (EDTA)

2.3   Hard and Soft Acids and Bases (Pearson)

While the simple Lewis theory enables qualitative predictions to be made about complex formation, it is the quantitative aspects of the theory that are lacking. Ideally, if it were possible to construct scales of Lewis acid and Lewis base strengths, the relative stabilities of all acid—base complexes could be predicted, and we would be in a position to predict the direction of chemical reactions of the type

$$ML + L' \rightleftharpoons ML' + L$$

where L and L′ represent competing ligands. However, while the strengths of bases can frequently be quantified with respect to a particular acid, the order of base strength so obtained is not independent of the acid. A useful series of base strengths can be constructed for aqueous solution by using the proton as reference acid, but a scale of Lewis acid strengths has not been constructed, partly because so many Lewis acids react with water. Broadly speaking it is found that acids as well as bases can be classified under two different headings. For the proton as reference acid, it is found that those bases containing nitrogen, oxygen, or fluorine as the donor atom are stronger than those containing phosphorus, sulphur, or iodine. The reverse is true for acids such as Hg^+, Cu^+, and Pt^{2+} which give more stable complexes with, for example, phosphorus ligands than with nitrogen ligands. The terms *hard* and *soft* have been used to describe these two classes of acids and bases. Donors having high electronegativity, low polarisability, and high resistance to oxidation are described as hard, while donors having low electronegativity, high polarisability, and which are readily oxidised are described as soft. The acids are then classified simply according to whether they prefer hard or soft bases. Some examples of hard and soft acids and bases are listed in Table 2.4. Some acids and bases can be considered as borderline cases; bases in this category include Br^-, py, and N_2, while Fe^{2+}, Co^{2+}, Ni^{2+}, Cu^{2+}, GaH_3, and BMe_3 can be considered as borderline acids.

   Although we have still not quantified our acid—base approach, we have clarified some of the anomalies found in the simple Lewis approach. The fact that hard acids prefer to bind to hard bases, and soft acids to soft bases, is not only useful in prediction but also in the correlation of inorganic facts.

TABLE 2.4   HARD AND SOFT ACIDS AND BASES

| Hard | | Soft | |
Acids	Bases	Acids	Bases
$BeMe_2$	H_2O	BH_3	R_2S
$AlCl_3$	R_2O	$GaCl_3$	R_3P
BF_3	ROH	Cu^+	R_3As
H^+	NH_3	Ag^+	CO
Be^{2+}	NR_3	Au^+	RNC
Mn^{2+}	N_2H_4	Hg^+	C_2H_4
Sc^{3+}	OH^-	Pd^{2+}	C_6H_6
Cr^{3+}	F^-	Pt^{2+}	CN^-
Fe^{3+}	Cl^-	Hg^{2+}	I^-
Th^{4+}	CH_3COO^-	Pt^{4+}	SCN^-

The factors affecting the hardness of acids and bases are size, charge or oxidation state, electronic structure, and the nature of attached groups already present. With metal ions, hardness as acids increases with oxidation state. In their zero oxidation states, metals behave as soft acids and are stabilised by carbon monoxide and phosphine ligands. For a given oxidation state, the transition metal ions become softer as we cross the series, all the really soft acceptor ions having at least a half-filled outer d shell. Indeed, the soft character of metals such as Cu(I), Ag(I), Pt(II), Pd(II), and Rh(I) appears to depend on the availability of metal d electrons for dative π-bonding with carbon ligands such as carbon monoxide and olefins. These ligands form stable complexes only with soft acids, and this fact can be used as a test for softness. In boron chemistry we see from Table 2.4 that when fluorine is bonded to boron(III) we have a hard acid, but when hydrogen is so bonded a soft acid results. This we might expect since the amount of positive charge on the boron atom is greater when it is bonded to the electronegative fluorines than when it is bonded to the soft hydride ions. Carbon monoxide thus forms borine carbonyl, BH_3CO, in its reaction with diborane but has no effect on boron trifluoride.

A slightly different method of classifying acids and bases was introduced somewhat earlier by Ahrland, Chatt, and Davies. From a survey of the relative affinities of ligand atoms for metal ions, they concluded that metals could be divided into two classes. In *class a* are metals that form more stable complexes with ligands whose donor atoms are first-row elements (N, O, F), while in *class b* are metals that prefer to bond to ligands having second-row elements as donor atoms (P, S, Cl). Examples of metals falling in these two classes are shown in Table 2.5; acceptors that do not show predominantly either character are listed as borderline. The degree of b-character of a metal is dependent on its oxidation state. In low oxidation states where the metals possess nonbonding electrons capable of being used in $d_\pi - p_\pi$ or $d_\pi - d_\pi$ bonding with the ligands, the degree of b-character is greatest. Ligands such as CO, C_2H_4, and RNC only co-ordinate strongly with class-b

TABLE 2.5   CLASS a AND b ACCEPTORS

	Class a	Class b
	H, Li, Na, K, Rb, Cs	Rh, Ir
	Be, Mg, Ca, Sr, Ba	Pd, Pt
	Sc, Y, Lanthanides	Ag, Au
	Actinides, Ti, Zr, Hf	Hg
	V, Nb, Ta, Cr, Zn	
	Al, Ga, In	
	Si, Ge, Sn	

Borderline behaviour
Mo, W, Mn, Tc, Re, Fe, Ru, Os, Co, Ni, Cu, Cd, Tl, Pb, Bi, Te, Po

metals and the borderline acceptors when their b-character is enhanced by a low oxidation state. The similarities between the 'a' and 'hard' acceptors and 'b' and 'soft' acceptors can be further seen by comparison of the data in Tables 2.4 and 2.5.

3 The Preparation and Stability of Complex Compounds

3.1 The Preparation of Complexes

A wide variety of methods are used in synthesising co-ordination compounds; we shall summarise here only the more general methods. In view of its ubiquitous nature, water has most frequently been used as a solvent for preparing complex compounds. However, since all cations form some kind of bond with water molecules, it is not often appropriate to consider the water as an inert solvent that plays no part in the chemical reaction. For this reason we shall now consider the species $M(H_2O)_6{}^{n+}$, known as the *hexaquo-ion*, and see how its reactions lead to other complex cations.

3.1.1 The Hexaquo-ion; Aqueous Substitution Reactions

For the alkali metal cations, and to some extent the alkaline earth metal cations, the bonding to water molecules is of a relatively weak ion—dipole type. That is to say, the cation is attracted electrostatically to the negative end of the water dipole. For many chemical reactions these cations can be regarded as reacting in the form of unsolvated species since, although a co-ordination sphere of water molecules surrounds each cation, these molecules do not take part in the overall reaction. However, with triply charged cations and transition metal cations, the water molecules are held by much stronger bonding forces, and these bonded molecules cannot be ignored if we are to understand the processes occurring in aqueous solution. In general, octahedral (or distorted octahedral) complexes are formed when water molecules co-ordinate to di- and tri-positive transition metal ions of the first transition series, so we may represent the hexaquo-ion formation as

There is much spectroscopic evidence for the presence of these ions in solution, and many have been shown by X-ray crystallography to occur as discrete species in the solid phase. We saw in chapter 1 that the alums contain the $[Al(H_2O)_6]^{3+}$ species, and many apparently simple salts such as $NiSO_4.7H_2O$ actually contain complex aquo-ions in the solid state, that is $[Ni(H_2O)_6]^{2+}$ (the extra water molecule is held by hydrogen bonds to the oxygen atoms of sulphate ions and the hydrogen atom of co-ordinated water molecules).

Two very important reactions of the hexaquo-ion, namely substitution and acid—base reactions, lead to new complex compounds. The most frequently used preparative method is that of substitution at the hexaquo-ion. The replacement of water molecules by other ligands occurs in a stepwise fashion and may or may not be complete in the presence of excess of the ligand. Thus, if we add an aqueous solution of nickel(II) bromide to an excess of concentrated aqueous ammonia, the hexamminenickel(II) ion is formed and this precipitates as a violet crystalline salt with the bromide ions

$$[Ni(H_2O)_6]^{2+} + 6NH_3 \rightarrow [Ni(NH_3)_6]^{2+} + 6H_2O$$

However, with copper(II), salts complete substitution of the water molecules with ammonia does not occur in aqueous solution. Instead the stepwise replacement occurs, and in concentrated ammonia solutions approximately four of the water molecules have been replaced

$$[Cu(H_2O)_6]^{2+} + 4NH_3 \rightleftharpoons [Cu(NH_3)_4(H_2O)_2]^{2+} + 4H_2O$$

If a solution of copper(II) sulphate is added to an excess of concentrated aqueous ammonia, the colour changes from blue to violet indicating that replacement of water molecules has occurred. If we now induce precipitation from this solution by the addition of ethanol, the precipitate does not have quite the composition $[Cu(NH_3)_4(H_2O)_2]SO_4$ expected. Instead the precipitate has the formula $[Cu(NH_3)_4(H_2O)]SO_4$ in which the co-ordination geometry is that of a square-based pyramid with the copper atom slightly out of the plane of the four nitrogen atoms. This is an example of a rather general point, namely, complexes that can be isolated from solution are not necessarily the same as the species present in the solution. The stepwise replacement of ligands will be discussed, along with the stability constants of the various species, later in this chapter. For the present, the failure of $[Cu(H_2O)_6]^{2+}$ to form $[Cu(NH_3)_6]^{2+}$ in aqueous solution can be attributed to the high concentration of water in the solution which is competing with the ammonia for the co-ordination sites. Other species such as $[Cu(NH_3)(H_2O)_5]^{2+}$, $[Cu(NH_3)_2(H_2O)_4]^{2+}$, and $[Cu(NH_3)_3(H_2O)_3]^{2+}$ will also be present in such solutions, although in strong ammonia their concentrations will be low and species such as $[Cu(NH_3)_5(H_2O)]^{2+}$ will become more important. The hexamminecopper(II) complexes such as $[Cu(NH_3)_6]Br_2$ are readily prepared by using liquid ammonia. Many neutral ligands are organic molecules rather insoluble in water; preparations then involve the dissolution of the ligand in alcohol, as in the preparation of bipyridyl complexes

$$[Fe(H_2O)_6]^{2+} + 3bipy \rightarrow [Fe(bipy)_3]^{2+} + 6H_2O$$

Substitution of the water molecules in aquo-ions by anions can lead to neutral and to anionic complexes. The acetylacetonates of the transition elements can frequently be prepared in aqueous solution by using a metal salt, acetylacetone, and a weak base (which deprotonates the diketone without precipitating the hydroxide of the metal), for example

$$[Cr(H_2O)_6]^{3+} + 3(acac)^- \rightarrow Cr(acac)_3 + 6H_2O$$

Compounds of this type frequently precipitate from water or can be crystallised from water. Anionic complexes tend to be more soluble but they will usually crystallise from solutions concentrated in the ligand anion, or they can be precipitated by ethanol. The preparation of potassium cyanoferrate(II) is thus achieved by reaction of an aqueous solution of iron(II) sulphate with an excess of hot potassium cyanide solution; cooling of the mixture causes the product to crystallise.

$$[Fe(H_2O)_6]^{2+} + 6CN^- \rightarrow [Fe(CN)_6]^{4-} + 6H_2O$$

As far as substitution reactions are concerned, complexes can be divided into two classes. *Labile* complexes are those that undergo rapid substitution reactions, while *inert* complexes undergo such reactions relatively slowly. In the examples we have seen so far, the displacement of water molecules from the copper(II) ion by ammonia occurs almost instantaneously on mixing since the copper(II) ion forms labile complexes. However, with the hexaquochromium(III) ions the substitution occurs slowly, and several hours may be required in a preparation from the 'inert' chromium(III) ion. It is important to realise that these terms refer to kinetic stability only, that is rate of reaching equilibrium conditions, and that they must not be confused with thermodynamic stability which is concerned with the concentrations of species when equilibrium has been reached. The kinetics and mechanisms of substitution reactions are discussed in chapter 8. Obviously preparative methods for complexes do depend upon the lability of the compounds as well as on the thermodynamics of the reaction (denoted by the stability constants of the complexes)

Substitution reactions are of course not confined to aquo-ions although many of them are most conveniently carried out in aqueous solution. Thus the pale yellow Zeise's salt is prepared (as a monohydrate) by shaking a solution of potassium tetrachloroplatinate(II) with ethylene

$$[PtCl_4]^{2-} + C_2H_4 \rightarrow [PtCl_3(C_2H_4)]^- + Cl^-$$

Substitution of nitro groups in hexanitrocobaltate(III) ions by ethylenediamine can be achieved by using aqueous solutions of the reagents at 70°

$$[Co(NO_2)_6]^{3-} + 2en \rightarrow cis\text{-}[Co(NO_2)_2(en)_2]^+ + 4NO_2^-$$

the brown *cis*-isomer precipitates from the solution on cooling. Substitution *by* water molecules (rather than *of* water molecules) is an important type of reaction known as *aquation*. Two kinetically well-studied aquation reactions are those of the inert chloropentamminecobalt(III) and trisoxalatochromate(III) ions

$$[Co(NH_3)_5Cl]^{2+} + H_2O \rightarrow [Co(NH_3)_5(H_2O)]^{3+} + Cl^-$$
$$[Cr(C_2O_4)_3]^{3-} + 2H_2O \rightarrow [Cr(C_2O_4)_2(H_2O)_2]^- + C_2O_4^{2-}$$

Sometimes a particular complex ion that is known to be formed in solution cannot be precipitated or crystallised from the solution by ordinary techniques. This is often the situation in substitution reactions in which two or more complex species may be present in solution and the required complex is more soluble than, or present in smaller amounts than, other complex species. In these cases we must increase the lattice energy effects of the required complex. A general method of isolation of such species is to use a precipitating ion of approximately the same size as the complex ion (usually large) and of equal but opposite charge. In the reaction of potassium cyanide with nickel(II) salts the species $[Ni(CN)_4]^{2-}$ and $[Ni(CN)_5]^{3-}$

are formed in solutions of high cyanide concentration. The predominating $[Ni(CN)_4]^{2-}$ crystallises readily as the orange potassium salt $K_2 Ni(CN)_4 . H_2 O$. The trivalent anion will crystallise if a large trivalent cation is used; for example, with $[Cr(en)_3]^{3+}$ the salt $[Cr(en)_3][Ni(CN)_5].1.5H_2 O$ is obtained.

This method is frequently of use in the stabilisation of ions of unusual co-ordination number. The unusual hexachloro-complexes of Cr^{III}, Mn^{III}, and Fe^{III} can be stabilised with the tris(propylenediamine)cobalt(III) cation, giving stable solids $[Co(pn)_3][MCl_6]$. Elements that normally form hexachloro-species can similarly have the five-co-ordinate species stabilised, for example $CuCl_5^{3-}$ by $[Cr(NH_3)_6]^{3+}$, VCl_5^- by PCl_4^+, and GeF_5^- by $[AsPh_4]^+$.

3.1.2   The Hexaquo-ion; Lowry–Brönsted Acidity

The second important reaction of aquo-ions is their Lowry–Brönsted acidity. Their power to act as proton donors according to the equation

$$[M(H_2 O)_6]^{n+} + H_2 O \rightleftharpoons [M(H_2 O)_5 (OH)]^{(n-i)+} + H_3 O^+$$

varies widely; however, the presence of this equilibrium must always be remembered even if the equilibrium constant for the forward reaction is small, since addition of a base will remove $H_3 O^+$ and encourage the hydrolysis reaction. In general terms, the greater the charge on the aquo-ion the greater its acidity. While dipositive transition metal ions are only weakly acidic, the tripositive ions are strong acids; ions of the type $[M(H_2 O)_6]^{4+}$ are not formed by elements in the first transition series because they are too strongly acidic and undergo hydrolysis. Whilst a knowledge of this hydrolytic reaction is not of great use in preparing complexes, its importance lies in informing us which complex salts may be difficult or impossible to prepare in aqueous solution. Let us exemplify these statements by considering the aqueous chemistry of iron. The pale violet $[Fe(H_2 O)_6]^{3+}$ undergoes the following hydrolytic equilibria to give hydroxo-species which are yellow because of a charge-transfer absorption that tails into the visible

$$[Fe(H_2 O)_6]^{3+} + H_2 O \rightleftharpoons [Fe(H_2 O)_5 (OH)]^{2+} + H_3 O^+$$
$$(K_1 = 10^{-3.05})$$
$$[Fe(H_2 O)_5 (OH)]^{2+} + H_2 O \rightleftharpoons [Fe(H_2 O)_4 (OH)_2]^+ + H_3 O^+$$
$$(K_2 = 10^{-3.26})$$

These solutions are thus strongly acidic; we can write the pK values for these equilibria as 3.05 and 3.26 respectively (where p$K = -\log_{10} K$). Any attempt to raise the pH above 2 causes these reactions to proceed to the right, and eventually, in the final equilibrium, $Fe(H_2 O)_3 (OH)_3$ [probable composition $FeO(OH)$aq] precipitates. Thus the addition of sodium carbonate solution to an iron(III) chloride solution does not result in the formation of iron(III) carbonate but rather in the precipitation of the 'hydroxide' with evolution of much carbon dioxide. Similarly, if magnesium ribbon is used in place of sodium carbonate, the solvated protons are removed by the reaction.

$$2H_3 O^+ + Mg \rightarrow Mg^{2+} + H_2 + 2H_2 O$$

effervescence occurs, hydrogen is evolved, and the 'hydroxide' eventually precipitates. On the other hand, the addition of noncomplexing acids, such as

HNO_3 and $HClO_4$, to these partially hydrolysed solutions causes the re-formation of the violet $[Fe(H_2O)_6]^{3+}$ ion. In addition to the mononuclear hydroxo-species, the formation of binuclear cations occurs by the *olation* reaction

$$2[Fe(H_2O)_6]^{3+} + 2H_2O \ \rightleftharpoons \ [Fe_2(H_2O)_8(OH)_2]^{4+} + 2H_3O^+$$
$$(K = 10^{-2.91})$$

This binuclear species, in which the iron atoms are bridged by OH groups (called *ol* bridges), has the structure

In the presence of base, further proton donation by this ion, followed by olation, leads to larger aggregates until eventually colloidal gels are formed and the iron(III) 'hydroxide' precipitates. The format.on of oxo bridges, known as *oxolation*, is probably an important process in this aggregation

$$[(H_2O)_4Fe(OH)_2Fe(H_2O)_4]^{4+} + 2H_2O \rightleftharpoons [(H_2O)_4Fe\underset{O}{\overset{O}{<}}Fe(H_2O)_4]^{2+} + 2H_3O^+$$

The exact nature of these species in solution is not known with certainty; for example, the binuclear iron species may be oxo-bridged, that is $[(H_2O)_5FeOFe(H_2O)_5]^{4+}$. However, the general ideas of olation and oxolation are useful concepts in understanding the aggregation of aquo-ions.

Aqueous solutions of iron(II) salts are hardly acidic ($K_1 = 10^{-9.5}$); iron(II) carbonate can be precipitated from them by the addition of sodium carbonate. The slight acidity of other divalent cations frequently results in the formation of *basic salts*. Thus nickel(II), cobalt(II), and copper(II) salts give basic carbonates unless a high pressure of carbon dioxide is maintained above the solutions. Basic salts frequently precipitate also when strong alkalis such as sodium hydroxide are added to solutions of transition metal cations, especially when the latter are in excess. For example, treatment of copper(II) sulphate solution with a little dilute sodium hydroxide solution results in the precipitation of basic sulphates having a composition that varies depending on the conditions used, that is concentrations of solutions and relative concentrations of the two reagents. Once the initial stage in the stepwise hydrolysis of these cations has proceeded, olation can occur and aggregates are formed that are large enough to precipitate with the sulphate anions. A typical product thus might be $[Cu(OH)(H_2O)_5]_2SO_4$ which in basic salt nomenclature would be written $Cu(OH)_2.CuSO_4.10H_2O$.

3.1.3   Preparation of Complexes by Direct Reaction

In theory this is the simplest method for preparing complexes; it is the reaction of a Lewis acid with a Lewis base. The reactions we have already seen indicate that since water is a base it is in competition with any other base we add for the acidic site.

Most 'direct' acid–base reactions are therefore carried out in the complete absence of water, that is in the gas phase, or in an inert solvent such as a hydrocarbon, or even by direct mixing if one or both of the reagents are liquid. Such methods frequently necessitate the use of high-vacuum apparatus or dry-box techniques.

When both reagents are gaseous, a carefully controlled flow of each gas is passed into a large vessel. With boron trifluoride and trimethylamine the product is deposited as a white powder

$$BF_3 + NMe_3 \rightarrow BF_3.NMe_3$$

More usually, at least one reagent is in the liquid or solid state, and then an inert solvent is used as diluent. Thus the trimethylamine adduct of boron trichloride is prepared by mixing the reagents in benzene from which it precipitates. Inert organic solvents such as hydrocarbons and halogenated hydrocarbons are widely used because they frequently dissolve both the acceptor (for example, covalent halides such as $SnCl_4$, VCl_4, and $TiCl_4$) and the donor (wide variety of organic ligands) and yet cause precipitation of the complexes, which often have polymeric structures. Just a few examples will suffice here; many others occur throughout this book.

$$VCl_4 + 2py \xrightarrow[-20°]{\text{toluene}} VCl_4.2py$$

$$SnCl_4 + 2Et_2O \xrightarrow{\text{benzene}} SnCl_4.2(OEt_2)$$

$$WOCl_4 + THF \xrightarrow{\text{dichloromethane}} WOCl_4.THF$$

Often the ligand itself conveniently serves as the solvent, and the complex is obtained by evaporation of the solvent. For example, ammines are prepared by condensing liquid ammonia on to a metal salt and then letting the ammonia (b.p. $-33°$) evaporate. Many ammonia complexes cannot be prepared in water because of the precipitation of metal hydroxides. Iron(II) chloride thus yields the hydroxide with aqueous ammonia whereas with liquid ammonia a series of ammoniates is formed of which the stable species at room temperature is the hexammine

$$FeCl_2 + 6NH_3 \rightarrow [Fe(NH_3)_6]Cl_2$$

Ammonia, like water, is a protonic solvent; in the same way that hydrolysis occurs in water, so ammonolysis occurs in ammonia. The more covalent halides thus do not form simple ammoniates but rather ammonolytic derivatives, for example

$$VBr_3 + 7NH_3 \rightarrow [V(NH_3)_5(NH_2)]Br_2 + NH_4Br$$
$$TiCl_4 + 6NH_3 \rightarrow TiCl(NH_2)_3 + 3NH_4Cl$$

Other ligands that react directly with Lewis acids include aliphatic amines, hydrazines, nitriles, alcohols, ethers, ketones, and organic sulphides. Again examples from the first transition series illustrate the types of complex formed

$$CoCl_2 + 6N_2H_4 \rightarrow [Co(N_2H_4)_6]Cl_2$$
$$VBr_3 + 3MeCN \rightarrow VBr_3(MeCN)_3$$
$$CoBr_2 + 2(acacH) \rightarrow CoBr_2(acacH)_2$$

3.1.4   Oxidation – Reduction Reactions

So far we have used examples in which the oxidation state of the metal remains unchanged throughout the preparations. However, many ligands can act as reducing agents, and for some metals stabilisation of high oxidation states is achieved by co-ordination. The most celebrated example is in cobalt chemistry. Here the 'simple' salts of cobalt(II) are stable while the 'simple' cobalt(III) salts are oxidising agents in aqueous solution; that is, with water as the ligand the cobalt(II) state is stabilised. However, with nitrogen ligands the cobalt(III) state is stabilised preferentially to the divalent state. In the preparation of cobaltammines therefore a cobalt(II) salt in aqueous ammonia is oxidised by, for example, bubbling air through the mixture. By use of cobalt(II) chloride, ammonium chloride, and a charcoal catalyst, the orange hexammine is obtained.

$$4[Co(H_2O)_6]Cl_2 + 4NH_4Cl + 20NH_3 + O_2 \rightarrow 4[Co(NH_3)_6]Cl_3 + 26H_2O$$

In the absence of charcoal, and with hydrogen peroxide instead of air, the aquopentamminecobalt(III) salt is formed; this is readily converted into the red chloropentamminecobalt(III) chloride by treatment with concentrated hydrochloric acid

$$2[Co(H_2O)_6]Cl_2 + 2NH_4Cl + 8NH_3 + H_2O \rightarrow 2[Co(NH_3)_5(H_2O)]Cl_3 + 12H$$
$$[Co(NH_3)_5(H_2O)]Cl_3 \rightarrow [Co(NH_3)_5Cl]Cl_2 + H_2O$$

It is particularly useful to start with cobalt(II) salts because cobalt(II) complexes being labile undergo rapid substitution reactions whereas cobalt(III) complexes are inert. The cobalt(III) complexes are thus produced by substitution of ligand molecules for co-ordinated water molecules around cobalt(II) followed by oxidation of the cobalt(II) ammine to a cobalt(III) ammine.

Preparative reactions in which the transition element is reduced are somewhat more common. Frequently the ligand itself is the reducing agent. In the preparation of the *cis-* and *trans*-isomers of potassium dioxalatodiaquochromate(III), potassium dichromate is reduced with oxalic acid

$$K_2Cr_2O_7 + 7H_2C_2O_4 \rightarrow 2K[Cr(C_2O_4)_2(H_2O)_2] + 6CO_2 + 3H_2O$$

Anhydrous copper(II) chloride is reduced by anhydrous hydrazine with the formation of bis(hydrazine)copper(I) chloride

$$2CuCl_2 + 8N_2H_4 \rightarrow 2Cu(N_2H_4)_2Cl + 2N_2H_4.HCl + 2NH_3 + N_2$$

Even ligands such as aliphatic amines, 2,2'-bipyridyl, pyridine, and nitriles, which one might not consider to be strong reducing agents, often cause reduction of transition metal salts in their higher oxidation states. Therefore, in studying reactions of such ligands, one of the co-ordination chemist's first objectives is to establish the oxidation state of the metal in the product. Vanadium(IV) chloride is especially prone to reduction and gives several interesting five-co-ordinate vanadium(III) compounds when treated with an excess of these ligands, for example $VCl_3(NMe_3)_2$ and $VCl_3(SMe_2)_2$; however, the six-co-ordinate complexes $VCl_3(py)_3$ and $VCl_3(MeCN)_3$ are produced when pyridine or acetonitrile is used as the reductant. Whether or not reduction occurs may depend on the conditions, particularly the temperature at which the reaction is carried out. We saw earlier that at $-20°$ in an inert solvent the vanadium(IV)–pyridine complex $VCl_4(py)_2$ can be isolated; in this preparation the vanadium(IV) chloride is deliberately kept in excess over the reducing ligand. It seems likely that in these reductions the initial step after co-ordination is that of electron transfer to the metal to give a radical anion $(VCl_4 \cdot)^-$ and a radical cation $(C_5 H_5 N \cdot)^+$. In the reductions of tungsten(VI) chloride and tungsten(V) bromide with pyridine, the oxidation product of the pyridine has been identified as the 1-(4-pyridyl)pyridinium ion.

Complexes in which the central metal ion is in an unusually low oxidation state are conveniently prepared in nonaqueous solvents. By virtue of its ability to dissolve the alkali metals as well as many inorganic complexes, liquid ammonia has been extensively used for this purpose. The solvent must be used in a vacuum apparatus to prevent oxidation of the products by the atmosphere, but this proves to be no real hindrance to its use. When potassium tetracyanonickelate(II) is treated with potassium in ammonia, a bright red precipitate forms while the nickel is in excess; this is believed to contain the binuclear nickel(I) anion $[Ni_2 (CN)_6]^{2-}$. With excess of potassium, a bulky yellow precipitate of the tetracyanonickelate(0) is obtained.

$$K_2 Ni(CN)_4 + 2K \rightarrow K_4 Ni(CN)_4$$

Similar low oxidation state cyanides which can be thus prepared include $K_4 Ti(CN)_4$, $K_4 Co(CN)_4$, and $K_6 Cr(CN)_6$. In all of these the oxidation state of the metal is formally zero and the compounds do behave as strong reducing agents, being rapidly oxidised in air and liberating hydrogen from water.

Similar low oxidation state complexes containing ligands such as carbon monoxide and phosphines can be prepared by using solutions of alkali metals in ammonia. Such ligands are capable of stabilising the low oxidation states by enabling charge to be transferred from the metal ion to the ligand. Ammonia has been considered incapable of this π-bonding, so it was somewhat surprising to find that reduction of certain metal halides in ammonia produces ammines of metals in their zero oxidation states. Tetrammineplatinum(0) is precipitated as a yellow-white solid when tetrammineplatinum(II) bromide is reduced with potassium in ammonia at the boiling point

$$[Pt(NH_3)_4] Br_2 + 2K \rightarrow Pt(NH_3)_4 + 2KBr$$

Pentammineiridium(0) $Ir(NH_3)_5$ is prepared similarly. These compounds are decomposed thermally to give ammonia and the metals only.

A rather special and recently systematised type of oxidation is the *oxidative addition* reaction of metal complexes having the d^8 configuration. In these reactions the metal increases its oxidation state by two units, that is it assumes a d^6 configuration and its co-ordination number is increased by co-ordination of the constituent parts of the oxidising molecule. The tendency to form stable adducts having the d^6 configuration increases as we descend the group (Figure 3.1) and

$$Fe^0 \quad Co^I \quad Ni^{II}$$
$$Ru^0 \quad Rh^I \quad Pd^{II}$$
$$Os^0 \quad Ir^I \quad Pt^{II}$$

Figure 3.1   Metals involved in oxidative addition reactions

also as we cross from right to left within group VIII; thus the relative ease of oxidation is

$$Pt_{(II)} > Pd_{(II)} > Ni_{(II)} ;$$
$$Os(0) > Ir_{(I)} > Pt_{(II)}$$

Vaska's compound *trans*-[Ir(PPh$_3$)$_2$(CO)Cl] is particularly versatile. In this, iridium(I) becomes oxidised to iridium(III) as illustrated by the reaction with hydrogen bromide

For many molecules, for example O$_2$, H$_2$, and SO$_2$, reversible uptake occurs

$$\text{Ir(PPh}_3\text{)}_2\text{(CO)Cl} \underset{\text{vacuum}}{\overset{\text{O}_2 \text{ in benzene}}{\rightleftharpoons}} \text{O}_2\text{Ir(PPh}_3\text{)}_2\text{(CO)Cl}$$

and it is this reversibility of the processes that makes this type of system so useful in organic synthesis. The reverse process is known for obvious reasons as *reductive elimination*.

Platinum(II) complexes resemble those of iridium(I) in going from square to octahedral

but the trigonal bipyramidal osmium(0) complexes dissociate a ligand in changing to octahedrally co-ordinated osmium(II)

3.1.5   Thermal Decompositions

When thermally decomposed, many complexes evolve any neutral and volatile ligand molecules in a stepwise fashion; by careful control of the temperature, new complexes can be obtained that may be difficult to prepare by other routes. In the reactions of nickel(II) salts with an excess of nitrogen ligands, complexes of the type [NiL$_6$]X$_2$ or NiL$_4$X$_2$ are commonly produced. By heating these the intermediate complexes NiL$_2$X$_2$ and NiLX$_2$ can frequently be isolated. Thus NiCl$_2$(NH$_3$)$_2$ is obtained from [Ni(NH$_3$)$_6$]Cl$_2$ at 100°, and the decomposition of the nickel(II) bromide—pyridine complex can be illustrated by the scheme.

$$\text{NiBr}_2\text{(py)}_4 \xrightarrow{140°} \text{NiBr}_2\text{(py)}_2 \xrightarrow{200°} \text{NiBr}_2\text{(py)} \xrightarrow{300°} \text{NiBr}_2$$

As is usual in thermal dissociation reactions, the co-ordination number of the nickel has remained constant (at six) in these reactions.

The study of these thermal dissociations is greatly aided by the techniques of thermogravimetric analysis (t.g.a.) and differential thermal analysis (d.t.a.). In t.g.a. a crucible containing the complex is attached to one pan of a balance and, by enclosing the crucible in a furnace, the weight of the complex can be recorded as the temperature is increased. The temperatures at which dissociation occurs for the various complexes are thus easily recorded together with the weight of ligand evolved. In d.t.a. two crucibles, one containing the complex and the other a thermally stable reference compound, are heated together in an enclosed space. The temperatures within each crucible are measured continuously, and when dissociation of the complex occurs the temperatures will differ depending on whether the process is exothermic or endothermic. From d.t.a. studies, the temperature and enthalpy of the dissociation are measured.

3.1.6   Metal Carbonyls and Organometallic Compounds

Compounds containing metal—carbon bonds have become exceedingly numerous in the last two decades and are now a very important part of co-ordination chemistry. The first metal carbonyl was discovered by Mond in 1890; he obtained nickel carbonyl $Ni(CO)_4$ by the reaction of carbon monoxide with nickel powder at atmospheric pressure and below $100°$. This carbonyl is prepared on an industrial scale today in the refining of nickel by using pressures of up to 15MPa and temperatures of up to $150°$. The reaction between carbon monoxide and iron or cobalt similarly yields carbonyls directly. For carbonyls of other metals the general preparative route is the reduction of metal halides in a solvent such as ammonia (see section 3.1.4), tetrahydrofuran, or diglyme under a $20-30$ MPa pressure of carbon monoxide and at up to $300°$. The various reducing agents used include alkali metals, magnesium, aluminium, and aluminium alkyls. We may exemplify these reactions by the preparation of vanadium hexacarbonyl

$$VCl_3 + CO + Na \xrightarrow[100°,\ 25\ \text{MPa}]{\text{diglyme}} [Na(\text{diglyme})_2]^+[V(CO)_6]^- \xrightarrow{H_3PO_4} V(CO)_6$$

In this reaction, the solvated sodium salt of the hexacarbonylvanadate(−I) ion is first formed, and this is converted into the carbonyl via thermal decomposition of the as yet unisolated $HV(CO)_6$. Carbon monoxide acts both as reducing agent and as ligand when it reacts under conditions of heat and high pressure with rhenium or technetium heptoxides

$$Re_2O_7 + 17CO \longrightarrow Re_2(CO)_{10} + 7CO_2$$

The metal carbonyls thus prepared are the starting point for the preparation of an enormous number and variety of derivatives. Just a few of the reactions of iron pentacarbonyl in Figure 3.2 must suffice to indicate the scope of such reactions.

Organometallic complexes contain metal—carbon σ bonds or metal—carbon π bonds (or both). The simple σ bonded alkyls and aryls MR_x are relatively rare; however, such M—C σ bonds are greatly stabilised if π bonding ligands such as CO, PPh_3, or cyclopentadienyl ($C_5H_5^-$) are present. Many preparative routes are used; three common methods are those involving the use of Grignard reagents,

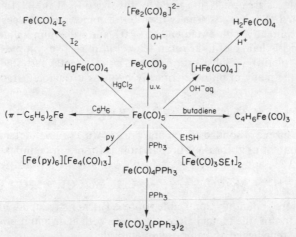

Figure 3.2   Reactions of iron pentacarbonyl

metathesis between an alkyl halide and a sodium carbonylmetallate, and the addition of unsaturated hydrocarbons to complex hydrides, for example

$$(Et_3P)_2PtCl_2 + 2MeMgBr \rightarrow (Et_3P)_2PtMe_2 + MgBr_2 + MgCl_2$$
$$Na[Co(CO)_4] + MeI \rightarrow MeCo(CO)_4 + NaI$$
$$(\pi\text{-}C_5H_5)Fe(CO)_2H + CF_2{=\!=}CF_2 \rightarrow (\pi\text{-}C_5H_5)Fe(CO)_2CF_2CF_2H$$

Complexes with unsaturated hydrocarbons bonded to transition metals are very much more stable and exceedingly numerous. Such complexes may contain, for example, olefins, acetylenes, cyclic unsaturated systems, or allyl groups π-bonded to the metal. We mentioned earlier the first olefin complex prepared, Zeise's salt. Perhaps the most celebrated complex is bis(cyclopentadienyl)iron, called ferrocene. This is an orange crystalline solid with remarkable thermal stability (it boils without decomposition at 230°), and it is not attacked by alkali or concentrated hydrochloric acid. It was the discovery of this compound in 1951 that sparked off the development of the now enormous organic chemistry of the transition metals. Ferrocene contains two cyclopentadienide rings which form a 'sandwich' with the iron atom between them. It is prepared in the laboratory from iron(II) chloride by treatment with sodium cyclopentadienide in tetrahydrofuran or ethylene glycol dimethyl ether

$$FeCl_2 + 2C_5H_5Na \rightarrow Fe + 2NaCl$$

This reaction is indeed a general route to π-cyclopentadienyl-metal compounds which may contain up to four cyclopentadienyl rings, for example

$$TiCl_4 + \tfrac{1}{2}Mg(C_5H_5)_2 \rightarrow \underset{Cl}{\overset{}{Ti}} \quad + \tfrac{1}{2}MgCl_2$$

$$UCl_4 + 4KC_5H_5 \rightarrow U(C_5H_5)_4 + 4KCl$$

Olefin complexes are usually prepared by the reaction of the olefin with a metal halide or carbonyl. A solvent is usually used but is sometimes unnecessary. Copper(I) halides combine directly with butadiene at $-10°$, and palladium halides combine with liquid straight-chain olefins at room temperature to give complexes of the formula $[PdCl_2(olefin)]_2$. Substitution by olefins into metal carbonyls can be effected either thermally or photochemically by using ultraviolet irradiation.

$$C_2H_4 + C_5H_5Mn(CO)_3 \xrightarrow{\text{u.v.}} (C_5H_5)Mn(CO)_2(C_2H_4) + CO$$

Complexes of benzene π-bonded to a metal are known as *arene* metal complexes. The carbonyl route is used for their preparation as for olefin complexes, for example

$$C_6H_6 + Mo(CO)_6 \xrightarrow{\text{u.v.}} C_6H_6Mo(CO)_3 + 3CO$$

However, the more commonly applicable route is Fischer's reducing Friedel – Crafts method. In this, the metal halide is reduced with aluminium powder in the presence of the arene ligand and aluminium chloride

$$3CrCl_3 + 2Al + AlCl_3 + 6C_6H_6 \rightarrow 3[Cr(C_6H_6)_2]^+[AlCl_4]^-$$

In this example bis(benzene)chromium, the first arene complex to be prepared, is obtained by reduction of the cation with aqueous sodium dithionite

$$2[Cr(C_6H_6)_2]^+ + S_2O_4{}^{2-} + 4OH^- \rightarrow 2Cr(C_6H_6)_2 + 2SO_3{}^{2-} + 2H_2O$$

3.2 The Stability of Complex Ions in Solution

We have already mentioned the need to differentiate between thermodynamic and kinetic stabilities. Indeed the word stability may take on even further connotations; a compound may be thermally stable yet react violently with water at room temperature, that is be hydrolytically unstable. The word 'stable' must always be qualified when talking about a compound; on its own it raises the immediate question: stable to what? In this section we are concerned with the thermodynamic stability of complex ions in solution; that is, the extent to which the complex ion will tend to be formed from, or dissociate into, its component species when the system has reached equilibrium. The kinetic stability of complexes, that is the rate at which the equilibria are reached, will concern us in chapter 8.

3.2.1 Stability Constants

When a metal ion reacts with a ligand in aqueous solution, stepwise replacement of the co-ordinated water molecules by the ligand occurs. Consider a metal ion M (for convenience we shall neglect the charge) and a ligand L. The first step in the replacement of water molecules is represented by

$$[M(H_2O)_n] + L \rightarrow [M(H_2O)_{n-1}L] + H_2O$$

If the equilibrium lies to the right we say a complex has been formed (despite the fact that a complex was already present on the left-hand side of the equation). For convenience such equilibria are usually written as

$$M + L \rightarrow ML$$

If we assume activity coefficients of unity, for this equilibrium we can write

$$K_1 = \frac{[ML]}{[M][L]}$$

For the stepwise replacement of further water molecules, we can write the stepwise equilibrium constants as follows

$$ML + L \rightleftharpoons ML_2 \qquad K_2 = \frac{[ML_2]}{[ML][L]}$$

$$ML_2 + L \rightleftharpoons ML_3 \qquad K_3 = \frac{[ML_3]}{[ML_2][L]}$$

$$ML_3 + L \rightleftharpoons ML_4 \qquad K_4 = \frac{[ML_4]}{[ML_3][L]}$$

These equilibrium constants K_1 to K_4 representing the formation of complexes are called the *stepwise formation constants* or *stability constants*. The numerical values of these constants nearly always decrease in the order $K_1 > K_2 > K_3 > K_4 > K_n$. For any particular reaction we may be more interested in the overall stability constant (denoted by the symbol β), as given by

$$\beta_4 = \frac{[ML_4]}{[M][L]^4}$$

Thus an alternative way of expressing formation constants is to represent the equilibria as

$$M + L \rightleftharpoons ML \qquad \beta_1 = \frac{[ML]}{[M][L]}$$

$$M + 2L \rightleftharpoons ML_2 \qquad \beta_2 = \frac{[ML_2]}{[M][L]^2}$$

$$M + 3L \rightleftharpoons ML_3 \qquad \beta_3 = \frac{[ML_3]}{[M][L]^3}$$

and so on. It is readily apparent that

$$\beta_4 = K_1 K_2 K_3 K_4$$

(substitute the values for K_1 to K_4 and multiply out if you do not see this immediately), or more generally that

$$\beta_n = \frac{[ML_n]}{[M][L]^n}$$

These terms β_n are called the *overall formation constants* or *stability constants*. Some numerical values of stability constants are listed in Table 3.1. Since these values occur over a very wide range, they are normally

TABLE 3.1   STEPWISE AND OVERALL STABILITY CONSTANTS

M	L	$\log K_1$	$\log K_2$	$\log K_3$	$\log K_4$	$\log K_5$	$\log K_6$	$\log \beta$
Cu^{2+}	NH_3	4.17	3.53	2.88	2.05			12.6
Ni^{2+}	NH_3	2.80	2.24	1.73	1.19	0.75	0.03	8.7
Ag^+	NH_3	3.14	3.82					7.0
Hg^{2+}	CN^-	18.00	16.70	3.83	2.98			41.5
Hg^{2+}	I^-	12.87	10.95	3.67	2.37			29.9
Cu^{2+}	en	10.55	9.05					19.6
Ni^{2+}	en	7.45	6.23	4.34				18.0

reported on a logarithmic scale. A large value of the stability constant for a particular complex indicates that the concentration of the complex is much greater than the concentrations of the species of which it is composed. An overall stability constant of around 10^8 or greater normally indicates the formation of what we would regard as a thermodynamically stable complex (the β values given in Table 3.1 are the products of the K values listed, that is β_4 for Cu^{2+}, β_6 for Ni^{2+}, etc.). The blank spaces in Table 3.1 are those for which the stability constants would be very low or negative; for example, in the formation of $[Cu(NH_3)_5(OH_2)]^{2+}$ $\log K_5$ is approximately -0.5.

For the simple substitution reactions in which no change of stereochemistry occurs, a steady decrease in the stepwise stability constants in any particular system, for example Ni^{2+}–NH_3, is generally observed. This decrease is to be expected solely on statistical grounds; the probability of exchanging a water molecule is greater in $[Ni(H_2O)_6]^{2+}$ than in $[Ni(H_2O)_5(NH_3)]^{2+}$, and so on. Other factors such as steric effects (the ligand being larger than a water molecule) are also important. When a stereochemical change occurs on substitution, a sudden change in the values of the stability constants may occur, and the magnitude may even increase in going from K_n to K_{n+1}. Thus the sudden drop from K_2 to K_3 for Hg^{2+} complexes with halide and cyanide ions can be attributed to a change from linear HgX_2 to tetrahedral $[HgX_3(H_2O)]^-$ with consequent change in hybridisation of the mercury from sp to sp^3. The increase in stability constant from K_1 to K_2 in the Ag^+–NH_3 system may indicate that a nonlinear species $[Ag(NH_3)(H_2O_n)]^+$ becomes linear in $[Ag(NH_3)_2]^+$. As long as n is greater than unity this reaction, that is

$$[Ag(NH_3)(H_2O)_n]^+ + NH_3 \rightleftharpoons [Ag(NH_3)_2]^+ + nH_2O$$

proceeds from left to right with an increase in entropy and consequently an increase in the equilibrium constant K_2.

So far we have used concentrations in our expressions for equilibrium constants, but this would be quantitatively acceptable only in very dilute solution. If a stability constant is measured by determining the concentrations of the various species in the expression for the equilibrium constant, it is called a *concentration constant*. Concentration constants are quantitatively valid only under the conditions at which they were measured, but they are very useful in a semiquantitative way. Stability constants expressed in terms of activities are called *thermodynamic constants*. In very dilute solutions concentrations are equal to activities, so thermodynamic constants could in theory be measured in such solutions. However, this is not convenient in practice so these are usually measured by determining the stability constants in a series of solutions containing different concentrations of a noncomplexing electrolyte such as sodium perchlorate. The thermodynamic constants are then obtained by extrapolation to zero ionic strength.

3.2.2   Stability Trends

A large amount of stability constant information is to be found in *Stability Constants* and *Stability Constants Supplement No. 1* (Special Publications Nos. 17 and 25 of The Chemical Society, London, 1964 and 1971). We shall now examine some of the trends to be found in stability constants, and the reasons for them.

If we use a simple electrostatic model for a complex compound we can fairly readily understand some of the trends in stabilities. Since the formation of a complex may be regarded as an interaction between a cation and either an anion or the negative end of a dipole, we should expect the magnitudes of the charges to be important as well as the sizes of the interacting species. Obviously, the smaller the interacting species or the larger the charge on them, the greater the electrostatic attraction and the more stable the resultant complex. It should be noted that this model is more appropriate to hard acids and bases; for soft acids and bases, other factors such as covalent bonding are important. Thus, for cations and hard bases the stability constants of complexes of a base with a metal increase as the oxidation state of the metal increases (for example, the stabilities of EDTA complexes in Table 3.2). In a series of ions of approximately the same ionic radius, the stability constants with a given ligand decrease as the charge decreases, for example $Th^{4+} > Y^{3+} > Ca^{2+} > Na^+$. Instead of just the charge or size, it is usual to talk about the ratio of charge to radius or the ionic potential of ions. For a given ligand and divalent ions of the first transition series, the stability constants roughly follow the order of the ionic potentials (and sizes) of the metal ions, that is

$$Mn^{2+} < Fe^{2+} < Co^{2+} < Ni^{2+} < Cu^{2+} > Zn^{2+}$$

This 'natural' order is often known as the Irving–Williams order. Copper(II) does not fit well into the series since it is reluctant to co-ordinate a fifth or sixth ligand very strongly. This order of size of the metal ions and the stability of the various complexes is more readily understood in terms of ligand field theory (chapter 5).

Entropy effects appear to make a major contribution to the stabilities of complexes. In order to understand this we must remember that the equilibrium constant is related to the standard free energy of a reaction by the equation

$$\Delta G^\circ = -RT \ln K$$

and that the free-energy change is related to the enthalpy and entropy changes by

$$\Delta G^{\circ} = \Delta H^{\circ} - T\Delta S^{\circ}$$

Thus we can see that the stability constant for the formation of a complex depends on the enthalpy and entropy changes in the complexing process. Enthalpy changes in reactions of the type

$$[M(H_2 O)_6]^{n+} + xL \rightleftharpoons [M(H_2 O)_{6-x} L_x]^{n+} + xH_2 O$$

are usually small, and entropy contributions may become significant. We saw earlier some examples of this affecting the stepwise stability constants when a stereochemical change occurs during the reaction. If, during a substitution reaction, we produce more free molecules than we started with, the increased randomness of the system results in an increase of entropy. This occurs in substitution reactions in which a unidentate ligand is replaced by a chelating ligand; the extra stability of chelate complexes over complexes of unidentate ligands having the same donor atoms is known as the *chelate effect*. In Table 3.1 we can see the stability constants for ethylenediamine complexes. The reactions

$$[Cu(H_2 O)_6]^{2+} + 3en \rightleftharpoons [Cu(en)_3]^{2+} + 6H_2 O$$
$$[Ni(H_2 O)_6]^{2+} + 3en \rightleftharpoons [Ni(en)_3]^{2+} + 6H_2 O$$

proceed from left to right with a net production of three extra free molecules; thus they have a positive entropy change. This multiplied by the absolute temperature T gives a substantial value of $T\Delta S^{\circ}$ and, because of the minus sign in front of this term, a negative contribution to ΔG° and hence a positive contribution to the equilibrium constant K. Therefore, not only do we expect the stability constants of these complexes to exceed those of the corresponding ammonia complexes, but also that reactions of the type

$$[Ni(NH_3)_6]^{2+} + 3en \rightleftharpoons [Ni(en)_3]^{2+} + 6NH_3$$

will be thermodynamically favourable. In this particular system it has been shown experimentally that the substitution reaction occurs with a large entropy change accompanied by a substantial change in enthalpy.

The stability of complexes containing chelate rings increases as the number of rings increases, but decreases as the ring size increases from five to seven-membered. Complexes containing eight-membered ring systems are unknown. Because of the stability of chelates they find many uses in industrial and analytical chemistry. The most widely used chelating agent is ethylenediaminetetraacetic acid (EDTA). The disodium salt of this acid reacts with metal ions to form very stable complexes containing up to five five-membered rings (Figure 2.1). The ligand is normally hexadentate, so 1 : 1 complexes are formed, that is, six water molecules are replaced by only one EDTA anion, resulting in a large entropy increase

$$[M(H_2 O)_6]^{2+} + EDTA^{4-} \rightleftharpoons [M(EDTA)]^{2-} + 6H_2 O$$

Table 3.2 shows the magnitude of the stability constants of some metal ions with EDTA. Commercially EDTA is known as Complexone or Sequestrene because of its ability to reduce the concentration of simple (that is hydrated) metal ions in solution by forming stable complexes. It can be used in water-softening, for example,

TABLE 3.2   STABILITY CONSTANTS OF EDTA COMPLEXES

Metal ion	$\log K_1$ for EDTA
Ag^+	7.3
Ca^{2+}	10.8
Cu^{2+}	18.7
Ni^{2+}	18.6
Fe^{2+}	14.3
Fe^{3+}	25.1
Co^{2+}	16.1
Co^{3+}	36.0
V^{2+}	12.7
V^{3+}	25.9

leaving no free calcium or magnesium ions to precipitate with soaps. In volumetric analysis a large number of metal ions can be titrated directly with EDTA using indicators that form weaker complexes with metals than does EDTA, and that possess a colour in the free state different from that in the complexed state.

Further reading

For the preparation of any specific complex the reader should consult *Inorganic Syntheses,* McGraw-Hill, New York, vol. I (1939) — vol. XIV (1973). More general reviews appear in *Preparative Inorganic Reactions,* Interscience, New York, vol. I (1964) — vol. 5 (1968). Brauer's *Handbuch der Praparativen Anorganischen Chemie,* Enke, Stuttgart (1962), will also be found useful.

4 Co-ordination Numbers and Stereochemistry

So far we have considered what types of species combine together to form complexes, and how stable the resultant complex may be. We must now consider the various known co-ordination numbers and how (that is in what shape) the ligands are arranged spatially around the central ion. Werner's prediction that certain metals have preferred co-ordination numbers, for example cobalt six and platinum four, and that the ligands surround the metal in a preferred definite shape enabled great advances to be made in the understanding of co-ordination compounds. Today a wide range of physical methods are available for determining the structure of complexes. Of these methods, the most direct is X-ray crystallography by which the positions of the various atoms are located and bond lengths and angles estimated. Other useful but less direct methods include electronic and vibrational spectroscopy, dipole-moment measurements, and magnetic-susceptibility studies. As a result of such measurements on a large number of compounds, the most commonly occurring stereochemistries for the various co-ordination numbers have been established. We shall deal with these in turn but first it will prove useful to consider, with the help of some very simple theory, the possible geometries that may occur.

4.1 Valence-Shell Electron-Pair Repulsion (VSEPR) Theory

The simple idea that localised electron pairs on atoms repel one another was first developed by Sidgwick and Powell around 1940 but has more recently been refined in detail by Gillespie and Nyholm. The theory has had particular success in explaining the shapes of molecules containing typical elements; the transition elements present some complications but the simple ideas are considered worthwhile at this stage.

In VSEPR theory we assume that the outermost or valency electrons surrounding the central atom in a molecule exist in localised pairs, that is occupying localised molecular orbitals. Each pair of electrons thus occupies a well-defined area of space, and these electron pairs behave as though they repel one another and get as far apart from each other as possible, that is they maximise the least distance apart. Thus if we have two pairs of valency electrons in a molecule they will be at $180°$ from each other, so the molecule will be linear. With three pairs, the angle between them will be $120°$ in a planar structure; four pairs will arrange themselves tetrahedrally at an angle of $109°28'$ apart. These deductions, together with those for five and six electron pairs, can be summarised diagrammatically if we assume that an element B contributes one electron to each bond and various elements A contribute one electron to each bond with B.

AB_2 linear   B——A——B

AB_3 trigonal planar   B——A$<$ ^B_B

AB₄ → AB_4 tetrahedral

AB₅ → AB_5 trigonal bipyramidal

AB₆ → AB_6 octahedral

As well as bonding pairs of electrons, we must consider nonbonding or lone pairs of electrons, and again we assume that they are localised in a doubly filled orbital. The ligands water and ammonia are therefore not linear and trigonal planar respectively because of the presence of the lone pairs of electrons. In each case the total number of electron pairs is four, so the structures are based on the tetrahedral shape. However, they are not regular tetrahedra since the lone pairs of electrons are more localised on the oxygen and nitrogen atoms than are the bonding electrons which rather occupy sausage-shaped orbitals between the bonded atoms. This results in greater repulsion forces between lone-pair electrons and bonding pairs of electrons than between two bonding pairs. In general we can say that repulsions between electron pairs decrease in the order

lone pair − lone pair > lone pair − bond pair > bond pair − bond pair

The bond angles in the water and ammonia molecules are thus reduced from the tetrahedral angle because of the effect of the lone pairs. This effect is best seen by comparing the bond angles in methane, ammonia, and water.

$HCH = 109°28'$ $HNH = 107°18'$ $HOH = 104°30'$

For complexes formed by main-group elements and by transition elements having d^0, d^5 (spin free), or d^{10} configurations, this approach is remarkably successful in predicting molecular shape. Transition-metal complexes having other configurations have not a symmetrical distribution of the d electrons and we must use ligand-field theory (chapter 5) to comprehend the resulting stereochemistries. Let us take some examples to illustrate the application of VSEPR theory to complex compounds.

In the $[Ag(NH_3)_2]^+$ ion silver is in oxidation state +1, so its ground-state electronic configuration is $[Kr]4d^{10}$. Each ammonia molecule is donating a pair of electrons to form the co-ordinate bonds, so we have only two bond pairs to consider (the d electrons being spherically symmetrical about the silver ion); the bond angle will be 180°; that is the ion is linear

$$[H_3 N - Ag - NH_3]^+$$

In $BF_3(NMe_3)$ boron provides three electrons, the three fluorines each provide one electron, and the nitrogen atom of the trimethylamine provides two electrons. We thus have eight electrons or four pairs of electrons, so the resulting structure is tetrahedral (around both boron and nitrogen)

In $TiCl_4(SMe_2)_2$ we have titanium in oxidation state +4, so its electronic configuration is $[Ar] 3d^0$. Each chloride ion and each sulphur atom is regarded as donating a pair of electrons to form the bonds, so we have six bonding pairs and consequently an octahedral structure

An alternative way of treating this example is to say that titanium ($3d^2 4s^2$) provides four electrons, each chlorine atom provides one electron, and each sulphur atom provides two electrons, to constitute the six bonding pairs. Although this is perhaps better chemically ($TiCl_4$ does not actually contain Ti^{4+} ions but is covalent like CCl_4), it does not readily indicate that we are dealing with what may be regarded as a d^0 system and consequently that VSEPR theory applies.

4.2   Co-ordination Numbers and Stereochemistries

4.2.1   Co-ordination Numbers Two and Three
Complexes of co-ordination number two are relatively uncommon and are confined principally to Cu^I, Ag^I, and Au^I; however, for these species it is the characteristic feature of their chemistry. The linear complexes may be cationic, neutral, or anionic, examples being $[H_3 N \rightarrow Cu-NH_3]^+$, $Et_3 P-Au-C{\equiv}CPh$, and $[Cl-Ag-Cl]^-$. Even bidentate ligands form linear complexes by bridging, for example $ClAgNH_2 CH_2 CH_2 NH_2 AgCl$. Mercury(II) also shows a considerable tendency toward two-co-ordination. Solutions of mercury(II) chloride contain largely $HgCl_2$ rather than $HgCl_3^-$ and $HgCl_4^{2-}$. The preferential stability of the linear two-co-ordinate species is indicated by the sudden fall in stability constants from K_2 to K_3 in mercury(II) complexes with, for example, NH_3, CN^-, and I^- (see Table 3.1).

Co-ordination number three occurs only extremely rarely in transition metal complexes. Trigonal planar three-co-ordination occurs in the NO_3^- and CO_3^{2-} ions but not in complexes such as $CsCuCl_3$ or $K_2 Ni(CN)_3$. In the former complex the

X-ray crystal structure shows each copper atom to be surrounded by four coplanar chlorines in a bridged structure ($Cl-MCl_2-Cl-MCl_2$); the diamagnetism of the nickel complex suggests that it exists as dimeric anions $[Ni_2(CN)_6]^{4-}$. While simple halides of the trivalent typical elements, for example BCl_3, often have the discrete trigonal planar structure, trivalent transition metals give halides, such as VF_3, having giant lattices in which the metal atoms are usually surrounded octahedrally by halide ions. Authentic examples of planar three-co-ordination in transition metal chemistry are the HgI_3^- ion in $[Me_3S]^+[HgI_3]^-$ and in $KCu(CN)_2$ where the anion exists in spiral polymeric chains with the copper atoms three-co-ordinate by virtue of cyanide bridges

The carbon and nitrogen atoms bound to copper are almost coplanar with the metal atom but the CCuC angle is about $134°$ (rather than $120°$). This structure is not found in the silver and gold cyano-ions; these contain the two-co-ordinate metals in discrete $[NC-M-CN]^-$ ions.

The compounds $M[N(SiMe_3)_2]_3$, in which $M = Ti$, V, Cr, or Fe, have recently been shown to contain three-co-ordinated metals but their structures are not yet known in detail.

4.2.2 Co-ordination Number Four

On VSEPR theory we expect only tetrahedral complexes for this co-ordination number unless lone-pair electrons are present as in, for example, XeF_4 and ICl_4^- which may be represented as square complexes

with the lone-pair electrons occupying the fifth and sixth octahedral positions. A large number of tetrahedral complexes are thus formed by the elements in the second row (Li to Ne) of the periodic table when they achieve their octet; beryllium and boron form especially numerous species of the types BeX_4^{2-}, BeX_2L_2, BX_4^-, and BX_3L (where L represents a unidentate ligand and X a univalent anion). Tetrahedral complexes are also predominant in four-co-ordinate complexes of the transition elements but the alternative square planar configuration is also common and is a characteristic feature of the chemistry of some metals, notably those with a d^8 electronic configuration such as Pt^{II}.

Tetrahedral complexes of transition elements are mostly anionic or neutral. Cationic complexes such as $[Cd(NH_3)_4]^{2+}$ are uncommon. The simple covalent molecules such as $TiCl_4$, VCl_4, and $Ni(CO)_4$ are tetrahedral. With some metals tetrahedral complexes are formed in solution even in the presence of an excess of ligand, the rather general aqueous substitution reaction of halide ions being

$$[M(H_2O)_6]^{n+} + 4X^- \quad \rightleftharpoons \quad [MX_4]^{(4-n)-} + 6H_2O$$

The addition of concentrated hydrochloric acid to solutions of iron(III) and cobalt(II), for example, results in the formation of $FeCl_4^-$ and $CoCl_4^{2-}$ complexes (rather than $FeCl_6^{3-}$ and $CoCl_6^{4-}$). With other metals, tetrahedral halide complexes can be prepared in the presence of an excess of halide ions but not in strongly co-ordinating solvents such as water. Salts containing species such as VX_4^-, MnX_4^{2-}, and NiX_4^{2-} can be crystallised from solvents such as acetonitrile and alcohol in the presence of an excess of halide ions. Other well-characterised tetrahedral anions include $[Co(NCS)_4]^{2-}$ and $[CoX_3(H_2O)]^-$. While tetrahedral species are common for cobalt(II) they are unknown in cobalt(III) chemistry; they are similarly unknown for chromium(III). Neutral tetrahedral complexes are formed particularly by cobalt(II) and sometimes by nickel(II); they are of the type MX_2L_2, examples being $CoCl_2(py)_2$ and $NiBr_2(Ph_2AsO)_2$. The factors involved in determining whether tetrahedral or octahedral complexes will be the more stable under any given set of conditions are not completely understood; some of these factors will be discussed in chapter 5. It is important to remember that the empirical formula of a compound does not give any information about stereochemistry; for example $NiCl_2(py)_2$ is not tetrahedral but six-co-ordinate with chloride bridges.

Square planar four-co-ordination occurs commonly for Rh^I, Ir^I, Pd^{II}, Pt^{II}, and Au^{III}; for these species it is the most frequently occurring stereochemical form. The occurrence of so many complexes of platinum(II) with this stereochemistry is of considerable historical importance since the occurrence of isomers of formula PtX_2L_2 led Werner to suggest the square structure for these complexes. In the solid state and in solution, cationic, neutral, and anionic complexes of platinum(II) are square, for example $[Pt(NH_3)_4]^{2+}$, $Pt(NH_3)_2Cl_2$, and $PtCl_4^{2-}$. Square planar complexes formed by the other d^8 elements are exemplified by $[Rh(CO)_2Cl]_2$ which contains chlorine bridges, $Ir(CO)Cl(PPh_3)_2$, $PdCl_4^{2-}$, and $AuCl_4^-$. As well as these elements, nickel(II) forms many square complexes but it is more commonly six-co-ordinate. Square nickel(II) complexes include the $[Ni(CN)_4]^{2-}$ ion in, for example, $Na_2Ni(CN)_4.3H_2O$, and the vivid red precipitate of bis(dimethylglyoximato)nickel(II) obtained when dimethylglyoxime is added to neutral nickel(II) solutions. For other metals the square planar configuration occurs only rarely.

4.2.3   Co-ordination Number Five

Complex compounds in this co-ordination number are considerably less numerous than those in co-ordination numbers four and six. Some apparently five-co-ordinate complexes are in fact polymers, and some are mixtures. The pentachlorides of niobium, tantalum, and molybdenum are five-co-ordinate (trigonal bipyramidal) in the gas phase but exist as dimeric molecules in the solid state with slightly distorted

octahedral co-ordination

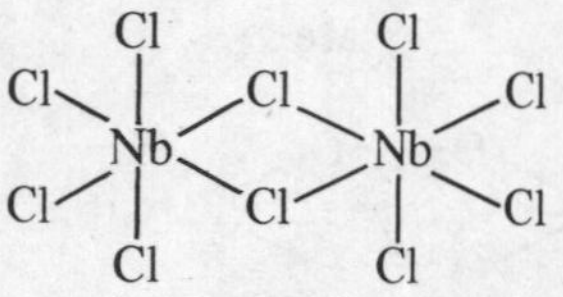

$$
\begin{array}{ccc}
& Cl & Cl \\
Cl\diagdown & | & | \diagup Cl \\
& Nb & \diagup Nb \\
Cl\diagup & | \diagup Cl \diagup & | \diagdown Cl \\
& Cl & Cl \\
\end{array}
$$

The caesium salt $Cs_2 CoCl_5$ contains the slightly distorted tetrahedral $CoCl_4{}^{2-}$ ions as well as Cl^- ions in the lattice. Of the several possible polyhedra for five-co-ordination, only two are commonly found, the trigonal bipyramid and the square-based pyramid.

While the trigonal bipyramidal structure is predicted on VSEPR theory (and is indeed found for the s and p block elements), the energy difference between this structure and the alternative square pyramidal structure is very small for complexes of the transition elements. Indeed the two structures can be readily interconverted; the displacements of the atoms necessary for this interconversion are illustrated in Figure 4.1. The consequence of this small energy difference and structural similarity is that both of the idealised structures are found as well as a large number of compounds with stereochemistries in between the two extremes indicated in Figure 4.1.

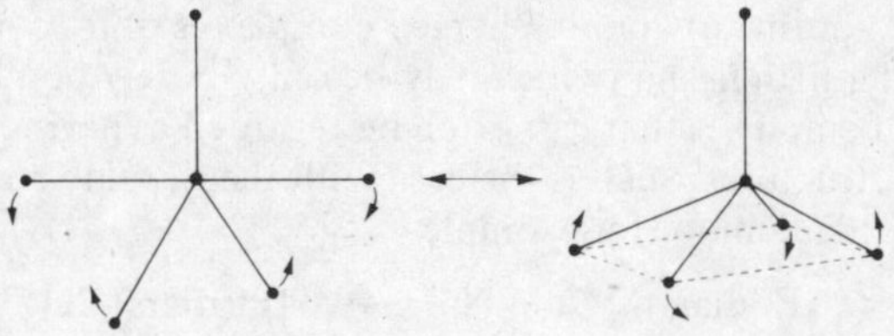

Figure 4.1   Interconversion of trigonal bipyramidal and square pyramidal structures

The trigonal bipyramidal structure is generally found for compounds having five identical ligands. The pentahalides PF_5, PCl_5, $TaCl_5$, $NbBr_5$, and $MoCl_5$ have this structure in the vapour phase, and $SbCl_5$ retains its five-co-ordination in the solid state. Iron pentacarbonyl $Fe(CO)_5$ and ionic species such as $[Mn(CO)_5]^-$, $[Pt(SnCl_3)_5]^{3-}$, $[Ni(CN)_5]^{3-}$, $CuCl_5{}^{3-}$, and $[Co(NCMe)_5]^+$ also have the trigonal bipyramidal structure. Substituted derivatives of iron carbonyl of the types $Fe(CO)_4 L$ and $Fe(CO)_3 L_2$ and several complexes of vanadium(III) halides of the general formula $VX_3 L_2$ have structures based upon the trigonal bipyramid. The complexes $MCl_3(NMe_3)_2$ (M = Ti, V, Cr) are isostructural, being *trans*-trigonal bipyramidal (the three chlorines occupy the terminal positions in the equatorial plane).

Square pyramidal arrangements in which the metal atom sits in the basal plane probably do not occur; distortion of such structures is expected even on simple VSEPR theory. The metal atom is thus normally out of the basal plane and inside the pyramid. In square pyramidal complexes the chemical nature of the apical ligand is frequently different from that of the four ligands in the plane. Examples include $VO(acac)_2$, $Co(NO)(S_2CNR_2)_2$, and $[NiBr(diars)_2]^+$

Such species as these do not always retain their five-co-ordinate structures in solution; for example, there is considerable spectroscopic evidence to suggest that, in solutions of $VO(acac)_2$ in co-ordinating solvents, a solvent molecule occupies the sixth or 'octahedral' position. The π-bonding such as is shown for $VO(acac)_2$ appears to be an important feature in stabilising the square pyramidal structure relative to that of the trigonal bipyramid.

Square (four-co-ordinate) doubly charged complexes of d^8 configurations may, by the addition of a fifth ligand (which may be only loosely bonded), produce square pyramidal complexes that can often be regarded as having only slightly perturbed square structures. Such is the case with the diarsine complexes of platinum(II) and palladium(II), for example

$$[Pt(diars)_2]^{2+} + X^- \rightleftharpoons [Pt(diars)_2 X]^+$$

4.2.4   Co-ordination Number Six

This is by far the most commonly occurring co-ordination number, and the most important for transition metal complexes. The great majority of six-co-ordinate complexes have structures based upon the octahedron. The octahedron has eight faces and six corners, and an atom at the centre is symmetrically surrounded by atoms placed at each of the corners; all the bond lengths are equal and all the bond angles are 90°. It is usual to draw octahedral structures as in the following examples

It is important to realise that each ammonia molecule in $[Co(NH_3)_6]^{3+}$ is in an environment identical to that of all the other ammonia molecules; that is, there is nothing special about the ammonia molecules we have drawn in the vertical bond positions. Let us consider the symmetry of the octahedron in more detail. It will be convenient here to introduce the concept of a symmetry axis. If rotation through an angle $360°/n$ about an axis produces a configuration indistinguishable from the original one, this axis is called a C_n axis. Some of the symmetry axes of an octahedron are shown in Figure 4.2. The axes of highest symmetry are the fourfold

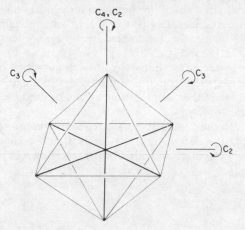

Figure 4.2 Symmetry axes of an octahedron

rotation axes (labelled C_4); there are three of these, passing through the centre of symmetry and through opposite corners of the octahedron. Four threefold axes C_3 pass through the centre and through the midpoints of opposite triangular faces. Twofold axes C_2 join the midpoints of opposite edges of the octahedron via the centre of symmetry; as well as these six C_2 axes, there is a set of C_2 axes coincident with the C_4 axes.

Although there are many fairly regular octahedral complexes, many complexes that are loosely described as octahedral have structures in which the octahedron has been distorted in some way. Two forms of such distortion are common. In *tetragonal distortion* the octahedron is either stretched or compressed along a C_4 (fourfold) axis. This results in the production of a structure with four coplanar bonds of the same length and two bonds at right-angles to these but longer or shorter than those in the set of four. The effects of these tetragonal distortions on bond lengths are illustrated in Figure 4.3. The new structures are less symmetrical than that of the

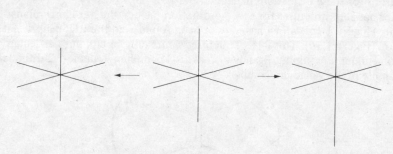

Figure 4.3 Tetragonal distortions of the octahedral structure

octahedron, they belong to the symmetry class D_{4h}. The effect of increasing tetragonal elongation of the octahedron is eventually to remove the axial ligands from the co-ordination 'sphere' of the central atom, leaving a square planar structure. There is therefore no sharp dividing line between tetragonal and square structures. Many copper(II) complexes, including $[Cu(H_2O)_6]^{2+}$, are tetragonal with two long bonds, and these have frequently been treated in elementary texts as being square, for example $[Cu(H_2O)_4]^{2+}$ and $[Cu(NH_3)_4]^{2+}$. In the nickel(II) dimethylglyoxime complex, which we considered (section 4.2.2) to contain nickel in a square planar environment, the X-ray crystal structure of the solid shows that the Ni(DMG)$_2$ units stack on top of each other, so the structure can be considered as tetragonal with long Ni–Ni bonds.

In *trigonal distortion* the octahedron is stretched or compressed along one of the C_3 (threefold) axes. In this deformation the octahedron becomes a trigonal antiprism of symmetry class D_{3d}. The effect of this is shown in Figure 4.4; the

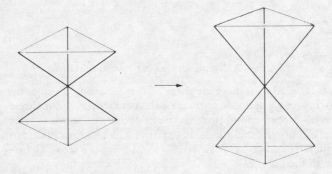

Figure 4.4   Trigonal distortion of the octahedral structure

octahedron is placed on a triangular base and then stretched (in this case) along the vertical C_3 axis. Examples of trigonal distortions are to be found in the hexaquo-complexes of some trivalent transition metal ions in the alums, for example $(NH_4^+)[V(H_2O)_6]^{3+}(SO_4^{2-})_2.6H_2O$. There is considerable X-ray spectroscopic and magnetic evidence to indicate that both tetragonal and trigonal distortions occur in the solid state as well as in solution. We shall consider the reasons for these distortions in chapter 5.

Other possible structures for six-co-ordination include the hexagonal planar structure (Figure 1.1*a*), which has not yet been found, and the trigonal prismatic structure (Figure 1.1*b*). This latter structure occurs only rarely, but particularly with some bidentate sulphur ligands such as *cis*-1,2-diphenylethane-1,2-dithiolate, for example in the vanadium complex

4.2.5 Co-ordination Numbers Greater than Six

Co-ordination numbers seven, eight, nine, ten, eleven, and twelve are known in complex compounds. These high co-ordination numbers are found only rarely with first-row transition elements but are more common with the second and third-row transition elements. The major factors responsible for this are the sizes of the heavier elements and the availability of closely energetic orbitals. The ligands that stabilise high co-ordination numbers are small electronegative unidentate ligands such as H^-, F^-, CN^-, and chelates of the electronegative oxygen and nitrogen atoms. The idealised shapes of species with high co-ordination number can be rationalised in terms of the VSEPR theory. However, as with five-co-ordination, the difference between these idealised shapes for any particular co-ordination number is very small; these differences appear to be comparable to the distortion imposed by vibrationally excited states. We shall consider only one or two examples here.

At least four types of structure are found in seven-co-ordination. The pentagonal bipyramid (Figure 4.5a) is found in the potassium salt of the

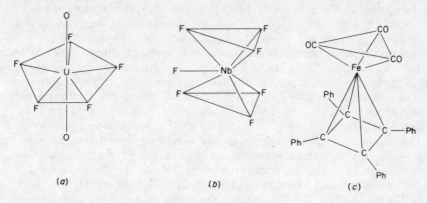

(a)   (b)   (c)

Figure 4.5 Some seven-co-ordinate structures

$[UO_2F_5]^{3-}$ ion. The addition of a seventh ligand above one of the rectangular faces of a trigonal prism gives the capped trigonal prism structure as found in NbF_7^{2-} and TaF_7^{2-} (Figure 4.5b). Similarly the capped octahedral structure has a seventh ligand added above the centre of one face of an octahedron; the $NbOF_6^{3-}$ ion has this structure. The tetragonal base—trigonal base structure is illustrated in Figure 4.5c for the iron carbonyl cyclobutadiene complex $Fe(CO)_3C_4(C_6H_5)_4$.

Eight-co-ordination is considerably more common. The cubic structure as found in ionic lattices such as caesium chloride is not commonly found for molecular species. Instead, two figures that can be constructed by distortion of the cube are observed; these are the square antiprism and the dodecahedron. The square antiprism is formed when one square face of a cube is rotated by $45°$ relative to the face opposite to it. A structure like this is found in ions such as TaF_8^{3-} (Figure 4.6) and ReF_8^{2-}, as well as in the tetrakis(acetylacetonates) of zirconium(IV), cerium(IV), and thorium(IV). The dodecahedral structure is typified

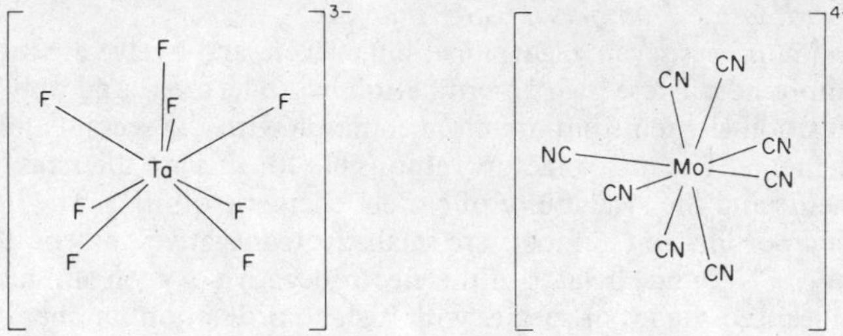

Figure 4.6   Some eight-co-ordinate structures

by the $[Mo(CN)_8]^{4-}$ ion (Figure 4.6); it occurs also in $Na_4Zr(C_2O_4)_4.3H_2O$ and $TiCl_4(diars)_2$. It is also found in polymeric eight-co-ordinate units; for example, K_2ZrF_6 contains zirconium surrounded by eight fluorines, four of which are shared. A further structural type of eight-co-ordination is that of the planar hexagonal bipyramid; this is found mainly in the actinide series of elements, and particularly for dioxouranium(VI) compounds. In $Rb[UO_2(NO_3)_3]$, for example, the collinear OUO group is bisected by a plane of three nitrate groups, each nitrate group being bidentate with respect to the uranium atom.

Perhaps the most remarkable nine-co-ordinate species are the complex hydrides TcH_9^{2-} and ReH_9^{2-}. These have the tricapped trigonal prismatic structure in which the basic trigonal prismatic structure has an extra hydride ion centred over each of its three rectangular faces. Nonahydrates of the trivalent rare earth ions, for example $[Nd(H_2O)_9]^{3+}$, also have this structure, as do a large number of ionic lattices containing these ions. In hexagonal $NaNdF_4$, for example, there are two Nd^{3+} sites, and in each the neodymium ions are nine-co-ordinated by fluoride ions.

Co-ordination numbers ten, eleven, and twelve are not commonly found for molecular species; again they occur principally in the lanthanide and actinide series. Ten-co-ordination is established in $La[(O_2CCH_2)_2NCH_2CH_2N-(CH_2CO_2H)CH_2CO_2].4H_2O$ in which a three-dimensional X-ray analysis has revealed two nitrogen atoms, four oxygen atoms from carboxylate groups, and four oxygen atoms from water molecules to be within bonding distance of the lanthanum. In the sulphate $La_2(SO_4)_3.9H_2O$, twelve sulphate oxygen atoms are bonded to lanthanum in the twelve-co-ordinate eicosahedral structure.

4.3   Isomerism in Co-ordination Compounds

In chapter 1 we mentioned briefly how Werner, using the phenomenon of isomerism, was able to conclude that the six-co-ordinate compounds he had been studying possessed the octahedral structure. The two most important types of isomerism found in co-ordination chemistry are geometrical and optical isomerism; there are many other kinds but we shall mention these only briefly.

4.3.1   Geometrical Isomerism
This kind of isomerism occurs in planar (but not tetrahedral) four-co-ordinate and in six-co-ordinate complexes. Planar compounds of the type MA_2B_2 may show

cis-trans isomerism; the classic examples of this occur in platinum(II) chemistry, for example

| *cis* | *trans* |

For complexes of formula MABCD, three isomers are possible (with B, C, or D *trans* to A), and all three have been isolated in several cases, for example $PtCl(Br)(py)(NH_3)$.

Octahedral complexes of this type MA_4B_2 similarly exhibit *cis-trans* isomerism

| *cis* | *trans* |

Even when compounds occur or can be isolated in only one form, the *cis* and *trans* nomenclature is useful as a method of describing a structure. Two (and only two) isomeric forms are possible for octahedral MA_3B_3 complexes; these isomers are distinguished by the prefixes *mer* and *fac* (meridional and facial). Both isomers can be isolated in the case of $RhCl_3(py)_3$

| *mer* | *fac* |

4.3.2 Optical Isomerism

Optically active substances are capable of rotating the plane of polarised light. If a complex is to show optical activity it must be asymmetric. The complex must have no plane of symmetry, and the structure and its mirror image must be different, that is nonsuperimposable. A useful analogy is left and right hands; the left hand has no plane of symmetry, and its mirror image (the right hand) cannot be superimposed on to the left hand because of the different orientations of the thumb and fingers. In organic chemistry many optically active compounds exist in which a central carbon atom is tetrahedrally surrounded by four different groups. Because of the much greater kinetic reactivity of tetrahedral inorganic complexes, it is difficult enough to prepare compounds of the type MABCD (tetrahedral) let alone resolve them into optical isomers. Tetrahedral complexes containing unsymmetrical chelate ligands can be more easily prepared, and some have been resolved. Their mirror images can best be visualised with the aid of models but can be represented for the general formula $M(A-B)_2$ by

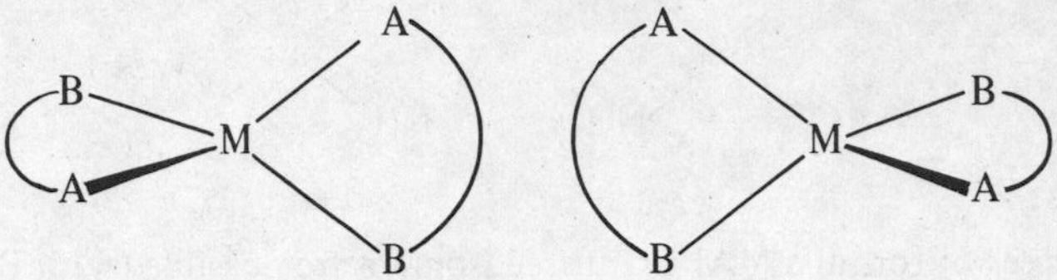

Beryllium(II), zinc(II), and boron(III) form resolvable chelates of this type. The letters A and B in the structures above do not represent necessarily different donor elements but rather the ends of unsymmetrical chelates. Thus a typical example is bis(benzoylacetonato)beryllium(II)

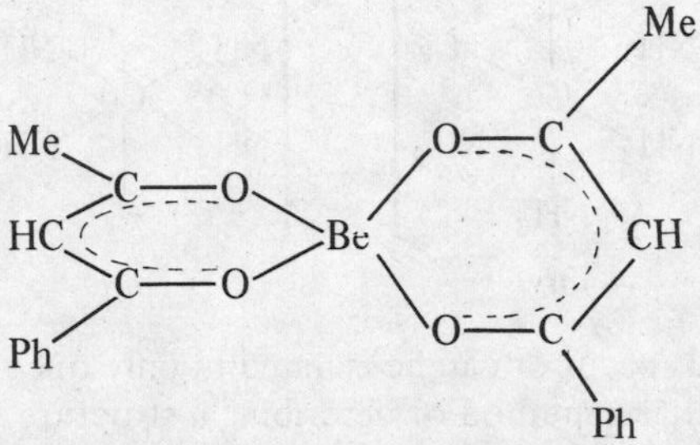

A much larger class of optically active complexes is that of octahedral complexes having chelate ligands (no complex of unidentate ligands MABCDEF has been resolved). The two classes that have been most extensively studied are the tris-bidentate $M(L-L)_3$ and bis-bidentate $M(L-L)_2 X_2$ complexes. These bidentate ligands frequently contain carbon; Werner, setting out to prove that their activity was not necessarily due to the presence of carbon, succeeded in resolving the hydroxo-bridged complex

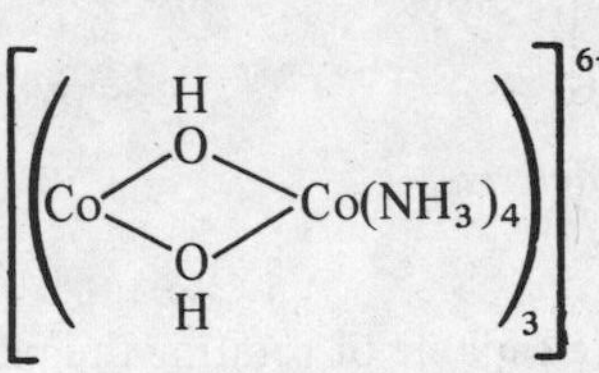

Complexes of the type $M(L-L)_3$ can be exemplified by the tris(oxalato)chromate(III) anion

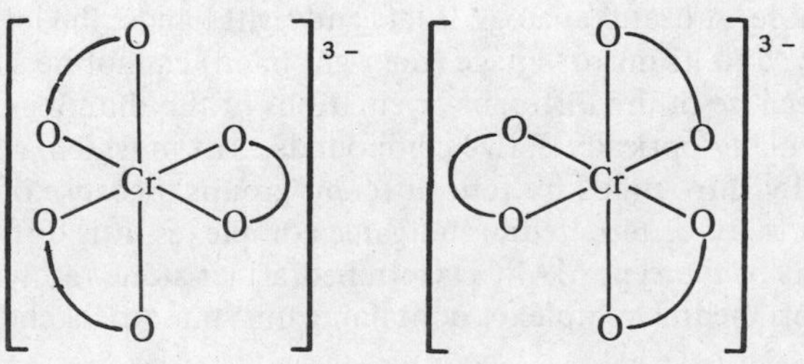

The fact that these compounds can be resolved into optical isomers is good evidence that they have the octahedral structure; neither the hexagonal planar nor the trigonal prismatic structure would show optical activity. Resolution of such compounds is not always possible owing to the speed at which racemisation takes place.

Optical activity in complexes of the type $M(L-L)_2 X_2$ proves the *cis* structure since the *trans* form has a plane of symmetry. Thus for $[Co(en)_2 Cl_2] Cl$ three forms exist, two optically active *cis* forms and an inactive *trans* form

cis forms; active   *trans* form; inactive

4.3.3   Other Types of Isomerism in Complexes

When two compounds having the same molecular formula dissolve to give different ionic species in solution, these compounds are called *ionisation isomers*. Typical examples are $[Pt(NH_3)_4 Cl_2] Br_2$ and $[Pt(NH_3)_4 Br_2] Cl_2$; the former gives bromide ions in aqueous solution while the latter gives chloride ions. A similar type is *hydration isomerism;* there are, for example, three isomers of $CrCl_3 .6H_2 O$ which can be formulated as $[Cr(H_2 O)_6] Cl_3$, $[Cr(H_2 O)_5 Cl] Cl_2 .H_2 O$, and $[Cr(H_2 O)_4 Cl_2] Cl.2H_2($ according to the number of ions they produce in solution and the number of chloride ions that can be precipitated immediately with silver nitrate solution. When ligands possess more than one donor atom, *linkage isomerism* may occur, for example thiocyanato (S-bonded) and isothiocyanato (N-bonded) complexes, $(Ph_3 P)_2 Pd(SCN)_2$ and $(Ph_3 P)_2 Pd(NCS)_2$. Nitro (N-bonded) and nitrito (O-bonded) complexes also show this kind of isomerism, for example $[Co(NH_3)_5 ONO]^{2+}$ and $[Co(NH_3)_5 NO_2]^{2+}$. In *co-ordination isomerism* the isomers consist of complex cations and anions in which the ligands are bound to different metals in each isomer, for example $[Co(NH_3)_6] [Cr(CN)_6]$ and $[Cr(NH_3)_6][Co(CN)_6]$. Substances with the same empirical composition but differing in the size of the smallest unit are called *polymerisation isomers,* for example $Pt(NH_3)_2 Cl_2$ and $[Pt(NH_3)_4] [PtCl_4]$.

5 Theories of Bonding in Complex Compounds

5.1 Magnetic Properties of Complexes

Studies of the magnetic properties of complex compounds have played a very important part in discussions of bonding theories. We must first consider elementary magnetism before the bonding theories are introduced. A more detailed account of the magnetic properties of transition metal ions will be met in chapter 7.

The magnetic properties of substances arise principally from the charge and from the spin and orbital angular momenta of the electrons. We need to distinguish initially between *diamagnetism* and *paramagnetism*. Diamagnetic materials are repelled by a magnetic field whereas paramagnetic materials are attracted by a magnetic field. All substances possess the property of diamagnetism. This effect derives from the presence of closed shells of electrons within substances, and results from an induced magnetic moment being set up in opposition to the applied magnetic field. Paramagnetism derives from the spin and orbital angular momenta of electrons. If all the electrons exist in pairs, their spin and orbital angular momenta cancel each other out, so paramagnetism occurs in substances having unpaired electrons. Because the paramagnetic effect is very much greater (about 10^3 times) than the diamagnetic effect, substances with unpaired electrons are usually paramagnetic. In calculating the extent of the paramagnetism we must correct for the diamagnetism due to the closed shells of electrons since these have the effect of reducing the paramagnetism due to the unpaired electrons. Paramagnetism is very easily measured in the laboratory (chapter 7); it is usual to express it in terms of the magnetic moment μ.

For ions in the first transition series the magnetic moment is given by

$$\mu = \sqrt{[4S(S+1) + L(L+1)]}$$

where S is the total spin quantum number, and L is the total orbital angular momentum quantum number. In practice it is found that the orbital contribution is often considerably less than the spin contribution, so a further approximation can be written

$$\mu = \sqrt{[4S(S+1)]}$$

Now, since $s = \pm\frac{1}{2}$ only and $S = \Sigma s$, the number of unpaired electrons in a system is given by $n = 2S$. Hence, by substitution for S in the above equation we get

$$\mu = \sqrt{[n(n+2)]}$$

This important result is known as the *spin-only formula*. It follows that we should expect:

for 1 unpaired electron, $\mu = \sqrt{3} = 1.73$ B.M.
for 2 unpaired electrons, $\mu = \sqrt{8} = 2.83$ B.M.
for 3 unpaired electrons, $\mu = \sqrt{15} = 3.87$ B.M.

and so on. The unit of magnetic moment in use here is the Bohr Magneton (B.M.)

$(=9.273 \times 10^{-24} A m^2$ molecule^{-1}). The experimental moments found for ions in the first transition series are often quite close to those given by the spin-only formula but in some cases they are considerably in excess of this theoretical value. It is these deviations (most commonly due to orbital contribution) from the spin-only value that enable magnetic measurements to be of value in determining the stereochemistry of metal complexes (chapter 7).

5.2   The Valence-Bond Theory

The ideas involved in the formation of complexes by donor–acceptor reactions were extended by Linus Pauling into the valence bond (VB) theory of the bonding in complex compounds. The basic idea is that pairs of electrons from the donor atoms are donated into empty orbitals on the metal ions. In order to receive these electron pairs, the atomic orbitals on the metal hybridise to give a set of equivalent orbitals having the required symmetry properties. The beryllium ion, for example, thus forms tetrahedral sp^3 hybrid orbitals, which can receive electron pairs from ligands. We can represent the electronic structure of $[Be(H_2O)_4]^{2+}$ thus

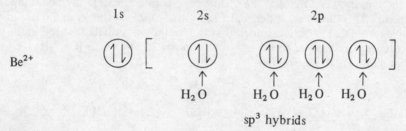

For square complexes the metal orbitals are dsp^2 hybridised, and for octahedral complexes d^{2}sp^3 hybridisation is assumed.

In an octahedral complex the set of six equivalent σ orbitals have their lobes pointing to the corners of a regular octahedron. These hybrids are thus constructed using the metal s, p$_x$, p$_y$, p$_z$, d$_{z^2}$, and d$_{x^2-y^2}$ orbitals (the d$_{xy}$, d$_{xz}$, and d$_{yz}$ orbitals do not point towards the corners of the octahedron). Since two of the metal d orbitals are used in the formation of the hybrid orbitals, only three are left in which to accommodate any electrons possessed by the metal ion. No problems arise for metal ions of configuration d^1, d^2, or d^3; we can illustrate the bonding in the hexaquotitanium(III) ion thus

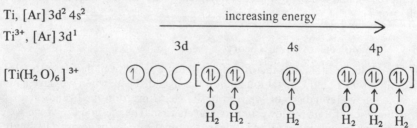

The $[Ti(H_2O)_6]^{3+}$ ion is indeed paramagnetic, with a magnetic moment.

corresponding to the presence of one unpaired electron per titanium ion.

If we now consider an ion with more than three d electrons, we see that we cannot always accommodate these electrons in the 3d orbitals if we wish to feed the electrons in with the maximum number of unpaired spins (that is applying Hund's rules). Cobalt(III) complexes can be used to illustrate the d^6 case. If the six electrons enter the three vacant 3d orbitals, the resulting complex would contain no unpaired electrons and thus be diamagnetic. Such is experimentally found to be the case for the cobaltammines, for example $[Co(NH_3)_6]^{3+}$ (bracketed orbitals contain electrons from the ligands)

$$d^2 sp^3 \text{ hybrids}$$

$[Co(NH_3)_6]^{3+}$ ⟨1↓⟩ ⟨1↓⟩ ⟨1↓⟩ [⟨1↓⟩ ⟨1↓⟩   ⟨1↓⟩   ⟨1↓⟩ ⟨1↓⟩ ⟨1↓⟩]

We might therefore predict that all cobalt(III) complexes will be diamagnetic, but it was found that in the ion CoF_6^{3-} four unpaired electrons are present, so the above description cannot apply. Initially this latter complex was regarded as being an ionic complex having Co^{3+} ions (six electrons to feed into five d orbitals, hence four remain unpaired) and F^- ions. The diamagnetic complexes were then described as 'covalent'. This theory was later replaced by the concept of inner and outer-orbital complexes. In CoF_6^{3-} it is assumed that the 'outer' 4d orbitals are used in the hybridisation as opposed to the use of the 'inner' 3d orbitals in $[Co(NH_3)_6]^{3+}$. We can illustrate this as follows

| 3d | 4s | 4p | 4d |

CoF_6^{3-} ⟨1↓⟩⟨1⟩⟨1⟩⟨1⟩⟨1⟩ [⟨1↓⟩ ⟨1↓⟩⟨1↓⟩⟨1↓⟩ ⟨1↓⟩⟨1↓⟩]○○○

While the valence-bond theory is still widely used in organic chemistry and to some extent in the chemistry of the typical (main group) elements, it has now been superseded in transition metal chemistry by the molecular orbital and ligand-field theories. The greater usefulness of these theories will become apparent as the reader becomes familiar with them. We can criticise the VB approach to complexes on the following grounds. Firstly, it assumes that all the 3d orbitals have the same energy in the complex; we shall see that this directly contradicts the other theories mentioned. Secondly, the rather arbitrary use of 3d and 4d orbitals for bonding when they are known to possess widely differing energies is, to say the least, unsatisfactory. Thirdly, the theory gives us no understanding of electronic spectra (chapter 6), and finally, it does not adequately explain the magnetic data (see chapter 7).

5.3 The Molecular-Orbital Theory

In molecular orbital (MO) theory we construct a series of molecular orbitals from atomic orbitals on the metal ion and on the ligand. Then, when we have a molecular orbital energy-level diagram, we feed in the electrons from the ligands and from the metal ion, filling the lowest energy levels first. The method by which we combine the atomic orbitals is the LCAO (linear combination of atomic orbitals) method. The overlap of atomic orbitals will occur only when these orbitals have the same symmetry;

we therefore need to consider the symmetry properties of the various metal and ligand orbitals.

Let us first consider the octahedral complex ML_6, in which we shall assume that only σ-bonding is important. We first pick out the valence shell atomic orbitals for the metal and the ligands. For a first-row transition metal ion these are the 3d, 4s, and 4p orbitals. Of these nine orbitals only six have lobes pointing towards the corners of an octahedron; these are the $3d_{x^2-y^2}$, $3d_{z^2}$, 4s, $4p_x$, $4p_y$, and $4p_z$ orbitals. The $3d_{xy}$, $3d_{xz}$, and $3d_{yz}$ orbitals have lobes lying between the cartesian co-ordinates and are therefore not suitably shaped for σ-bonding (but they are for π-bonding) in an octahedral complex. In group theoretical terminology we classify the σ-bonding orbitals on the metal as follows

$$3d_{x^2-y^2}, \ 3d_{z^2} \qquad\qquad e_g$$
$$4s \qquad\qquad a_{1g}$$
$$4p_x, \ 4p_y, 4p_z \qquad\qquad t_{1u}$$

These new labels refer to symmetry classes of the orbitals. The label a_{1g} represents a single orbital that is totally symmetrical, e_g represents a pair of orbitals that are equivalent except for their spatial orientations, while t_{1u} represents a set of three orbitals that are equivalent except for their spatial orientations. The symmetry of these orbitals is shown in Figure 5.1.

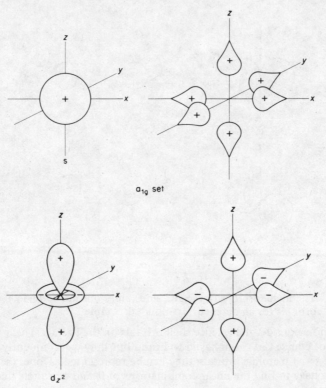

a_{1g} set

d_{z^2}

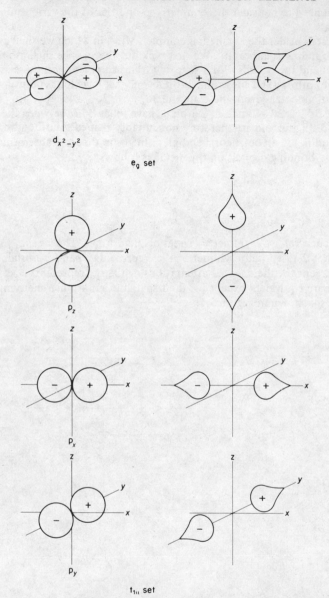

Figure 5.1   The metal orbitals (left-hand side) and their matching ligand group
orbitals for an octahedral σ-bonded complex

We must now consider the valence shell orbitals on the ligands. The σ valence
orbitals will of course vary from ligand to ligand but they will frequently be
composed of s and p atomic orbitals; they can be represented as lobes (see Figure
5.1). We now have to find the linear combinations of ligand σ orbitals that may

bond with the metal a_{1g}, e_g, and t_{1u} orbitals. This is readily done by writing down the linear combination of σ orbitals that has the same symmetry properties as each of the metal orbitals. Thus the linear combination of ligand σ orbitals that has the same symmetry as the $d_{x^2-y^2}$ orbital has a plus sign in the $+x$ and $-x$ directions and a minus sign in the $+y$ and $-y$ directions. This is the combination $(\sigma_x + \sigma_{-x} - \sigma_y - \sigma_{-y})$. The combinations for the other matching ligand orbitals are found similarly; these with their appropriate normalisation constants are listed in Table 5.1 under the heading 'ligand group orbitals'. The only one that presents any difficulty is the d_{z^2} matching combination. The d_{z^2} orbital is a shorthand notation for $d_{2z^2-x^2-y^2}$, and hence the ligand group orbital. Note that the d_{z^2} orbital (Figure 5.1) has a negative collar surrounding the positive lobes along the z axis.

In order to construct an energy-level diagram for our complex ML_6 we must make some assumptions about the energies of the atomic orbitals (3d, 4s, 4p, and σ_L) from which we are forming molecular orbitals. For transition metal complexes the most usual order of energies of metal orbitals is $nd < (n+1)s < (n+1)p$, so

TABLE 5.1   METAL AND LIGAND ORBITAL COMBINATIONS FOR A
σ-BONDED OCTAHEDRAL COMPLEX

Symmetry	Metal orbitals	Ligand-group orbitals
a_{1g}	s	$(1/\sqrt{6})(\sigma_x + \sigma_{-x} + \sigma_y + \sigma_{-y} + \sigma_z + \sigma_{-z})$
e_g	d_{z^2}	$(1/\sqrt{12})(2\sigma_z + 2\sigma_{-z} - \sigma_x - \sigma_{-x} - \sigma_y - \sigma_{-y})$
	$d_{x^2-y^2}$	$(1/2)(\sigma_x + \sigma_{-x} - \sigma_y - \sigma_{-y})$
t_{1u}	p_z	$(1/\sqrt{2})(\sigma_z - \sigma_{-z})$
	p_y	$(1/\sqrt{2})(\sigma_y - \sigma_{-y})$
	p_x	$(1/\sqrt{2})(\sigma_x - \sigma_{-x})$

for first-row complexes we shall write 3d $<$ 4s $<$ 4p. For most ligands, for example H_2O, NH_3, and F^-, the σ_L orbitals used for bonding to the metal are lower in energy, that is more stable, than the metal valence orbitals. The energy-level diagram thus resulting from an octahedral σ-bonded complex is shown in Figure 5.2. For each combination of metal and ligand orbitals we get a lower-energy bonding molecular orbital and a higher-energy antibonding (denoted by an asterisk) molecular orbital. The t_{2g} orbitals on the metal, that is the d_{xy}, d_{xz}, and d_{yz} orbitals, are nonbonding in a σ-bonded octahedral complex, so their energy remains unchanged.

In order to apply this diagram to any particular complex, we feed the six pairs of electrons from the ligands into the lowest energy levels, that is the a_{1g}, t_{1u}, and e_g levels, which thus all become filled (each horizontal line in Figure 5.2 represents an orbital that can contain up to two electrons). Notice the similarity here to the VB approach. The molecular orbitals here that 'receive' the donated electron pairs correspond to the sp^3d^2 orbitals on VB theory. Any valence shell electrons in the

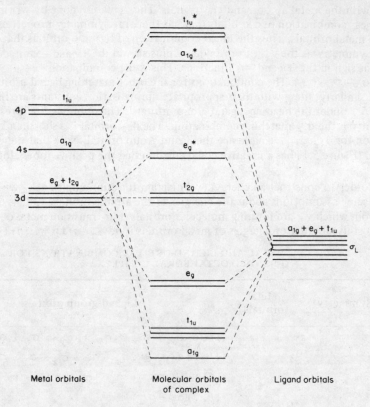

Metal orbitals   Molecular orbitals   Ligand orbitals
                         of complex

Figure 5.2   Molecular-orbital energy-level diagram for a σ-bonded octahedral
             complex ML_6

metal ion are placed in the t_{2g} and e_g* orbitals. These orbitals, because they lie
closer in energy to the metal orbitals from which they were derived, have largely
the character of the pure metal orbitals. Similarly, the a_{1g}, t_{1u}, and e_g molecular
orbitals are largely ligand orbitals and, as we have seen, accommodate the ligand
electrons. The result of forming molecular orbitals from the metal d orbitals has
thus been to separate the initially degenerate d levels into a set of two (e_g*) and a
set of three (t_{2g}).

Let us now see how simple magnetic and spectroscopic properties of transition
metal complexes can be explained by using this MO method. The simplest case of a
d^1 complex ion available in aqueous solution is the hexaquotitanium(III) ion
$[Ti(H_2O)_6]^{3+}$. We can construct the energy-level occupation as follows for this ion

Ti $[Ar]\, 3d^2\, 4s^2$

Ti^{3+} $[Ar]\, 3d^1$

$[Ti(H_2O)_6]^{3+}$ $\underbrace{(a_{1g})^2\ (t_{1u})^6\ (e_g)^4}\ (t_{2g})^1$

                    paired electrons from water molecules

We are thus left with one unpaired electron in the t_{2g} orbital, and this
satisfactorily accounts for the paramagnetism of this ion
($\mu \approx 1.7$ B.M. at room temperature). The ions of d^2 and d^3 configuration follow
naturally with the outermost electronic configurations of $(t_{2g})^2$ and $(t_{2g})^3$. With
the d^4 configuration we have a choice; either we can begin to pair up electrons
in the t_{2g} level to give the configuration $(t_{2g})^4$, or we can put the fourth electron
in the e_g^* level to give the configuration $(t_{2g})^3 (e_g^*)^1$. Which of these two
configurations is actually obtained will depend on the energy separation between
the t_{2g} and the e_g^* orbitals. This energy gap is usually referred to as Δ or 10 Dq.
If Δ is small, the electrons will be expected to fill the levels so as to leave the
maximum number of electrons with unpaired spins (Hund's rules). If Δ is large,
the energy required to put an electron in the higher-energy e_g^* orbitals may be
greater than that required to pair electrons in the t_{2g} levels. We thus predict two
possible types of complexes with metal ions of configurations d^4, d^5, d^6, and d^7.
We can now use the example that we used on VB theory, that is the d^6 Co^{3+}
ion, to illustrate this point. The diamagnetic complexes of cobalt(III) are those
having all the d electrons paired in the t_{2g} levels. The paramagnetic complexes
have the six electrons distributed between the two levels so as to give the
maximum number of unpaired electrons. We can write this as follows

Co $\quad$ [Ar] $3d^7 4s^2$

Co^{3+} $\quad$ [Ar] $3d^6$

$[Co(NH_3)_6]^{3+}$ $\qquad$ $(a_{1g})^2 (t_{1u})^6 (e_g)^4 (t_{2g})^6$

$[CoF_6]^{3-}$ $\qquad$ $(a_{1g})^2 (t_{1u})^6 (e_g)^4 (t_{2g})^4 (e_g^*)^2$

The CoF_6^{3-} ion thus has four unpaired electrons (two in the e_g^* level and two in
the t_{2g} levels). These complexes possessing the maximum number of unpaired
electrons are called *high-spin* or *spin-free* complexes, while those in which the d
electrons are paired as far as possible are referred to as *low-spin* or *spin-paired*
complexes. In the cobalt example, $[CoF_6]^{3-}$ is a high-spin complex and
$[Co(NH_3)_6]^{3+}$ is a low-spin complex. We shall be considering the factors favouring
the formation of high and low-spin complexes in more detail shortly. For the
moment we may just observe that the energy separation Δ is less for CoF_6^{3-} (and
hence the electrons remain unpaired) than for $[Co(NH_3)_6]^{3+}$ where a lower-energy
state is obtained by pairing the electrons in the t_{2g} orbitals.

As well as giving a satisfactory explanation of the magnetic data, the molecular
orbital diagram accounts satisfactorily for the electronic spectra of complexes. For
many electron systems we must use spectroscopic states in place of orbitals when
discussing spectra (see chapter 6) but d^1 systems can be simply described by using
the molecular orbital energy-level diagram. The best-characterised octahedral d^1
complexes are the ions $[Ti(H_2O)_6]^{3+}$ and $[VCl_6]^{2-}$. The violet hexaquotitanium(III)
ion has the visible absorption spectrum shown in Figure 5.3. The ground state of
this ion has the outermost electron configuration $(t_{2g})^1$. The first excited state
corresponds to $(t_{2g})^0 (e_g^*)^1$. The absorption of visible light of a frequency
$\nu = \Delta / h$ results in the excitation of the electron from the $(t_{2g})^1$ ground state to the
$(e_g^*)^1$ excited state. We say that the violet colour of the $[Ti(H_2O)_6]^{3+}$ ion is due

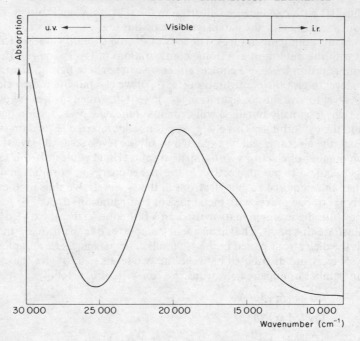

Figure 5.3   The visible absorption spectrum of the $[Ti(H_2O)_6]^{3+}$ ion

to the $t_{2g} \rightarrow e_g{}^*$ transition. Because we have measured the frequency at which absorption occurs and Planck's constant h is known, the energy separation between the t_{2g} and e_g levels has been determined by experiment. For convenience the spectra are usually measured in units of reciprocal centimetres or wavenumbers, cm^{-1}, and by convention this unit is also used as the unit of energy. The spectrum of $[Ti(H_2O)_6]^{3+}$ thus shows an absorption maximum at 20 000 cm^{-1} and the $t_{2g} - e_g{}^*$ separation $\varDelta$ thus equals 20 000 cm^{-1} (note that 20 000 $cm^{-1} \equiv 5000$ Å $\equiv 239$ kJ $mol^{-1} \equiv 2.479$ eV). The rising absorption towards the ultraviolet shown in Figure 5.3 is due to charge-transfer transitions (chapter 6). The spectrum of the $[VCl_6]^{2-}$ ion is very similar in general shape to that of the $[Ti(H_2O)_6]^{3+}$ ion; the absorption maximum occurs at 15 400 cm^{-1}, so for $VCl_6{}^{2-}$ $\varDelta = 15 400$ cm^{-1}. We shall consider more detailed aspects of these spectra in chapter 6.

Molecular orbital energy-level diagrams can be constructed in a similar way for other stereochemistries. However, when the effects of π-bonding are included some of these diagrams become very complicated and hence their general utility is diminished.

5.4   The Crystal-Field Theory

In the MO method we considered the overlap of orbitals, that is covalent bonding, as being of paramount importance. Crystal field (CF) theory starts at

the opposite extreme, that is it is an electrostatic theory. We assume that the
bonding in a complex is a result of electrostatic interaction between the positive
nucleus of the metal cation and the negatively charged electrons of the ligands.
Electrons contained in the cation exert a repulsion force on the ligand electrons.

5.4.1   Octahedral Complexes

Let us consider an octahedral complex which we represent on CF theory as
shown in Figure 5.4; that is we choose the cartesian axes to be the fourfold axes

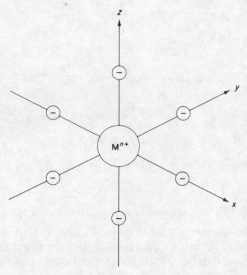

Figure 5.4   Crystal-field model of an octahedral complex

of the octahedron. The ligand electrons are assumed to constitute an electrostatic
field around the metal ion (just as in ionic crystals). We now consider the effect of
this field on the energies of the metal orbitals. In view of the spherical symmetry of
s orbitals it is apparent that an electron in, for example, the metal 4s orbital is
affected equally by the electrostatic field in all six directions. The energy of the 4s
orbital is merely raised by the presence of the electrostatic field. Similarly the p
orbitals, which lie along the cartesian co-ordinates are equally affected by the
field, so they remain degenerate (that is they all have the same energy) in the
complex although again their energy is raised compared with that in the free ion.
It is with the d orbitals that the presence of the crystal field becomes most
important.

The five d orbitals have the approximate electron-density distribution indicated
in Figure 5.5. The e_g set, that is the d_{z^2} and the $d_{x^2-y^2}$ orbitals, point directly along
the cartesian co-ordinates and consequently directly at a ligand atom. The t_{2g}
set, that is the d_{xy}, d_{xz}, and d_{yz} orbitals, have their lobes pointing between the
axes and therefore do not point directly at the ligand atoms in an octahedral
complex. It follows that an electron in an e_g orbital is repelled more strongly by
the ligand electrons than is an electron in a t_{2g} orbital. An electron in a t_{2g}

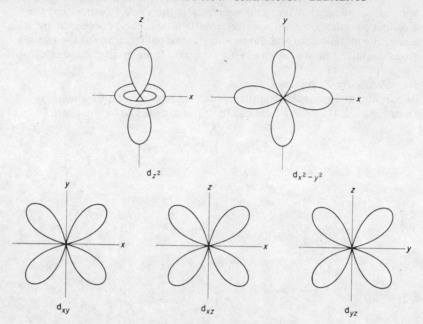

Figure 5.5 The electron-density distribution in the d orbitals

orbital is thus in a lower-energy state than one in an e_g orbital. We therefore deduce from simple electrostatics that in an octahedral field the five d orbitals are split into a group of two orbitals (e_g, higher energy) and a group of three orbitals (t_{2g}, lower energy). This is represented diagrammatically in Figure 5.6. In this figure we have chosen to ignore the increase in energy of all the d orbitals that results upon application of the crystal field; we have focused our attention on the splitting of the orbitals since this is the informative part of the diagram. We therefore draw the free-ion d orbital energies as though they are at the centre of gravity of the energies of the orbitals in the complex.

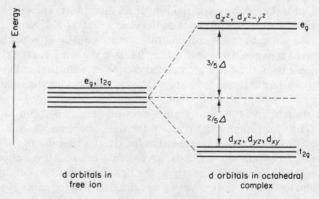

Figure 5.6 Crystal-field splitting diagram for an octahedral complex

We have thus arrived at the same conclusion as when using MO theory, that is, the degeneracy of the d orbitals is removed in an octahedral complex, the e_g set (denoted e_g* on MO theory) being at a higher energy than the t_{2g} set. Further, the separation between these levels, that is Δ, is justified experimentally. The CF theory enables the splitting of the d orbitals to be obtained very simply, regardless of the stereochemistry. However, it does start with the extreme assumption that only electrostatic bonding is important; there is now ample spectroscopic and magnetic evidence that in most complexes at least some covalent bonding occurs. In order to solve this difficulty, most authors use (as we shall here) the term *ligand field theory* to encompass the two extremes of CF and MO. Ligand-field theory can thus be regarded as derived from CF theory by the admission of covalent bonding, or from just the d orbital part of the complete MO diagram. Because the d orbital energy-level diagrams are most easily obtained by using CF theory, we shall now consider the splitting of the d orbitals in crystal fields other than octahedral.

5.4.2   Tetrahedral Complexes

In order to derive the d orbital energy-level diagram for a tetrahedral complex, we must first visualise the tetrahedral structure as being related to a cube. This is shown in Figure 5.7 which illustrates that the tetrahedral structure is formed by joining

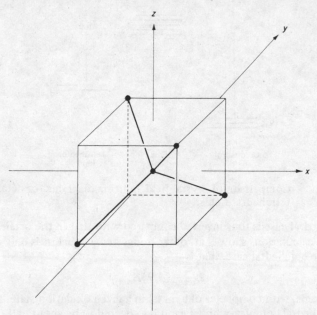

Figure 5.7   The relationship between the tetrahedral structure and a cube.

opposite corners of the top and bottom sides of a cube to the centre. With the cartesian co-ordinates drawn as before, we can now envisage the crystal-field interactions with electrons in the d orbitals in a tetrahedral environment. The d_{z^2}

and $d_{x^2-y^2}$ orbitals now do not point in the direction of the ligand charges and hence are at a lower energy than the d_{xy}, d_{xz}, d_{yz} set which point more closely to the ligand positions. Unlike the situation in the octahedral case, notice that in the tetrahedral field none of the d orbitals points exactly at the ligands. The d_{xy}, d_{xz}, and d_{yz} orbitals have lobes that can be imagined to intersect the edges of the cubes at a distance equal to half a cube edge away from a ligand position, while the lobes of the d_{z^2} and $d_{x^2-y^2}$ orbitals intersect the cube faces at a distance equal to half a diagonal away from a ligand position. The energy separation between the two sets of orbitals is thus less than it was in the octahedral case. This is illustrated in Figure 5.8 which compares the splitting in octahedral and tetrahedral fields. The splitting

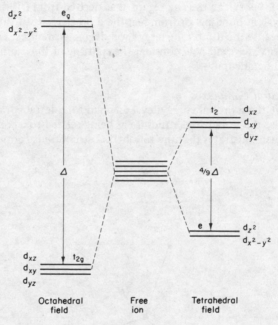

Figure 5.8   Comparison of crystal-field splitting diagrams for octahedral and tetrahedral fields.

in the tetrahedral case is thus inverted compared with that in the octahedral case. and detailed calculations show that, other things being equal, it is only four-ninths of that in the octahedral case, that is

$$\Delta_{tet} = (4/9)\Delta_{oct}$$

In general terms, with complexes of a metal in a given oxidation state, the value of Δ for a tetrahedral complex with a ligand is expected to be about half that for an octahedral complex with the same ligand. This is found experimentally to be the case.

In the tetrahedral case we feed metal electrons into the lower-energy doublet first. Notice that this orbital doublet carries the label e and the upper orbital triplet carries the label t_2. The subscripts g and u do not apply in the tetrahedral case since this does not possess a centre of symmetry (the g and u subscripts refer to symmetry properties with respect to inversion about a centre of symmetry). For the simplest case of a d^1 ion we can illustrate the crystal field approach by using the VCl_4 molecule (despite the obvious covalency in this compound). The configuration of this molecule is represented by $(e)^1(t_2)^0$. The absorption of light by this molecule results in the electronic transition $e \rightarrow t_2$; this absorption is a fairly broad band centred around 8000 cm^{-1}, that is Δ_{tet} for $VCl_4 \approx 8000$ cm^{-1}. This is in reasonable agreement with Δ_{oct} for the $VCl_6{}^{2-}$ ion which we saw earlier is $15\,400$ cm^{-1}, that is about twice the value for Δ_{tet} in VCl_4.

5.4.3 Crystal-field Splittings in Stereochemistries Other than Octahedral or Tetrahedral

The energy-level splittings for the d orbitals in other stereochemical environments can be deduced by using the same ideas of electrostatics and symmetry that were used in the octahedral and tetrahedral cases.

The square planar crystal-field diagram can be obtained in one of two ways. Either we can imagine the square plane as being derived from the octahedral shape by removal of two *trans*-axial ligands to infinity, or we can start with the square planar structure. The former method is appropriate to tetragonal complexes, and we shall see the energy-level diagrams obtained by this method later in this chapter. In the square planar structure, the ligands are attached to the metal ion along the x and y axes (using the same model as in the octahedral case). The orbital of highest energy is thus the $d_{x^2-y^2}$ orbital since this points directly at the ligands. The d_{xy} orbital is expectedly the next highest in energy because it lies in the same plane as the ligands. The d_{xz} and d_{yz} orbitals will be degenerate and may be higher in energy than the d_{z^2} orbital which has its major component pointing along the z axis away from the ligands. Thus we arrive at the qualitative square planar energy-level diagram as shown in Figure 5.9. While this diagram is believed to be correct in the case of, for example, the $[PtCl_4]^{2-}$ ion, for other combinations of metal

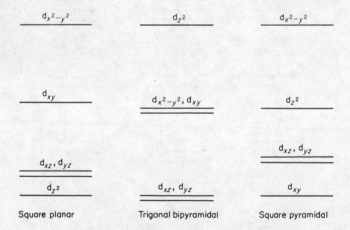

Figure 5.9   Qualitative crystal-field splitting diagrams for different stereochemistries.

ion and ligand the d_{z^2} level may rise above that of the d_{xz} and d_{yz} orbitals. This is also the case when there are ligands in the z axes that are too far away to be normally considered as being bonded, and of course in tetragonal complexes.

The two crystal-field splitting diagrams for the pure five-co-ordinate geometries are also included in Figure 5.9; their main features follow from the arguments that we have made for other stereochemistries.

5.5   Ligand-Field Theory in Application

5.5.1   High- and Low-spin Complexes
We have already seen that on MO and CF theories the magnetic and spectral properties of transition metal complexes are concerned principally with nonbonding electrons. We shall now examine the number of these electrons in each of the configurations from d^1 to d^{10} for octahedral complexes.

For the configurations d^1, d^2, d^3, d^8, d^9, and d^{10} there is only one possible way to distribute the electrons among the t_{2g} and e_g orbitals so as to obtain the lowest-energy configurations. We obtain the lowest-energy configurations for d^1, d^2, and d^3 species by feeding electrons first into the lower-energy t_{2g} levels with the electron spins unpaired. With d^8, d^9, and d^{10} configurations the t_{2g} level is necessarily filled with electrons and the e_g level fills by maintaining the maximum number of unpaired electron spins. These configurations are illustrated in Figure 5.10 together with the magnetic moment expected for each configuration if the spin-only formula is assumed to apply.

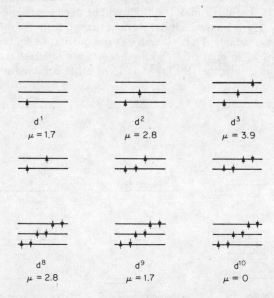

Figure 5.10   Ligand-field diagrams showing the ground state electron occupancy and spin-only magnetic moments (in B.M. units) for six of the configurations from d^1 to d^{10} for octahedral complexes.

For the configurations d^4, d^5, d^6 and d^7 there are two possibilities as was pointed out in section 5.3. The most important single factor in determining whether high-spin or low-spin complexes arise is the ligand-field splitting parameter Δ. When Δ is larger than the pairing energy P for the electrons, the electrons pair in the t_{2g} orbitals as far as is possible. If the energy required to pair up the spins of the electrons and place them in the same orbital (electrostatic repulsion) is greater than Δ, the electrons will distribute themselves between the t_{2g} and e_g levels so as to have the maximum number of unpaired spins. In the former case we have the *strong-field* ($\Delta > P$) arrangement with low-spin complexes, while in the latter we have the *weak-field* ($\Delta < P$) arrangement with high-spin complexes. These configurations are illustrated in table 5.2 together with the number of unpaired electrons and hence the *spin-only* magnetic moment for each type of complex.

Complexes of configurations d^1, d^2, d^3, d^8, and d^9 have the same number of unpaired electrons as in the free ion, while those of configurations d^4, d^5, d^6, and d^7 may or may not have the maximum number of unpaired electrons. Two examples will suffice here to illustrate that this is found experimentally. We saw the electronic structures of the d^6 complexes $[Co(NH_3)_6]^{3+}$ and $[CoF_6]^{3-}$ described

TABLE 5.2.   HIGH AND LOW-SPIN COMPLEXES OF OCTAHEDRAL
d^4, d^5, d^6, AND d^7 CONFIGURATIONS

| No. of d electrons | High-spin | | No. of unpaired electrons | μ (B.M.) | Low-spin | | No. of unpaired electrons | μ (B.M.) |
| | Weak-field arrangement | | | | Strong-field arrangement | | | |
	t_{2g}	e_g			t_{2g}	e_g		
4	↑↑↑	↑	4	4.9	↑↓↑↑		2	2.8
5	↑↑↑	↑↑	5	5.9	↑↓↑↓↑		1	1.7
6	↑↓↑↑	↑↑	4	4.9	↑↓↑↓↑↓		0	0
7	↑↓↑↓↑	↑↑	3	3.9	↑↓↑↓↑↓	↑	1	1.7

on VB and MO theories, and for comparison we describe them here using ligand field notation

$[CoF_6]^{3-}$ μ(experimental) = 4.3 B.M.   high-spin $t_{2g}^4 e_g^2$

$[Co(NH_3)_6]^{3+}$ μ(experimental) = 0   low-spin t_{2g}^6

Iron(III) complexes provide an example from the d^5 configuration

$[FeF_6]^{3-}$ μ(experimental) = 5.9 B.M.   high-spin $t_{2g}^3 e_g^2$

$[Fe(CN)_6]^{3-}$ μ(experimental) = 2.4 B.M.   low-spin t_{2g}^5

In these two examples the deviations of the experimentally determined magnetic moments at room temperature from the spin-only moments are considerable, but the moments leave no doubt about the existence of the two separate classes of complex.

In tetrahedral fields complexes of the first-row transition elements are always high-spin. This is to be expected in view of the lesser splitting in the tetrahedral case.

5.5.2   *The Magnitude of Δ; the Spectrochemical Series*

Studies of the electronic spectra of transition-metal complexes enable the ligand-field splitting parameter Δ $(10Dq)$ to be determined experimentally. From the values so obtained (table 5.3 shows typical values) certain generalisations can be made.

TABLE 5.3 SOME TYPICAL VALUES OF Δ (cm^{-1}) FOR OCTAHEDRAL COMPLEXES

Metal ion	Ligands			
	$6F^-$	$6H_2O$	$6NH_3$	$6CN^-$
Ti^{3+}	17 500	20 100		22 100
V^{3+}	16 100	18 500		23 400
Cr^{3+}	15 100	17 400	21 600	26 600
Fe^{3+}	14 000			35 000
Co^{3+}	13 000	19 100	22 900	34 800
Fe^{2+}		10 400		33 800
Co^{2+}		9 300	11 000	
Ni^{2+}		8 500	10 800	

(1) For a given ligand, Δ does not vary much among ions of the first transition series in the same oxidation state. For example, values of Δ for hydrates of M^{2+} lie within the region $7\,500 - 12\,000$ cm^{-1}

(2) For a given ligand Δ increases rapidly with increase in oxidation state of the metal; for example values of Δ for hydrates of M^{3+} are in the $14\,000 - 25\,000$ cm^{-1} region.

(3) For a given ligand and stereochemistry the metal ions can be arranged in order of increasing Δ, and this order is more or less independent of the nature of the ligand.

(4) For a given metal ion the ligands can be arranged in order of increasing Δ; this order is more or less independent of the nature of the metal ion.

These last two series are known as the *spectrochemical series*. The spectrochemical series for common metal ions in the first transition series is

$$Mn^{2+} < Ni^{2+} < Co^{2+} < Fe^{2+} < V^{2+} < Fe^{3+} < Cr^{3+} < V^{3+} < Co^{3+}$$

For the more common ligands the spectrochemical series is

$$I^- < Br^- < Cl^- < S^{2-} < F^- < OH^- < CH_3COO^- < C_2O_4^{2-} \approx O^{2-}$$
$$< H_2O < py \approx NH_3 < en < bipy < phen < PR_3 < CO \approx CN^-$$

This spectrochemical series for ligands is found to be useful in a semiquantitative way. The ligands are arranged in order of increasing ligand field strength. For example, the CN^- ion is known as a strong-field ligand while the I^- ion is known as a weak-field ligand. Ligands such as water and ammonia are said to have medium field strengths. If we know the position of the absorption maxima in the spectrum of a metal ion with one ligand, we can make approximate predictions about the positions of the bands for complexes of this ion with other ligands. Similarly, if a metal forms low-spin complexes with, for example, ethylenediamine, it will form low-spin complexes with, for example, bipyridyl or cyanide ions. However, the series must be used with some caution since minor variations sometimes occur. The order in the series is very difficult to understand on crystal-field theory; for example, it is hard to envisage why the negatively charged F^-ion exerts a weaker field than a neutral water molecule. Ligand-field theory, which takes covalent bonding into account, does account for these apparent anomalies.

A certain amount of rationalisation of the spectrochemical series of ligands can be achieved through the consideration of π-bonding. In the molecular orbital diagram for an octahedral complex (Figure 5.2) we deliberately, for the sake of simplicity, included only σ-bonding. However, the value of Δ is affected by π-bonding. The t_{2g} orbitals on the metal ion are nonbonding with σ-ligands but become bonding orbitals with ligands having π symmetry. The ligand π orbitals may be p or d orbitals or π^* molecular orbitals. Some suitable combinations of metal and ligand π-bonding orbitals are shown in Figure 5.11. (For simplicity only

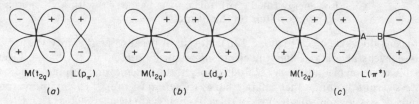

Figure 5.11   Some suitable combinations of metal and ligand π-bonding orbitals.

one ligand orbital is drawn in each example.) The effects that π-bonding of these types have upon the value of Δ can now be considered. For simplicity we shall consider two extreme but important cases.

In the first case we have ligand orbitals of π symmetry filled with electrons and low in energy. Such a case is represented by filled p orbitals on F^- or Cl^- ions, and the overlap by Figure 5.11a. The π-bonding molecular orbital that is formed has a lower energy than the p_π atomic orbitals from which it was formed (Figure 5.12a), and contains the electrons from the filled ligand π orbitals. This could thus be represented as L→M π interaction in which the ligand is regarded as a π electron donor. The π-antibonding molecular orbital has a higher energy than the original t_{2g} metal orbital, and since the bonding π orbital is filled, the π^* t_{2g} orbital must contain the metal electrons. The net effect is to decrease Δ, the separation between the orbitals that have largely the character of the original d orbitals on the metal. We thus expect to find ligands such as halide ions at the weak-field end of the spectrochemical series.

The second case to consider is that of ligands possessing vacant high-energy π orbitals. These are the so-called π-acceptor ligands which give rise to M→L π interaction. The effect of the π interaction here is to stabilise the t_{2g} level, thus

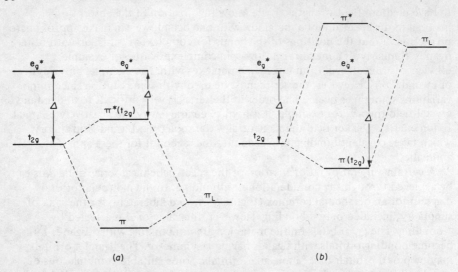

(a)   (b)

Figure 5.12   The effect of π-bonding on the magnitude of Δ: (a) with
low-energy filled π orbitals on the ligand; (b) with high-energy
empty π orbitals on the ligand

increasing the value of Δ (Figure 5.12b). Note that because π_L is empty the
π (t_{2g}) level still contains only the electrons originally present on the metal. Ligands
capable of behaving in this way include phosphorus, arsenic, and sulphur donors which
possess vacant d orbitals that can overlap as shown in Figure 5.11b, and polyatomic
ligands possessing multiple bonds. In this latter class are ligands such as CO, CN⁻,
and ethylene which possess empty π^* molecular orbitals; these are represented as
A—B in Figure 5.11c. These ligands cause a large splitting of the d levels and are thus
the strong-field ligands found at the strong-field end of the spectrochemical series.
The ligands of intermediate field strength have little or no π-bonding effects.

5.5.3   Crystal-field Stabilisation Energies

In an octahedral complex of a d^1 ion the d electron resides in a t_{2g} orbital. This
orbital lies (see Figure 5.6) at an energy $(2/5)\Delta$ below the arbitrary zero of
energy that we assign to the unsplit d orbitals in the presence of a ligand field. This
extra stability arising out of the splitting of the d orbitals is known as the
crystal-field stabilisation energy (CFSE). Some authors refer to this as ligand-field
stabilisation energy, but since it had its origin in crystal-field theory we retain here
the more commonly used term CFSE.

Every electron that resides in a t_{2g} orbital of an octahedral complex contributes
0.4Δ to the CFSE. However, an electron in an e_g orbital is at a higher energy than
that of the unsplit d orbitals, and thus it destabilises the system by $(3/5)\Delta$. Every
electron in an e_g orbital therefore contributes -0.6Δ to the total CFSE. It is thus
easy to work out the total CFSE for all octahedral configurations; the results are
shown in table 5.4. Since values of Δ are in the region $10-30\,000$ cm⁻¹ it follows
that CFSEs will typically lie in the range $5\,000-40\,000$ cm⁻¹, that is $60-480$

TABLE 5.4. CRYSTAL-FIELD STABILISATION ENERGIES FOR OCTAHEDRAL COMPLEXES

Number of d electrons	High-spin Configuration		CFSE (Δ)	Low-spin Configuration		CFSE (Δ)
d^1	t_{2g}^{1}	e_g^{0}	0.4			
d^2	t_{2g}^{2}	e_g^{0}	0.8			
d^3	t_{2g}^{3}	e_g^{0}	1.2			
d^4	t_{2g}^{3}	e_g^{1}	0.6	t_{2g}^{4}	e_g^{0}	1.6
d^5	t_{2g}^{3}	e_g^{2}	0.0	t_{2g}^{5}	e_g^{0}	2.0
d^6	t_{2g}^{4}	e_g^{2}	0.4	t_{2g}^{6}	e_g^{0}	2.4
d^7	t_{2g}^{5}	e_g^{2}	0.8	t_{2g}^{6}	e_g^{1}	1.8
d^8	t_{2g}^{6}	e_g^{2}	1.2			
d^9	t_{2g}^{6}	e_g^{3}	0.6			
d^{10}	t_{2g}^{6}	e_g^{4}	0.0			

kJ mol^{-1}. These amounts of energy are not insignificant, and we shall now try to illustrate how their existence adds to the power of ligand-field theory in interpreting the properties of transition-metal complexes.

5.5.4 Stabilities of Hexaquo-ions
Let us consider the heats of hydration of divalent metal ions in the first transition series, that is the heats of the reactions

$$M^{2+}(g) + H_2O \text{ (excess)} \rightarrow [M(H_2O)_6]^{2+}aq$$

If the ratio of charge to radius were the only factor affecting the bond strength, we should expect the heats of hydration of the metal ions to increase steadily from Ca^{2+} to Zn^{2+} as the ions become progressively smaller. The actual values are those illustrated by circles in Figure 5.13. However, if the value of the CFSE for each hydrated ion (all high-spin) is deducted from the observed value of the hydration energy, we get the expected smooth curve passing through the crosses in Figure 5.13. Notice that for the Ca^{2+} (d^0), Mn^{2+} (d^5), and Zn^{2+} (d^{10}) ions which have no CFSE the experimental values already lie on this smooth curve. The idea that CFSE is playing an important part in the energy cycle is thus justified.

5.5.5 Octahedral or Tetrahedral Co-ordination?
The three most commonly found stereochemistries for first-row transition metal complexes are octahedral, tetrahedral, and square planar. Of these the octahedral structure, which uses the largest number (six) of the available σ-bonding orbitals on

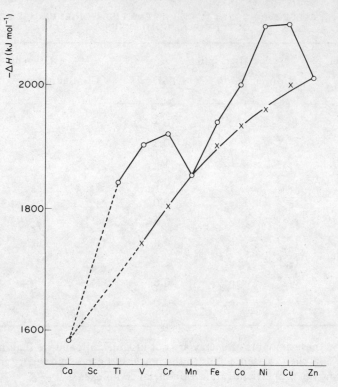

Figure 5.13   Hydration energies of divalent metal ions: O, experimental values; X, values corrected for CFSE.

the metal, is the most stable unless significant steric or electronic effects are present.

The steric effects that may be important in determining the shapes of complex ions concern ligand—ligand repulsion forces. Repulsion between bonded ligands in complexes occurs through coulombic and van der Waals type forces. Large negatively charged one-atom ions such as I^- and Br^- will experience greatest repulsion, while small neutral molecules such as H_2O and NH_3 will exhibit the smallest interligand repulsion forces. Large molecules and small ions will be in the intermediate repulsion range. Since the splitting of the d orbitals and hence the CFSE is least with the weak-field ligands, it follows that the tetrahedral structure is most likely to be preferred with weak-field ligands that have large mutual repulsion effects. In general terms we observe that, while water and ammonia form predominantly octahedral complexes, it is the iodide and bromide ions that form a large number of tetrahedral complexes. These ions combine weak ligand fields with maximum interligand repulsions. Cyanide ion, which exerts less interligand repulsion, has also one of the strongest ligand fields, so the difference between the octahedral and tetrahedral CFSE for any particular metal is at a maximum for this ion. This results in a considerable preference for the octahedral shape in most cases.

In the first-row transition series, certain metal ions have a greater tendency than others to form tetrahedral complexes. The number of electrons in the d orbitals in each case is responsible for the electronic effects that contribute to this tendency. The CFSE for tetrahedral complexes can be calculated in the same way as for octahedral complexes. In tetrahedral complexes the metal d electrons enter the e levels first (Figure 5.8) and each e electron contributes $(3/5)\Delta_{tet}$ to the CFSE. Electrons in t_2 levels contribute $-(2/5)\Delta_{tet}$ to the total CFSE. Since $\Delta_{tet} = (4/9)\Delta_{oct}$ we can compare the CFSEs for tetrahedral and octahedral fields by using Δ_{oct} as a common unit. Thus for d^1 in a tetrahedral field the CFSE is

$$(3/5)\Delta_{tet} = (3/5) \times (4/9)\Delta_{oct}$$
$$= 0.266\Delta_{oct}$$

The values of CFSE thus obtained for all the d^0 to d^{10} configurations are plotted in Figure 5.14 (lower graphs) along with the values for octahedral high-spin

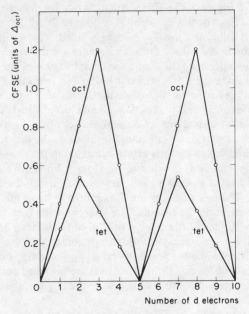

Figure 5.14   Crystal-field stabilisation energies for d^0 to d^{10} high-spin ions in octahedral and tetrahedral fields.

configurations (upper graphs). The figure shows that in terms of CFSE the octahedral field is preferred most by the d^3 and d^8 configurations. Chromium(III) (d^3) and nickel(II) (d^8) do indeed show an exceptional preference for octahedral as opposed to tetrahedral co-ordination; tetrahedral chromium(III) complexes are virtually unknown. The d^4 and d^9 configurations also show a strong preference for octahedral stereochemistry, for example Mn(III) and Cu(II), but, as we shall see shortly, distortion is common in these cases. For the d^1, d^2, d^6, and d^7 ions the preference for octahedral stereochemistry is not so great, and tetrahedral complexes are expected to be more common. Thus, unlike Cr(III), vanadium(II) (d^2) forms tetrahedral $VX_4{}^-$ species (X = Cl, Br, I), and unlike nickel(II), cobalt(II) (d^7) forms a very large number of tetrahedral complexes with neutral ligands as well as with anions. For the configurations d^0, d^5 (high-spin), and d^{10}, the d electrons have no influence on stereochemistry since there is no CFSE in either case.

5.5.6   Distortions from Perfect Symmetry; the Jahn–Teller Effect

So far we have considered the electronic effects that contribute towards the stability of octahedral as opposed to tetrahedral stereochemistry. A very important electronic effect tells us that in many cases these pure stereochemistries will not be obtained but that distortion will occur. This effect was first formulated by Jahn and Teller in 1937 and is named after them. We shall begin by considering the copper(II) ion which perhaps provides the best illustration of this effect.

In an octahedral field the copper(II) ion (d^9) has the electronic structure $t_{2g}{}^6 e_g{}^3$ as shown in Figure 5.15a. Since the e_g levels are degenerate, we do not know whether the $d_{x^2-y^2}$ or the d_{z^2} orbital is doubly filled. Let us assume that the d_{z^2} orbital is doubly filled and the $d_{x^2-y^2}$ singly filled. The result of this orbital occupancy is that the ligands along the x and y axes are drawn closer to the metal. This is because they experience the electrostatic attraction of the Cu^{2+} ion more than do the ligands along the z axis which are shielded from the Cu^{2+} ion by an extra electron. We thus have a tetragonal distortion in which the octahedron has been contracted along the x and y axes. If we had chosen the $d_{x^2-y^2}$ as the doubly filled orbital, again a tetragonal distortion would occur but this time with a contraction along the z axis. For copper(II) compounds the distortion frequently observed experimentally is in fact the contraction along the x and y axes resulting in a structure having four short bonds and two longer bonds. We thus have the $d_{x^2-y^2}$ orbital singly filled as shown in the tetragonal energy-level diagrams in Figure 5.15b and c. The pair of electrons thus prefer to be in the d_{z^2} orbital where they feel less repulsion from the ligands. The energy of the d_{z^2} orbital is thus no longer degenerate with that of the $d_{x^2-y^2}$ orbital but is instead below it in energy. As the tetragonal distortion increases with elongation along the z axis, so the level of the d_{z^2} orbital drops while the levels of the $d_{x^2-y^2}$ and d_{xy} orbitals rise (Figure 5.15c). Eventually, if the ligands along the z axis are removed completely, we have the square planar structure (Figure 5.9) in which the d_{z^2} level may even drop below that of the d_{xz} and d_{yz} doublet level.

A close examination of Figure 5.15 will enable the reader to see that, in this d^9 case, distortion has resulted in a lower energy for the system, that is distortion has caused stabilisation. Since the *centre of gravity* rule applies to the tetragonal splittings, the energy of the system in Figure 5.15b is less than that in Figure 5.15a by an amount equal to $0.5\delta_{e_g}$ (there is no net energy change in the t_{2g} orbitals since the increase in energy of the two electrons in the d_{xy} orbital is equal to twice the amount by which the four electrons in the d_{xz} and d_{yz} orbitals are decreased in energy). Therefore, with copper(II) complexes, tetragonal rather than octahedral complexes are found even when six identical ligands are present as in $[Cu(H_2O)_6]^{2+}$. In many copper(II) compounds the distortion is so great (represented by Figure 5.15c) that the co-ordination is almost square.

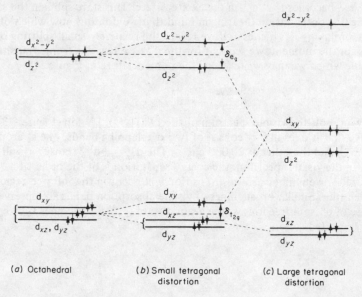

(a) Octahedral   (b) Small tetragonal distortion   (c) Large tetragonal distortion

Figure 5.15   Tetragonal distortion of an octahedral d^9 complex; bracketed levels are degenerate.

In general terms this Jahn–Teller effect, or theorem as it is sometimes known, can be stated: *If, as first represented, a molecular system gives a degenerate electronic state, it will be found to have distorted itself so as to split the degenerate state.* The theorem tells us only that a distortion will occur; it does not indicate the direction or magnitude of the distortion. As well as the $t_{2g}^6 e_g^3$ configuration we should expect the $t_{2g}^6 e_g^1$ configuration to undergo distortion with similar stabilising effect. Low-spin cobalt(II) and nickel(III) have this configuration and their 'octahedral' complexes are indeed found to have undergone distortion. Such distortions would also be correctly predicted for ions of the $t_{2g}^3 e_g^1$ configuration, that is high-spin chromium(II) and manganese(III). For ions of configurations

$$t_{2g}^3; \quad t_{2g}^3 e_g^2; \quad t_{2g}^6; \quad t_{2g}^6 e_g^2; \quad t_{2g}^6 e_g^4$$

no distortions are expected; these are the configurations most likely to give rise to perfectly octahedral ML_6 complexes.

Interesting cases arise with complexes containing partly filled t_{2g} levels. The extent of the splitting of the t_{2g} levels (δt_{2g}; Figure 5.15b) is considerably less than that of the e_g levels because the t_{2g} orbitals do not point directly at the ligand positions. Thus, while we should expect distortion for the configurations $t_{2g}^1, t_{2g}^2, t_{2g}^4$, and t_{2g}^5, it is more difficult to detect experimentally; however, these small distortions can be detected by magnetic measurements (chapter 7). In the first excited states of these configurations, electrons are present in e_g orbitals; these excited states undergo the larger Jahn–Teller distortion, and it is these distortions that are readily observed in the electronic spectra. For the d^1

case $t_{2g}{}^1 e_g{}^0$ becomes $t_{2g}{}^0 e_g{}^1$ in the excited state. This state splits in the same way as the d^9 configuration (bearing in mind that we do not know which of the levels d_{z^2} or $d_{x^2-y^2}$ lies highest). If we ignore the relatively small splitting of the t_{2g} levels for the moment, we see that for the d^1 case two electronic transitions are possible which we may write (using the nomenclature in Figure 5.15) as

$$d_{xz,yz} \rightarrow d_{z^2}$$
$$d_{xz,yz} \rightarrow d_{x^2-y^2}$$

If we now consult the visible spectrum of the $[Ti(H_2O)_6]^{3+}$ ion (Figure 5.3), we observe that it does in fact consist of two overlapping bands, one at around $17\,000$ cm^{-1} and the other at $20\,000$ cm^{-1}. The $d_{xz,yz} \rightarrow d_{xy}$ transition will not be observed in the visible spectrum; the energy separation is of the order of a few hundred wavenumbers, so this transition will occur in the infrared region of the spectrum. Similar broadenings of visible absorption bands are observed in the case of d^6 spin-free ions whose ground state $t_{2g}{}^4 e_g{}^2$ becomes $t_{2g}{}^3 e_g{}^3$ in the first excited state.

6 Electronic Spectra of Transition-metal Complexes

6.1 Introduction

We have, to a very limited extent, already discussed the magnetic and spectral properties of complexes in chapter 5. We now look at these topics in greater detail. The theory behind this material is difficult; in order that some progress can be made towards understanding the spectra and magnetism of complexes, it will be assumed that the student has some elementary knowledge of spectroscopic term symbols. Some of the more theoretical aspects of electronic spectra and a discussion of term symbols are to be found in the references cited in the bibliography. The study of the spectral and magnetic properties of complexes provides us with much information that we have seen can shed considerable light on bonding theories. Spectral properties are concerned with differences between the ground state and the excited states of molecules, whereas the magnetic properties are concerned with the nature of states very close in energy to the ground state. Since electronic spectra can be measured directly in a few minutes, information is often quickly forthcoming from this technique. Because these electronic spectral bands are usually very broad, the spectra are not often used as 'fingerprints' or in looking for functional groups as is done in infrared spectroscopy. However, a combination of spectral and magnetic studies often enables the stereochemistry of a complex to be determined, and may also indicate something about the extent of distortion therein.

The visible and ultraviolet regions of the spectrum, that is $10\,000-50\,000$ cm^{-1}, are those in which electronic excitations usually occur. Light of these wavelengths may be absorbed by a complex for a variety of reasons. These possible interactions can be classified as follows.

(1) *Ligand spectra.* Ligands such as water or organic molecules possess characteristic absorption bands that are normally in the ultraviolet. These bands remain in the spectra of the complexes but may be shifted somewhat from their original position.

(2) *Counter-ion spectra.* A complex ion must be associated with a counter-ion; a knowledge of the spectrum of this counter-ion must be known in order to interpret the spectrum due to the complex ion.

(3) *Charge-transfer spectra.* These spectra arise from transitions between orbitals that are principally those of the metal and orbitals that are largely ligand orbitals.

(4) *Ligand-field spectra.* These arise from transitions between the d orbitals of the metal that have been split in a ligand field; they are otherwise known as d–d spectra.

It is principally this last category with which we shall be most concerned, but a brief indication of the importance of each of the other spectral types will now be given.

6.1.1   Ligand Spectra

Some organic ligands possess visible absorption, and virtually all absorb in the ultraviolet region of the spectrum. There are three principal types of ligand absorption which we now consider in turn.

(1)   n→σ* *transitions*. When atoms in molecules possess lone-pair electrons that are not involved in the internal bonding, the transition of lowest energy is n→σ*. The σ-bonding orbitals are filled and lower in energy than the level containing the nonbonding electrons. These transitions are found in molecules such as water, alcohols, amines, and alkyl halides. This ultraviolet absorption limits the use of these substances as solvents in which to measure spectra. For example, it becomes difficult to balance spectroscopic cells with water above $50\,000$ cm^{-1} or with chloroform above $40\,000$ cm^{-1} because of the intense absorption above these frequencies.

(2)   n→π* *transitions*. These transitions can occur in molecules containing atoms that are involved in π-bonding and that also contain nonbonding electron pairs. Such molecules are aldehydes and ketones which contain the $>C\!\!=\!\!O$ group. Again the nonbonding electrons are in a level higher than the σ and π-bonding levels but now the lowest unoccupied level is the π* level. Hence the transition of lowest energy is n→π*; this occurs in the ultraviolet region. The next transition is π→π*; this normally occurs above $50\,000$ cm^{-1}, in what is described as the vacuum ultraviolet.

(3)   π→π* *transitions*. Molecules possessing double or triple bonds but no atoms with nonbonding electrons have the π-bonding orbital as the highest occupied level and the π* orbital as the lowest unoccupied level. The transition of lowest energy is thus π→π*. Such transitions occur in olefins, dienes, and aromatic systems.

Some ligands possess more than one of these types of band. Pyridine, for example, has n→π* and π→π* transitions. In complexes containing these ligands, the ligand absorption does not change significantly and is frequently recognisable by its characteristic band shape.

6.1.2   Counter-ion Spectra

There are many simple anions that do not absorb radiation in the visible or ultraviolet regions, for example Cl^-, SO_4^{2-}, and ClO_4^-. However, of these only perchlorate does not form stable complexes with metal ions. Therefore, if we wish to measure the spectrum of a hexaquo-ion, the solution must be acidic to prevent hydrolysis, so perchloric acid would be a suitable acid to add. The spectrum of $[Ti(H_2O)_6]^{3+}$ (figure 5.3) is thus obtained in perchloric acid, but if hydrochloric acid is used the positions of the band maxima move to lower wavenumbers as species such as $[TiCl(H_2O)_5]^{2+}$ begin to become important. Many oxo-anions such as NO_3^- and NO_2^- have intense absorptions in the ultraviolet, and others have absorption bands that either tail into the visible, for example CrO_4^{2-}, or actually absorb in the visible, for example MnO_4^-.

6.1.3   Charge-transfer Spectra

The chromate and permanganate ions mentioned above owe their absorptions to charge-transfer transitions. These transitions take place between molecular orbitals of which one is largely a metal orbital and the other is largely a ligand orbital. They thus correspond either to a transfer of an electron from the metal to the ligand (*metal oxidation*) or to a transfer of an electron from the ligand to the metal (*metal reduction*). They are intense absorptions usually located in the ultraviolet but frequently tailing into the visible. They account for the colours of d^0 complexes such as $TiCl_6{}^{2-}$ and $CrO_4{}^{2-}$, and of d^{10} compounds such as the red HgI_2. The transitions are not of course limited to these configurations; they occur widely in the spectra of transition metal complexes. Because they are so much more intense than the ligand-field bands, they may obscure the ligand-field spectra if the two types of transition are close to each other in energy. We shall now consider the various types of charge-transfer spectra.

(1)   *Ligand-to-metal charge transfer.* Since this corresponds to metal reduction, the more easily the metal is reduced and the ligand oxidised, the lower the energy of the transition. Thus iodide, being a readily oxidisable ligand, frequently forms compounds that have charge-transfer absorption in the visible region; cations that are reducible are thus expected to form coloured iodides, for example TiI_4 violet-black, HgI_2 red, AgI yellow, etc.

The spectra of the hexahalo-complexes $MX_6{}^{n-}$ have been studied in some detail. Figure 6.1 shows a simplified molecular orbital energy-level diagram for this type

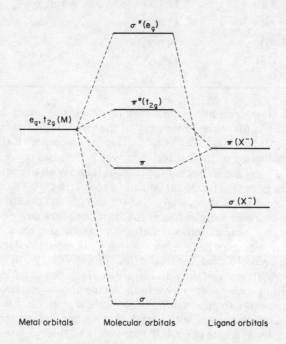

Figure 6.1   Simplified molecular orbital energy-level diagram for an octahedral $MX_6{}^{n-}$ complex (X^- = a halide ion)

of complex. The σ and π levels are filled with bonding electrons while the π^* and σ^* levels are vacant. Four transitions are thus expected; in order of increasing energy these are

$$\pi \to \pi^* \quad (t_{2g})$$
$$\pi \to \sigma^* \quad (e_g)$$
$$\sigma \to \pi^* \quad (t_{2g})$$
$$\sigma \to \sigma^* \quad (e_g)$$

Notice that each of these represents transfer of charge from an orbital that has largely the character of the ligand to an orbital that has largely the character of the metal. The spectra of some d^0 and ^{1} complexes MX_6^{2-} are listed in table 6.1. The spectra are not quite as simple as the table implies, and they are not yet

TABLE 6.1   SPECTRA OF HEXAHALO-SALTS (in cm^{-1})

Complex	$\pi \to \pi^*$ (t_{2g})	$\pi \to \sigma^*$ (e_g)
$TiCl_6^{2-}$	31 850	42 500
$TiBr_6^{2-}$	25 200	36 500
$ZrCl_6^{2-}$	42 400	
$ZrBr_6^{2-}$	38 900	
VCl_6^{2-}	21 400	

fully understood. However, some general features become apparent from the limited assignments in the table. The energy of the $\pi \to \pi^*$ (t_{2g}) transition decreases in the order $ZrCl_6^{2-} > TiCl_6^{2-} > VCl_6^{2-}$. This is in accord with the established order of ease of reduction of these metal ions, which increases in the series $Zr^{4+} < Ti^{4+} < V^{4+}$. A similar logical sequence is found with the bromides. Only the first transition is located (up to 50 000 cm^{-1}) for $ZrBr_6^{2-}$. Titanium(IV) is a better oxidising agent than zirconium(IV), so for $TiBr_6^{2-}$ the resulting shift of the absorption bands to lower energies means that two bands are now observed for the titanium(IV) complex. Vanadium(IV) is such a strong oxidising agent to bromide ions that no VBr_6^{2-} complex has yet been prepared. It should be mentioned that other bands have been observed in the spectrum of the VCl_6^{2-} ion but these have not been assigned with any certainty and so are not included in table 6.1.

Strongly oxidising cations often have ligand-to-metal charge-transfer bands that are low enough in energy to encroach into the visible region. Thus the iron(III) ion forms many coloured complexes with ligands that act as good electron donors (reducing agents), for example phenols, halide ions, and thiocyanate ions. In the first-row tetrahedral complexes of formula MO_4^{n-}, the trend observed in their colours follows that expected from the oxidising power (oxidation state) of the metal ion. Thus the transition of lowest energy (oxygen to metal) decreases in energy in the series

$$VO_4{}^{3-} > CrO_4{}^{2-} > MnO_4{}^{-}$$

white   yellow   purple

(2) *Metal-to-ligand charge transfer.* In this case it is necessary for the metal to be readily oxidisable and the ligand to be readily reducible. Ligands that have low-lying vacant orbitals suitable for receiving an electron from the metal include pyridine, 2,2'-bipyridyl, and 1,10-phenanthroline. These have vacant π^* orbitals and consequently give strongly coloured complexes with readily oxidised cations such as Ti^{3+}, V^{2+}, Fe^{2+}, and Cu^+. Spectra of this M→L type have not been very thoroughly studied. In order to be observed the charge-transfer bands must lie lower in energy than the $\pi \rightarrow \pi^*$ transitions in the ligand. We can represent the charge-transfer transitions diagrammatically as in figure 6.2. Two transitions are possible and are represented as ν_1 and ν_2; however, assignments of this class of spectra have so far been only tentative.

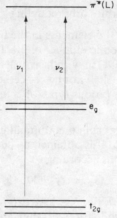

Figure 6.2   A simple representation of metal-to-ligand charge-transfer transitions.

(3) *Metal-to-metal charge transfer.* Whenever an inorganic compound contains a metal in two different oxidation states, 'intravalence' charge transfer can occur, one metal ion acting as the ligand, so to speak. Compounds in which this occurs are typically intensely coloured. Well-known examples include Prussian blue $KFe^{III}[Fe^{II}(CN)_6]$, molybdenum blue which contains Mo^{IV} or Mo^V in MoO_3, and the black gold compound $Cs_2 Au^I Au^{III} Cl_6$. In this latter compound intravalence electron transfer occurs between $[Au^{III}Cl_4]^-$ and $[Au^I Cl_2]^-$ groups.

6.2   Ligand-Field Spectra

6.2.1   Terms and Russell–Saunders States

So far we have discussed electronic spectra as though one electron is being excited into a higher energy level, and we have ignored the effect of other electrons that might interact with this electron. For the d^1 case as in $[Ti(H_2O)_6]^{3+}$ we have only one electron in the valence shell, so complications from other electrons do not arise. However, when more than one valence electron is present we must take into account the interactions or *couplings* between the quantum numbers for the individual electrons. An electron in an ion can initially be assigned to a set of four one-electron quantum numbers, which are designated as n (the principal quantum number), l (the azimuthal or orbital angular momentum quantum number), m_l (the magnetic orbital quantum number), and s (the spin quantum number). If we now consider a two-electron system, for example a d^2 ion, the interactions that can occur are of three types. These are spin–spin coupling in which the spin angular momenta of the two electrons couple, orbit–orbit coupling in which the orbital angular momenta couple, and spin–orbit coupling. In spin–orbit coupling we are concerned with coupling of spin and orbital angular momenta on the same electrons; the coupling of the spin of one electron with the orbital angular momentum of a different electron is so small that it is ignored. In the Russell–Saunders scheme it is assumed that

$$\text{spin–spin coupling} > \text{orbit–orbit coupling} > \text{spin–orbit coupling}$$

This is found to be the case for the elements in the first transition series, and in general for elements up to an atomic number of 30. Thereafter spin–orbit coupling becomes more important, so with heavy elements a different scheme known as *jj* coupling applies; we shall not be concerned with this here. It will now be appropriate to consider these various types of coupling in turn.

Spin–spin coupling. The resultant spin quantum number for a system of electrons is denoted by the capital letter S. It is obtained by coupling the spin quantum numbers for the separate electrons according to

$$S = (s_1 + s_2), (s_1 + s_2 - 1), \ldots, (s_1 - s_2)$$

Thus for two electrons $S = 1$ or 0. For $S = 1$ we can represent the spins as being coupled in parallel, that is ↑↑, while for $S = 0$ they are coupled with their spins opposed, that is ↑↓. For three electrons we can have $S = {}^3/2$ (↑↑↑) and $S = ½$ (↑↑↓) only, while for four electrons $S = 2$ (↑↑↑↑), 1 (↑↑↑↓), or 0 (↑↑↓↓).

Orbit–orbit coupling. For two electrons whose orbital angular momenta we can represent as l_1 and l_2, the total orbital angular momentum quantum number L is obtained by adding l_1 and l_2 vectorially. Thus

$$L = (l_1 + l_2), (l_1 + l_2 - 1), \ldots, (l_1 - l_2)$$

For two p electrons, $l_1 = 1$ and $l_2 = 1$, so L can have values 2, 1, and 0. For two d electrons, $l_1 = l_2 = 2$, so L can have values 4, 3, 2, 1, and 0. Now, just as for single electrons the value of l defines an orbital (that is $l = 0$ means we have an s orbital, $l = 1$ a p orbital, and so on), so the value of L defines a quantum number or energy state for a system of electrons. We use capital letters for these states or *term letters* as follows

L	0	1	2	3	4	5	6
Term letter	S	P	D	F	G	H	I

Spin—orbit coupling. The total angular momentum quantum number J, for a system of electrons, is obtained by coupling the resultant spin and orbital momenta according to

$$J = (L + S), (L + S - 1), \ldots, (L - S)$$

Terms of different L values have appreciably different energies; for given values of L and S we get several *levels* close together. The number of these levels is called the *multiplicity*. The multiplicity is given by the formula $(2S + 1)$. This is readily seen by considering examples

$$S = 0 \quad J = L \qquad\qquad\qquad \text{multiplicity 1 (singlet)}$$
$$S = \tfrac{1}{2} \quad J = (L + \tfrac{1}{2}),\ (L - \tfrac{1}{2}) \quad \text{multiplicity 2 (doublet)}$$
$$S = 1 \quad J = (L + 1), L, (L - 1) \quad \text{multiplicity 3 (triplet)}$$

Note that if $L = 0$ the multiplicity can only be 1, that is S terms can only have one value of J.

All this information that we have been considering can be conveyed in one symbol, known as a *term symbol*.

$$\text{Term symbol} = {}^{(2S+1)}L_J$$

For example, if $L = 2$ and $S = 1$ the term symbol is ^3D and the three states of the triplet are $^3D_3, ^3D_2, ^3D_1$.

Terms for d *electron systems.* In the d^1 case we have only one electron, so $L = l = 2$ and $S = \tfrac{1}{2}$. The only term arising is thus ^2D.

The d^2 case gives rise to L values of 4, 3, 2, 1, and 0, and so gives G, F, D, P, and S terms and multiplicities of 1 and 3. Not all of these terms are allowed since some of them correspond to a contravention of the Pauli exclusion principle. The terms that are allowed are $^3P, ^3F, ^1S, ^1D$, and ^1G. We can decide the lowest-energy or *ground* term with the help of Hund's rules.

(1) The most stable state is the one with maximum multiplicity.
(2) For a group of terms with the same multiplicity, the one with the largest value of L lies lowest in energy.

We thus have ^3F as the ground term for d^2; the order of the higher terms may also be indicated, but less reliably, by these rules.

We are now in a position to summarise the Russell—Saunders coupling scheme for a d^2 ion in one diagram, that is figure 6.3. The energy separations in this figure are not drawn to scale. The separations between the levels obtained by spin—orbit coupling are measured in terms of a spin—orbit coupling constant λ; the separation between the levels of J values J and $(J + 1)$ is $(J + 1)\lambda$. Thus for the ^3F term, $L = 3$ and $S = 1$, so $J = (3 + 1)$, 3, and $(3 - 1)$, that is the three states are $^3F_4$, $^3F_3$, and $^3F_2$. These have separations of 4λ between $^3F_4$ and $^3F_3$, and 3λ between $^3F_3$ and $^3F_2$.

In addition to these couplings, a further splitting of the energy levels can occur under the influence of an external magnetic field (the Zeeman effect). The J levels are split into $(2J + 1)$ equally spaced levels corresponding to the number of values that can be assumed by the magnetic quantum number m. These values are

$-J$, ..., 0, ..., $+J$; the separation between them is $g\beta H$ where g is known as the Landé splitting factor, β is the Bohr magneton, and H is the applied magnetic-field strength. The Landé formula relates g to L, S, and J.

$$g = \frac{3}{2} + \frac{S(S + 1) - L(L + 1)}{2J(J + 1)}$$

It can be seen from this equation that g varies between 1 and 2 as J varies between L ($S = 0$) and S ($L = 0$).

The terms arising from the other d^n configurations can similarly be deduced and Russell–Saunders coupling applied. Since many terms arise from the d^3, d^4, d^5, d^6, and d^7 configurations, these are normally consulted in reference books rather than committed to memory. The terms arising from all the d^n configurations are listed in table 6.2 (numbers in parentheses refer to the number of times the term occurs).

TABLE 6.2   TERMS ARISING FROM THE d^n CONFIGURATIONS

Configuration	Terms
d^1, d^9	^2D
d^2, d^8	^3F, ^3P, ^1G, ^1D, ^1S
d^3, d^7	^4F, ^4P, ^2H, ^2G, ^2F, $^2D(2)$, ^2P
d^4, d^6	^5D, ^3H, ^3G, $^3F(2)$, ^3D, $^3P(2)$, ^1I, $^1G(2)$, ^1F, $^1D(2)$, $^1S(2)$
d^5	^6S, ^4G, ^4F, ^4D, ^4P, ^2I, ^2H, $^2G(2)$, $^2F(2)$, $^2D(3)$, ^2P, ^2S

It is useful to know the ground term arising from each d^n configuration. Fortunately this can readily be deduced as follows. Since the ground term is the one with the largest value of S, we can write down the multiplicity, bearing in mind that there are five degenerate d orbitals in the free ion, so pairing of spins necessarily begins after d^5. The multiplicities of the ground terms in going from d^1 to d^9 are thus 2, 3, 4, 5, 6, 5, 4, 3 and 2. We now need to find the maximum value of M_L ($M_L = L, L-1, ..., -L$) corresponding to the maximum number of unpaired spins and obeying the Pauli exclusion principle. This operation is readily performed as shown in table 6.3. Apart from the d^5 case all the configurations give rise to D or F ground terms. This fact, together with the fact that the terms arising from the configuration d^n are the same as those from d^{10-n}, considerably simplifies our treatment of spectral and magnetic data.

Energies of terms above ground terms. The energy separation between the various terms can be described in terms of two *electron-repulsion* parameters known as *Racah parameters* B and C. If we are concerned only with d electrons, the energy differences between states of different spin multiplicities are given by sums of multiples of B and C. Energy differences between states of the same spin multiplicity are often multiples of B only. For example, in the d^3 V^{2+} ion

$$^4F - ^4P \quad \text{separation} = 15B$$
$$^4F - ^2G \quad \text{separation} = 4B + 3C$$

TABLE 6.3   GROUND TERMS FOR d^n CONFIGURATIONS

Configuration	Example	m_1					M_L	S	Ground term
		2	1	0	−1	−2			
d^1	Ti^{3+}	↑					2	1/2	^2D
d^2	V^{3+}	↑	↑				3	1	^3F
d^3	Cr^{3+}	↑	↑	↑			3	3/2	^4F
d^4	Cr^{2+}	↑	↑	↑	↑		2	2	^5D
d^5	Mn^{2+}	↑	↑	↑	↑	↑	0	5/2	^6S
d^6	Fe^{2+}	↑↓	↑	↑	↑	↑	2	2	^5D
d^7	Co^{2+}	↑↓	↑↓	↑	↑	↑	3	3/2	^4F
d^8	Ni^{2+}	↑↓	↑↓	↑↓	↑	↑	3	1	^3F
d^9	Cu^{2+}	↑↓	↑↓	↑↓	↑↓	↑	2	1/2	^2D

Ions of the first transition series have a C/B ratio of about 4, with $B \approx 1000$ cm^{-1}. These parameters are determined experimentally from the spectra.

6.2.2   Selection Rules

We saw in figure 5.3 that the intensity of the absorption by the $[Ti(H_2O)_6]^{3+}$ ion due to the d–d transition was much less than that due to charge transfer, that is the rising absorption in the ultraviolet. This is because the d–d transition is *forbidden* under the quantum-mechanical selection rules for light absorption. These rules may be stated as follows.

(1) *Spin forbidden.* Transitions in which there is a change in the number of unpaired electron spins are forbidden; that is, for a transition to give optical absorption, $\Delta S = 0$. Transitions for which $\Delta S \neq 0$ are said to be spin-forbidden

(2) *Orbitally forbidden (Laporte rule).* Transitions involving the redistribution of electrons in a single quantum shell are forbidden. Thus d→d and p→p transitions are forbidden but s→p and p→d transitions are allowed. Transitions should only involve one electron, so that for a transition to be allowed $\Delta L = \pm 1$. Transitions of the type g→g and u→u are described as being *parity forbidden.*

If these selection rules held strictly, we should neither observe ligand field spectra nor see the colours of transition metal ions. Indeed, when both rules apply, for example in the d^5 case Mn^{2+}, the intensity of the absorption and hence the colour is very weak. The spin selection rule is relaxed somewhat by spin–orbit coupling. The intensities of spin-forbidden bands relative to those of spin-allowed bands increase with increase in the spin–orbit coupling constants. However, spin-forbidden bands are extremely weak.

The Laporte selection rule may be relaxed as follows. If a complex contains a centre of symmetry it must use a vibronic mechanism to show optical absorption. For an octahedral complex, for example, there are a number of normal modes of vibration, some of which are antisymmetric with respect to the inversion centre, that is they are u-type vibrations. On mixing the vibrational and electronic parts of the wave function (so-called *vibronic coupling*), the ground term may become mixed with a g-type vibration and the excited term with a u-type vibration. The transition thus instead of being g→g becomes partly g→u and is allowed. If the complex already lacks a centre of symmetry, mixing of p and d orbitals can occur to some extent. Thus transitions can occur between d orbitals containing different amounts of p character. Using the molecular orbital picture, covalent bonding utilising p orbitals of the ligand and d orbitals of the metal causes mixing. In a tetrahedral complex the g character of the bonding d orbitals, that is the t_2 set, is partly lost. The nonbonding e levels (π-bonding ignored) retain all the g character of the pure d orbitals. Transitions of the type e→t_2 thus occur between states having different amounts of u character and become partially allowed. It is thus commonly found that tetrahedral complexes have absorption bands of the order of 10^2 times as intense as those of similar octahedral complexes.

Finally, there is the phenomenon known as *intensity stealing*. When a ligand-field transition occurs close to a charge-transfer band, its intensity often increases markedly. This is believed to be due to mixing of the electronic wave functions of the forbidden excited term with the allowed level, resulting in electronic transitions to the excited term becoming more allowed. This gaining of intensity decreases rapidly as the separation between the allowed and forbidden bands increases.

Experimentally the intensities of absorption bands are measured in terms of the *molar absorption coefficient* (ϵ) Spectrophotometers measure optical density (d) as a function of wavelength; d is given by

$$d = \log_{10} \frac{I_0}{I}$$

where I_0 is the intensity of the light incident on the sample and I the intensity of the emergent light. The molar absorption coefficient is then given by

$$\epsilon = \frac{d}{cl}$$

where c is the concentration of the solution in gram-moles per litre and l is the path length in centimetres (1-cm cells are frequently used). Typical values of the molar extinction coefficient for the types of absorption that we have discussed are indicated in table 6.4.

Band breadths. If the transitions occurred between one discrete energy level and other discrete levels, we would expect to find sharp absorption peaks corresponding to the energy differences between the levels. In practice electronic spectral bands are usually very broad with band widths commonly of the order $1\,000-3\,000$ cm^{-1}. The principal reasons for this broadening are the accompanying vibrational excitations, the occurrence of the Jahn—Teller effect, and spin—orbit coupling effects.

TABLE 6.4   MOLAR ABSORPTION COEFFICIENTS FOR VARIOUS
TYPES OF TRANSITION METAL COMPLEX

Type of transition	Example	Typical value of ϵ
Spin forbidden, Laporte forbidden	$[Mn(H_2O)_6]^{2+}$	0.1
Spin allowed, Laporte forbidden	$[Ti(H_2O)_6]^{3+}$	10
Spin allowed, Laporte 'partially allowed' by d–p mixing	$[CoCl_4]^{2-}$	5×10^2
Spin allowed, Laporte allowed (charge transfer)	$[TiCl_6]^{2-}$	10^4

The effect of vibrations in complex molecules is to modulate the ligand-field strength. On the crystal-field model we should expect the magnitude of Δ ($10Dq$) to vary rapidly as the metal–ligand bond distance changes. Thus the energy separations between the terms is spread over a range of energies corresponding to the values of Δ obtained during the bond vibrations. This broadening can be overcome to some extent by measuring spectra at low temperatures where fewer vibrational levels are occupied and the probability of transition is reduced.

Broadening due to the Jahn–Teller effect has already been discussed in chapter 5. Similar broadening and splitting of absorption bands occurs when the ligands surrounding the metal ion are not all the same. The ligand-field strength then varies in different directions and the symmetry is lower than when all the ligands are identical.

When terms are split by spin–orbit coupling, the separation between the resulting states (figure 6.3) is a multiple of the spin–orbit coupling constant and is normally of the order of several hundreds of wavenumbers. We might therefore expect to see fine structure in bands due to this coupling. However, this is normally observed only when the spin–orbit coupling constants are very large, notably with some second and third-row transition metal ions. Usually this fine structure is obscured by the vibrational broadening.

6.2.3   Terms arising in Ligand Fields

We have so far considered the terms arising from free ions. Before we can interpret the spectra of complex ions we must consider the effect of ligand fields on the terms in the Russell–Saunders scheme. Two cases need to be considered. In the *weak-field case* we assume that the crystal-field perturbation is small compared with the interelectronic repulsion forces but larger than the spin–orbit coupling forces. We thus derive the terms arising for the free ion and then consider the effect of the crystal field on these terms. In the *strong-field case* electron pairing may occur, so here the crystal field is more important than the interelectronic repulsions. In this case therefore it is more appropriate to consider the ligand-field splitting first and then to superimpose effects due to the

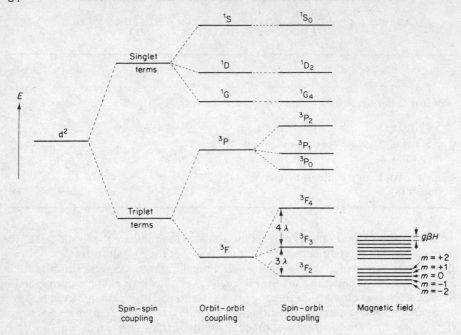

Figure 6.3   The effect of Russell–Saunders coupling on the d^2 configuration.

interelectronic repulsions. We shall confine ourselves to the weak-field case here and deal with the spectra of high-spin ions.

The terms arising from free-ion d^n configurations in octahedral and tetrahedral fields are listed in table 6.5. In the octahedral case the subscript g must be added to the crystal-field terms. These terms are known as Mulliken symbols; they are of group theoretical origin and their derivation is described in Urch's book (see bibliography, page 207). While they can be regarded merely as labels, some cursory examination of their meaning may be helpful.

TABLE 6.5   TERMS ARISING FROM d^n CONFIGURATIONS IN OCTAHEDRAL AND TETRAHEDRAL FIELDS

Free-ion term	Terms arising in cubic fields
S	A_1
P	T_1
D	$E + T_2$
F	$A_2 + T_1 + T_2$
G	$A_1 + E + T_1 + T_2$

The symbols for the crystal-field spectroscopic states have meanings similar to those of the small letters that we used in molecular orbital theory. Thus A and B represent an orbital singlet state, E a doublet, and T a triplet. The S state of the free ion can be likened to an s orbital. In a crystal field an s orbital is not split and is totally symmetric, so the state arising from the S term is A_1. Similarly, p orbitals are not split in a cubic crystal field, so they remain degenerate and the state from a P term is thus T_1. As we have already seen, the d orbital splits into a doublet and a triplet and thus the D term gives rise to E and T_2 states. Of the seven f orbitals, one is unique in that it has eight lobes directed to the corners of a cube. The other six f orbitals are of two kinds. Of one set of three, each orbital has lobes pointing along one of the x, y, or z axes (that is like the three p orbitals) with two rings one above and one below the nodal plane, that is the xy plane in the case of the orbital lying along the z axis. The other set of three orbitals each have eight lobes. They can be imagined to be constructed by splitting the lobes of, for example, a $d_{x^2-y^2}$ orbital so that each original lobe gives a lobe above the xy plane and one below it. In a crystal field therefore the degeneracy of the f orbitals is removed to an orbital singlet and two orbital triplets, that is an F term gives rise to $A_2 + T_1 + T_2$ states. Similarly, there are nine g orbitals, which give rise to an orbital singlet, a doublet, and two triplets in a cubic field.

6.2.4   Spectra of d^1 and d^9 Ions

d^1 *ions in octahedral and tetragonal fields.* The d^1 electronic configuration gives rise to only one term, that of ^2D. In octahedral complexes this term gives rise to an upper doublet ^2E_g and a lower triplet $^2T_{2g}$. The separation between these states is dependent on the ligand-field strength. This dependence is usually illustrated in an energy-level diagram (figure 6.4). Thus, as we saw in figure 5.3, the spectrum of

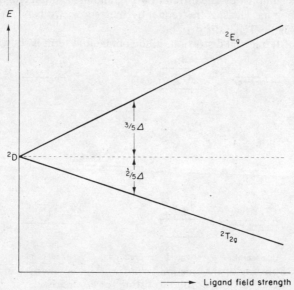

Figure 6.4   Energy-level diagram for a d^1 ion in an octahedral field.

$[Ti(H_2O)_6]^{3+}$ shows a peak at 20 100 cm^{-1} that we can now assign to the $^2T_{2g} \rightarrow {}^2E_g$ transition. The transition is spin-allowed but Laporte-forbidden, so $\epsilon \approx 6$ 1 mol^{-1} cm^{-1}. If we prepare a series of complexes $[TiL_6]^{3+}$ and measure their visible spectra, we can determine the spectrochemical series with Ti^{3+} as the reference ion. When this is done we obtain the following series (numbers in brackets after the ligand are the values of Δ)

$$H_2O\,(20\,100) \quad > \quad urea\,(17\,550) \quad > \quad Cl\,(12\,750) \quad > \quad Br\,(11\,750)$$

We can increase the number of ligands in the series if we use Jørgensen's rule of average environment. This states that the position of the peak in a complex TiX_3L_3 will be midway between that for TiL_6^{3+} and TiX_6^{3-} provided that all three complexes have the same stereochemistry. Thus, for example, since $TiCl_3.3CH_3COCH_3$ absorbs at 15 400 cm^{-1} and $TiCl_6^{3-}$ at 12 750 cm^{-1}, we can calculate Δ for $[Ti(CH_3COCH_3)_6]^{3+}$ (which has not been prepared) as follows

$$\Delta[TiCl_3.3CH_3COCH_3] \quad = \quad \tfrac{1}{2}\Delta\,[TiCl_6^{3-}] + \tfrac{1}{2}\Delta\,[Ti(CH_3COCH_3)_6{}^{3+}]$$

$$15\,400 \quad = \quad 6\,375 + \tfrac{1}{2}\Delta\,[Ti(CH_3COCH_3)_6{}^{3+}]$$

$$\Delta[Ti(CH_3COCH_3)_6{}^{3+}] \quad = \quad 18\,050\,cm^{-1}$$

The rule can also be used, with some caution, to indicate the nature of species present in solution. However, complexes containing a mixture of ligands frequently have low-symmetry components to the ligand field, so considerable band splitting occurs and the rule becomes inapplicable.

Tetragonal distortion of the octahedral structure occurs commonly for d^1 complexes, particularly those of vanadium(IV) of the type VCl_4L_2. In these compounds, for example *trans*-VCl_4L_2, the tetragonal distortion is already 'built in' to the complex by the different M–Cl and M–L bond lengths and ligand-field strengths. The spectra frequently then show two bands rather than one band with Jahn–Teller distortion.

The effect of tetragonal distortion on the octahedral term is shown in figure 6.5

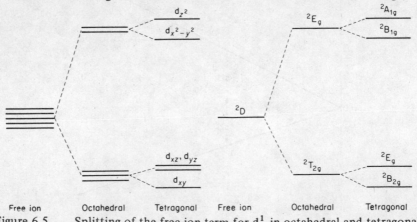

Free ion   Octahedral   Tetragonal   Free ion   Octahedral   Tetragonal

Figure 6.5   Splitting of the free-ion term for d^1 in octahedral and tetragonal fields, with the orbital splitting for comparison.

with the corresponding orbital diagram for comparison. This diagram is also appropriate to many titanium(III) complexes. The E_g (in octahedral field) level is always split in these complexes either by the Jahn—Teller effect or by other low-symmetry fields as well as tetragonal, so the spectra alone cannot be used to diagnose stereochemistry.

The tetragonal distortion in figure 6.5 represents that in which the M—L bonds in the z axis are shorter than the bonds in the x and y axes, that is an octahedron compressed along the z axis. Notice that this results in a different orbital splitting diagram from that represented in figure 5.15 which is for the octahedron elongated along the z axis. Which of these types of tetragonal distortion is actually present in any complex cannot be deduced from the electronic spectra, so assignments of tetragonal spectra are tentative until the actual stereochemistry is known from, for example X-ray crystallographic studies. We shall assume that figure 6.5 is applicable to d^1 systems (figure 5.15 is certainly more appropriate for d^9 Cu^{2+} complexes).

Because only one d electron is present, the symbols used for the tetragonal field states can be understood by comparison with the orbital splitting diagram in figure 6.5. Since $trans$-VCl_4L_2 has a centre of symmetry, the subscripts g and u apply. All the states are even *(gerade)* with respect to inversion through the centre, since the sign of the wave function in the d orbital does not change on inversion (this can be seen for the t_{2g} orbitals in figure 5.11, and it is also true for the e_g set). They thus all carry the subscript g. The only level corresponding to an orbital doublet has the symbol E. The other levels corresponding to orbital singlets carry the labels A or B depending on whether or not the orbitals with which they can be equated are symmetrical with respect to rotation about the principal axis of the molecule. The d_{z^2} orbital is symmetrical with respect to rotation by an angle $360°/4$ about the C_4 axis (the z axis), and thus the state carries the label A. The $d_{x^2-y^2}$ and d_{xy} orbitals are antisymmetric with respect to this rotation; for example, rotation of the d_{xy} orbital (figure 5.11) by $90°$ about the z axis causes + and − signs to change; the states corresponding to an electron in those orbitals thus carry the label B. The subscripts 1 and 2 relate to whether or not the orbitals are symmetrical with respect to rotation about a C_2 axis perpendicular to the principal axis, in this case the C_4 (z) axis. Both the d_{z^2} and $d_{x^2-y^2}$ orbitals do not change when they are rotated by $180°$ about a C_2 axis, for example the x axis. Hence the states corresponding to these carry the subscript 1. However, the d_{xy} orbital on $180°$ rotation about the x or y axes is not symmetrical; the signs on the lobes are interchanged. It thus gives rise to a B_2 state. Note that the spin multiplicities of the states are unchanged as a result of the ligand field splittings, and all the states carry the superscript 2, being derived from ^2D.

The splitting of the ^2E_g and $^2B_{2g}$ terms is relatively small (about $1\,000\ cm^{-1}$), with a somewhat larger splitting of the upper $^2A_{1g}$ and $^2B_{1g}$ terms. Two excitations are thus expected to be observed in the visible and ultraviolet region, corresponding to $^2B_{2g} \rightarrow {}^2B_{1g}$ and $^2B_{2g} \rightarrow {}^2A_{1g}$ transitions. Table 6.6 lists some typical spectra of d^1 compounds that we have interpreted according to this scheme.

TABLE 6.6   ELECTRONIC SPECTRA (cm^{-1}) OF SOME TETRAGONAL d^1 COMPLEXES

Complex	$^2B_{2g} \rightarrow {}^2B_{1g}$	$^2B_{2g} \rightarrow {}^2A_{1g}$
$TiCl_3 \cdot 3THF$	13 500	14 700
$TiCl_3 \cdot bipy$	13 500	15 750
$TiCl_3 \cdot 3MeCN$	14 700	17 100
$VCl_4 \cdot 2THF$	13 600	18 100
$VCl_4 \cdot 2PhCOPh$	14 300	20 400
$VCl_4 \cdot bipy$	17 400	21 300

d^9 *ions in octahedral and tetragonal fields.* The d^9 configuration also gives rise to the ^2D ground term. However, in an octahedral field the odd electron resides in an e_g orbital, so the ground term now becomes ^2E_g rather than $^2T_{2g}$. In other words, the spectroscopic levels are inverted relative to d^1. This is an example of a more general phenomenon, that is that a d^{10-n} configuration has the same behaviour in the crystal field as the d^n configuration except that inversion occurs. This is sometimes known as the *hole formalism* in which the shortage of one electron, that is one hole, is regarded as the presence of one positron. On application of crystal-field theory we see that the positron is most stable in those orbitals in which the electrons are least stable. Thus the positron will occupy the e_g set of orbitals and on absorption of radiation will be excited into the t_{2g} set.

In octahedral copper(II) complexes we might expect a single absorption band in the visible spectrum corresponding to the $^2E_g \rightarrow {}^2T_{2g}$ transition. The energy-level diagram for this configuration is shown in figure 6.6. In fact, as we have already

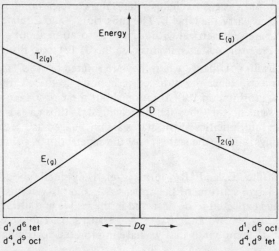

Figure 6.6   The splitting of free-ion D terms in octahedral and tetrahedral fields.

discussed (chapter 5), the Jahn–Teller distortion in the d^9 case is considerably greater than that in the d^1 case. Thus, instead of one band in the spectrum, six-co-ordinate copper(II) complexes show spectra having broad bands resulting from several overlapping bands. For $[Cu(H_2O)_6]^{2+}$ the broad band has a long tail into the near infrared; Δ is above $12\,000$ cm^{-1}.

Tetrahedral fields. The orbital splitting diagram for d^1 in tetrahedral fields is inverted relative to that in octahedral fields. That is, the ^2E state corresponding to the $e^1 t_2{}^0$ configuration is the ground state, and the first excited state corresponding to $e^0 t_2{}^1$ is the $^2T_2$ state. The spectra of tetrahedral d^1 complexes are thus expected to show a band due to the transition $^2E \rightarrow {}^2T_2$. This band will be at a lower energy $[\Delta_{tet} = (4/9)\,\Delta_{oct}]$ than in corresponding octahedral complexes. It is thus apparent that the D term in d^1 tetrahedral splits in the same way as in d^9 octahedral (both inverted with respect to d^1 octahedral). In more general terms we have the following relationships.

(1) Octahedral d^n splits in the same way as tetrahedral d^{10-n}.
(2) Octahedral d^n splits inversely to octahedral d^{10-n}.

These relationships are exemplified in figure 6.6 which shows the splitting of D terms for the d^1, d^4, d^6, and d^9 configurations in octahedral and tetrahedral fields. The spin multiplicities are omitted since they are 2 for d^1, d^9 and 5 for d^4, d^6. The d^4 and d^6 configurations give rise to the ^5D ground term. The general rule relating the splitting patterns from these configurations is

(3) d^{n+5} splits in the same way as d^n.

Thus d^6 splits as for d^1, the spherically symmetrical half-filled d shell accounting for this relationship. Similarly, d^4 and d^9 configurations are related as shown in figure 6.6.

Because of these relationships we are able to discuss the spectra arising from the configurations d^2, d^3, d^5, d^7, and d^8 in the pairs d^2, d^8 and d^3, d^7, but d^5 is unique. For this configuration, applying the rules above, we have

(4) d^5 octahedral splits in the same way as d^5 tetrahedral.

6.2.5 Spectra of d^2 and d^8 Ions

d^2 *ions in octahedral fields*. The d^2 configuration gives rise to the ^3F ground term with higher-energy terms ^3P, ^1G, ^1D, and ^1S. In an octahedral field these terms split to give the states listed in table 6.5. The energy-level diagram for a d^2 ion is thus considerably more complicated than those we have seen for d^1 and d^9 ions. The variation in energies of the various levels with increasing ligand field strength has been calculated by Orgel for the weak-field case and by Tanabe and Sugano for the strong-field case. The resulting energy-level diagrams are named after them. The Orgel diagram for the d^2 ion is shown in figure 6.7. This diagram is only of qualitative application; for semiquantitative work the Tanabe–Sugano diagrams must be consulted (these can be found in the books by Lever, Sutton, or Figgis). The continuous lines in the diagrams are those representing spin triplet states while the broken lines represent the spin singlet states. Since the ground state at all ligand field strengths is $^3T_{1g}(F)$, the only spin-allowed transitions will be those to

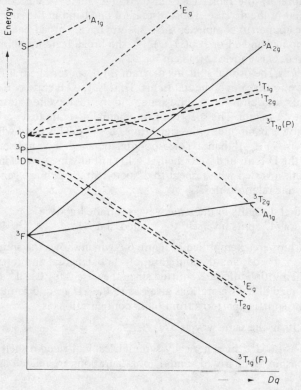

Figure 6.7   Orgel diagram for a d^2 ion (V^{3+}) in an octahedral field

the other spin triplet states. Transitions from $^3T_{1g}(F)$ to the singlet states will be spin-forbidden and observed at the most as very weak bands in the spectra.

The electronic spectrum of the d^2 ion $[V(H_2O)_6]^{3+}$ (as the perchlorate salt) is illustrated in figure 6.8. It shows two low-intensity bands in the visible and near-ultraviolet regions; further into the ultraviolet only high-intensity (charge-transfer) bands are observed. Since no other bands are observed at lower energies (until the infrared spectrum is recorded) we can tentatively assign these two observed bands to the two spin-allowed transitions of lowest energy indicated on the energy-level diagram, thus

$$^3T_{1g}(F) \rightarrow {}^3T_{2g} \qquad \nu_1 = 17\,250 \text{ cm}^{-1}$$

$$^3T_{1g}(F) \rightarrow {}^3T_{1g}(P) \qquad \nu_2 = 25\,000 \text{ cm}^{-1}$$

Now, if the diagram (figure 6.7) has been constructed specifically for V^{3+} (with a $^3P-{}^3F$ separation of $15B$, using the free-ion value for B), we should be able to deduce the expected position of the $^3T_{1g}(F) \rightarrow {}^3A_{2g}$ (ν_3) transition. To do this we need to calculate the observed ratio of ν_1 to ν_2 and then find the vertical line that fits this ratio in the diagram. The point at which this vertical line cuts the

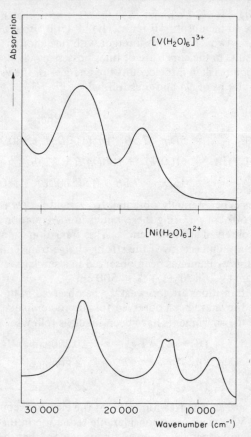

Figure 6.8   Electronic spectra of the ions $[V(H_2O)_6]^{3+}$ (d^2) and $[Ni(H_2O)_6]^{2+}$ (d^8)

$^3A_{2g}$ level enables the ratio of ν_1 to ν_3 or of ν_2 to ν_3 to be found, and hence ν_3 can be calculated. This method corresponds to a pure crystal-field approach and assumes that the value of the free-ion Racah parameter B is maintained in the complex. In practice it is found that this does not usually give very satisfactory results. A better fit between theory and experiment is obtained if B is treated as a variable parameter. In the more quantitative Tanabe–Sugano diagrams, the energy (ordinate) is plotted as a ratio E/B while the abscissa is plotted in units of Dq/B. The ratio method then enables Dq, B, and ν_3 to be determined from the two experimentally observed bands. The original assignments of ν_1 and ν_2 then become justified if physically realistic values result for $10Dq$ and B. Notice that it will be slightly more convenient to use $10Dq$ in place of Δ in discussing spectra, but these symbols are interchangeable throughout the text; $Dq = 0.1\Delta$.

Alternatively, we can calculate these parameters directly from the mathematical equations. In an octahedral field the ^3F term of a d^2 complex splits into the three terms $^3T_{1g}$ ($-6\ Dq$), $^3T_{2g}$ ($2\ Dq$), and $^3A_{2g}$ ($12\ Dq$), with the energies given in parentheses relative to the unsplit ^3F term. The ^3P term also gives rise to a $^3T_{1g}$ term, and these two $^3T_{1g}$ terms interact with one another; it is this interaction that results in the curvature of lines possessing identical designations. When this interaction is taken into account the energies of the various transitions can be described in terms of Dq and B as follows

Transition   Energy

$$\nu_1, {}^3T_{1g}(F){\to}{}^3T_{2g} \quad = \quad 5Dq - 7.5B + 0.5(100Dq^2 + 180Dq.B + 225B^2)^{\frac{1}{2}}$$

$$\nu_2, {}^3T_{1g}(F){\to}{}^3T_{1g}(P) \quad = \quad (100Dq^2 + 180Dq.B + 225B^2)^{\frac{1}{2}}$$

$$\nu_3, {}^3T_{1g}(F){\to}{}^3A_{2g} \quad = \quad 15Dq - 7.5B + 0.5(100Dq^2 + 180Dq.B + 225B^2)^{\frac{1}{2}}$$

From the values experimentally obtained for ν_1 and ν_2, Dq is calculated to be 1850 cm^{-1} and B 602 cm^{-1}. Using these values in the equation for ν_3, we find that ν_3 should be observed at 35 700 cm^{-1}. This transition is not observable for the hexaquovanadium(III) ion because of the strong charge-transfer absorption in this region. However, this ν_3 band has been observed in a similar system, that of vanadium corundum (V^{3+} in Al_2O_3) at 34 500 cm^{-1}.

Spin-forbidden transitions are too weak to be observed in the spectrum of $[V(H_2O)_6]^{3+}$, but they have been observed for fluoro-complexes, for example $VF_6{}^{3-}$, for which the assignments have been made as follows

$$^3T_{1g}(F){\to}{}^1E_g, {}^1T_{2g} \qquad 10\ 200\ cm^{-1}$$

$$^3T_{1g}(F){\to}{}^3T_{2g} \qquad 14\ 800\ cm^{-1}$$

$$^3T_{1g}(F){\to}{}^3T_{1g}(P) \qquad 23\ 000\ cm^{-1}$$

The nephelauxetic effect. The free-ion value of the electron repulsion parameter B is 860 cm^{-1} for V^{3+}, so we see a considerable reduction in the value of B in the complex. This is a very generally observed phenomenon, that the value of B in a complex is approximately 70 per cent of the free-ion value. Similarly, the Racah parameter C becomes reduced by about 70 per cent on complex formation (we have not been concerned here with this parameter because we have been dealing with transitions between states of the same spin multiplicity). These results show that the interelectronic repulsions of the d electrons in a complex are less than in the free ion since it is these repulsions that are responsible for the separations between the Russell–Saunders states. If the interelectronic repulsions have decreased it would suggest that the electrons are further apart in the complex, that is that the d orbital electron clouds have expanded to some extent. The effect is thus known as the *nephelauxetic effect* (Greek nephelauxetic = cloud expanding). The major part of the reduction in the value of B is believed to be a direct result of covalency within the complex. We saw on MO theory (chapter 5) that in an octahedral complex the e_g electrons become σ-antibonding and the t_{2g} electrons may be involved in π-bonding. The electrons are thus no longer pure metal electrons but are to some extent delocalised on to the ligands. This delocalisation increases the average separation between the various d electrons and thus reduces B. It can then be argued that the greater the reduction in the value of B the greater the covalency in the metal—ligand bond. The ratio

$$\frac{B \text{ in complex}}{B \text{ in free ion}} = \beta$$

It is found that ligands can be arranged in order of decreasing β, that is increasing covalency. This order is more or less independent of the metal ion and is known as the *nephelauxetic series*. The series for the more common ligands is

$$F^- > H_2O > NH_3 > en \approx C_2O_4{}^{2-} > NCS^- > Cl^- \approx CN^- > Br^- > S^{2-} \approx I^-$$

Ionic ligands such as F^- ions give β values close to unity, indicating ionic complexes, while the more covalently bonding ligands such as sulphur donors and I^- ions give values of β as low as 0.3. We can similarly draw up a nephelauxetic series for metal ions with a constant ligand. The more polarising the metal ion, the more covalent the metal—ligand bond and hence the lower the value of β. The order for some metals is

$$Mn^{2+} \approx V^{2+} > Ni^{2+} \approx Co^{2+} > Mo^{3+} > Cr^{3+} > Fe^{3+} > Co^{3+} > Pt^{4+} \approx Mn^{4+}$$

d^8 *ions in octahedral fields*. The relative energies of the free-ion terms for d^8 are the same as for d^2, so spin-allowed transitions will result from transitions between levels derived from ^3F and ^3P. As with d^1 and d^9, however, we get inversion of the crystal-field states derived from the free-ion terms. The simplified energy-level diagram for d^8 ions (showing only states giving rise to spin-allowed transitions) is given in figure 6.9. This diagram is useful in that it summarises all the spin-allowed

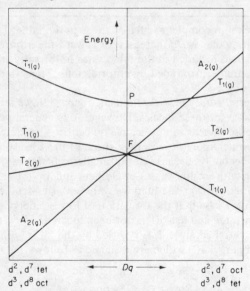

Figure 6.9   Simplified energy-level diagram for spin triplet terms of d^2, d^8 and spin quartet terms of d^3, d^7 ions in octahedral and tetrahedral fields.

transitions for d^2, d^3, d^7, and d^8 ions in octahedral and tetrahedral fields. In conjugation with figure 6.6 which covers the d^1, d^4, d^6, and d^9 ions, these diagrams represent all the octahedral and tetrahedral d^n configurations except d^5 which presents a special case.

The octahedral d^8 transitions are given on the left-hand side of figure 6.9 (note that the d^2 octahedral diagram on the right-hand side of this figure is a simplified picture of figure 6.7). Again three transitions are expected from the ground term (which is now $^3A_{2g}$), and these are frequently observed in octahedral complexes of nickel(II). The spectrum of the hexaquonickel(II) ion is illustrated in figure 6.8. The three bands are assigned for some representative $[NiL_6]^{2+}$ complexes in table 6.7. The trends of the spectrochemical series are readily

TABLE 6.7 ELECTRONIC SPECTRA (cm^{-1}) OF SOME OCTAHEDRAL NICKEL(II) COMPLEXES

Complex	$^3A_{2g} \rightarrow {}^3T_{2g}$	$^3A_{2g} \rightarrow {}^3T_{1g}(F)$	$^3A_{2g} \rightarrow {}^3T_{1g}(P)$
$NiCl_2$	6 800	11 800	20 600
$[Ni(H_2O)_6]^{2+}$	8 500	13 800	25 300
$[Ni(NH_3)_6]^{2+}$	10 750	17 500	28 200
$[Ni(en)_3]^{2+}$	11 200	18 350	29 000
$[Ni(bipy)_3]^{2+}$	12 650	19 200	obscured

discernible from nickel(II) complexes such as those in the table. The three bands move toward the ultraviolet as Dq increases, or toward the infrared on lowering Dq. Nickel(II) chloride is included in the table since in the solid state it contains a nickel(II) ion octahedrally surrounded by chloride ions. Transitions to spin singlet levels are sometimes observable.

d^2 *and* d^8 *ions in tetrahedral fields.* Again using figure 6.9, we see that three transitions are possible for these species. However, two general points must be noted. Firstly, because these complexes have no centre of symmetry, their absorptions will be much more intense than those of the corresponding octahedral species. Secondly, because of the smaller splitting in the tetrahedral field, we are dealing with smaller values of Dq here and are thus closer to the centre of figure 6.9 on the Dq scale than we were with the octahedral complexes. It is therefore often the case that the transition of lowest energy gives rise to absorption in the infrared and has often not been located.

Vanadium(III) in tetrahedral fields is typified by the VCl_4^- ion. This has two absorption bands (figure 6.10) which are assigned as follows

$$^3A_2 \rightarrow {}^3T_2 \qquad \text{not found}$$
$$^3A_2 \rightarrow {}^3T_1(F) \qquad 8\ 900\ cm^{-1}$$
$$^3A_2 \rightarrow {}^3T_1(P) \qquad 15\ 000\ cm^{-1}$$

The band of lowest energy is calculated to occur around $5\ 500\ cm^{-1}$ but has not been found experimentally.

Tetrahedral nickel(II) complexes absorb light from the red part of the spectrum

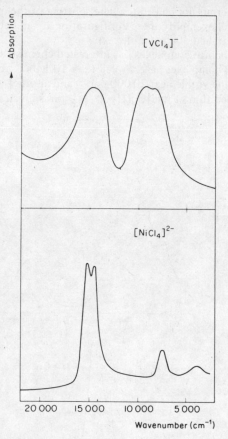

Figure 6.10   Electronic spectra of the tetrahedral anions $[VCl_4]^-$ (d^2) and $[NiCl_4]^{2-}$ (d^8)

and hence appear intensely blue. A typical spectrum, that of $NiCl_4{}^{2-}$, is shown in figure 6.10, and assignments are made using figure 6.9

$$^3T_1 \rightarrow {}^3T_2 \qquad\qquad 4\,000\ cm^{-1}$$

$$^3T_1 \rightarrow {}^3A_2 \qquad\qquad 7\,500\ cm^{-1}$$

$$^3T_1 \rightarrow {}^3T_1(P) \qquad\quad 15\,000\ cm^{-1}$$

In general, tetrahedral nickel(II) complexes show a strong multiple absorption band $(\epsilon \approx 10^2\ l\ mol^{-1}\ cm^{-1})$ near $15\,000\ cm^{-1}$, with a weaker band near $8\,000\ cm^{-1}$; the lowest-energy transition is not very often observed. Weak bands sometimes observed on the high and low-energy sides of the $15\,000\ cm^{-1}$ band have been assigned to spin-forbidden bands to levels derived from ^1D and ^1G free-ion terms.

6.2.6   Spectra of d^3 and d^7 Ions
The high-spin d^3 and d^7 ions give rise to two spin quartet free-ion terms ^4F and ^4P

(table 6.2) and numerous doublet states. The splitting of the quartet terms is shown in figure 6.9. We shall exemplify these configurations from cobalt(II) complexes (there is a lack of tetrahedral d^3 complexes).

The colour of cobalt(II) complexes has interested chemists for many years. The pale pink octahedral complexes have a weak ($\epsilon \approx 10$ 1 mol^{-1} cm^{-1}) multiple absorption band in the visible near 20 000 cm^{-1} and another weak band around 8 000 cm^{-1}. The spectrum of the $[Co(H_2O)_6]^{2+}$ ion is illustrated in figure 6.11.

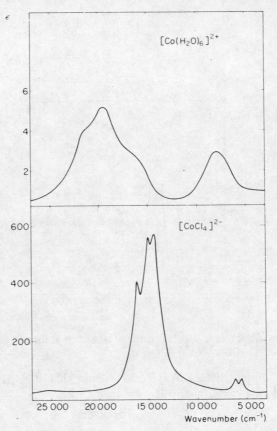

Figure 6.11   Electronic spectra of $[Co(H_2O)_6]^{2+}$ and $[CoCl_4]^{2-}$

Its assignment is not straightforward partly because of the poor resolution of the bands around 20 000 cm^{-1}. The lowest-energy band at 8 000 cm^{-1} is unambiguously assigned to the $^4T_{1g} \rightarrow ^4T_{2g}$ transition. This leaves three bands, at around 16 000, 19 400, and 21 600 cm^{-1}. It is apparent that we are on that part of the energy-level diagram where the $^4A_{2g}$ and $^4T_{1g}(P)$ states are close together, that is near the intersection, and this, coupled with the poorly resolved spectrum, means that assignments are still somewhat tentative. Most authors favour the following

$$^4T_{1g} \rightarrow {}^4A_{2g} \qquad 16\,000 \text{ cm}^{-1}$$

$$^4T_{1g} \rightarrow {}^4T_{1g}(P) \qquad 19\,400 \text{ cm}^{-1}$$

with the extra band (shoulder on the high-frequency side) being attributed to spin—orbit coupling effects or to transitions to doublet states.

Tetrahedral cobalt(II) complexes are frequently an intensely blue colour, and as such are readily distinguishable from octahedral complexes. The intensity of the visible band of $[CoCl_4]^{2-}$ can be compared with that of the octahedral $[Co(H_2O)_6]^{2+}$ in figure 6.11. There is no ambiguity about the principal assignments in tetrahedral complexes because only two bands are to be expected, the lowest-energy band being in the infrared. The $[CoCl_4]^{2-}$ assignments are thus

$$^4A_2 \rightarrow {}^4T_1(F) \qquad 5\,800 \text{ cm}^{-1}$$

$$^4A_2 \rightarrow {}^4T_1(P) \qquad 15\,000 \text{ cm}^{-1}$$

With these assignments the $^4A_2 \rightarrow {}^4T_2$ transition is expected to lie at $3\,300 \text{ cm}^{-1}$. The reasons for the fine structure of the bands have been variously debated. Spin—orbit coupling effects almost certainly contribute to the fine structure as well as transitions to doublet states.

6.2.7 Spectra of d⁵ Ions

The d^5 configuration gives rise to only one sextet term in the free ion, that is ^6S (table 6.2), as well as many quartet and doublet terms. The result of this is that any transitions between crystal field states will be spin-forbidden (the ^6S term gives rise to only one term, that is $^6A_{1g}$, in octahedral fields). Consequently the spectra of d^5 ions are very weak, as are the colours of d^5 complexes (which are free from charge-transfer absorption); for example the pale violet iron(III) alum and the pale pink manganese(II) salts. Since we are dealing with d^5 cases, we need consider only one stereochemistry (octahedral splitting $\equiv$ tetrahedral splitting for d^5).

The spectrum of the hexaquomanganese(II) ion in perchlorate media is illustrated in figure 6.12. Several features of this spectrum are unusual where it is

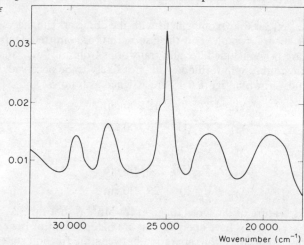

Figure 6.12   Electronic spectrum of the ion $[Mn(H_2O)_6]^{2+}$

compared with the spectra that we have seen previously. Firstly, there is the weakness of the bands; compare these with the bands for cobalt(II) complexes in figure 6.11. Secondly, there is the relative sharpness of some of the bands. Spin-allowed bands are invariably broad whereas we see here that spin-forbidden bands may be broad or sharp.

The Orgel diagram for octahedral Mn^{2+} is shown in figure 6.13. In this diagram

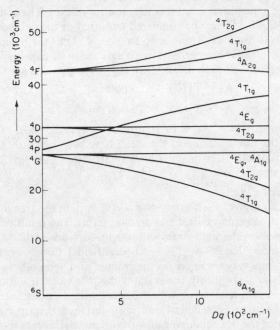

Figure 6.13   Orgel diagram for octahedral Mn^{2+} (d^5)

the ground state $^6A_{1g}$ is drawn coincident with the abscissa. The diagram is also applicable to tetrahedral complexes if the g subscripts are omitted. Only the quartet terms have been included because transitions from the $^6A_{1g}$ state to spin doublet states are doubly spin-forbidden and not likely to be observed. Using the diagram, the bands shown in figure 6.12 are assigned as follows

$$^6A_{1g} \rightarrow {}^4T_{1g} \qquad 18\,900\ cm^{-1}$$

$$^6A_{1g} \rightarrow {}^4T_{2g}(G) \qquad 23\,100\ cm^{-1}$$

$$^6A_{1g} \rightarrow {}^4E_g, {}^4A_{1g}\ \ 24\,970, 25\,300\ cm^{-1}$$

$$^6A_{1g} \rightarrow {}^4T_{2g}(D) \qquad 28\,000\ cm^{-1}$$

$$^6A_{1g} \rightarrow {}^4E_g(D) \qquad 29\,700\ cm^{-1}$$

An interesting feature of the Orgel diagram for Mn^{2+} is that the $^4A_{1g}$, $^4E_g(G)$, $^4E_g(D)$, and $^4A_{2g}(F)$ terms have energies that are independent of the crystal field. It is thus very easy to measure the separation between the free-ion states at zero

Dq. By comparison of the assignments above and the energies of the free-ion states in figure 6.13, the usual reduction of the interelectron repulsion parameters can be seen to have occurred.

The sharpness of the bands is connected with the ratio of the slope of the curve for the upper state to that of the ground state. During vibrations of the complex ion, the ligand field strength $10Dq$ varies about some mean value. If the energy separation between two levels is not dependent on ligand-field strength, a sharp absorption band is expected. However, if the slope of the line for the upper state is considerable (bearing in mind the zero slope of the $^6A_{1g}$ line as we have drawn it), the transition energy will vary as the ligand field varies during the vibrations. The sharp bands are thus due to the transitions to the ^4E_g, $^4A_{1g}(G)$, $^4T_{2g}(D)$, and $^4E_g(D)$ states, while transitions to the heavily sloping $^4T_{1g}(G)$ and $^4T_{2g}(G)$ states give rise to broader bands.

7 Magnetic Properties of Transition-metal Complexes

7.1 Introduction'

In this chapter we continue from the very brief introduction given to magnetism in chapter 5. We begin by defining the terms and quantities that are used in magnetochemistry.

Magnetic susceptibility and paramagnetism. When a compound is placed inside a magnetic field of strength H, the magnetic field inside the compound, that is the magnetic induction of flux B, is given by

$$B = H + 4\pi I$$

where I is the intensity of magnetisation. Chemists use a quantity known as susceptibility, the *volume susceptibility* K being given by

$$K = \frac{I}{H}$$

The *gram susceptibility* is then given by

$$\chi = \frac{K}{\rho}$$

where ρ is the density. The *molar susceptibility* χ_M is defined as ($\chi \times M$) where M is the molecular weight of the compound appropriate to one metal atom. If B is smaller than H the substance is *diamagnetic;* it is repelled by the field and its susceptibility is negative. When B is larger than H the substance is *paramagnetic;* it is attracted into the field and has a positive susceptibility.

Compounds in which the paramagnetic centres are separated from one another by numbers of diamagnetic atoms (from ligands, etc.) are said to be *magnetically dilute.* If the ligands or other diamagnetic species are removed from such a system, the neighbouring paramagnetic centres interact with one another. These interactions give rise to *ferromagnetism* (neighbouring magnetic dipoles aligned in the same direction) and *antiferromagnetism* (neighbouring magnetic dipoles aligned in alternate directions). Ferromagnetic materials have a greatly enhanced paramagnetism; the phenomenon is rare and confined largely to iron and other metals and alloy systems. Antiferromagnetism is much more common; its effect is to reduce the susceptibility and hence the magnetic moment of the compound. It is commonly found to occur in transition-metal halides and oxides where a reduction in effective spin is observed. Thus at room temperature the following magnetic moments are obtained for $TiCl_3$ and VCl_2

$$\alpha\text{-}TiCl_3 \quad \mu = 1.31 \text{ B.M. } (\mu_{\text{spin only}} = 1.73 \text{ B.M.})$$

$$VCl_2 \quad \mu = 2.42 \text{ B.M. } (\mu_{\text{spin only}} = 3.87 \text{ B.M.})$$

These various forms of paramagnetism exhibit characteristic relationships between susceptibility and temperature. These are shown qualitatively in figure 7.1.

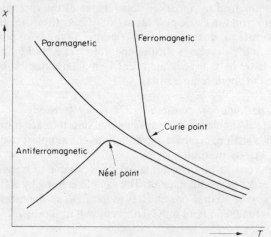

Figure 7.1   Temperature dependence of susceptibility for paramagnetic, ferromagnetic, and antiferromagnetic materials

Normal paramagnetic substances obey *Curie's Law* which states that paramagnetic susceptibilities depend inversely on temperature, that is

$$\chi \propto \frac{1}{T} = \frac{C}{T}$$

where C is the Curie constant. Thus a plot of $1/\chi$ against T gives a straight line of slope $1/C$ passing through the origin (0 K). The constant C is given by the Langevin expression which relates the susceptibility χ_M to the magnetic moment μ in Bohr magnetons

$$\chi_M = \frac{N\mu^2}{3kT}$$

where N is the Avogadro number, k is the Boltzmann constant, and T is the absolute temperature. Thus

$$C = \frac{N\mu^2}{3k}$$

Many substances give straight lines that intersect the temperature axis a little above or below 0 K. These compounds are said to obey the *Curie–Weiss Law*, that is

$$\chi = \frac{C}{T + \theta}$$

in which θ is known as the Weiss constant.

Ferromagnetic materials behave as normal paramagnetics at high temperatures, but at some temperature known as the Curie point (figure 7.1) there is a marked increase in χ accompanied by a marked dependence of the susceptibility on the field strength. For pure iron the Curie point is at 768 ˇC. Antiferromagnetic materials also behave as normal paramagnetics until (figure 7.1) the Néel point is reached when the susceptibility decreases with temperature. The Néel point is often below 100K although, as we have already mentioned, there are many substances that are antiferromagnetic at room temperature.

7.1.1   Determination of Susceptibility; The Gouy Method

Many methods are available for the experimental determination of magnetic susceptibilities. We shall confine ourselves here to the simplest and most widely used method, the Gouy method.

The apparatus consists of an accurate balance (capable of weighing to five decimal places) and a powerful magnet. The sample is hung by a fine thread from one pan of the balance so that one end lies in the field of the magnet and the other end in the earth's magnetic field only. This is shown in figure 7.2. The sample is

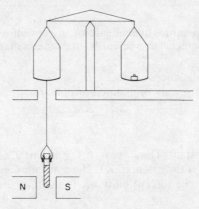

Figure 7.2   Diagrammatic representation of the Gouy method.

contained in a cylindrical tube (usually about the size of a long 'ignition tube') and is weighed in the magnetic field and out of the magnetic field. The out-of-field weight is obtained either by switching off the current if an electromagnet is used, or by moving the magnet away if a permanent magnet is used. The force on the sample in the magnetic field is given by

$$F = \tfrac{1}{2}(K_1 - K_2)(H_2{}^2 - H_1{}^2)A$$

where K_1 is the volume susceptibility of the sample, K_2 is the volume susceptibility of air, H_2 is the magnetic field strength, H_1 is the earth's magnetic field strength, and A is the cross-sectional area of the sample.

Of these terms, F and A are measured and K_2 is known. The magnetic field strength is usually calibrated by using a substance of known K_1. The above expression then reduces to

$$F = \tfrac{1}{2}(K_1 - K_{air}) \times \text{constant}$$

Thus, by measuring F for the unknown in a calibrated apparatus (that is of known constant), K_1 is calculated. The density of the sample is required in order to convert K into χ

$$\chi = \frac{K}{\rho}$$

We convert χ into χ_M by multiplying by the molecular weight. The magnetic moment of the sample is then given (from the Langevin expression) by

$$\mu = \sqrt{\left(\frac{3RT\chi_M}{N^2}\right)} = 2.83 \sqrt{(\chi_M T)}$$

In order to obtain the magnetic moment of the metal ion alone, we must correct for the diamagnetic effects of the ligands (and strictly also for the inner-core diamagnetism of the metal ion). These diamagnetic parts of the sample were in fact exerting a force away from the magnetic field, thus making the sample appear lighter in the magnetic field than out of it, whereas the paramagnetic ions were exerting a force into the magnetic field, hence giving an apparently increased weight of sample in the magnetic field. Tables for such diamagnetic corrections are available in the literature (see the bibliography). These are normally added on to χ_M thus

$$\chi'_M = \chi_M \text{ corrected for diamagnetism of ligands.}$$

The effective magnetic moment of the metal ion is then given by

$$\mu_{\text{eff}} = 2.83 \sqrt{(\chi'_M T)}$$

7.2 Orbital Contribution to Magnetic Moments

We saw in chapter 5 that the spin-only formula often gives a reasonable experimental fit between the magnetic moment and the number of unpaired electrons present. Whether or not a good agreement results in this way depends largely on whether or not the orbital contribution to the moment is significant. We shall now see how deviations (or lack of them) from the spin-only formula are useful in diagnosing the stereochemistry of complexes.

In order that an electron can have orbital angular momentum, it must be possible to transform the orbital that it occupies into an exactly equivalent and degenerate orbital by rotation. The electron is then effectively rotating about the axis used for the orbital rotation. In an octahedral complex, the degenerate t_{2g} orbitals (d_{xz}, d_{yz}, d_{xy}) can be interconverted by $90°$ rotations. For example, the d_{xz} orbital is transformed into the d_{yz} orbital by a rotation of $90°$ about the z axis. During this rotation the electron is in part orbiting the nucleus. Thus an electron in a t_{2g} orbital does contribute orbital angular momentum. However, the e_g orbitals (d_{z^2} and $d_{x^2-y^2}$) cannot be interconverted by rotation about any axis because of their different shapes; an electron in the e_g set therefore cannot contribute to orbital angular momentum. Electrons in t_{2g} orbitals will not always contribute orbital angular momentum. Consider the d^3, t_{2g}^3 case for example.

An electron in the d_{xz} orbital cannot by rotation be placed in the d_{yz} orbital because this orbital already contains an electron with the same spin as the incoming electron. In other words, in the d^1 case we have three possible arrangements while in the d^3 case there is only one, that is

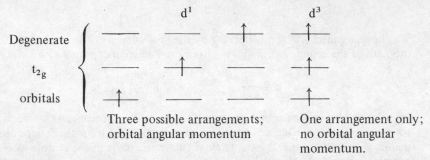

Three possible arrangements; orbital angular momentum

One arrangement only; no orbital angular momentum.

Tetrahedral complexes can be treated in the same way except that we now fill the e orbitals first and electrons in these do not contribute orbital angular momentum. It becomes apparent that, for many configurations, octahedral and tetrahedral complexes differ in whether or not they contribute to orbital angular momentum. This fact is useful in enabling magnetic measurements to distinguish between these two stereochemistries. In table 7.1 are listed all the d^1 to d^9 configurations, including high and low-spin complexes, with the ground term for the complex and a statement of whether or not orbital contribution to the magnetic moment is expected.

Examination of table 7.1 shows that the configurations corresponding to A_1 (from free-ion S term), E (from D term), or A_2 (from F term) are those for which there is no orbital angular momentum. Configurations that give rise to either a T_2 term (from D term) or a T_1 term (from F term) as the lowest terms are those for which orbital contribution to the magnetic moment is expected. We shall now examine these two groups of magnetic behaviour.

7.2.1   Magnetic Properties of Complexes with A and E Ground Terms

These complexes possess no orbital contribution in the ground state, so the magnetic moment is expected to follow the spin-only formula

$$\mu = \sqrt{[4S(S + 1)]} = \sqrt{[n(n + 2)]}$$

We now see why this formula was apparently so successful in interpreting magnetic data. It so happens that no orbital contribution is expected for many very common configurations, for example d^3 (Cr^{3+}), d^5 (Mn^{2+}, Fe^{3+}), d^8 (Ni^{2+}), and d^9 (Cu^{2+}). Since the spin-only formula contains no temperature term, the magnetic moment μ is expected to be independent of temperature. In table 7.2 are listed some salts containing octahedral $[M(H_2O)_6]^{n+}$ cations or tetrahedral $[MCl_4]^{n-}$ anions, all of which possess A or E ground terms (compare table 7.1). Their magnetic moments at two widely different temperatures show in every case that the moment does not vary with temperature. However, the fit with the spin-only moment is not always good; this is because spin–orbit coupling, although it cannot

TABLE 7.1   ORBITAL CONTRIBUTION FOR d^1 TO d^9 IONS IN
OCTAHEDRAL AND TETRAHEDRAL STEREOCHEMISTRIES

Number of d electrons	Octahedral complexes			Tetrahedral complexes		
	Configuration	Ground term	Orbital contribution	Configuration	Ground term	Orbital contribution
1	t_{2g}^1	$^2T_{2g}$	yes	e^1	^2E	no
2	t_{2g}^2	$^3T_{1g}$	yes	e^2	$^3A_2$	no
3	t_{2g}^3	$^4A_{2g}$	no	$e^2 t_2^1$	$^4T_1$	yes
4	$t_{2g}^3 e_g^1$	^5E_g	no	$e^2 t_2^2$	$^5T_2$	yes
	t_{2g}^4	$^3T_{1g}$	yes			
5	$t_{2g}^3 e_g^2$	$^6A_{1g}$	no	$e^2 t_2^3$	$^6A_1$	no
	t_{2g}^5	$^2T_{2g}$	yes			
6	$t_{2g}^4 e_g^2$	$^5T_{2g}$	yes	$e^3 t_2^3$	^5E	no
	t_{2g}^6	$^1A_{1g}$	no			
7	$t_{2g}^5 e_g^2$	$^4T_{1g}$	yes	$e^4 t_2^3$	$^4A_2$	no
	$t_{2g}^6 e_g^1$	^2E_g	no			
8	$t_{2g}^6 e_g^2$	$^3A_{2g}$	no	$e^4 t_2^4$	$^3T_1$	yes
9	$t_{2g}^6 e_g^3$	^2E_g	no	$e^4 t_2^5$	$^2T_2$	yes

occur as a first-order effect, is able to mix in some orbital contribution from higher
T terms of the same multiplicity as the ground term. This occurs because the spin
and orbital angular momenta of the free ion are coupled by spin—orbit coupling and
the ligand field is unable to effect a perfect separation of the terms of the same
multiplicity according to their different orbital angular momenta. The ground term
thus contains some orbital angular momentum.

The A_1 ground terms present no problems. The d^5 (Mn^{2+}) configuration gives
rise to no higher T terms of the sextuplet spin multiplicity, so no mixing occurs and
manganese(II) complexes therefore give the spin-only moment (table 7.2). The
low-spin d^6 configurations give rise to diamagnetism only.

For the A_2 (from F) and E (from D) terms there is necessarily a higher T term of
the same multiplicity as the ground state. For these terms the magnetic moment is
given by

$$\mu_{eff} = \mu_{spin\ only}\left(1 - \frac{a\lambda}{10Dq}\right)$$

where $a = 2$ for an E term or 4 for an A_2 term. The spin—orbit coupling constant
λ has a positive sign for d shells less than half filled and is negative (*J* values inverted

TABLE 7.2   MAGNETIC MOMENTS OF SOME COMPLEXES WITH A
OR E GROUND TERMS

Number of d electrons	Compound	Stereochemistry	μ_{eff}		
			80K	300K	spin-only
1	VCl_4	tetrahedral	1.6	1.6	1.73
3	$KCr(SO_4)_2.12H_2O$	octahedral	3.8	3.8	3.87
4	$CrSO_4.6H_2O$	octahedral	4.8	4.8	4.90
5	$K_2Mn(SO_4)_2.6H_2O$	octahedral	5.9	5.9	5.92
7	Cs_2CoCl_4	tetrahedral	4.5	4.6	3.87
8	$(NH_4)_2Ni(SO_4)_2.6H_2O$	octahedral	3.3	3.3	2.83
9	$(NH_4)_2Cu(SO_4)_2.6H_2O$	octahedral	1.9	1.9	1.73

for d^6 to d^9) for d shells more than half filled. We can thus get magnetic moments that are slightly less than, or slightly more than, the spin-only moment depending on the sign of λ. The effect is normally of the order of a few tenths of a Bohr magneton. We can exemplify the effect with the help of table 7.2. The d^1 (V^{4+}), d^3 (Cr^{3+}), and d^4 (Cr^{2+}) ions have λ positive and hence $\mu_{eff} < \mu_{spin\ only}$. The d^7 (Co^{2+}), d^8 (Ni^{2+}), and d^9 (Cu^{2+}) ions have λ negative, and hence, by substitution in the above equation, $\mu_{eff} > \mu_{spin\ only}$. By using values of λ and $10Dq$ obtained from the electronic spectra, a good fit is obtained between the experimental and calculated moments. Because the difference in energy between the ground term and the term being mixed in is very large (approximately 10^4 cm^{-1}) compared with kT (approximately 200 cm^{-1} at room temperature) this spin—orbit contribution to the moment is independent of temperature.

7.2.2   Magnetic Properties of Complexes with T Ground Terms

The situation for these ground terms is very much more complex and we can indicate only the results here. The T terms are split by spin—orbit coupling to produce levels whose energy differences are frequently of the order of kT. Temperature will thus have a direct effect on the population of the levels arising in a magnetic field, so for these terms we expect magnetic moments to vary with temperature. A study of magnetic properties over a temperature range thus enables a distinction to be made between these two types of ground terms, and hence permits the deduction of stereochemistry of complexes. Notice (table 7.1) that only for the d^5 case do the octahedral and tetrahedral stereochemistries give rise to the same type of ground term.

Some representative magnetic moments for complexes with T ground terms are listed in table 7.3. The room-temperature moment is in excess of the spin-only moment for all except the V^{3+} case which we shall discuss shortly. This increase in magnetic moment is as expected on the simple picture of orbital contribution. Notice that this contribution is significantly greater than that usually observed for complexes of the same element having an A or E ground term, that is in a different stereochemistry.

The extent of orbital contribution in these complexes can be deduced only when the extent of the splittings of the various levels is known. Calculations have thus been performed to determine the effect of the spin—orbit coupling on the orbitally degenerate ground term. The results are best expressed in terms of a graph of μ_{eff} against the parameter kT/λ which is a convenient scale of temperature. These plots are sometimes known as *Kotani plots* after the originator of the equation on which they are based. Two graphs can be drawn for each term, one for which λ is positive and one for which λ is negative. The graphs for all the T terms are to be found in the works by Figgis, or Figgis and Lewis (see bibliography).

TABLE 7.3   MAGNETIC MOMENTS OF SOME COMPLEXES WITH
T GROUND TERMS

Number of d electrons	Compound	Stereochemistry	μ_{eff} 80K	μ_{eff} 300K	spin-only
1	$Cs_2 VCl_6$	octahedral	1.4	1.8	1.73
2	$(NH_4)V(SO_4)_2 . 12H_2O$	octahedral	2.7	2.7	2.83
4	$K_3[Mn(CN)_6]$	octahedral	3.1	3.2	2.83
5	$K_3[Fe(CN)_6]$	octahedral	2.2	2.4	1.73
6	$(NH_4)_2 Fe(SO_4)_2 . 6H_2O$	octahedral	5.4	5.5	4.90
7	$(NH_4)_2 Co(SO_4)_2 . 6H_2O$	octahedral	4.6	5.1	3.87
8	$(Et_4 N)_2 NiCl_4$	tetrahedral	3.2	3.8	2.83

We shall show here the general variations by using the curves for the T terms derived from D configurations; these are drawn in figure 7.3. The moments of appropriate individual ions are marked on the curves using the free-ion values of λ and $T = 300$ K. The bottom curve in the figure represents the variation of magnetic moment with temperature for the octahedral d^1 case. At absolute zero the magnetic moment is zero because the orbital and spin moments ($l = 1, s = \frac{1}{2}$) cancel each other out, that is the lowest level corresponds to the orbital and spin contributions being in opposite directions. As the temperature is increased, higher levels become occupied corresponding to an alignment of the spin and orbital moments. The spin-only value (1.73 B.M.) is reached around $kT/\lambda = 1$. In the limit when $kT/\lambda \to \infty$, the formula $\mu = \sqrt{[4S(S + 1) + L(L + 1)]}$ applies and a moment of $\sqrt{5}$, that is 2.24 B.M., is expected. We can see that the success of the spin-only formula for T terms, such as in Ti^{3+} and V^{3+}, is rather fortuitous; it so happens that at room temperature the values of λ and T sometimes give moments close to the spin-only value.

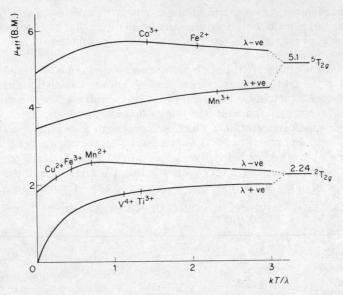

Figure 7.3   The variation of magnetic moment with temperature for d electron
T terms

In practice, measurements of magnetic moments over a temperature range do
not fit these curves very well. In many cases, for example V^{3+} (see table 7.3),
the variation of the moment with temperature is much less than that predicted.
Several factors are responsible for this. Firstly, we have used the free-ion value
for λ in the plots, and this differs from the actual values found in the complexes;
no allowance has been made for electron delocalisation. If covalent bonding is
present it is unlikely that the unpaired electron(s) remain permanently on the
metal. In the presence of σ or π-bonding, an electron can be removed from the
metal to the ligand, in which position it can no longer contribute to orbital
angular momentum around the metal. The reduction in the value of the spin–
orbit coupling constant in the complex compared with that in the free ion is best
interpreted in terms of this covalency. Even when this delocalisation is taken into
account a poor fit is still found with, for example, vanadium and titanium(III)
alums. We have assumed cubic (that is pure octahedral or tetrahedral) symmetry
for the complexes throughout, and this is evidently an unjustified assumption.
The $[Ti(H_2O)_6]^{3+}$ and $[V(H_2O)_6]^{3+}$ ions which are present in the alums
contain orbitally degenerate ground levels as first written, and as such are subject
to Jahn–Teller distortion. In this distortion the ground T level is split into orbital
singlet and doublet levels (figure 5.15). The ground term is now not T but A or E,
and the moment is thus expected to be invariant with temperature. The separation
between the singlet and doublet levels produced varies from zero, that is pure T
level, to around 2000 cm^{-1}, but is commonly of the order of a few hundred cm^{-1}.
Compared with the primary ligand-field splitting ($10Dq$), this splitting is thus very
small and difficult to detect spectroscopically. However, magnetic properties of

complexes are very sensitive to these small ground-state variations and are thus useful in detecting these small distortions. The magnitude of this distortion can be discovered by attempting to fit the experimental plot of μ against T with various ratios of δ/λ where δ is the separation between the singlet and doublet levels (figure 5.15). The best fit gives the value of δ, a value of $\delta = 0$ corresponding to octahedral symmetry.

7.3   High-spin—low-spin Equilibria

We have so far assumed that each complex occurs in a given structural form with a given high or low-spin configuration. Complications arise when a complex is in a given stereochemical form but in high-spin—low-spin equilibrium or when a complex readily changes stereochemistry (and hence the number of unpaired electrons) in different chemical environments. Each of these types of equilibrium is now discussed briefly.

7.3.1   Thermal Equilibria between Spin States

Ions of the d^4 to d^7 configurations in octahedral fields may give high or low-spin complexes depending on the strength of the ligand field. For any metal ion the change from high-spin to low-spin configuration might be expected to occur abruptly at a certain value of the ligand-field splitting parameter Δ. If we can choose a ligand field close to this *cross-over point*, a chemical equilibrium of the two forms might occur. Further, if the energy separation between the two spin states is of the order of kT, temperature will have a great effect on the position of the high-spin—low-spin equilibrium.

Such equilibria are well known, occurring particularly in iron complexes. The d^6 iron(II) systems are especially suitable for experimental studies because the low-spin state is diamagnetic and the change in the number of unpaired electrons in going to the high-spin state is four. In figure 7.4 is shown the variation of magnetic moment

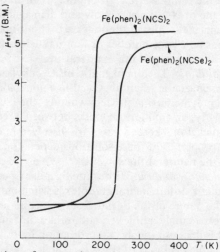

Figure 7.4   Variation of magnetic moment with temperature for some iron(II) complexes near the $^5T_2 - {}^1A_1$ cross-over

with temperature for two octahedral iron(II) phenanthroline complexes. At room temperature $Fe(phen)_2(NCS)_2$ and $Fe(phen)_2(NCSe)_2$ exist in the high-spin $^5T_2$ ground state (table 7.1) with magnetic moments in excess of the spin-only value for four unpaired electrons. At 174 K the moment of $Fe(phen)_2(NCS)_2$ suddenly drops to below 1 B.M., and a similar drop occurs for $Fe(phen)_2(NCSe)_2$ at 232 K. Below these temperatures the complexes have the $^1A_{1g}$ ground state, and changes in the electronic spectra support this.

Similar magnetic behaviour is shown by iron(III) surrounded by six sulphur ligands as in the dialkyldithiocarbamates $[Fe(S_2CNR_2)_3]$. Such complexes have magnetic moments at room temperature that lie between those expected for high and low-spin octahedral complexes; this behaviour persists in benzene in which the complexes are monomeric. The thermal equilibrium between spin states $S = \frac{1}{2}$ and $S = ^5/_2$ is thus occurring.

7.3.2   Structural Equilibria between Spin States

When the energy difference between two stereochemistries for a particular combination of metal ion plus ligand is small, a slight change of conditions may bring about a structural change and consequently a magnetic change. In this class are a large number of nickel(II) complexes that were for many years regarded as 'anomalous'. These are conveniently classified under three headings.

Octahedral–square-planar equilibria in solution. The solutions of a large number of nickel(II) complexes in organic solvents have magnetic moments the magnitude of which depends on factors such as the concentration, the temperature, and the nature of the solvent. In donor solvents, square (diamagnetic) and octahedral (paramagnetic) complexes are often interconvertible by way of the equilibrium

$$NiL_4 + 2L' \rightleftharpoons trans\text{-}[NiL_4L'_2]$$

The Lifschitz salts, that is nickel(II) complexes of substituted ethylenediamines, can be represented by the formula NiL_2X_2 where X is an anion. These often occur in hydrated form and are of two types: yellow and diamagnetic (usually square planar); blue and paramagnetic (usually tetragonal). The form that is obtained in any reaction depends on a large number of factors such as the nature of the solvent, the nature of the anion, and the temperature.

Monomer–polymer equilibria. The diamagnetic bis-(N-alkylsalicylaldimine)nickel(II) complexes become paramagnetic when dissolved in a nonco-ordinating solvent. This is attributed to a polymerisation process whereby the nickel becomes at least five-co-ordinate. Similarly, an equilibrium exists between a red monomer and a green polymer when many nickel(II) β-ketoenolates are dissolved in organic solvents. The colours and magnetic properties of these solutions depend on the concentration, the temperature, and the nature of the solvent.

Tetrahedral–square-planar equilibria. With some ligands, especially phosphines, nickel(II) salts form square or tetrahedral complexes when only minor changes occur in the conditions or the nature of the ligand. Solution equilibria often occur between square and tetrahedral complexes, and unlike that in the octahedral–square equilibrium, this does not involve co-ordination by solvent molecules. The solid square planar complex $NiBr_2[Pr^iP(Ph)_2]_2$ is stable at $0\,°C$ but isomerises to the tetrahedral form in about one day at $25\,°C$. Perhaps even more remarkable is the occurrence of $(PhCH_2PPh_2)_2NiBr_2$ in two forms. The red form is diamagnetic and square planar; the green form has a magnetic moment of 2.7 B.M. at room temperature. This green form has been shown by an X-ray structural analysis to contain nickel in two different environments, both the square and the tetrahedral. Its structure can be represented by the formula (writing L for the phosphine) NiL_2Br_2(square).$2NiL_2Br_2$(tetrahedral). If allowance is made for one-third of the molecules being diamagnetic, a reasonable recalculated moment of 3.3 B.M. is obtained for the tetrahedral form.

8 Mechanisms of Complex-ion Reactions

8.1 Introduction

In chapter 3 we mentioned that substitution reactions of metal aquo-ions, for example

$$[M(H_2O)_6]^{n+} + L \rightleftharpoons [M(H_2O)_5L]^{n+} + H_2O$$

do not all occur instantaneously even though the stability constant may show that the equilibrium lies well over to the right-hand side. Those complexes for which equilibrium is attained rapidly, for example in less than one minute, are called *labile* while those for which ligand substitution is a slow process are called *inert*. These terms labile and inert are concerned with kinetics and must not be confused with stability which, when unqualified, refers to thermodynamic stability. It is thus possible to have complexes which are unstable but inert, or conversely stable but labile. For example, the cobaltammines are thermodynamically unstable in acid solution with respect to the formation of $[Co(H_2O)_6]^{2+}$ and oxygen. Such solutions can however be kept for days at room temperature without noticeable decomposition, that is, the rate of decomposition is very low because the complexes are inert.

Measurements of the rates of reaction of complex compounds can provide information concerning the mechanisms of the reactions. Many kinetic studies on such systems have been performed but here we shall concern ourselves with only the general conclusions to be gained from these studies; more detailed discussions are to be found in the books recommended in the bibliography. Inert complexes are easily studied kinetically by so-called 'static' methods. The reactants are mixed and the rate of reaction is measured with time, using, for example, a spectro-photometric method of observing the increase or decrease in concentration of particular species. Labile complexes are more difficult to study. By using a flow system reactants can be mixed in shorter time than by conventional 'static' mixing, and observations can be made on moderately fast reactions. However, very fast reactions require very specialised techniques such as relaxation methods. In these methods the position of the equilibrium is suddenly changed by, for example, a temperature or pressure jump, and the speed with which the new equilibrium is reached is measured.

The results of such rate studies give information from which some broad generalisations emerge. Taube was the first to point out the existence of a relationship between the rate of reaction of complexes and their electronic structure.

8.1.1 Inert and Labile Complexes

Taube suggested that for inner-orbital complexes (valence bond nomenclature) at least one vacant d orbital needed to be present to confer lability. The substitution process is envisaged as the addition of the new ligand to form a seven-co-ordinate activated complex which then evolves one of the original ligands. Thus, if no vacant orbital is available the activation energy for the formation of the activated complex will be high, and reaction rates correspondingly slow.

Basolo and Pearson have used the crystal-field approach to correlate lability with d electron configurations. In this method the crystal field stabilisation energy of the complex is compared with that of its activated complex (that is the configuration of reactant molecules from which reaction can proceed without further addition of energy). If the CFSE of the initial complex is much greater than that of the activated complex, the complex will be inert, but if the difference between the CFSEs is small the complex will be labile. Assuming, for example, that an octahedral complex reacts via a square pyramidal five-co-ordinate activated complex, the difference in CFSEs for the various d electron configurations can be calculated. This simple approach is only a crude approximation but nevertheless has led to useful correlations.

For d^0, d^5, and d^{10} ions the CFSE is zero in both the initial complex and activated complex, so complexes of these configurations are labile. Similarly, in the d^1, d^2, and high-spin d^6, d^7, d^9 cases the change in CFSE is negligible and labile complexes result. Inert complexes are expected on CFSE argument for d^3 $(t_{2g}{}^3)$ and d^8 $(t_{2g}{}^6 e_g{}^2)$ and low-spin d^4 $(t_{2g}{}^4)$, d^5 $(t_{2g}{}^5)$, and d^6 $(t_{2g}{}^6)$ configurations. The predicted order of decreasing reaction rates for inert complexes is $d^5 > d^4 > d^8 \approx d^3 > d^6$. In general the experimental rates found support this order except for the d^8 configuration. Nickel(II) complexes of the $t_{2g}{}^6 e_g{}^2$ configuration are generally labile although their reaction rates are usually slower than those of Mn(II) $(t_{2g}{}^3 e_g{}^2)$, Co(II) $(t_{2g}{}^5 e_g{}^2)$, Cu(II) $(t_{2g}{}^6 e_g{}^3)$, and Zn(II) $(t_{2g}{}^6 e_g{}^4)$.

As a general rule, four-co-ordinate complexes react more rapidly than similar six-co-ordinate systems. They are consequently often difficult to study kinetically. Their greater lability is attributed to the ease of formation of five-co-ordinate intermediates which then break down with liberation of one of the originally bound ligands. We shall now consider substitution reactions of six and four-co-ordinate complexes.

8.2   Substitution Reactions of Metal Complexes

The reactions with which we are concerned here are of the general type

$$MX_5 Y + L \rightarrow MX_5 L + Y$$

For these reactions two limiting mechanisms can be defined.

If the intermediate formed in the rate-determining step is of lower co-ordination number than the original complex, this is called a *dissociative* mechanism (symbol D). In other words the mechanism can be written in the form

$$MX_5 Y \xrightarrow{\text{slow}} MX_5 + Y$$

followed by

$$MX_5 + L \xrightarrow{\text{fast}} MX_5 L$$

In this mechanism the rate-determining step involves the first process; the rate of disappearance of $MX_5 Y$ is proportional to the concentration of this species and, as only one reactant is involved in the rate-determining step, the process is unimolecular. An alternative label for this overall process is $S_N 1$, that is substitution, nucleophilic, unimolecular.

The alternative mechanism is that in which the rate-determining step involves the formation of an intermediate of higher co-ordination number than the starting compound. This is the *associative* mechanism (symbol A). The process can be written

$$MX_5Y + L \xrightarrow{\text{slow}} MX_5YL$$

followed by

$$MX_5YL \xrightarrow{\text{fast}} MX_5L + Y$$

The rate of disappearance of MX_5Y is now proportional to the concentrations of both MX_5Y and L, the process is bimolecular, and the mechanism is labelled S_N2, that is *substitution, nucleophilic, bimolecular.*

These two mechanistic routes represent extreme cases, and any particular reaction may proceed with say S_N1 character. This would imply that M—Y bond breaking is of primary importance but the ligand L may be present in the co-ordination sphere of M before the five-co-ordinate intermediate is actually formed.

In the first transition series, substitution reactions of the inert d^3 Cr(III) ions and d^6 Co(III) ions occur slowly enough to permit easy measurement of rates. The consequence is that an enormous amount of information is available on the substitution reactions of these elements. However, with the advent of fast-reaction techniques, rate studies on complexes of the more labile ions have become possible. It is impossible to give an adequate treatment to all this work in this book; we shall look only at a few important generalisations.

8.2.1 *Substitution Reactions of Octahedral Complexes*

Substitution in aquo-ions. The most fundamental substitution reaction of aquo-ions is that of water exchange between hydrated cations and the solvent water. Studies on a number of aquo-cations have been performed using relaxation methods. Except for a few cations, notably Cr(III) and Rh(III), these reactions are very fast. The general trends are as follows. Within a periodic group of ions of the same charge, for example Be^{2+}, Mg^{2+}, Ca^{2+}, Sr^{2+}, Ba^{2+}, the rates of exchange increase with the size of the cation (Be^{2+}, $k \approx 10^2$ s^{-1}; Ba^{2+}, $k \approx 10^9$ s^{-1}). The more highly charged M^{3+} ions undergo water exchange more slowly than M^{2+} and M^+. Now, the M—OH_2 bond strength is expected to increase with charge on M and decrease with size of M for ions of the same charge. It follows therefore that the breaking of the M—OH_2 bond is of greater importance than the making of a new bond in the transition state. The exchange reactions thus appear to occur largely by dissociative or S_N1 processes.

The rates of replacement of water molecules by other ligands in general depend on the concentration of metal ion but not on the concentration of the added ligand, that is, they follow a first-order rate law. Surprisingly, the rates of substitution for a given metal aquo-ion do not vary much with the nature of the added ligand.

Aquation reactions. Substitution reactions carried out in aqueous solution appear to proceed always via the intermediate formation of the aquo-complex. This aquation reaction, that is

$$[MX_5Y]^{n+} + H_2O \rightleftharpoons [MX_5(H_2O)]^{n+} + Y$$

is thus of prime importance in the study of substitution reactions, and many studies have been performed on cobalt(III) complexes. The reaction is often given the inappropriate name of *acid hydrolysis;* this term is meant to imply that the hydration occurs under acid conditions. The reaction

$$[MX_5Y]^{n+} + OH^- \rightleftharpoons [MX_5(OH)]^{n+} + Y^-$$

is called *base hydrolysis.*

In the aquation reactions (pH < 4) the entering ligand H_2O is present in such a high and effectively constant concentration ($\approx 55.5M$) that the rate law cannot help in deciding whether an associative or a dissociative mechanism is in operation. The observed rates are of course dependent on the concentrations of $[MX_5Y]^{n+}$ only. In order to decide which mechanism is the more important, evidence from the hydration of closely related ions must be sought.

The dissociative process would be expected to be favoured if bulky groups are attached to the cobalt, causing some steric strain and discouraging the possibility of association to a seven-co-ordinate intermediate. To this end, studies have been made on the rates of 'hydrolysis' of $[Co(LL)_2Cl_2]^+$ in which LL represents a chelating ligand derived from ethylenediamine. The rate of substitution for the process

$$[Co(LL)_2Cl_2]^+ + H_2O \rightarrow [Co(LL)_2Cl(H_2O)]^{2+} + Cl^-$$

is indeed found to increase with the size of the ligand LL, and a dissociative process is indicated. Steric effects may not be the only important ones here however; dissociation might be assisted by inductive effects from the more highly substituted ligands, increasing the electron density on the cobalt ion and facilitating the dissociation of the chloride ion.

Further evidence that a dissociative process predominates comes from a comparison of the rates of aquation of complexes having similar ligands but different charges. The aquation of $[Co(NH_3)_4Cl_2]^+$, for example, is of the order of 10^3 times faster than that of $[Co(NH_3)_5Cl]^{2+}$. The increase of charge on a complex is expected to increase the metal–ligand bond strengths and hinder metal–ligand bond cleavage. It should however facilitate the bonding of a seventh ligand and thereby enhance the associative process. The fact that substitution occurs more slowly with the more highly charged ions is indicative of a D process.

8.2.2   *Substitution Reactions of Square Complexes*

The complexes of platinum(II) are square planar and fairly inert, so these have received most detailed study. The rates of substitution reactions in platinum(II) chemistry are insensitive to the charge on the complex ion, indicating that both bond breaking and bond formation are important. Further, the overall kinetics is second-order with a two-term rate law. For example, for the reaction

$$[Pt(NH_3)_3Cl]^+ + Y^- \rightarrow [Pt(NH_3)_3Y]^+ + Cl^-$$

the rate is given by

$$\text{Rate} = k\,[\text{Pt(NH}_3)_3\text{Cl}^+] + k'\,[\text{Pt(NH}_3)_3\text{Cl}^+][\text{Y}^-]$$

Such a rate law implies that the reaction proceeds by two parallel paths, one of which involves Y^- in the rate-determining step, and one involving the solvent. These mechanisms are illustrated below.

$$[\text{Pt(NH}_3)_3\text{Cl}]^+ + \text{Y}^- \xrightleftharpoons{\text{slow}, k'} [\text{Pt(NH}_3)_3\text{ClY}]$$

$$\downarrow$$

$$[\text{Pt(NH}_3)_3\text{Y}]^+ + \text{Cl}^-$$

$$[\text{Pt(NH}_3)_3\text{Cl}]^+ + \text{H}_2\text{O} \xrightleftharpoons{\text{slow}, k} [\text{Pt(NH}_3)_3\text{H}_2\text{O}]^{2+} + \text{Cl}^-$$

$$\downarrow \text{Y}^-, \text{fast}$$

$$[\text{Pt(NH}_3)_3\text{Y}]^+ + \text{H}_2\text{O}$$

Both paths are believed to proceed via the associative process A, involving a trigonal bipyramidal intermediate.

The trans *effect.* One of the most remarkable features of substitution reactions of platinum(II) complexes is the so-called *trans* effect. This concerns the effect of *trans* substituents on the lability of the leaving group in substitution reactions. As an example let us consider the following reaction schemes:

In these schemes the order in which the PtCl_4^{2-} ion is treated with ammonia and nitrite ion determines which of the two isomers of $[\text{PtCl}_2(\text{NO}_2)(\text{NH}_3)]^-$ is predominantly produced. It is apparent that the NO_2^- ion has labilised the chloride ion *trans* to itself in the reaction of $[\text{PtCl}_3(\text{NO}_2)]^{2-}$ with ammonia, whereas in the reaction of $[\text{PtCl}_3(\text{NH}_3)]^-$ with NO_2^- it is the chloride ion which has labilised a chloride ion *trans* to itself. The order of *trans* effect for these ligands is thus $\text{NO}_2^- > \text{Cl}^- > \text{NH}_3$. By studying a vast number of these substitution reactions it is possible to draw up a series of ligands of decreasing *trans* effect

$$\text{CO} \approx \text{CN}^- \approx \text{C}_2\text{H}_4 \approx \text{PR}_3 \approx \text{H}^- > \text{NO}_2^- \approx \text{I}^- \approx \text{SCN}^- > \text{Br}^-$$

$$> \text{Cl}^- > \text{NH}_3 \approx \text{py} > \text{OH}^- > \text{H}_2\text{O}$$

This order can then be used to predict preparative methods for particular isomers. Consider the preparation of [PtClBr(NH$_3$)py] from PtCl$_4$$^{2-}$. One isomer can be prepared as follows.

In this sequence, the reaction with Br$^-$ occurs in the manner shown since Cl$^-$ has a greater labilising effect than ammonia; the reaction with pyridine proceeds as shown because Br$^-$ has a greater labilising effect than either Cl$^-$ or NH$_3$. Obviously by altering the starting material and order of reactants the other two possible isomers of the product (that is, with NH$_3$ and Cl *trans* to Br) can be synthesised.

Theories which attempt to explain the observed order of *trans* effect have been generally debated and no conclusive explanation has been reached. However, those ligands of high *trans* effect are those having either high π-acceptor properties, for example CO, CN$^-$, and C$_2$H$_4$, or strong σ-donor properties (to metal p orbitals), for example H$^-$ and PR$_3$. The NO$_2$$^-$ ion probably has a large π-acceptor contribution but a poor σ-donor effect in comparison with CO and CN$^-$.

8.3 Oxidation–Reduction Reactions

Reactions in which an electron is transferred from one complex to another are mechanistically of two main types. These are called the *outer-sphere* and *inner-sphere* processes. In the outer-sphere process each complex maintains its own co-ordination shell during the electron transfer. This process is the one occurring between inert complexes; in these, substitution reactions are slower than electron transfer reactions. Inner-sphere processes are those in which electron transfer takes place through a bridging atom or group which is common to the co-ordination spheres of both complexes. They occur between reactants which are labile to substitution processes.

8.3.1 Outer-sphere Reactions

These reactions have a rate law that is first order in both reactants. Usually both reactants are inert. Many examples of these reactions are to be found in electron exchange reactions which do not represent overall chemical reactions. Consider, for example, the electron-exchange reaction between [Fe(CN)$_6$]$^{4-}$ and [Fe(CN)$_6$]$^{3-}$. The rate of this exchange can be studied by labelling one of the complexes with a radioactive isotope of iron

$$[\overset{*}{Fe}(CN)_6]^{4-} + [Fe(CN)_6]^{3-} \rightarrow [\overset{*}{Fe}(CN)_6]^{3-} + [Fe(CN)_6]^{4-}$$

The hexacyanoferrate(III) is a low-spin d^5 system and hexacyanoferrate(II) is a low-spin d^6 system, both of which are inert to substitution, and hence the loss or exchange of cyanide ions is very small. The electron-exchange rate is very fast and thus the possibility of a bridged activated complex being important is ruled out since the mere formation of such a complex amounts to a substitution process. Other 'reactions' proceeding similarly are included in table 8.1; the rate constants indicate the wide range of rates exhibited, but accurate values are not quoted since the reactions were often measured under different conditions.

The direct electron transfer process between complexes presents some interesting theoretical problems. The Franck–Condon principle states that electronic transitions occur virtually instantaneously as compared to the time of atomic rearrangements. In other words we can expect the electron to move very much faster than the atoms in the complexes. In the hexacyanoferrate complexes,

TABLE 8.1 SOME OUTER-SPHERE REACTIONS

Reactants	Rate constant $(dm^3 \ mol^{-1} s^{-1})$
$[W(CN)_8]^{3-}$, $[W(CN)_8]^{4-}$	10^5
$[IrCl_6]^{2-}$, $[IrCl_6]^{3-}$	10^5
$[Mo(CN)_8]^{3-}$, $[Mo(CN)_8]^{4-}$	10^4
MnO_4^-, MnO_4^{2-}	10^3
$[Fe(CN)_6]^{3-}$, $[Fe(CN)_6]^{4-}$	10^2
$[Co(phen)_3]^{3+}$, $[Co(phen)_3]^{2+}$	1
$[Co(en)_3]^{3+}$, $[Co(en)_3]^{2+}$	10^{-4}
$[Co(NH_3)_6]^{3+}$, $[Co(NH_3)_6]^{2+}$	10^{-12}

the Fe–C bond length is shorter in $[Fe(CN)_6]^{3-}$ than it is in $[Fe(CN)_6]^{4-}$. If no movement of the atoms occurs during electron transfer, then in the newly formed $[Fe(CN)_6]^{3-}$ the Fe–C bond will be longer than the equilibrium value for this ion, and the newly formed $[Fe(CN)_6]^{4-}$ will have an Fe–C bond shorter than the equilibrium value. Both of these new ions are thus in excited states, that is with higher energy, so it appears that energy has been created. Since this cannot be the case (conservation of energy principle) the activation energy that needs to be supplied to the reaction is that necessary to produce a transition state in which each species has the same dimensions. Since the bond lengths are not very different in $[Fe(CN)_6]^{3-}$ and $[Fe(CN)_6]^{4-}$, only a small activation energy is required, so the electron transfer is fairly rapid. The exchange between $[Co(NH_3)_6]^{3+}$ and $[Co(NH_3)_6]^{2+}$ (table 8.1) is extremely slow. These complexes however do not have very dissimilar bond lengths, so an alternative explanation must be sought for the remarkable rate differences. In the cobalt ammine case the complexes differ widely in electronic structure; $[Co(NH_3)_6]^{3+}$ has the t_{2g}^6 ground state while $[Co(NH_3)_6]^{2+}$ has the ground state $t_{2g}^5 e_g^2$. It is therefore apparent that changes in electronic arrangement as well as bond lengths must occur before electron transfer in such systems.

8.3.2   Inner-sphere Reactions

In these reactions the mechanism involves a bridged complex formed by substitution into one of the reactants of a ligand from the other reactant. Provided that the substituting ligand remains bonded to the original metal ion as well as to the new ion, a bridge is formed. This bridging group is usually transferred along with the electron, and thus its presence in the product of the new ion (that is, not that to which it was originally bonded) is evidence for the bridge mechanism.

There are not very many systems that are suitable for studying this mechanistic process. An ideal reducing agent is Cr^{2+} because it is labile to substitution and forms chromium(III) complexes, which are inert. Any transfer of ligand must occur therefore in the activated complex. Suitable inert oxidising agents include Cr(III), Co(III), and Pt(IV). Taube has demonstrated the generality of the reaction

$$[Co(NH_3)_5X]^{2+} + [Cr(H_2O)_6]^{2+} + 5H_3O^+ \rightarrow [Cr(H_2O)_5X]^{2+} + [Co(H_2O)_6]^{2+} + 5NH_4^+$$

for a wide range of inorganic and organic anions. Thus the reaction between $[Co(NH_3)_5Cl]^{2+}$ and $[Cr(H_2O)_6]^{2+}$ results in quantitative transfer of the chloride ion from cobalt to chromium. Further, if labelled $^{36}Cl^-$ is used in the $[Co(NH_3)_5Cl]^{2+}$ ion and unlabelled chloride ion is also present in the solution which reacts with Cr^{2+}, the product contains only labelled Cl^-, proving that the Cl^- had come only from the cobalt cation. The bridged intermediate in these reactions can thus be written as $[(NH_3)_5Co^{III}-X-Cr^{II}(H_2O)_5]^{4+}$. Sometimes such bridged intermediates can be isolated in the solid state. For example, in the oxidation of $[Co(CN)_5]^{3-}$ (this ion may be hydrated) with $[Fe(CN)_6]^{3-}$, the barium salt of the ion $[(NC)_5Fe^{II}CNCo^{III}(CN)_5]^{6-}$ has been separated from the reaction mixture by fractional crystallisation.

9 Introduction to the First-row Transition Elements

9.1 Introduction

The elements of the first transition series are those for which the 3d electron shell contains between one and nine electrons. Copper is included because, although its outer electronic configuration is $3d^{10} 4s^1$, it has the $3d^9$ configuration in its commonly occurring +2 oxidation state. Zinc is not normally considered a transition element since in both the element and its compounds the 3d electron shell remains filled. It therefore does not show the characteristic properties of coloured compounds and paramagnetism shown by the other elements in at least one of their oxidation states. Scandium is included by the definition, but so far only the +3 oxidation state has been established with certainty. Its compounds are thus diamagnetic with no colour from d−d transitions, and its chemistry thus resembles that of aluminium rather than that of the other transition elements. We shall not deal specifically with its chemistry here; interested readers should consult Remy.

The elements with their outermost electronic configurations are listed in table 9.1.

TABLE 9.1   ELECTRONIC CONFIGURATIONS AND IONISATION ENERGIES OF THE FIRST-ROW TRANSITION ELEMENTS

Element	Configuration	Ionisation energies (kJ mol^{-1})			
		1st	2nd	3rd	4th
Sc	$3d^1 4s^2$	631	1235	2389	7130
Ti	$3d^2 4s^2$	656	1309	2650	4173
V	$3d^3 4s^2$	650	1414	2828	4600
Cr	$3d^5 4s^1$	653	1592	3056	4900
Mn	$3d^5 4s^2$	717	1509	3251	5020
Fe	$3d^6 4s^2$	762	1561	2956	5510
Co	$3d^7 4s^2$	758	1644	3231	5114
Ni	$3d^8 4s^2$	737	1752	3489	5404
Cu	$3d^{10} 4s^1$	745	1958	3545	5683

In the atoms of these elements the 3d and 4s orbitals have similar energies. The 4s orbitals are conventionally denoted as the outermost orbitals in table 9.1. This is because the radial probability plots of the 4s and 3d orbitals indicate that there is a greater probability of the 4s electron being furthermost from the nucleus. Further, on ionisation, the neutral atoms usually lose a 4s electron first. In other words, for charged atoms the 3d orbitals are the most stable. Calcium has the outer electron configuration $4s^2$, so the first transition element is scandium with the configuration $3d^1 4s^2$. Along the transition series the 3d shell is then filled up regularly with the exceptions of Cr $3d^5 4s^1$ and Cu $3d^{10} 4s^1$ whose configurations are attributable to the stabilities of half-filled and filled d shells respectively. However, no great practical importance is attached to the configurations of the atoms; it is the ions with which we shall more often be concerned. Thus the divalent state corresponding to the ionisation of the two s electrons is well-established for all the elements except scandium. Oxidation states less than two are usually found (except in the case of Cu^1) only with π-acceptor ligands. Copper(I) has the extra stability associated with the $3d^{10}$ closed-shell configuration.

The trends in the ionisation energies of the elements (table 9.1) illustrate some features of transition metal chemistry. In general the ionisation energies double as the oxidation state increases by one unit, at least up to the fourth ionisation energy. The fourth ionisation energy of scandium indicates that Sc(IV) is unlikely to occur normally, whereas Ti(IV) and V(IV) for example do occur commonly. Unfortunately many factors are important in determining the stability of oxidation states, so we cannot decide which states will occur on the basis of ionisation energies alone. Unipositive ions of these elements are not normally stable in aqueous solution because of the significant sublimation energy term in the process

$$M(s) \xrightarrow{\Delta H_{subl}} M(g) \xrightarrow{I} M^+(g) \xrightarrow{\Delta H_{hydration}} M^+(aq.)$$

As we go to dipositive ions the sublimation energy term remains the same, while the ionisation and hydration energies increase markedly. The sublimation energy thus becomes less significant and the formation of hydrated dipositive ions is favourable. The increased third ionisation energies of nickel and copper, compared with those of the other elements, are significant in the instability of Ni(III) and Cu(III) in aqueous solution.

9.2   Oxidation States

The factors that influence the stability of oxidation states are many. As with compounds the word 'stable' must be modified when applied to the term oxidation state. Thus copper(I) is a thermally stable oxidation state of copper but it is an unstable state with respect to disproportionation in aqueous solution. If we ignore for the moment the complexes of the elements with π-acceptor ligands such as carbon monoxide, which give the very low oxidation states (that is $0, -1, -2$), we can tabulate the known oxidation states as in table 9.2.

In the first half of the series, that is up to manganese, the highest oxidation state corresponds to the periodic group number. Thereafter the high oxidation states are difficult to obtain and the maximum oxidation state reached no longer corresponds to the removal of all the 3d and 4s electrons. The oxidation state of +2 is strongly reducing for Ti, V, and Cr, but much more redox stable for Mn and the elements that follow. Indeed, the existence of any particular oxidation state for a metal with a given ligand or anion depends largely on the redox properties of the metal—ligand combination. For example, vanadium in its reactions with the halogens gives a

TABLE 9.2   KNOWN OXIDATION STATES OF THE FIRST-ROW
TRANSITION ELEMENTS

Element	Oxidation states						
Sc			3				
Ti		2	3	4			
V		2	3	4	5		
Cr		2	3	4	5	6	
Mn		2	3	4	5	6	7
Fe		2	3	4	5	6	
Co		2	3	4	5		
Ni		2	3	4			
Cu	1	2	3				

fluoride VF_5, a chloride VCl_4, and a bromide VBr_3. In other words vanadium(V) is too strong an oxidising agent to exist in sole combination with chloride or bromide ions; they become oxidised to the free halogens. We shall now consider the effects of the nature of the ligand and of pH on the stability of high and low oxidation states.

9.2.1   High Oxidation States

These are exhibited when the metals are in combination with fluoride ions or oxide ions. These are the small, highly electronegative, and hence most difficultly oxidised, anions. Fluorine is often said to bring out the maximum covalency of any element, and we have already seen one example of this in vanadium chemistry. With oxide ions we get the stabilisation of the highest oxidation states of V, Cr, and Mn in the vanadate VO_4^{3-}, chromate CrO_4^{2-}, and permanganate MnO_4^- ions. The high oxidation states of iron and cobalt (that is greater than +3) and of nickel and copper (that is greater than +2) almost invariably occur in compounds containing oxygen or fluorine. Examples of these compounds will be found under the individual elements concerned in the chapters that follow.

In aqueous solution high oxidation states are favoured in alkaline solution; indeed, oxidation of transition metal hydroxides by atmospheric oxygen is a characteristic feature of these elements. The addition of sodium hydroxide solution to an iron(II) salt, for example, gives a pale green precipitate of $Fe(OH)_2$ which turns brown on standing in the air owing to oxidation to iron(III). We can understand such oxidations simply in terms of the equilibria

$$[M(H_2O)_6]^{n+} \underset{H^+}{\overset{OH^-}{\rightleftharpoons}} [M(OH)_6]^{(6-n)-}$$

It is easier to remove an electron from (that is oxidise) the negatively charged hydroxo-species than from the positively charged aquo-ion. It follows that, in order to stabilise lower oxidation states in aqueous solution, nonoxidising acids must be used.

9.2.2   Low Oxidation States

The ligands that favour low oxidation states are those that are capable of bonding by both σ-donor and π-acceptor bonds. In order to do this the ligands must possess both lone-pair electrons and vacant orbitals of π symmetry. The electrons from the ligand are donated along the σ bond; the build-up of negative charge on the metal is then delocalised back on to the ligand via π-bonding. Since the ligands are accepting electron density in the π-bond they are sometimes called π-*acid ligands.* Typical π-acid ligands are CO, N_2, NO, PR_3, RNC, bipy, and phen.

Since carbon monoxide is uncharged, the binary metal carbonyls, for example $Cr(CO)_6$ and $Fe(CO)_5$, contain the metals in zero oxidation states. The bonding in these carbonyls consists of an M—C σ bond formed by donation of the lone pair of electrons on carbon and an M—C π bond formed by donation of electron density from metal d orbitals into the vacant antibonding molecular orbitals (π*p) on carbon monoxide. These two types of bond are illustrated in figure 9.1. In many

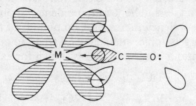

Figure 9.1   The σ and π-bonding in metal carbonyls; only the M—C bonding orbitals are included; shaded orbitals are filled and unshaded orbitals vacant.

complexes the metal has a formally negative oxidation state. In the sodium salts $Na[Co(CO)_4]$ and $Na[Mn(CO)_5]$, for example, the metals are present in uninegative anions, and since there is no charge on the CO ligands, the metals are in the —1 oxidation state. The +1 oxidation state occurs in the carbonyl halides such as $Mo(CO)_5Cl$.

Complexes containing molecular nitrogen (usually called dinitrogen complexes) resemble the carbonyls in that the M—N_2 bond involves σ donation and π acceptance by the nitrogen. Typical dinitrogen complexes are $Co(N_2)H(PPh_3)_3$ and $Fe(N_2)H_2(Ph_2EtP)_3$ which contain linear M—N—N groupings. Some complexes containing bridging rather than terminal dinitrogen groups are known, for example $[(NH_3)_5RuN=NRu(NH_3)_5]^{4+}$. However, the range of dinitrogen complexes is much less extensive than that of the carbonyl complexes (the first N_2 complex was obtained in 1965), and N_2 complexes are less stable than those of CO.

Complexes of nitrogen monoxide usually contain the ligand bonded as NO^+. In the carbonyl nitrosyl complexes we have the isoelectronic series $MnCO(NO)_3$, $Fe(CO)_2(NO)_2$, $Co(CO)_3NO$ [and $Ni(CO)_4$]. Since the physical evidence indicates that these compounds contain NO^+, the oxidation states of the metals in this series are Mn(–III), Fe(–II), Co(–I) [and Ni(0)].

While phosphorus and arsenic ligands can stabilise low oxidation states because of their vacant d orbitals (which can form π bonds with the metal d orbitals), simple nitrogen donors cannot act effectively in this way. The d orbitals on nitrogen are too high in energy, and the antibonding molecular orbitals on ammonia are of σ symmetry. One or two compounds having only ammonia bonded to a metal have been prepared, for example tetrammineplatinum(0), $Pt(NH_3)_4$. However, compounds of this type have low thermal stability and have not been obtained for the first-row transition elements. If the donor nitrogen atom is connected to a delocalised π system as in bipy and phen, the π^* orbitals of the ring can be used in π-bonding to the metal. Typical of such compounds are $Ti(bipy)_3$ and $Li[Ti(bipy)_3] . 3.5THF$ which formally contain $Ti(0)$ and $Ti(-I)$ respectively. However, these oxidation state assignments assume uncharged bipy ligands; an alternative assignment involves the bipy being reduced to radical anions bipy$^-\cdot$ with the extra electrons occupying π^* orbitals in the ring.

9.2.3   *Ionic and Basic Character*

The lower oxidation states in table 9.2 are those in which the elements show their most pronounced ionic character. As the oxidation state increases the charge density on the metal increases and hence the tendency to polarisation of the anion charge cloud and covalency increases. Compounds of these elements in the divalent state are thus largely ionic, and a general trend to covalency as the oxidation number increases is observed. This is shown for example in the halides

$TiCl_2$	$TiCl_3$	$TiCl_4$	
VCl_2	VCl_3	VCl_4	$VOCl_3$
Solids, insoluble in benzene; weak acids		Liquids, soluble in benzene; strong acids	

$$\xrightarrow{\hspace{4cm}}$$

increasing covalent character

Accompanying this increase in charge on the metal, as the oxidation state increases, is a corresponding increase in acidity. The more covalent halides are strong Lewis acids and hydrolyse violently in water. The more ionic halides have weaker Lewis acidity; they form aquo-ions in water, $[M(H_2O)_6]^{2+}$ being almost neutral while $[M(H_2O)_6]^{3+}$ is acidic. Similar trends are to be observed in the more common oxides

TiO	Ti_2O_3	TiO_2	
VO	V_2O_3		V_2O_5
CrO	Cr_2O_3		CrO_3
basic	amphoteric	acidic	

The oxides of the metals (Ti to Co) in their maximum oxidation states are decidedly acidic and give oxo-anion salts with alkalis. Thus titanium(IV) oxide forms titanates(IV), for example TiO_4^{4-}, vanadium(V) oxide forms vanadates(V), for example VO_4^{3-}, and chromium(VI) oxide forms chromates(VI), for example

$CrO_4{}^{2-}$. At the other extreme the oxides of metals in the +2 oxidation state show only basic properties, dissolving in acids to form aquo-ions but having little reaction with alkalis.

9.3 Physical and Chemical Properties of the Elements

9.3.1 Physical Properties

All the first-row transition elements are typical metals, many of which find great use as engineering materials. They are characterised by high melting and boiling points, high densities, and low atomic volumes. Some of these properties are listed in table 9.3.

TABLE 9.3   SOME PHYSICAL PROPERTIES OF FIRST-ROW TRANSITION ELEMENTS

Element	M.p. (K)	B.p. (K)	Density (g/cm^3)	Sublimation energy $(kJ\ mol^{-1})$ (298 K)
Sc	1810	3000	3.0	326
Ti	1950	3530	4.5	473
V	2160	3270	6.0	515
Cr	2160	2750	7.2	397
Mn	1520	2370	7.2	281
Fe	1810	3270	7.9	416
Co	1760	3170	8.9	425
Ni	1730	3010	8.9	430
Cu	1356	2868	8.9	339

The horizontal relationships between transition metal ions in the same oxidation state are related to the very similar sizes, and hence charge-to-size ratios, found along the series. Although one cannot strictly define the size of an atom, it is useful to have a relative scale of sizes of the various atoms for comparative purposes. In table 9.4 are listed some atomic and ionic radii obtained by the method of Pauling. A general decrease in size is observed as the transition series is traversed. This is due to contraction of the electron cloud, all the electrons being in the same shell and hence not adequately screening each other from the effects of the increased nuclear charge. The overall decrease in size along the series is not large, and there are some striking size similarities. The chemistry of Ti^{3+} resembles that of V^{3+} in many respects, as is to be expected from their similar ionic radii. In the so-called *transition triads* of group VIII a very close similarity is observed between the metals iron, cobalt, and nickel (see table 9.3). There are considerable similarities also in the properties of their M^{2+} ions, particularly in those properties dependent primarily on charge-to-size ratio rather than electronic configuration, for example solubilities of salts and acidity of the $[M(H_2O)_6]^{2+}$ ions.

TABLE 9.4   ATOMIC AND IONIC RADII (nm)

Element	Atomic radius	Ionic radii M^{2+}	M^{3+}
Ti	0.132	0.090	0.076
V	0.122	0.088	0.074
Cr	0.117	0.084	0.069
Mn	0.117	0.080	0.066
Fe	0.116	0.076	0.064
Co	0.116	0.074	0.063
Ni	0.115	0.072	0.062
Cu	0.117	0.072	

9.3.2   Chemical Properties

The first-row transition elements are reactive metals which undergo reaction with a very large number of the chemical elements. A characteristic feature of their chemistry is alloy formation between themselves as well as with nontransition metals. With the light nonmetallic elements, H, B, C, N, and O, interstitial solid solutions are formed. In these 'compounds' the light atoms occupy positions in the interstices of the metal lattices. The structures of the metals are often preserved but may be distorted somewhat; the formulae of the compounds are determined by solubilities, so these compounds are nonstoichiometric. The ratio of the atomic radii of the nonmetal to the metal needs to be less than 0.59 for interstitial compound formation to occur. Stoichiometric compounds do arise when the available interstitial holes are all filled by the light element. Thus, if all the 'octahedral holes' in a face-centred cubic close-packed structure are filled, a 1:1 stoichiometry results as in TiB, TiC, TiN, and TiO. Such compounds have the rock salt structure. These interstitial compounds are usually hard, of high melting point, and chemically inert. Some of them are of great commercial importance, for example steel. When carbon dissolves in iron it can occur as cementite Fe_3C or austenite. In cementite the carbon atoms are at the centres of nearly regular trigonal prisms of iron atoms, while in austenite there are insufficient carbon atoms to form a regular structure and they probably occupy octahedral holes in the γ-Fe lattice. Other interstitial compounds, for example TiB_2, have extreme hardness coupled with inertness to chemical attack.

The elements are extremely reactive in an excess of oxygen; they are often pyrophoric in air especially when in powdered form. The free energies of formation of the oxides in the maximum oxidation states of the metals are among the highest known. The reaction of iron with oxygen in the presence of water, that is rusting, is of great economic importance. With the halogens reaction again occurs very exothermically to produce the highest oxidation state halides, that is the ones which are stable at the temperature of preparation. The hydrogen halides are less reactive toward the transition metals but at elevated temperatures usually react to give halides of lower oxidation states. Thus, for example, the reaction of fluorine with titanium gives TiF_4 while that of hydrogen fluoride leads to TiF_3.

Towards aqueous acids the metals behave in varying ways. Hydrochloric acid dissolves them, Ti and V giving aquo (or chloroaquo) ions of M^{3+}, while Cr, Mn, Fe, Co, and Ni give M^{2+} aquo (or chloroaquo) ions. Oxidising acids such as concentrated nitric acid often have a passivating effect, for example on Fe and Cr. The metals can then be dissolved only with difficulty in other acids, presumably because of a protective oxide film on the surface of the metal. Metals that react with concentrated nitric acid give hydrous oxides (or oxo-ions) if the maximum oxidation state of the metal is +4 or over; thus Ti gives TiO_2.aq. The formation of oxides from nitric acid reactions on metals is due to the great acidity of, for example, $[M(H_2O)_6]^{4+}$ ions which, in the presence of water as a base, become $[M(OH)_4(H_2O)_2]$, that is MO_2.aq. Copper is rather unique in the series in requiring nitric acid to dissolve it rapidly.

10 General Preparative Methods

10.1 Introduction

The preparation of compounds will always be an important part of chemistry whether it be industrial production or the synthesis of new compounds on the research scale. Occasionally the synthesis of a new compound has an enormous impact on our progress and understanding in chemistry. Such was the case with ferrocene $Fe(\pi\text{-}C_5H_5)_2$, discovered in 1951, and xenon tetrafluoride XeF_4, first prepared in 1962.

In this chapter we consider the preparation of the metals and of their simple compounds. The general methods will be illustrated so that in studying the chemistry of each element in the subsequent chapters there will be no need to repeat preparative details of simple compounds. With a knowledge of general methods the reader should then be able to devise syntheses for a large number of simple compounds without a detailed knowledge of the particular element involved.

10.2 Occurrence and Extraction of the Transition Metals

10.2.1 Occurrence of Elements in the Earth's Crust
The approximate abundances of the elements of the first transition series in the earth's crust are indicated in table 10.1. The table includes some other elements

TABLE 10.1 ABUNDANCE OF THE FIRST-ROW TRANSITION ELEMENTS IN THE EARTH'S CRUST

Element	Abundance (wt %)	Element	Abundance (wt %)
O	46	Cl	0.1
Si	28	Cr	0.04
Al	8.1	C	0.03
Fe	5.1	V	0.02
Ca	3.6	Ni	0.008
Na	2.9	Cu	0.007
K	2.6	Co	0.002
Mg	2.1	Pb	0.001
Ti	0.6	Sc	0.0005
Mn	0.1	Zn	0.0001

for comparison purposes. One can see by inspection of the table that iron and titanium are the most plentiful transition elements and that all the other first-row transition elements are more plentiful than such 'common' elements as zinc. Since oxygen is the most plentiful of all the elements and is also electronegative, the elements are most likely to occur as oxides often in combination with silica, silicon being the next most plentiful element. Some typical ores with their approximate compositions are listed in table 10.2.

The extraction of a metal from its ore involves in the first place a concentration of the ore. Various processes are used to separate the ore from sundry unwanted rock etc. These processes include gravity separation and flotation methods as well as aqueous leaching processes (hydrometallurgy). We shall not elucidate them here but shall confine ourselves to the extraction of the metals from their purified ores. In general the ore must be reduced to the metal. The choice of reducing agent is governed by economic as well as chemical factors and is also dependent on the purity of metal required and the use to which it is to be put. The most important methods involve the reduction of oxides with carbon, hydrogen, or aluminium, the reduction of chlorides with active metals, and electrolysis of salts either molten or in aqueous solution. We consider these methods below.

TABLE 10.2   SOME COMMON ORES OF THE TRANSITION ELEMENTS

Element	Ore	Typical composition
Sc	thortveitite	$ScSi_2O_7$
Ti	ilmenite	$FeTiO_3$
	rutile	TiO_2
V	patronite	$V_2S_5.nS$
	vanadinite	$3Pb_3(VO_4)_2.PbCl_2$
Cr	chromite	$FeCr_2O_4$
Mn	pyrolusite	MnO_2
	rhodocrosite	$MnCO_3$
Fe	haematite	Fe_2O_3
	magnetite	Fe_3O_4
	siderite	$FeCO_3$
	pyrites	FeS_2
Co	smaltite	$CoAs_2$
	cobaltite	$CoAsS$
Ni	pentlandite	$(Ni,Fe)_9S_8$
	garnierite	$(Ni,Mg)_6Si_4O_{10}(OH)_8$
Cu	chalcocite	Cu_2S
	chalcopyrite	$CuFeS_2$

10.2.2   Extraction from Oxides by Thermal Methods

The thermal reduction of a metal oxide with carbon is frequently the preferred route to a metal on economic grounds. Because the general reaction, that is

$$MO + C \rightarrow M + CO$$

proceeds from left to right with the net production of carbon monoxide, a gas of large entropy, the reaction will become thermodynamically feasible for all metals at some temperature or other. This is because in the equation

$$\Delta G^\circ = \Delta H^\circ - T\Delta S^\circ$$

the ΔS term is positive; thus, provided that T is large enough, ΔG° becomes negative and hence favourable for reaction. In practice temperatures of over $2000^\circ C$ are uneconomical and other methods are used to extract metals that form very stable oxides. Metals that are normally extracted by carbon reduction of oxides include lead and zinc.

However, with the transition elements there is a complication, interstitial compound formation. If titanium(IV) oxide is heated with an excess of carbon, for example, the reaction

$$TiO_2 + 3C \rightarrow TiC + 2CO$$

proceeds and titanium carbide is the product. For titanium, vanadium, and chromium this method cannot be used. However, the later transition elements can be obtained in impure form by this method, for example

$$Mn_3O_4 + 4C \rightarrow 3Mn + 4CO$$

$$Co_2O_3 + 3C \rightarrow 2Co + 3CO$$

Iron is the most important example economically. The reduction of iron oxides by coke in the blast furnace gives 'pig iron' which contains about 4.6 per cent carbon (the solubility limit of carbon in iron). This excess of carbon is removed by oxidation using air or oxygen with the molten metal in a Bessemer converter or an open hearth. Steels rather than pure iron are obtained in this way; pure iron has no industrial importance. Despite the fact that we have written carbon as the reducing agent in the equations, the effective reducing agent is probably carbon monoxide, the carbon serving to reduce the carbon dioxide produced back to the monoxide, for example

$$CO + FeO \rightarrow Fe + CO_2$$

$$CO_2 + C \rightarrow 2CO$$

Hydrogen reduction of metal oxides is not so cheap or effective as carbon reduction, but when contamination from carbon is undesirable it may be a useful process. Thus pure iron can be obtained by heating high purity Fe_2O_3 in hydrogen; cobalt, molybdenum, and tungsten are similarly obtained free from carbides by, for example, the reaction

$$Co_2O_3 + 3H_2 \rightarrow 2Co + 3H_2O$$

Alternatively, if the temperature for carbon reduction is too high and/or carbon contamination is a problem, reduction by an active metal such as calcium, magnesium, or aluminium can be used. The 'thermit' process is used in the extraction of chromium and of carbon-free ferrovanadium and ferrotitanium

$$Cr_2O_3 + 2Al \rightarrow 2Cr + Al_2O_3$$

Vanadium is obtained from the oxide by reduction with silicon or calcium

$$2V_2O_5 + 5Si \rightarrow 4V + 5SiO_2$$

In the pyrometallurgical process for extracting copper from sulphide ores, the sulphur in the ores is used as reducing agent after about half of the copper has been oxidised, that is

$$2Cu_2O + Cu_2S \rightarrow 6Cu + SO_2$$

A rather special process is used for refining nickel; this is known as the Mond or carbonyl process. The impure nickel obtained by carbon reduction of NiO is treated with carbon monoxide at $50°$, and the volatile carbonyl so produced is decomposed to the pure metal at $180°$

$$Ni + 4CO \underset{180°}{\overset{50°}{\rightleftharpoons}} Ni(CO)_4$$

Pure iron can similarly be obtained by thermal decomposition of $Fe(CO)_5$.

10.2.3 Extraction from Halides

Transition-metal halides have been reduced by alkali metals and alkaline earth metals, by hydrogen, and by thermal decomposition. The most important production from halides is by the Kroll process which accounts for virtually all the titanium manufactured commercially. In this process, titanium(IV) chloride is reduced by magnesium or sodium at around $800°$ in an atmosphere of argon

$$TiCl_4 + 2Mg \rightarrow Ti + 2MgCl_2$$

This rather expensive process ($TiCl_4$ must be manufactured from $TiO_2 + C + Cl_2$) is used for many reasons. Firstly, titanium is required (unlike iron) in a pure form for use as an engineering metal, and impurities lower the ductility of the metal. Secondly, titanium has a large affinity for oxygen and carbon, so carbon reduction of the oxide cannot be used. Thirdly, titanium forms interstitial compounds with nitrogen and hydrogen, and hence the more expensive argon has to be used as the protective atmosphere. The magnesium chloride is removed from the titanium by sublimation in a vacuum or by leaching with water.

Scandium is obtained in a similar process, using the reduction of the fluoride by calcium under an inert atmosphere at a temperature in excess of $1000°$

$$2ScF_3 + 3Ca \rightarrow 2Sc + 3CaF_2$$

Laboratory methods for preparing titanium include passing the chloride $TiCl_4$ in hydrogen through a red-hot tube or over a heated filament

$$TiCl_4 + 2H_2 \rightarrow Ti + 4HCl$$

A purification process used for refining some of the transition metals (notably Ti and Zr) on a laboratory scale is that due to van Arkel and de Boer. The impure metal is allowed to react with iodine at a moderate temperature (about $200°$), and the volatile iodide produced is then decomposed on a hot filament (about $1300°$), for example

$$\text{Ti(crude)} + 2I_2 \xrightarrow{\;200°\;} \text{TiI}_4 \xrightarrow{\;1300°\;} \text{Ti(pure)} + 2I_2$$

10.2.4 Electrolytic Methods

Electrolysis of molten salts has not been used widely for the preparation of first-row transition elements, since the later elements in the series can be deposited by electrolysis of aqueous solutions. Methods similar to that used commercially for aluminium have been developed for titanium. The oxide can be electrolysed in a bath of molten calcium, magnesium, or alkali metal fluorides; alternatively, the lower chlorides can be electrolysed in molten alkali metal chloride baths. Scandium was initially produced by electrolysis of its chloride in a molten salt bath, but the product is not as pure as that obtained by calcium reduction.

Aqueous electrolysis is widely used as a method of refining impure samples of metals such as chromium, nickel, and copper. The crude metal is cast into anodes and used in electrolytic cells containing a solution of the metal sulphate or chloride. On electrolysis, the metal dissolves from the anode and is deposited on the pure metal cathode; the impurities remain as the 'anode sludge'.

10.3 Preparation of Simple Transition-metal Compounds

The distinction between simple and complex compounds is a very arbitrary one, and we shall concern ourselves here largely with compounds formed between the metals and simple anions (there being no ligand present other than the anion). In chapter 3, the hexaquo-ion was discussed, and at this point the reader should be reminded that reactions in aqueous solution are not likely to lead to simple salts. The products of the dissolution of a transition-metal oxide in an acid, for example, are not simple salts. Rather, they contain complex ions with water and/or anions from the acid in the co-ordination sphere of the metal. The consequence of this is that, in transition metal chemistry, preparative procedures for simple compounds must usually be carried out in the absence of water. Thus, reactions frequently involve vacuum-line techniques, inert atmospheres, or nonaqueous solvent systems. Because of the variable oxidation states shown by these elements, the absence of oxygen from preparations is also frequently desirable.

10.3.1 Preparation of Halides

Direct reaction with halogens. The reaction between the metals and the halogens is the most straightforward way of preparing the more covalent metal halides. If the pure metal and halogen are available, and the product can be separated easily from excess halogen, a product free from impurities is readily obtained. The reactions are exothermic and once started will usually continue without further heat, but a temperature of a few hundred degrees is often necessary to start the reaction. When the halide is relatively involatile, continuous heating of the reaction zone may be necessary to effect a sublimation of the halide away from the metal.

For fluorides and chlorides, the undiluted halogen is passed over the heated metal in a combustion tube, and any products are collected in a cooled vessel as they emerge from the tube. Simple fractionation or evacuation is then necessary to separate the halide from adsorbed or dissolved halogen. Some halides typically prepared in this way are indicated in the equations

$$Ti + 2F_2 \rightarrow TiF_4$$

$$V + 2Cl_2 \rightarrow VCl_4$$

$$2Fe + 3Cl_2 \rightarrow 2FeCl_3$$

With bromine and iodine, a carrier gas such as nitrogen must be used if the halogen vapour is to be transported over the heated metal. Alternatively, the reaction can be carried out using a solvent or with the halogen itself as solvent. Titanium(IV) bromide is most easily prepared, for example, by adding granules of titanium, in small amounts, to an excess of liquid bromine. The metal burns with incandescence; the bromine may eventually boil and the vapour is returned to the reaction vessel by a reflux condenser. The excess of bromine is finally distilled off and the bromide remaining is distilled in a vacuum. Titanium(IV) iodide is most easily prepared by heating titanium and iodine under reflux in a solvent such as benzene.

Direct reaction with hydrogen halides. Metals react with hydrogen halides at high temperatures to produce a low oxidation state halide. Typical reactions and conditions are given in the equations

$$V + 2HCl \xrightarrow{950^\circ} VCl_2 + H_2$$

$$Cr + 2HBr \xrightarrow{750^\circ} CrBr_2 + H_2$$

$$Cu + HCl \xrightarrow{900^\circ} CuCl + \tfrac{1}{2}H_2$$

From metal oxides. When the metal is not readily available, or is expensive compared with the price of its oxide, the halogen can be caused to react in a flow system with a mixture of the metal oxide and carbon. Examples are

$$TiO_2 + 2C + 2Cl_2 \xrightarrow{500^\circ} TiCl_4 + 2CO$$

$$Cr_2O_3 + 3C + 3Cl_2 \xrightarrow{800^\circ} 2CrCl_3 + 3CO$$

Alternatively, carbon tetrachloride can be passed over the hot oxide; CCl_4 decomposes into carbon and chlorine above 400°.

Some metals form oxohalides as well as the binary halides; for these metals, problems of contamination may arise if oxides are used as starting materials. Thus chlorination of a mixture of vanadium(V) oxide and carbon can produce both

VCl_4 and $VOCl_3$

$$V_2O_5 + 5C + 4Cl_2 \xrightarrow{800°} 2VCl_4 + 5CO$$

$$V_2O_5 + 3C + 3Cl_2 \xrightarrow{350°} 2VOCl_3 + 3CO$$

Since both VCl_4 and $VOCl_3$ are dark red liquids with similar boiling points, contamination of one chloride by the other is likely. In the corresponding bromine reaction, however, the reaction products are VBr_3 and $VOBr_3$

$$V_2O_5 + 5C + 3Br_2 \rightarrow 2VBr_3 + 5CO$$

$$V_2O_5 + 3C + 3Br_2 \rightarrow 2VOBr_3 + 3CO$$

Since VBr_3 is a black relatively involatile solid and $VOBr_3$ is a brown liquid, no contamination problems arise; the $VOBr_3$ distils out of the hot zone and the VBr_3 remains as a condensate on the cooler parts of the combustion tube.

Reduction of higher oxidation state halides. The lower oxidation state halides of titanium, vanadium, and chromium are conveniently prepared by reduction of the higher halides. Reducing agents include hydrogen, aluminium, and the element itself

$$2VBr_3 + H_2 \xrightarrow{450°} 2VBr_2 + 2HBr$$

$$3TiBr_4 + Al \xrightarrow{200°} 3TiBr_3 + AlBr_3$$

$$2CrCl_3 + Cr \xrightarrow{900°} 3CrCl_2$$

Apart from the reductions with hydrogen, these reactions are carried out in sealed tubes; glass 'Carius tubes' are satisfactory for the lower-temperature reactions (for example $TiBr_3$ preparation) while metal bombs must be used for reactions that involve high vapour pressures of the metal halide.

Sometimes internal oxidation–reduction occurs when a transition metal halide is heated. The preferred method of preparing VCl_3, for example, is to heat VCl_4 under reflux (b.p. 150°) in an inert atmosphere. The brown liquid decomposes smoothly leaving the violet trichloride

$$2VCl_4 \rightarrow 2VCl_3 + Cl_2$$

In this reaction the chloride ion has been oxidised to chlorine and the vanadium(IV) reduced to vanadium(III). Many similar reactions occur with halides in high oxidation states; this redox process is of course the reason for the instability of compounds such as VCl_5, VBr_4, and $CrCl_6$. Some other examples of thermal decompositions are

$$2VOBr_3 \xrightarrow{100°} 2VOBr_2 + Br_2$$

$$2FeBr_3 \xrightarrow{120°} 2FeBr_2 + Br_2$$

$$2CuBr_2 \xrightarrow{150°} 2CuBr + Br_2$$

When the metallic element undergoes both oxidation and reduction, we have disproportionation occurring. If the products are readily separated, the reaction can be of preparative use. The involatile halides of titanium(II) and vanadium(II) are conveniently prepared in this way because of the volatility of the higher halides, for example

$$2TiBr_3 \xrightarrow{400°} TiBr_2 + TiBr_4$$

$$2VCl_3 \xrightarrow{450°} VCl_2 + VCl_4$$

Exchange reactions. Because the reaction of fluorine with the later transition elements tends to produce ionic fluorides that coat the surface of the metals and hence prevent further reaction, exchange reactions are often found preferable for preparing fluorides. In these reactions, another halide is caused to react with fluorine, a halogen fluoride, or hydrogen fluoride. The ionic starting materials are used in a flow system; the more covalent and volatile halides can be treated in solution. Examples illustrating the preparation of fluorides are

$$2CoCl_2 + 3F_2 \xrightarrow{250°} 2CoF_3 + 2Cl_2$$

$$FeCl_3 + ClF_3 \xrightarrow{500°} FeF_3 + 2Cl_2$$

$$NiCl_2 + 2HF \xrightarrow{600°} NiF_2 + 2HCl$$

Similar exchange reactions can be used for the other halides. The most useful are those in which a halide is obtained directly from the metal oxide. Typical of the reagents that can effect this exchange are thionyl chloride, boron tribromide, and aluminium iodide, for example

$$V_2O_3 + 3SOCl_2 \xrightarrow{200°} 2VCl_3 + 3SO_2$$

$$Fe_2O_3 + 2BBr_3 \xrightarrow{100°} 2FeBr_3 + B_2O_3$$

$$V_2O_3 + 2AlI_3 \xrightarrow{330°} 2VI_3 + Al_2O_3$$

From hydrated halides. The reader may have noticed that in the examples given so far the elements titanium, vanadium, and chromium have featured frequently. The halides of these elements are best prepared under nonaqueous conditions; their M^{3+} and M^{2+} halide hydrates are not easily dehydrated thermally without hydrolysis occurring, for example

$$VBr_3(H_2O)_6 \xrightarrow[\text{vacuum}]{200°} VOBr + 2HBr + 5H_2O$$

The elements toward the other end of the first transition series, that is iron, cobalt, nickel, and copper, readily form hydrated halides in aqueous solution, from which the anhydrous halides can frequently be obtained by dehydration.

Some hydrated salts evolve the water of co-ordination when heated to above $100°$ in a vacuum. Manganese(II), iron(II), cobalt(II), and nickel(II) halides can be dehydrated in this way, for example

$$MnCl_2(H_2O)_4 \xrightarrow{\ 210° \ } MnCl_2 + 4H_2O$$

In order to be certain that no hydrolysis has occurred, it is usually preferable to dehydrate the halides in a stream of the dry hydrogen halide gas or sublime the crude anhydrous halide in an atmosphere of the hydrogen halide

$$CuCl_2(H_2O)_2 \xrightarrow[HCl]{\ 150° \ } CuCl_2 + 2H_2O$$

Chemical methods of dehydration are also used. Thionyl chloride has particular advantage in that, in its reaction with water, only volatile products result

$$SOCl_2 + H_2O \rightarrow SO_2 + 2HCl$$

A standard method for chlorides is thus to treat the hydrated chloride with an excess of thionyl chloride and heat under reflux if necessary. Effervescence is observed as the water reacts. No protection from water vapour in the atmosphere is necessary during the dehydration since thionyl chloride reacts more readily with water than does the anhydrous metal chloride. This method has been used successfully for the dehydration of $MCl_3 . 6H_2O$ (M = Cr, Fe) and $MCl_2 . xH_2O$ (M = Mn, Fe, Co, Ni, Cu). The disadvantage of the method is that the final traces of thionyl chloride (and possibly other sulphur compounds) are difficult to remove from the metal chloride even by continued evacuation on a vacuum line. Other chemical dehydrating agents include acetyl halides and 2,2-dimethoxypropane

10.3.2   *Preparation of Nitrates*

Anhydrous nitrates of the first-row transition elements were unknown some fifteen years ago. Thermal dehydration of the hydrated salts cannot be used because decomposition of the nitrate sets in before all the water has been removed.
As with so many transition metal salts, isolation of the anhydrous nitrates has proved possible by using nonaqueous solvents. The ideal solvent for preparing metal nitrates is dinitrogen tetroxide. This solvent undergoes the autoionisation

$$N_2O_4 \rightarrow NO^+ + NO_3^-$$

This ionisation is greatly increased if a donor solvent such as acetonitrile or ethyl acetate is used. Copper metal will dissolve with effervescence in such solvent mixtures

$$Cu + 2N_2O_4 \rightarrow Cu(NO_3)_2 + 2NO$$

The nitrate is isolated as a blue-green solid of composition $Cu(NO_3)_2 . N_2O_4$; when this is heated in a vacuum to $120°$ the blue nitrate remains. This nitrate, unlike the hydrated salt, can be sublimed in a vacuum at $200°$. Manganese(II) and cobalt(II) nitrates can be prepared similarly starting from the metals. Vanadium, however, gives $VO_2(NO_3)$ under this treatment.

The other first-row transition elements are insoluble in N_2O_4 or its mixtures with donor solvents. Other methods must therefore be used to prepare their nitrates. One method is to solvolyse a metal halide or carbonyl. Nickel(II) nitrate is conveniently prepared, for example, by adding acetonitrile to a suspension of nickel(II) chloride in N_2O_4

$$NiCl_2 \;+\; 2N_2O_4 \;\xrightarrow{\;\;MeCN\;\;}\; Ni(NO_3)_2 \;+\; 2NOCl$$

The nitrate is isolated as the adduct $Ni(NO_3)_2.3MeCN$ which is desolvated at $170°$ to give lime green $Ni(NO_3)_2$.

The early transition elements in their high oxidation states show a tendency to form oxide nitrates in the reactions of their halides with N_2O_4; titanium(IV) chloride forms $TiO(NO_3)_2$ for example. The more highly nitrated product can usually be obtained using dinitrogen pentoxide N_2O_5 instead of N_2O_4. Thus the reaction of vanadium(V) oxide or oxide trichloride with N_2O_5 yields the oxide trinitrate

$$V_2O_5 \;+\; 3N_2O_5 \;\rightarrow\; 2VO(NO_3)_3$$

[compare $VO_2(NO_3)$ from $V + N_2O_4$]. When chromium carbonyl is added to N_2O_5 in carbon tetrachloride solution, the green nitrate is formed as a precipitate.

$$Cr(CO)_6 \;+\; 3N_2O_5 \;\rightarrow\; Cr(NO_3)_3 \;+\; 6CO \;+\; 3NO_2$$

This nitrate cannot be readily obtained using N_2O_4 since the product $Cr(NO_3)_3.2N_2O_4$ is difficult to desolvate without decomposition of the $Cr(NO_3)_3$. Dinitrogen pentoxide is the anhydride of nitric acid and is capable of removing water from hydrated metal nitrates. Hydrated titanium(IV) nitrate is thus converted into $Ti(NO_3)_4$ by reaction with N_2O_5; sublimation at $100°$ in a vacuum gives the pure nitrate.

10.3.3 Preparation of Other Oxo-salts

Salts such as sulphates, phosphates, acetates, etc. are more thermally stable than nitrates, so they can frequently be obtained by thermal dehydration of the salts crystallising from aqueous solution, for example

$$CuSO_4.5H_2O \;\xrightarrow{\;\;200°\;\;}\; CuSO_4 \;+\; 5H_2O$$

This is most commonly true for salts of the later transition elements, that is manganese to copper. The early transition elements in their more covalent oxidation states yield oxo-anion salts, for example $TiOSO_4$ and $VO(OCOCH_3)_2$, if aqueous solutions are used in preparation, owing to the great acidity of $[M(H_2O)_6]^{4+}$ ions. Nonaqueous solvents or reactions must therefore be used. For sulphates, sulphuric acid or sulphur trioxide is used, for example

$$TiCl_4 \;+\; 6SO_3 \;\rightarrow\; Ti(SO_4)_2 \;+\; 2S_2O_5Cl_2$$

Acetates are best prepared in acetic acid to which a little acetic anhydride has been added to remove any water from the system. Thus vanadium(III) acetate is prepared by dissolving vanadium diboride (vanadium does not react) in acetic acid

$$2VB_2 + 6CH_3COOH \rightarrow V_2(O_2CCH_3)_6 + 3H_2 + 4B$$

10.3.4   *Preparation of Amides, Imides, and Nitrides*
Just as the solvent used to prepare hydroxides is water, so for amides the solvent used is liquid ammonia (transition metal amides are hydrolysed by water). The general route to transition-metal amides involves the addition of the stoichiometric quantity of potassium amide to an ammonia-soluble salt of the metal. The amides, like hydroxides, are polymeric and are precipitated, for example

$$[Mn(NH_3)_6](SCN)_2 + 2KNH_2 \xrightarrow{\text{liq.NH}_3} Mn(NH_2)_2 + 2KSCN + 6NH_3$$

$$[Co(NH_3)_6](NO_3)_3 + 3KNH_2 \xrightarrow{\text{liq.NH}_3} Co(NH_2)_3 + 3KNO_3 + 6NH_3$$

Amides are sometimes produced in the ammonolysis (using ammonia only) of covalent compounds of the early transition elements, but frequently only partial ammonolysis occurs, for example

$$VCl_4 + 6NH_3 \rightarrow VCl(NH_2)_3 + 3NH_4Cl$$

Thermal decomposition of amides produces imides and nitrides (compare hydroxides → oxides), for example

$$Co(NH_2)_3 \rightarrow CoN + 2NH_3$$

Nitrides can also be produced directly by passing a mixture of ammonia and a volatile metal halide (such as $TiCl_4$) over an incandescent filament.

Transition-metal amides are usually amphoteric in liquid ammonia, dissolving in an excess of the added base to form amido- or imido-complexes

$$Cr(NH_2)_3 + KNH_2 \rightarrow KCr(NH_2)_4$$

$$Ti(NH_2)_3 + KNH_2 \rightarrow KTi(NH)_2 + 2NH_3$$

11 Titanium

11.1 The Element

Titanium is the most naturally abundant transition metal after iron, and is widely distributed throughout the earth's surface. Despite this fact, the metal has been of commercial importance only during the last two or three decades. Although it has some very desirable physical properties as an engineering metal, the high cost of production of the pure metal (Kroll process, chapter 10) still hinders its utilisation. It is used in supersonic aircraft, in which its high strength-to-weight ratio is important, and in the construction of chemical plant where its corrosion resistance (particularly to moist chlorine) is its important feature.

Titanium (atomic number 22) is a silver-grey metal with a density of 4.5 and a melting point around 1670°. It is about as strong as steel yet only half as dense; however, its strength and resistance to corrosion drop rapidly above 800°. When finely divided, the metal is pyrophoric in air at room temperature; at higher temperatures the massive metal reacts with oxygen, nitrogen, carbon, and boron to give TiO_2, TiN, TiC, and TiB_2 as limiting products (in excess of the nonmetal). Hydrogen is absorbed reversibly. Dilute mineral acids have little effect in the cold, but the metal is dissolved in hot dilute hydrochloric acid or cold dilute hydrofluoric acid. Hot concentrated nitric acid slowly attacks the metal with the formation of insoluble $TiO_2 .nH_2O$. Titanium is not attacked by hot aqueous alkali.

The ground state outer electronic configuration of titanium is $3d^2 4s^2$. The maximum oxidation state is +4 and this is the commonest state found; compounds in lower oxidation states are readily oxidised. Certain similarities are shown to other group IV metals, notably tin; for example, the chlorides MCl_4 are covalent colourless liquids readily hydrolysed in moist air and forming complexes such as $MCl_6{}^{2-}$. We shall discuss now the compounds of titanium in the various oxidation states, but for convenience will consider the organometallic compounds separately at the end.

11.2 Compounds of Titanium(IV) (d^0)

11.2.1 Aqueous Chemistry
The +4 oxidation state for titanium gives rise to largely covalent compounds. The ion $[Ti(H_2O)_6]^{4+}$ is unknown; its high charge-to-size ratio leads to great acidity and hence only hydrolysed derivatives are obtained in aqueous solution. In dilute noncomplexing acids such as $HClO_4$, the hydrolytic equilibrium probably lies in the region indicated by

$$[Ti(H_2O)_4(OH)_2]^{2+} + H_2O \rightleftharpoons [Ti(H_2O)_3(OH)_3]^+ + H_3O^+$$

The addition of a base causes the precipitation of the hydrous oxide, that is the equilibrium

$$[Ti(H_2O)_3(OH)_3]^+ + H_2O \rightleftharpoons [Ti(H_2O)_2(OH)_4] + H_3O^+$$

lies to the right. The hydrous oxide is normally regarded as $TiO_2 . nH_2O$ [$n = 4$ in the formula $Ti(H_2O)_2(OH)_4$], there being no evidence for $Ti(OH)_4$. The hydrolytic equilibrium is of course (see chapter 3) complicated by olation and oxolation. However, the simple monomeric species $[TiO(H_2O)_5]^{2+}$ does not appear to be important; contrast this with the aqueous chemistry of vanadium(IV) in which $[VO(H_2O)_5]^{2+}$ is a predominant species. In subsequent chemical equations we shall for simplicity write TiO^{2+} as the species present, it being obviously preferable to Ti^{4+} as a realistic species. Oxo-species of titanium(IV) are usually polymeric with $-Ti-O-Ti-O-$ chains. Such a structure has been found, for example, in the solid $TiOSO_4 . H_2O$ in which each titanium atom is approximately octahedrally co-ordinated by oxygen atoms. The TiO group is believed to occur in salts of the $TiOCl_4{}^{2-}$ ion which have a tetragonal pyramidal structure for the anion.

In acids containing anions capable of co-ordinating to the titanium, complexes containing the anions are formed. The exact nature of these depends on the concentration of the acid (that is the ligand). Thus the hydrolysis of titanium(IV) chloride in water does not lead to the precipitation of much $TiO_2 . aq$; rather, the titanium is kept in solution as anionic complexes of the type $[Ti(OH)Cl_5]^-$ and $[TiCl_6]^{2-}$. Salts of this hexachlorotitanate(IV) anion can be precipitated from solutions of $TiCl_4$ in saturated hydrochloric acid.

In excess of aqueous alkali the initially precipitated $TiO_2 . aq$ dissolves to form titanates; such titanates are more readily formed in the fused state between TiO_2 and metal oxides. These titanates are basically of two types, the metatitanates $M^I_2 TiO_3$ and $M^{II} TiO_3$, and the orthotitanates $M^I_4 TiO_4$ and $M^{II}_2 TiO_4$. In most cases they do not contain discrete titanate anions but have mixed oxide type structures, that is the perovskite $CaTiO_3$, ilmenite $FeTiO_3$, and spinel $M^{II}_2 TiO_4$ structures.

A very useful and characteristic reaction of aqueous titanium(IV) solutions is that with hydrogen peroxide; the orange coloration produced in mildly acidic solutions can be used to estimate titanium colorimetrically. The nature of the species responsible for the colour is not known but various peroxo-complexes have been isolated in the solid state from these solutions. These include $Ti(O_2)SO_4 . 3H_2O$ obtained by the addition of hydrogen peroxide to a solution of $TiOSO_4$ in concentrated sulphuric acid, $M^I_3[Ti(O_2)F_5]$, and $M^I_2[Ti(O_2)(SO_4)_2] . nH_2O$. A yellow solid, probably $TiO_3 . 2H_2O$, is obtained when a solution of $TiOSO_4$ containing hydrogen peroxide is precipitated at pH 8.6 by the addition of ethanol.

11.2.2 *Titanium(IV) Oxide*, TiO_2

This oxide exists in three crystalline forms, rutile, anatase, and brookite. In all three forms the titanium atoms are co-ordinated to six oxygen atoms. Rutile is the more commonly occurring natural form but all three forms have been prepared synthetically. Titanium dioxide is commercially very important as a white pigment. It is nontoxic, chemically inert, and gives an opaque white finish when coating other colours. It is prepared industrially by two processes. In one process, ilmenite is dissolved in sulphuric acid and the sulphate $TiOSO_4$ allowed to hydrolyse in hot water. The $TiO_2 . aq$ that precipitates is ignited at 900° to give the anhydrous oxide. In the other process, rutile is converted into $TiCl_4$ by heating with carbon and chlorine. The tetrachloride is then combusted with dry oxygen at around 700°

$$TiCl_4 + O_2 \rightarrow TiO_2 + 2Cl_2$$

Titanium(IV) oxide is dissolved with difficulty in hot concentrated sulphuric acid but more readily in fused alkali metal bisulphates. With fused alkali metal carbonates, titanates are formed. Because of its constant composition and high thermal stability it is used in the gravimetric estimation of titanium. All the metal is precipitated from solution with aqueous ammonia, and the precipitate is ignited to constant weight at 900° and weighed as TiO_2.

11.2.3 Titanium(IV) Halides

The physical properties of the titanium(IV) halides are listed in table 11.1. The fluoride differs from the other halides in having a relatively high boiling point and

TABLE 11.1   PHYSICAL PROPERTIES OF TITANIUM(IV) HALIDES

Halide	Colour	M.p. (°C)	B.p. (°C)
TiF_4	white	–	284
$TiCl_4$	colourless	−23	136
$TiBr_4$	yellow	39	230
TiI_4	violet-black	150	377

in being insoluble in nonpolar solvents. It is probably a fluorine-bridged polymer with six-co-ordinate titanium atoms. The solid is hygroscopic but can be readily purified by vacuum sublimation. It dissolves in aqueous hydrogen fluoride to form the $TiF_6{}^{2-}$ ion; salts of this ion are stable to hydrolysis in water.

The chloride, bromide, and iodide behave as covalent monomers in nonpolar solvents. They are strong Lewis acids, and fume copiously in moist air owing to hydrolysis. Their reactions with Lewis bases can be considered in two categories, those in which only adduct formation occurs and those in which adduct formation is followed by the elimination of hydrogen halide, that is solvolysis. Solvolytic reactions, for example hydrolysis, ammonolysis, etc., occur when the ligand has protonic hydrogen atoms adjacent to the donor atom or close to it. Some of these reactions are illustrated in figure 11.1. Even in an excess of the base, solvolysis is usually incomplete. The mechanism of these reactions is believed to occur via the intermediate formation of adducts, for example

$$TiCl_4 + 2NH_3 \longrightarrow Cl_3Ti^{\delta+} \!\!-\!\! Cl^{\delta-}$$

$$Cl_3TiNH_2 + HCl$$

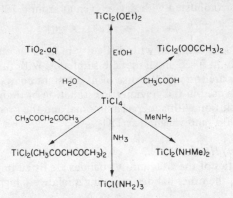

Figure 11.1   Some solvolytic reactions of titanium(IV) chloride.

In this representation we draw in one molecule of the base for simplicity (most known adducts are six-co-ordinate). The donation of electron density from the ligand donor atom does two things. Firstly, it partially neutralises the charge on the titanium atom, and secondly, it increases the charge on the protons adjacent to the donor atom (by the electron drift towards the titanium). The protonic hydrogen can then form a hydrogen bond with the adjacent chloride 'ions'. When this interaction is strong, the N–H bond breaks and HCl is eliminated. This HCl will normally be found in combination with the base, that is as NH_4 Cl in the example quoted. The titanium compound so produced may then co-ordinate another molecule of base which again may eliminate HCl, and so on to complete solvolysis. Each time a chlorine is replaced by a less electronegative group (such as NH_2), the electron drift away from the titanium atom, and hence its fractional positive charge, is decreased. The Lewis acidity of the substitution products is thus progressively reduced until such a time as co-ordination of further ligand occurs only weakly, and only weak hydrogen-bonding occurs, so no elimination of the HCl results. In liquid ammonia, for example, $TiCl_4$ forms $TiCl(NH_2)_3(NH_3)_2$ + $3NH_4$ Cl. On warming to room temperature, the weakly bonded ammonia molecules are evolved leaving the amido-chloride $TiCl(NH_2)_3$.

The extent of solvolysis can be increased by base catalysis. Alkoxides, for example, are prepared by the reaction with alcohols in the presence of ammonia

$$TiCl_4 + 4ROH + 4NH_3 \rightarrow Ti(OR)_4 + 4NH_4 Cl$$

These solvolytic products are usually polymeric, containing six-co-ordinate titanium, and have low solubilities in solvents that do not destroy them. In the solid state $Ti(OEt)_4$, for example, is tetrameric with approximately octahedral co-ordination around each titanium; in benzene, however, this and many other alkoxides are trimeric.

The adducts formed between titanium(IV) halides and an excess of nitrogen, phosphorous, arsenic, oxygen, and sulphur donor molecules are in general six-co-ordinate. Although Ti(IV) is a $3d^0$ system, the adducts are usually coloured yellow, red, or black through charge-transfer absorption bands reaching into the visible region. They are generally insoluble in nonpolar solvents and their structures are mostly not known with certainty. The range of adducts is illustrated for the chloride in figure 11.2. X-ray studies on the phosphorus oxochloride adducts

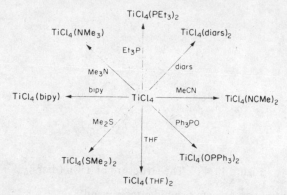

Figure 11.2   Some adducts of titanium(IV) chloride.

$TiCl_4(OPCl_3)_2$ and $TiCl_4(OPCl_3)$ have established the octahedral structure for titanium in both cases. The bis-complex is *cis*-octahedral while $TiCl_4(OPCl_3)$ is dimeric with halogen bridges (figure 11.3). However, the 1 : 1 adduct with

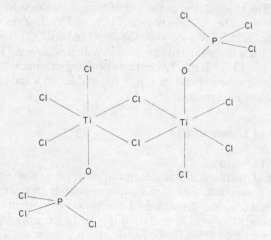

Figure 11.3   The structure of $[TiCl_4(OPCl_3)]_2$

trimethylamine is monomeric in benzene and hence five-co-ordinate; its infrared spectrum has been interpreted on the basis of a trigonal bipyramidal structure.

Bidentate ligands usually form 1 : 1 adducts containing six-co-ordinate titanium. However, with certain bidentate ligands eight-co-ordinate complexes are formed. Typical of these are $TiCl_4[o\text{-}C_6H_4(PMe_2)_2]_2$ and the arsine analogue $TiCl_4(diars)_2$. The dodecahedral structure of the latter (figure 11.4) has been established by X-ray crystallography. A similar eight-co-ordinate structure occurs in $Ti(NO_3)_4$

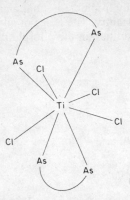

Figure 11.4   The structure of $TiCl_4(diars)_2$

in which each nitrate group acts as a bidentate ligand.

Anionic complexes are formed in the reactions of the halides with halide ions. The octahedral hexahalotitanates(IV) containing the $TiCl_6{}^{2-}$ and $TiBr_6{}^{2-}$ ions can be precipitated from hydrohalic acid solutions of the appropriate halide in the presence of a halide of a univalent cation. Salts such as yellow $(NH_4)_2TiCl_6$ and black $(NH_4)_2TiBr_6$ (unlike the salts of $TiF_6{}^{2-}$) undergo extensive hydrolysis in aqueous solution. The charge-transfer spectra of these complexes have been considered in chapter 6. The reaction between PCl_5 and $TiCl_4$ gives a compound of formula $(PCl_4TiCl_5)_n$. This does not contain $TiCl_5{}^-$ ions but instead has octahedrally co-ordinated titanium in halogen-bridged $[Ti_2Cl_{10}]^2$ anions. The anion $[Ti_2Cl_9]^-$ similarly contains bridging chlorines, the two $TiCl_6$ octahedra being joined at one face; it is isolated as the tetraethylammonium salt from dichloromethane solution.

11.3 Compounds of Titanium(III) (d^1)

11.3.1 Aqueous Chemistry
In its +3 oxidation state, titanium has the ground state outer electronic configuration $3d^1$. Unlike Ti(IV) compounds therefore Ti(III) compounds are paramagnetic ($\mu_{eff} \approx 1.7$ B.M.) and coloured through d–d absorption even when charge-transfer absorptions are absent. This oxidation state is less acidic than the +4 state, and its compounds show a considerably greater degree of ionic character. The hexaquo-ion $[Ti(H_2O)_6]^{3+}$ is readily prepared by reduction of solutions containing Ti(IV) with zinc and hydrochloric acid or electrolytically. There is, however, some substitution of water molecules by chloride ions in high acid concentrations, species such as $[Ti(H_2O)_5Cl]^{2+}$ and $[Ti(H_2O)_4Cl_2]^+$ being predominant in these solutions. Solutions containing Ti(III) are also readily obtained by dissolving the metal in hydrochloric acid.

The violet aquo-ion is strongly acidic

$$[Ti(H_2O)_6]^{3+} + H_2O \rightleftharpoons [Ti(OH)(H_2O)_5]^{2+} + H_3O^+ \ (pK_1 = 1.4)$$

The addition of salts of weak acids to titanium(III) solutions thus results in the precipitation of the hydrous oxide. The reaction with sodium carbonate solution, for example, results in evolution of carbon dioxide

$$[Ti(H_2O)_6]^{3+} + 3H_2O \rightleftharpoons Ti(OH)_3(H_2O)_3 + 3H_3O^+$$

$$2H_3O^+ + CO_3{}^{2-} \rightarrow CO_2 + 3H_2O$$

Water-stable titanium(III) salts are thus limited to salts of strong nonoxidising acids, that is halides and sulphate. The dark purple hydrous oxide is often written as $Ti(OH)_3$ but since there is considerable evidence for 'polymerisation', that is olation and oxolation, in the formation of this product, it is probably best considered as $Ti_2O_3.aq$.

In acidic solution Ti^{3+} is a strong reducing agent (slightly stronger than Sn^{2+})

$$TiO^{2+} + 2H^+ + e \rightarrow Ti^{3+} + H_2O \quad (E^\circ = 0.1 \cdot V)$$

The solution is oxidised by air and must be kept under nitrogen. It is used in volumetric analysis as a one-electron reductant; in organic chemistry it is used to estimate nitro-compounds since these are reduced to amines

$$RNO_2 + 6Ti^{3+} + 4H_2O \rightarrow RNH_2 + 6TiO^{2+} + 6H^+$$

In alkaline media titanium(III) is an even stronger reducing agent, $Ti_2O_3.aq$ being rapidly oxidised in air.

Complexes of titanium(III) with neutral donor molecules are normally hydrolysed in water and therefore must be prepared under anhydrous conditions. The $[TiCl_2(NH_3)_4]^+$ ion, for example, can be prepared in liquid ammonia but when added to water substitution and hydrolysis occur

$$[TiCl_2(NH_3)_4]^+ + 6H_2O \rightleftharpoons \underbrace{[Ti(H_2O)_6]^{3+} + 4NH_3 + 2Cl^-}_{\downarrow}$$
$$Ti_2O_3.aq$$

A remarkably stable complex (both oxidatively and hydrolytically) is that with urea, $Ti[OC(NH_2)_2]_6I_3$. This can be prepared in aqueous solution; its crystal structure contains titanium atoms octahedrally co-ordinated by oxygen atoms from the ligand.

Anionic complexes containing halide ions can be obtained from strongly acidic solutions. When titanium is dissolved in hot hydrobromic acid and the resulting solution is cooled, violet crystals of $TiBr_3(H_2O)_6$ form. Spectroscopic evidence indicates that these, as well as $TiCl_3(H_2O)_6$, contain the *trans*-octahedral $[TiX_2(H_2O)_4]^+$ ion ràther than the hexaquo-ion. In the presence of added cations, salts such as $Cs_2[TiCl_5(H_2O)]$ and $Cs_2[TiBr_5(H_2O)].2H_2O$ can be crystallised. Some cyanide and thiocyanate complexes have also been prepared, for example $K_4Ti(CN)_7.KCN$ (in liquid ammonia) and $NH_4[Ti(SCN)_4(H_2O)_2]$.

The green sulphate $Ti_2(SO_4)_3$ is obtained by electrolytic reduction of titanium(IV) solutions in sulphuric acid and is precipitated from the hot concentrated acid. It is insoluble in water but gives violet solutions in dilute sulphuric and hydrochloric acids. The alums such as $CsTi(SO_4)_2.12H_2O$ have been shown by X-ray studies to contain the octahedral $[Ti(H_2O)_6]^{3+}$ ion in the solid state; they give this ion also in aqueous solution.

11.3.2   Titanium(III) Halides

The blue crystalline fluoride TiF_3 is prepared by passing a mixture of hydrogen and hydrogen fluoride over titanium at $700°$. It is magnetically dilute (μ_{eff} = 1.75 B.M.) at room temperature. In the crystal each titanium is surrounded by a slightly distorted octahedral array of fluoride ions. Very little of its chemistry has been studied. It is insoluble in water and stable in air at room temperature. Above $950°$ disproportionation occurs

$$4TiF_3 \rightarrow Ti + 3TiF_4$$

The violet-black iodide has also been little studied. Unlike the fluoride it is antiferromagnetic; it is prepared by the reaction of stoichiometric quantities of titanium and iodine in a sealed tube at $700°$.

The chloride and bromide are more readily prepared, for example by reduction of the tetrahalides with hydrogen. The chloride exists in at least two different modifications. The commonly found violet α-form results from the high-temperature reductions of the tetrachloride. It has a layer lattice with octahedrally co-ordinated titanium. The brown β-form arises in the relatively low-temperature reductions of the tetrachloride with, for example, organoaluminium compounds. It has a fibrous structure with single chains of $TiCl_6$ octahedra each sharing two faces. The halides disproportionate above $400°$ in a vacuum into the di- and tetra-halides, and at higher temperatures the involatile dihalides disproportionate further

$$2TiCl_3(s) \xrightarrow{\;450°\;} TiCl_4(g) + TiCl_2(s)$$
$$\downarrow 700°$$
$$\tfrac{1}{2}Ti(s) + \tfrac{1}{2}TiCl_4(g)$$

The halides $TiCl_3$, $TiBr_3$, and TiI_3 are readily oxidised in moist air (to TiO_2.aq). They are insoluble in nonpolar solvents, but dissolve readily in the aqueous hydrogen halides, forming halo-aquo-complexes such as $[TiCl_2(H_2O)_4]^+$. They are less Lewis acidic than their TiX_4 analogues but react with a variety of donor molecules. Unidentate ligand molecules form mainly pseudo-octahedral complexes $TiX_3.3L$ (L = CH_3CN, CH_3COCH_3, THF, py) and five-co-ordinate complexes $TiX_3.2L$ (L = Me_3N, α-picoline). The five-co-ordinate complexes have the trans-trigonal bipyramidal structure. The dimethyl sulphide adducts $TiX_3.2SMe_2$, however, are dimeric with octahedral co-ordination around titanium. Neutral bidentate ligands form complexes of the types $Ti(LL)_3X_3$ (LL = en, propylenediamine) and $TiX_3.1.5LL$ (LL = bipy, phen). While compounds of the former type are cationic, for example $[Ti(en)_3]Cl_3$, the structures of the latter type are as yet unknown. Anionic complexes $TiCl_6^{3-}$ and $TiBr_6^{3-}$ occur in chloride ion melts, the pyridinium salts can be isolated by reaction between the pyridinium halide and $TiX_3.3MeCN$ in chloroform. At high temperatures in molten LiCl–KCl the tetrahedral tetrachlorotitanate(III) ion probably exists in equilibrium with the octahedral ion.

$$TiCl_6^{3-} \rightleftharpoons TiCl_4^- + 2Cl^-$$

11.4 Compounds of Titanium in Lower Oxidation States

Oxidation state +2 is a strongly reducing state for titanium. Compounds of Ti^{II} reduce water to hydrogen, so scarcely any aqueous chemistry of Ti^{2+} is known. The halides $TiCl_2$, $TiBr_2$, and TiI_2 evolve hydrogen from water, as do their blue-black complexes with donor molecules, for example $TiCl_2.2DMF$ and $TiCl_2.2MeCN$. The complex chlorides $RbTiCl_3$ and Rb_2TiCl_4 have been prepared by fusing together the stoichiometric quantities of the constituent chlorides under an inert atmosphere. TiO is obtained from TiO_2 and Ti at $1600°$; it has the rock-salt structure but tends not to be stoichiometric, with vacant sites in the lattice.

Oxidation states of 0 and -1 occur only in complexes with bipyridyl. The reduction of $TiCl_4$ by lithium in tetrahydrofuran in the presence of an excess of bipyridyl results in the formation of the green-blue $Ti(bipy)_3$ ($\mu_{eff} = 0$) and the blue-violet $LiTi(bipy)_3.3.5THF$ ($\mu_{eff} = 1.72$ B.M). These compounds are extremely sensitive to air and moisture, and oxidation-state titrations with iodine confirm that they at least formally contain titanium in oxidation states 0 and -1 respectively.

Titanium does not form any stable binary carbonyls or nitrosyls but a few complexes are known containing other π-bonded ligands in addition to these ligands, for example $(\pi-C_5H_5)_2Ti(CO)_2$ and $[(\pi-C_6H_6)Ti(CO)_4]Br$.

11.5 Organometallic Compounds of Titanium

Compounds in which titanium is bonded directly to carbon via a σ or π bond were scarcely known prior to 1960. The discovery by Ziegler that a mixture of an organoaluminium compound and a titanium halide was a successful catalyst for the polymerisation of α-olefins led to intense activity in the field of organotitanium chemistry. A typical heterogeneous catalyst is a mixture of $AlEt_3$ and $TiCl_4$ in heptane; when ethylene is bubbled through this mixture at room temperature and atmospheric pressure 'linear' polyethylene is produced. While the mechanism of such polymerisation processes is incompletely understood, many intermediate complexes have been isolated from such systems. The reaction between $(\pi-C_5H_5)_2TiCl_2$ and $AlEt_3$ gives as one of the products the blue crystalline compound $(\pi-C_5H_5)_2TiCl_2AlEt_2$ (figure 11.5). Solutions of this compound polymerise ethylene.

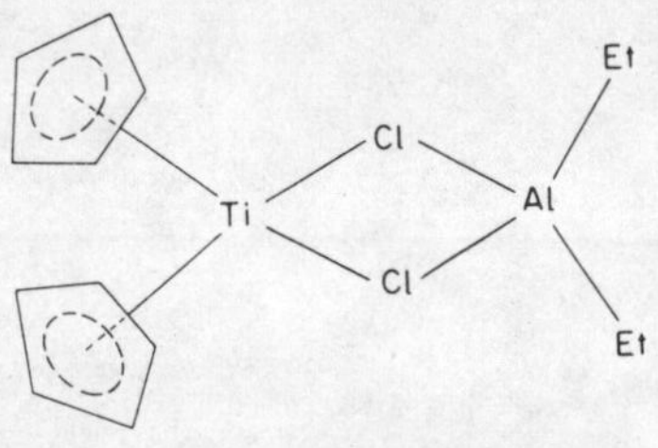

Figure 11.5   The structure of $(\pi-C_5H_5)_2TiCl_2AlEt_2$

The σ-bonded alkyls of titanium are very unstable to oxidation, and most are thermally unstable at room temperature. The tetramethyl $TiMe_4$, for example, can be prepared by the slow addition of lithium methyl in ether solution to a suspension of $TiCl_4(OEt_2)_2$ in ether at $-80°$. The yellow crystalline $TiMe_4$ begins to decompose above $-78°$. The alkyl titanium halides such as $MeTiCl_3$ and Me_2TiCl_2 behave as Lewis acids and form adducts with donor molecules. The π-cyclopentadienyl compounds of titanium are very much more stable to oxidation and heat. Titanocene $Ti(\pi\text{-}C_5H_5)_2$ forms dark green pyrophoric crystals, which decompose at $200°$ in a vacuum. More stable are the roughly tetrahedral compounds $(\pi\text{-}C_5H_5)_2TiCl_2$ and $(\pi\text{-}C_5H_5)TiCl_3$. These are orange diamagnetic solids, stable to dry air, which can be sublimed in a vacuum without decomposition. They have consequently been used as the starting points for the synthesis of a large number of other organotitanium compounds. One unusual property of $(\pi\text{-}C_5H_5)_2TiCl_2$ is that, when mixed with a Grignard reagent, it is capable of 'fixing' molecular nitrogen. The mixture absorbs nitrogen at room temperature, and subsequent hydrolysis produces ammonia. The mechanism of this process is not yet understood.

12 Vanadium

12.1 The Element

Unlike titanium, vanadium shows a wide range of oxidation states (+5 to +2) in aqueous solution, as well as the low oxidation states of +1, 0, and −1. As a result of this, vanadium is perhaps the most colourful of the transition elements, showing compounds of every colour and even with considerable colour variation within each oxidation state. It was because of the beauty of these colours that Sefström in 1830 named the element vanadium in honour of the Scandinavian goddess Vanadis.

Vanadium (atomic number 23) is a silver-grey metal having a body-centred cubic crystal structure. The metal itself does not find any great engineering applications and most of the metal produced is in the form of the iron–vanadium alloy ferrovanadium; this is used in the production of vanadium steels. Vanadium dissolves in oxidising acids such as HNO_3 and H_2SO_4, and also in HF. It combines with most nonmetals on heating; interstitial and nonstoichiometric products often result. Oxygen gives the lower oxides (which sometimes coat the metal at room temperature) and ultimately V_2O_5.

The outer electronic configuration in the ground state is $3d^3 4s^2$. The maximum oxidation state is thus +5, corresponding to a d^0 configuration. Compounds in this oxidation state are consequently diamagnetic and frequently colourless. In the +4, +3, and +2 oxidation states compounds are normally coloured and paramagnetic with magnetic moments corresponding to one, two, and three unpaired electrons per vanadium respectively. The chemistry of vanadium shows little resemblance to that of the elements in group VB, As, Sb, and Bi.

12.2 Compounds of Vanadium(V) (d^0)

12.2.1 Vanadium(V) Oxide and its Aqueous Chemistry

Vanadium(V) is a strongly oxidising state, so its thermally stable compounds are largely limited to those containing the electronegative ligands oxide and fluoride. The oxide V_2O_5 is the most important oxide and most vanadium chemicals are derived from it. It is a brown poisonous solid obtained on complete oxidation of the metal or its sulphides, nitrides, or lower oxides. It is also formed when vanadium(V) oxohalides are hydrolysed and when ammonium vanadate is heated

$$2NH_4VO_3 \rightarrow V_2O_5 + 2NH_3 + H_2O$$

It is used industrially as a catalyst particularly in the oxidation of SO_2 to SO_3 in the contact process for the manufacture of sulphuric acid. Some reactions of the pentoxide are summarised in figure 12.1.

V_2O_5 is an acidic oxide; it is sparingly soluble in water to give a feebly acidic solution, and in alkalis to give vanadates. In reducing acids, the vanadium is reduced. Thus boiling of V_2O_5 in hydrochloric acid results in the evolution of chlorine and the formation of blue vanadium(IV) species ($[VO(H_2O)_5]^{2+}$ in dilute solutions). V_2O_5 is also easily dissolved in dilute sulphuric acid in the presence

149

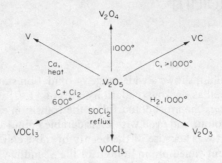

Figure 12.1   Some reactions of V_2O_5

of sulphite ions, $[VO(H_2O)_5]^{2+}$ again being formed. The nature of the solutions in alkali, that is the vanadates, depends on the pH. These solutions have been extensively studied but we can present only the most general findings here. The orthovanadate ion VO_4^{3-} is a colourless mononuclear tetrahedral ion existing in the most basic solutions (pH > 12.6). As the acidity of the solution is increased, condensed oxo-anions begin to form and the initially colourless solution becomes yellow and then red. Typical equilibria present in these solutions are indicated in the equations

$$VO_4^{3-} \;+\; H^+ \;\rightleftharpoons\; [VO_3(OH)]^{2-}$$
$$2[VO_3(OH)]^{2-} +\; H^+ \;\rightleftharpoons\; [V_2O_6(OH)]^{3-} + H_2O$$
$$3[VO_3(OH)]^{2-} +\; 3H^+ \;\rightleftharpoons\; V_3O_9^{3-} \;+\; 3H_2O$$
$$10V_3O_9^{3-} + 12H^+ \;\rightleftharpoons\; 3V_{10}O_{28}^{6-} + 6H_2O$$
$$V_{10}O_{28}^{6-} \;+\; H^+ \;\rightleftharpoons\; HV_{10}O_{28}^{5-}$$
$$HV_{10}O_{28}^{5-} \;+\; H^+ \;\rightleftharpoons\; H_2V_{10}O_{28}^{4-}$$
$$H_2V_{10}O_{28}^{4-} \;+\; 14H^+ \;\rightleftharpoons\; 10VO_2^+ \;+\; 8H_2O$$

Salts of the $(VO_3^-)n$ ion are known as metavanadates. The decavanadate ions occur naturally in minerals such as pascoite $Ca_3V_{10}O_{28}.16H_2O$. In strongly acidic solutions (pH < 1) the dioxovanadium ion VO_2^+ is the stable species (the degree of hydration in this and the other species is unknown).

12.2.2   Vanadium(V) Halides
Because the V^{5+} state is so strongly oxidising, the only stable binary halide is the fluoride VF_5. This is a white solid (m.p. $19.5°$) which is extremely readily hydrolysed by moist air. In the vapour phase it is monomeric with the trigonal bipyramidal structure. It is a violent oxidising and fluorinating agent, and with ligands often gives adducts of vanadium(IV) fluoride, such as $VF_4.NH_3$, $VF_4.py$, and $V(en)_3F_4$. Salts of the VF_6^- ion have been prepared by the reaction of BrF_3 on a mixture of VCl_3 and an alkali metal halide; KVF_6 is obtained by direct reaction between KF and VF_5 at room temperature.

The thermal stabilities of the oxide halides VOF_3, $VOCl_3$, and $VOBr_3$ decrease rapidly from the fluoride to the bromide; $VOBr_3$ begins to liberate bromine at room temperature, and VOI_3 is yet to be prepared. This is in accord with the relative reducing power of the halide anions. $VOCl_3$ and $VOBr_3$ behave as covalent monomers; they are soluble in nonpolar solvents, and $VOCl_3$ is known, from electron-diffraction studies, to have the tetrahedral structure in the vapour phase. These halides are very readily hydrolysed in moist air, giving copious brown fumes of V_2O_5. The acceptor properties of VOF_3 have not been studied; three types of reaction may occur in the reactions of $VOCl_3$ and $VOBr_3$ with ligands. These are addition reactions, substitution (solvolytic) reactions, and reduction. The chemistry of $VOBr_3$ is, as may be expected, dominated by reduction reactions; it is reduced to V^{IV} derivatives by most ligands. With bromide ions, for example, in acetonitrile solution, salts of the $VOBr_3{}^-$ and $VOBr_4{}^{2-}$ ions are formed depending on the nature of the cation. $VOCl_3$ is reduced by amines and thioethers; reduction by chloride ions to $VOCl_4{}^{2-}$ occurs in aqueous but not ethanolic media. Some reactions of $VOCl_3$ are summarised in figure 12.2.

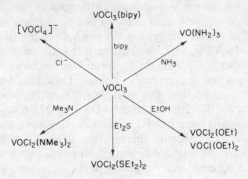

Figure 12.2   Some reactions of $VOCl_3$

12.3 Compounds of Vanadium(IV) (d^1)

12.3.1 *Aqueous chemistry and* VO^{2+} *compounds*
The +4 oxidation state is the most important one for vanadium in aqueous solutions; it is neither strongly oxidising nor reducing, and acidified solutions are stable to atmospheric oxidation

$$[V(OH)_4]^+ + 2H^+ + e \rightarrow VO^{2+} + 3H_2O \quad (E^\circ = 1.0 \text{ V})$$

V(IV) is quantitatively oxidised to V(V) by acid permanganate at 70°, and this is used in the volumetric estimation of vanadium. The vanadium compound is first boiled with an excess of acidified $KMnO_4$ to convert all the vanadium into the +5 state which is then readily reduced by sodium sulphite to V(IV). Excess sulphur dioxide is removed by boiling, and the V(IV) solution is then titrated with permanganate. The hexaquo-ion is far too acidic to exist in water, and the blue solutions of vanadium(IV) in noncomplexing acids contain the oxovanadium(IV) or vanadyl ion $[VO(H_2O)_5]^{2+}$. The aqueous chemistry of vanadium(IV) is in fact dominated by that of the VO^{2+} group; literally hundreds of compounds containing this group have been studied. The VO^{2+} ion is one of the most stable diatomic cations known, and since it has the $3d^1$ configuration its compounds have attracted widespread spectroscopic interest. While cationic species of V^{4+} are virtually unknown ($[V(en)_3]F_4$ may be an exception), oxovanadium(IV) derivatives may be cationic, anionic, or neutral.

The $[VO(H_2O)_5]^{2+}$ ion is weakly acidic

$$[VO(H_2O)_5]^{2+} + H_2O \rightleftharpoons [VO(OH)(H_2O)_4]^+ + H_3O^+ \quad (pK_1 = 6.0)$$

The yellow-brown $VO(OH)_2$ aq is precipitated by the addition of sodium carbonate solution; in strong alkalis, soluble vanadates(IV) are formed, and from these solutions crystalline compounds of the type $M^I_2[V_4O_9].7H_2O$ have been isolated. Aerial oxidation to vanadium(V) occurs much more readily in these alkaline solutions.

The structures of several oxovanadium(IV) salts have been determined by X-ray methods. In general the compounds have either five or six-co-ordination around the vanadium. The structures are usually based on that of the tetragonal pyramid with the oxygen atom of the $V{=\!=}O$ bond at the apex of the pyramid and with the site *trans* to this atom either vacant as in the five-co-ordinate structures or filled when a six-co-ordinate structure results. An example of the five-co-ordinate structure is shown by $VO(acac)_2$, the structure of which is illustrated in section 4.2.3. In the sulphate $VOSO_4.5H_2O$ the six-co-ordination involves one oxygen atom, four water molecules, and one oxygen from sulphate. These six-co-ordinate structures are distorted from octahedral by the short $V{=\!=}O$ bond; in $VOSO_4.5H_2O$, $V{=\!=}O$ is 159 pm, $V-OH_2$ is 198–205 pm, and $V-OSO_3$ is 222 pm. The multiple VO bond involves π-bonding from filled $2p_\pi$ orbitals on oxygen and vacant 3d orbitals on vanadium. Evidence for this multiple VO bonding is also gained from infrared spectroscopy. All oxovanadium(IV) compounds show a strong sharp band in the region 985 ± 50 cm^{-1} corresponding to the metal–oxygen stretching frequency. The highest frequencies are shown by those compounds that are known from X-ray studies to possess the shortest V–O bond lengths.

The five-co-ordinate compounds can become six-co-ordinate by accepting a ligand molecule in the 'vacant' site *trans* to the oxygen atom of the $V{=\!=}O$ bond. Thus $VO(acac)_2$ dissolves in nonpolar solvents (such as C_6H_6 and CS_2) to give solutions that show $\nu(V-O)$ around 1006 cm^{-1}, that is very close to the 996 cm^{-1} observed for the solid state. When bases such as pyridine, ethylenediamine, or thiocyanate ion are added to these solutions, $\nu(V-O)$ is decreased by about 50 cm^{-1}, indicating that co-ordination has occurred, accompanying shifts, etc. also observed in the electronic spectra. Six-co-ordination is also achieved in polymeric complexes in which the vanadium atoms are joined through V–O–V bonding. Thus the grey-green acetate $VO(OCOCH_3)_2$ is a remarkably air and water-stable compound (it is insoluble in water) showing $\nu(V-O)$ at 898 cm^{-1} in accord with a polymeric structure involving not only oxygen but also acetate bridging.

Oxovanadium(IV) complexes are often most conveniently prepared from V_2O_5 since this is reduced by many anions in acid solution. Thus salts of the $[VOCl_4]^{2-}$ and $[VO(C_2O_4)_2]^{2-}$ ions are isolated from solutions of V_2O_5 in hot hydrochloric and oxalic acids respectively. Examples of cationic complexes are $[VO(DMSO)_5](ClO_4)_2$ (DMSO = dimethyl sulphoxide) and $[VO(bipy)_2](ClO_4)_2$, while neutral complexes include $VO(C_2O_4)_2 . 2H_2O$ and $VOCl_2(NMe_3)_2$. The last-named complex differs from most vanadyl compounds in that its structure approximates more closely to that of a trigonal bipyramid with the Me_3N groups occupying the *trans*axial positions.

The high $V{=}O$ bond strength thus results in all complexes of vanadium(IV) being obtained from aqueous solution containing VO^{2+} as well as being obtained from other systems such as reduction reactions of $VOCl_3$. The $V{=}O$ bond can, however, be broken by reaction with thionyl chloride, for example in the preparation of hexachlorovanadates(IV)

$$(pyH)_2 VOCl_4 + SOCl_2 \rightarrow (pyH)_2 VCl_6 + SO_2$$

The electronic spectra of oxovanadium(IV) complexes have been studied very extensively. Unfortunately, although it is the easiest of the d^1 systems to use experimentally, the interpretation of the spectra has been complicated by the deviation of the complexes from regular octahedral stereochemistry. Three bands are normally observed in the spectra, in the regions $11\,000-15\,000\ cm^{-1}$, $14\,500-20\,000\ cm^{-1}$, and $21\,000-31\,000\ cm^{-1}$. On a simple crystal-field model these would be interpreted (figure 6.5) as the transitions from the $^2B_2$ ground term to the ^2E, $^2B_1$, and $^2A_1$ terms. However, such a treatment does not give a good quantitative fit to the spectra, and in the more complex molecular orbital treatment of the $[VO(H_2O)_5]^{2+}$ ion the axial π-bonding is taken into account. It is still not possible to assign the spectra of these complexes with confidence.

12.3.2 Vanadium(IV) Halides

Vanadium(IV) fluoride is a lime-green solid that hydrolyses in moist air and forms 1 : 1 adducts with SeF_4, NH_3, and py. It undergoes disproportionation slowly even at room temperature and rapidly at $100°$

$$2VF_4 \rightarrow VF_3 + VF_5$$

The tetrabromide VBr_4 has only recently been isolated by careful condensation of the vapours from the thermal decomposition of VBr_3 at $325°$. The purple VBr_4 is condensed at $-78°$; it is unstable above $-23°$, decomposing into VBr_3 and bromine. It is therefore not surprising that the chloride VCl_4 is the only vanadium(IV) halide to be studied in detail. The iodide has not been prepared.

VCl_4 is a dark red-brown liquid that fumes in moist air and readily forms the blue $[VO(H_2O)_5]^{2+}$ ion in water. It decomposes slowly at room temperature, and rapidly under reflux, into VCl_3 and chlorine. Electron-diffraction studies on the vapour phase show the molecule to be tetrahedral. It resembles $TiCl_4$ in its Lewis acidity, forming a large number of adducts and substitution products all of which are neutral or anionic. It is considerably more susceptible to reduction than $TiCl_4$; these properties are illustrated in figure 12.3. The products from these reactions are commonly six-co-ordinate but five-co-ordination is common when reduction to vanadium(III) occurs. Eight-co-ordination occurs with *o*-phenylenebisdimethylarsine, the compound $VCl_4(diars)_2$ being isomorphous with the titanium analogue. The 1 : 1 compound formed between PCl_5 and VCl_4 has the ionic structure $[PCl_4^+][VCl_5^-]$ with a five-co-ordinate VCl_5^- anion.

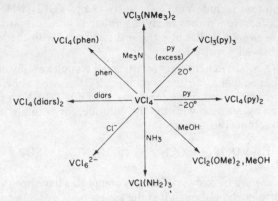

Figure 12.3   Some reactions of VCl$_4$

12.4   Compounds of Vanadium(III) (d^2)

In this oxidation state vanadium has the outer electron configuration 3d^2 in the ground state; the spectroscopic properties arising from this configuration have been considered in section 6.2.1. Magnetically dilute vanadium(III) compounds have magnetic moments close to the spin-only value, that is around 2.8 B.M. at room temperature. The only important simple compounds of V^{III} are the halides, all of which are known, and the oxide V$_2$O$_3$. However, a large number of complexes can be isolated from aqueous as well as nonaqueous solutions. In many respects the chemistry of vanadium(III) closely resembles that of titanium(III).

12.4.1   Aqueous Chemistry

Solutions containing the blue [V(H$_2$O)$_6$]$^{3+}$ ion are obtained by dissolution of the black basic oxide V$_2$O$_3$ in noncomplexing acids or by electrolytic reduction of V(IV) or V(V) in sulphate media. The V(III) state can also be obtained in aqueous solution by reduction of the higher states with an excess of bromide ions or by dissolution of the halides VX$_3$.

The hexaquo-ion is acidic and can exist only in aqueous solutions under acid conditions

$$[V(H_2O)_6]^{3+} + H_2O \rightleftharpoons [V(OH)(H_2O)_5]^{2+} + H_3O^+ \quad (pK_1 = 2.9)$$

It occurs in the solid state in the blue-violet alums such as NH$_4$V(SO$_4$)$_2$.12H$_2$O which can be crystallised from sulphuric acid solutions. The olation reaction

$$2[V(H_2O)_6]^{3+} + 2H_2O \rightleftharpoons [(H_2O)_4V \underset{O \atop H}{\overset{H \atop O}{<>}} V(OH_2)_4]^{4+} + 2H_3O^+$$

for which pK = 3.9, is characterised by a strong charge-transfer band at 22 930 cm^{-1} (compare the spectrum of [V(H$_2$O)$_6$]$^{3+}$ in figure 6.8). This same binuclear vanadium(III) species is formed as an intermediate in the reaction between V^{II} and V^{IV} in acid perchlorate solutions. In alkaline solutions the hydrous oxide V_2O_3.aq precipitates; V^{III} complexes with nitrogen donors similarly give V_2O_3.aq when they are hydrolysed in water.

Vanadium(III) solutions are unstable to aerial oxidation especially under alkaline conditions; the reduction potential in acid solutions is given by

$$VO^{2+} + 2H^+ + e \rightarrow V^{3+} + H_2O \quad (E^\circ = 0.36 \text{ V})$$

In the presence of added ligands, cationic complexes are more commonly formed than anionic ones. The halide hydrates $VCl_3.6H_2O$ and $VBr_3.6H_2O$, which can be crystallised from strongly acidic solutions, contain the $[VX_2(H_2O)_4]^+$ cations (compare their Ti analogues), and even complex halides such as $Cs_2VBr_5.5H_2O$ probably contain this cation, that is $[VBr_2(H_2O)_4]^+$, $2Cs^+, 3Br^-, H_2O$. In alcohols too, cations of the type $[VX_2(ROH)_4]^+$ predominate. Other cationic complexes include those with urea, such as $V[OC(NH_2)_2]_6^{3+}$, while typically anionic complexes include the octa-hedral oxalate $[V(C_2O_4)_3]^{3-}$ and thiocyanate $[V(NCS)_6]^{3-}$ complexes. The addition of an excess of potassium cyanide to aqueous V^{III} solutions results in the formation of an intense blue colour; on standing in the absence of air at room temperature, the deep blue solution changes to a wine-red colour. From the red solution, red crystals of the potassium salt $K_4V(CN)_7.2H_2O$ can be precipitated by the addition of methanol. This complex contains seven-co-ordinate vanadium(III), the $[V(CN)_7]^{4-}$ ion having the pentagonal bipyramidal structure. The deep-blue complex has been isolated at low temperatures and is believed to contain vanadium in two different oxidation states, one of which is V^{II}.

12.4.2 Vanadium(III) Halides

All four halides are thermally stable solids at room temperature. The yellow-green fluoride VF_3 is particularly stable; it is insoluble in water and melts without decomposition at 1400°. The other halides VCl_3 (violet), VBr_3 (black), and VI_3 (black) are hydrolysed in water and undergo hydrolysis and oxidation in moist air. They are unstable at elevated temperatures, for example

$$2VCl_3 \xrightarrow{400^\circ} VCl_2 + VCl_4$$

$$2VI_3 \xrightarrow{300^\circ} 2VI_2 + I_2$$

With the exception of the fluoride, these halides dissolve in co-ordinating solvents such as alcohol or acetonitrile with the formation of complexes, but unlike the V^{IV} halides they are insoluble in nonpolar solvents. Co-ordination compounds (figure 12.4) are thus prepared by direct reaction or by ligand exchange on, for example, $VCl_3.3MeCN$. Cationic complexes are formed only rarely in these reactions. Ethylenediamine forms $[V(en)_3]Cl_3$ which is probably cationic, but the compounds originally believed to be hexammines, for example $[V(NH_3)_6]Cl_3$, are in fact mixtures such as $VCl_2(NH_2).4NH_3 + NH_4Cl$. Neutral complexes are very similar to the analogous compounds of titanium(III), being either five or six-co-ordinate. As well as the octahedral VX_6^{3-} ions,

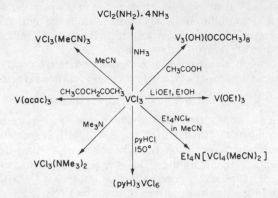

Figure 12.4   Complexes derived from VCl_3

which must be prepared in the absence of ligands other than X^-, complex salts containing anions of the type $[VX_4L_2]^-$ ($L = MeCN$, CH_3COOH) can be isolated from the solvents denoted by L; these evolve the solvent molecules when heated to above $100°$ in a vacuum, to give salts of the tetrahedral VCl_4^- or VBr_4^- ions.

12.5 Compounds of Vanadium(II) (d^3)

Compounds of vanadium(II) are strong reducing agents and difficult to work with in aqueous solution because of their rapid oxidation. The oxide VO is a basic oxide with the rock-salt structure; it is usually not stoichiometric. It dissolves in acids to give the $[V(H_2O)_6]^{2+}$ ion. This ion is also obtained as the final product in the reduction of NH_4VO_3 by zinc and acid. In this interesting reaction the colours of each oxidation state can be observed in turn; for example, if HCl is used, we have vanadate(V) yellow, $[VO(H_2O)_5]^{2+}$ blue, V^{3+} (for example $[VCl_2(H_2O)_4]^+$) green, and $[V(H_2O)_6]^{2+}$ violet. If sulphuric acid is used and the final solution evaporated in a vacuum over phosphorus pentoxide, the violet sulphate $VSO_4.6H_2O$ is obtained. This liberates hydrogen from water but is stable in acid solutions. The double sulphates $M_2V(SO_4)_2.6H_2O$ (Tutton's salts) contain the regularly octahedral $[V(H_2O)_6]^{2+}$ ion; they are somewhat more stable to oxidation than the simple sulphate. Electrolytic reduction from hydrochloric acid solutions of V(IV) gives $VCl_2(H_2O)_4$ which has the *trans*-octahedral (that is tetragonal) structure; $VBr_2(H_2O)_6$, $VI_2(H_2O)_6$, and $RbVCl_3.6H_2O$ contain the hexaquo-cations.

Cationic complexes containing, for example, $[V(bipy)_3]^{2+}$ are obtained from $VSO_4.7H_2O$ by the addition of a methanolic solution of the ligand, using KI or NH_4ClO_4 to precipitate the cation as the iodide or perchlorate salt. They arise also in the reactions of bases with vanadium hexacarbonyl, in which vanadium undergoes disproportionation, for example

$$3V(CO)_6 + 6CH_3COCH_3 \rightarrow [V^{II}(CH_3COCH_3)_6][V^{-I}(CO)_6]_2 + 6CO$$

Anionic complexes like the orange $K_4V(CN)_6$ (which contains octahedral

$[V(CN)_6]^{4-}$ ions) crystallise from solutions saturated in the anion.

The anhydrous halides VCl_2, VBr_2, and VI_2 are obtained by thermal decomposition or disproportionation of the higher halides. The chloride dissolves in hot pyridine to form $VCl_2(py)_4$ and reacts slowly with monomethylamine to form $VCl_2(MeNH_2)_6$. Complex chloride species such as VCl_3^-, VCl_4^{2-}, and VCl_6^{4-} are formed in melts of VCl_2 with alkali metal chlorides.

12.6 Compounds of Vanadium-(I), -(0), and -(−I)

12.6.1 Carbonyl and Nitrosyl Complexes

Vanadium hexacarbonyl $V(CO)_6$ forms blue-green sublimable crystals (for preparation see section 3.1.6). It is air-sensitive and decomposes above $60°$ with evolution of carbon monoxide. Being a monomeric molecule one electron short of the inert-gas configuration, it is paramagnetic (μ_{eff} = 1.7 B.M. at room temperature) both in the solid state and in benzene solution. The ease of formation of the hexacarbonylvanadate(−I) ion $[V(CO)_6]^-$, which has the noble-gas configuration, is a characteristic feature of its chemistry. Thus, with many oxygen and nitrogen bases the vanadium carbonyl disproportionates to give $[VL_6]^{2+}$ cations and $[V(CO)_6]^-$ anions (compare with section 12.5). However, some substitution products can be obtained. Phosphines give red-brown monomeric complexes of the type $V(CO)_4(PR_3)_2$ which are paramagnetic and, on the basis of their infrared spectra, are assigned the *trans* configuration.

The violet-red unstable nitrosyl carbonyl $V(CO)_5(NO)$ is formed when nitrogen monoxide is passed through a cyclohexane solution of the carbonyl at $0°$; the orange diamagnetic $[V(CO)_4(NO)(PPh_3)]$ is similarly obtained by starting from $V(CO)_4(PPh_3)$. The nitrosyl cyanide complex $K_3[V(CN)_5NO].2H_2O$ contains discrete distorted octahedral $[V(CN)_5NO]^{3-}$ anions; the N—O stretching frequency occurs at 1530 cm^{-1}, indicating that NO is present as NO^+ and hence the oxidation state of vanadium is formally +1 in this ion.

12.6.2 Other Low Oxidation-state Complexes

2,2′-Bipyridyl and 1,10-phenanthroline form compounds of V(0) and V(−I) that are analogous to the Ti–bipy complexes. $V(bipy)_3$ and $V(phen)_3$ have magnetic moments indicating the presence of one unpaired electron, that is they are low-spin complexes. The lithium salts $Li[V(bipy)_3].4THF$ and $Li[V(phen)_3].3.5THF$ are diamagnetic and also low-spin therefore; they are highly air-sensitive. Metal-to-ligand $d_\pi - p_\pi$ bonding is probably important in the stabilisation of these low oxidation state compounds. The phosphine $Me_2PCH_2CH_2PMe_2$ is also capable of stabilising V(0); black $V(Me_2PCH_2CH_2PMe_2)_3$ has $\mu_{eff} \approx 2.1$ B.M., and oxidises rapidly in air.

Bidentate sulphur donors form an interesting series of compounds of the types $[VS_6C_6R_6]^z$ (for example R = CN, $z = -1$ or -2; R = C_6H_5, $z = -2, -1$, or 0). These compounds have the trigonal prismatic six-co-ordinate structure around vanadium (see section 4.2.4); this has been established with certainty in the X-ray crystal structure determination on $[VS_6C_6Ph_6]$. The oxidation state formalism cannot be applied satisfactorily to these complexes; neither the +6 state (the vanadium being assumed to be bonded to $[S_2C_2Ph_2]^{2-}$ anions) nor the 0 state (with vanadium bonded to the neutral ligand PhCSCSPh) provides a reasonable explanation of the properties of these complexes. Vanadium is probably in some intermediate oxidation state with the ligands bonded as radical anions.

12.7   Organometallic Compounds of Vanadium

The most stable organovanadium compounds are the π-bonded, cyclopentadienyls. The violet-black vanadocene $V(C_5H_5)_2$ (m.p. 167°) has the 'sandwich' structure with the metal atom lying between the two cyclopentadienyl rings (as in ferrocene, chapter 15). Bis(benzene)vanadium $V(C_6H_6)_2$ (m.p. 277°) similarly has the π-bonded sandwich structure [as in $Cr(C_6H_6)_2$, chapter 13]. These compounds are fairly stable thermally, not decomposing below their melting points, but are unstable to aerial oxidation. σ-Bonded aryls are obtained by reactions of cyclopentadienylvanadium halides with lithium aryls in ether solvents at low temperatures, for example

$$(\pi\text{-}C_5H_5)_2\,VCl + LiPh \rightarrow (\pi\text{-}C_5H_5)_2\,VPh + LiCl$$

Treatment of $VCl_3.3THF$ with a tenfold excess of lithium phenyl in ether gives the remarkable purple complex $Li_4[VPh_6].3.5Et_2O$. This has a magnetic moment of 3.85 B.M., consistent with V(II); it is decomposed by water.

Organovanadium compounds are less effective as catalysts for polymerisation reactions than their titanium analogues, and consequently have received considerably less attention.

13 Chromium

13.1 The Element

Like vanadium, chromium shows a wide range of oxidation states and colours of compounds; its name derives from the Greek 'chromos' meaning colour. The green colour of emeralds is due to the presence of chromium in the mineral beryl, and the colour of ruby is due to the substitution of Cr^{III} ions for Al^{III} ions in the structure of α-Al_2O_3. Chromium is the first element of group VIA, lying above molybdenum and tungsten. The ground state outer electronic configuration is $3d^5 4s^1$, and like Ti and V it shows the highest oxidation state corresponding to the loss of all these outer electrons, that is +6 for Cr. Like V(V), however, Cr(VI) is a strongly oxidising state, chromium being bonded to fluorine or oxygen in its compounds. All oxidation states down to -2 are known in chromium compounds but the +3 state is by far the most stable and common.

Chromium (atomic number 24) is a light silvery grey lustrous metal that has considerable inertness to chemical attack. It is therefore widely used to plate other metals to give them a shiny corrosion-resistant coating. Another important use is in stainless steel, which contains 12—15 per cent chromium. Chromium is inert to attack by nitric acid, phosphoric acid, and aqua regia at room temperature; these reagents render the metal passive. However, it dissolves in nonoxidising acids such as dilute hydrochloric and sulphuric acids. When not in the passive state, chromium readily enters solution in salts of copper, tin, and nickel with precipitation of these metals. However, since many oxidising agents including oxygen of the air passivate the metal to some extent, the reactivity of any chromium sample is difficult to predict. At high temperatures the element reacts with alkali metal hydroxides, hydrogen halides, water, and many nonmetals including oxygen and the halogens.

13.2 Compounds of Chromium(VI) (d^0)

In the +6 oxidation state chromium has the $3d^0$ configuration; however, the compounds are usually yellow or red because of charge-transfer absorption. This is a strongly oxidising and acidic state; its complexes are predominantly anionic.

13.2.1 Aqueous Chemistry

The most important chromium(VI) compounds are the chromates and dichromates. The chromates are produced commercially by the oxidation of chromite by air at above 1000° in the presence of sodium carbonate

$$2FeO.Cr_2O_3 + 4Na_2CO_3 + {}^{7}\!/_2 O_2 \rightarrow 4Na_2CrO_4 + Fe_2O_3 + 4CO_2$$

The +6 oxidation state is formed in aqueous solution by the oxidation of Cr(III) with, for example, peroxide in alkaline solution. The poisonous yellow chromate CrO_4^{2-} and orange dichromate $Cr_2O_7^{2-}$ ions are the species present in aqueous solutions containing no added ligands. These two ions coexist in equilibrium over quite a wide pH range, the chromate species being predominant in alkaline

159

solution. When acid is added to a solution of chromate ions, protonation and olation occur

$$\left[\begin{array}{c} O \\ \diagdown \\ O == Cr - O \\ \diagup \\ O \end{array}\right]^{2-} + H^+ \rightleftharpoons \left[\begin{array}{c} O \\ \diagdown \\ O == Cr - OH \\ \diagup \\ O \end{array}\right]^{-}$$

$$\Updownarrow$$

$$H_2O + \left[\begin{array}{c} O \quad\quad O \\ \diagdown \quad \diagup \\ == Cr - O - Cr == O \\ \diagup \quad\quad \diagdown \\ O \quad\quad O \end{array}\right]^{2-} \rightleftharpoons \left[\begin{array}{c} O \quad H \quad\quad O \\ \diagdown \quad O \quad \diagup \\ O == Cr \diagdown O \diagup Cr == O \\ \diagup \quad H \quad\quad \diagdown \\ O \quad\quad\quad O \end{array}\right]^{2-}$$

The $CrO_4{}^{2-}$ ion is tetrahedral while the $Cr_2O_7{}^{2-}$ ion is composed of two tetrahedral CrO_4 groups joined by a common oxygen atom with the $Cr\widehat{O}Cr$ angle of $115°$. The addition of solutions of Ag^+, Pb^{2+}, or Ba^{2+} ions to solutions of potassium dichromate causes complete precipitation of these metals as chromates since these are insoluble and because the $CrO_4{}^{2-}$–$Cr_2O_7{}^{2-}$ equilibrium is so quickly re-established once disturbed. In the presence of high concentrations of complexing anions such as $SO_4{}^{2-}$ or Cl^-, substitution products of the chromate ion are formed

$$[CrO_3(OH)]^- + H^+ + Cl^- \rightleftharpoons [CrO_3Cl]^- + H_2O$$

A surprising feature of Cr^{VI} chemistry is the stability of these oxo-halo-compounds, especially $[CrO_3Br]^-$ and $[CrO_3I]^-$, which we might expect to undergo internal redox; the stability may be due to the kinetically slow reaction between $Cr(VI)$ and halide ions.

The addition of concentrated sulphuric acid to a solution of a dichromate results in precipitation of the red crystalline CrO_3. This is a very water-soluble acidic oxide which acts as a powerful oxidising agent. It reacts explosively with some organic substances, being reduced to Cr_2O_3.

The addition of hydrogen peroxide solutions to acid dichromate solutions gives a deep violet-blue colour, which rapidly fades with evolution of oxygen. However, if ether is added, the blue colour becomes concentrated in the ether layer where it is more stable; this is the basis of the so-called 'blue lake' qualitative test for chromium. If pyridine is added to the ethereal solution, the explosive solid 'pyridine perchromate' $(py)CrO(O_2)_2$ is obtained. The structure of this compound (figure 13.1) has been determined by X-ray crystallography. The chromium atom is in the centre of an approximately pentagonal pyramid in which the oxide oxygen occupies the apical position. The blue compounds are thus all believed to be adducts (with water, ether, or pyridine) of the chromium(VI) peroxide $CrO(O_2)_2$. Its formation in water can be represented thus

$$[CrO_3(OH)]^- + 2H_2O_2 + H^+ \rightarrow CrO(O_2)_2 + 3H_2O$$

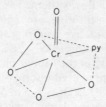

Figure 13.1   The structure of $(py)Cr(O_2)_2$

Many other peroxochromium species are known. Among these are the red-brown peroxochromates $M^I_3CrO_8$ formed in alkaline solution. These are paramagnetic (with one unpaired electron per chromium atom) and form mixed crystals with the niobium(V) and tantalum(V) compounds $M^I_3M^VO_8$; they are thus formulated as containing chromium(V), that is $[Cr(O_2)_4]^{3-}$.

13.2.2   Halides and Oxide Halides

The only known hexahalide is the unstable lemon-yellow CrF_6 prepared from the elements at 400° and 200 atm; at low pressures it decomposes above $-100°$ into CrF_5 and fluorine.

The 'chromyl' halides CrO_2X_2 (X = F, Cl, Br) are dark liquids or low-melting solids rapidly hydrolysed in water, for example

$$CrO_2Cl_2 + 2H_2O \rightarrow H_2CrO_4 + 2HCl$$

Their thermal stability decreases in the order $F^- > Cl^- > Br^-$; the bromide is unstable at room temperature and the iodide has not been isolated. The chloride is readily prepared by the addition of concentrated sulphuric acid to a mixture of potassium dichromate and potassium chloride

$$K_2Cr_2O_7 + 4KCl + 6H_2SO_4 \rightarrow 2CrO_2Cl_2 + 6KHSO_4 + 3H_2O$$

The evolution of the red-brown vapour on heating this mixture has been used as a distinguishing test between chloride and bromide ions, since CrO_2Br_2 is unstable under these conditions. CrO_2Cl_2 is a covalent, roughly tetrahedral, monomer in the vapour and in most nonpolar solvents. It is a violent oxidising agent with some organic compounds; however, some apparently Lewis base adducts have been isolated, for example $CrO_2Cl_2.2D$ (D = acetone or Me_2SO), but these have been shown to contain chromium(IV). Reaction with N_2O_4 gives $Cr(NO_3)_3.2N_2O_4$, but with N_2O_5 chromyl nitrate is formed as a brown volatile liquid

$$CrO_2Cl_2 + 2N_2O_5 \rightarrow CrO_2(NO_3)_2 + 2NO_2Cl$$

This is a violent oxidising agent, causing benzene to inflame on contact.

13.3 Compounds of Chromium(V) (d^1) and Chromium(IV) (d^2)

The +4 and +5 oxidation states are relatively rarely found in chromium compounds. In most of these compounds, which usually undergo ready disproportionation, chromium is bound to oxygen only or to a mixture of oxygen and halide ligands.

The pentafluoride CrF_5 is a red solid (m.p. 30°) formed in the direct reaction between the elements at 400° in a flow system. The green solid CrF_4 is obtained at a slightly lower temperature. Both fluorides are rapidly hydrolysed by water but, while the pentafluoride is a very reactive oxidising agent, the tetrafluoride is rather inert and does not react with, for example, NH_3, SO_2, or BrF_3 at room temperature. Evidence for the presence of $CrCl_4$ in the vapour phase (for example over $CrCl_3$ by disproportionation) has been presented but no other CrX_5 or CrX_4 compounds have been isolated. The oxochloride $CrOCl_3$ is a dark red solid made by the reaction of CrO_3 with thionyl chloride. From concentrated hydrochloric acid, complexes $M^I_2[CrOCl_5]$ can be crystallised starting from CrO_3 and M^ICl.

Some dark blue solid alkali metal chromates(V), for example Na_3CrO_4, are known. These are paramagnetic and believed to contain discrete tetrahedral anions. In water they disproportionate to Cr^{III} and Cr^{VI} Ca_2CrO_4Cl is known to contain distorted CrO_4 tetrahedra. Chromium(V) species have been shown to be present in solutions of CrO_3 in phosphoric acid and of chromates(VI) in 65 per cent oleum. Some simple chromium(IV) compounds include the green $Cr(NEt_2)_4$ (prepared from $CrCl_3$ and $LiNEt_2$) and the blue alkoxides $Cr(OR)_4$ obtained by alcoholysis of the tetrakis(diethylamide). The alkoxides are paramagnetic ($\mu_{eff} \approx 2.8$ B.M.), monomeric, and tetrahedral.

13.4 Compounds of Chromium(III) (d^3)

This is the most common and stable oxidation state of chromium especially in aqueous solution. Complexes of chromium(III) have the d^3 configuration and are kinetically inert (see chapter 8), undergoing only slow ligand substitution reactions. This has resulted in the study and isolation of a very large number of these complexes, virtually all of which possess octahedral co-ordination, which is a characteristic feature of chromium(III) chemistry. Three, four, and five-co-ordination do occur rarely as in, for example, $Cr(NPr^i_2)_3$, $Li[Cr(OBu^t)_4]$, and $CrCl_3(NMe_3)_2$.

13.4.1 Aqueous Chemistry

The violet hexaquo-ion $[Cr(H_2O)_6]^{3+}$ occurs in the solid state in salts such as $Cr_2(SO_4)_3.18H_2O$ and chrome alum $KCr(SO_4)_2.12H_2O$, as well as in solutions of $Cr(OH)_3$ in noncomplexing acids. The green chromium(III) salts, for example $Cr_2(SO_4)_3.6H_2O$, do not contain the hexaquo-ion; rather, they have anions in the co-ordination sphere of the metal. The acidity of the aquo-ion, that is

$$[Cr(H_2O)_6]^{3+} + H_2O \rightleftharpoons [Cr(OH)(H_2O)_5]^{2+} + H_3O^+ \qquad (pK_1 = 1.6)$$

$$2[Cr(OH)(H_2O)_5]^{2+} \rightleftharpoons [(H_2O)_4Cr(OH)_2Cr(OH_2)_4]^{4+} + 2H_2O\ (K \approx 10^4)$$

prohibits the formation of salts of weak acids such as CO_3^{2-} and S^{2-} in aqueous solution. The addition of alkali results in the precipitation of the green $Cr(OH)_3$ aq. This is soluble in an excess of alkali to give species that may be $[Cr(OH)_6]^{3-}$. In ammonia, solution of the hydroxide is slow but soluble ammines are eventually formed. A very large number of these ammines have been characterised, ranging through $[Cr(NH_3)_6]^{3+}$ to $[Cr(NH_3)_2(H_2O)_4]^{3+}$, and with anionic substitution products such as $[Cr(NH_3)_5X]^{2+}$ and $[Cr(NH_3)_2(H_2O)_3X]^{2+}$ (X = univalent anion). Reinecke's salt $NH_4[Cr(NH_3)_2(NCS)_4].H_2O$ as often been used as a source of a large univalent anion in the precipitation of large cations as 'reineckates'. Complete replacement of co-ordinated water or ammonia molecules by anions leads to complexes of the types $[CrX_6]^{3-}$, such as $[CrF_6]^{3-}$ and $[Cr(CN)_6]^{3-}$, and $[Cr(AA)_3]^{3-}$ such as $[Cr(C_2O_4)_3]^{3-}$. Many polynuclear complexes are also known. These are usually hydroxo-bridged, for example $[(NH_3)_5Cr(OH)Cr(NH_3)_5]^{5+}$ (the so-called normal rhodo-ion), or oxo-bridged as in the basic rhodo-ion, that is $[(NH_3)_5CrOCr(NH_3)_5]^{4+}$. Trinuclear chromium(III) complexes also occur commonly. In the basic acetates, which contain, for example, the ion $[Cr_3O(O_2CCH_3)_6(H_2O)_3]^+$, the oxygen atom is at the centre of an equilateral triangle of chromium atoms each pair of which is bridged by two acetate groups, the octahedron around each chromium being completed by a terminal water molecule.

13.4.2 Chromium(III) Oxide

The green Cr_2O_3 results when chromium is burned in air, when dichromates are reduced by heating with, for example, carbon or sulphur, and in the spectacular thermal decomposition of ammonium dichromate (in which reaction the ammonium ion acts as the reducing agent)

$$(NH_4)_2Cr_2O_7 \rightarrow Cr_2O_3 + N_2 + 4H_2O$$

The crystallised form, made at high temperatures, is black with a metallic lustre. It has the corundum structure; indeed, isomorphous replacement of Al^{III} by Cr^{III} in corundum gives the mineral ruby. When Cr_2O_3 is ignited strongly, that is in its preparation from '$Cr(OH)_3$', it becomes inert to acids and bases. On fusion with basic oxides, compounds of the spinel type, that is $M^{II}O.Cr_2O_3$, are obtained. Cr_2O_3 has been used in a variety of green pigments.

13.4.3 Chromium(III) Halides

All four halides have been prepared, the green fluoride by the high-temperature reaction of the metal with HF, and the violet $CrCl_3$, black $CrBr_3$, and CrI_3 by direct combination of the elements at various temperatures. Only the chemistry of the chloride has been studied in any detail. A large number of adducts are known with donor molecules; these are usually of the six-co-ordinate type, for example $CrCl_3.3py$ and $CrCl_3.3THF$. The trimethylamine adduct $CrCl_3.2NMe_3$ has the five-co-ordinate trans-trigonal bipyramidal structure. With alkali metal chlorides in melts, two series of complex halides arise, $M^I_3CrCl_6$ and $M^I_3Cr_2Cl_9$; the latter series of compounds are also conveniently prepared in thionyl chloride. These halides are magnetically dilute; in $Cr_2Cl_9^{3-}$ each chromium ion is bonded to three terminal chloride ions and shares three bridging chloride ions in an approximately octahedral array about each chromium.

Chromium(III) chloride dissolves at a negligible rate in water but becomes rapidly soluble in the presence of Cr^{II} or a reducing agent capable of producing Cr^{II} in the solution. The Cr^{II} ions can undergo electron-transfer reactions with the inert Cr^{III} ions in the solid $CrCl_3$ via a chloride bridge, thus enabling Cr^{II} sites to be formed which can leave the solid and cause further solubilisation of the $CrCl_3$. The hexahydrate $CrCl_3.6H_2O$ exists in three isomeric forms; the normally purchased green form has the formula $[CrCl_2(H_2O)_4]Cl.2H_2O$, while the less common violet and pale green isomers have the formulae $[Cr(H_2O)_6]Cl_3$ and $[CrCl(H_2O)_5]Cl_2$ respectively.

13.5 Compounds of Chromium(II) (d^4)

The +2 state is a strongly reducing one for chromium. Aqueous solutions of chromium(II) are best obtained by dissolving pure chromium in dilute nonoxidising mineral acids such as HCl. They are also readily obtained by the reduction of dichromate by zinc and acid. These preparations must be carried out under a protective atmosphere. The aquo-ion $[Cr(H_2O)_6]^{2+}$ is very readily oxidised to the green $[Cr(H_2O)_6]^{3+}$ in air

$$Cr^{3+} + e \rightarrow Cr^{2+} \qquad (E^\circ = -0.41 \text{ V})$$

Several hydrated salts can be obtained from aqueous solution; these frequently contain the hexaquo-ion in the solid state. The sulphates resemble those of iron(II). The blue $CrSO_4.7H_2O$ is isomorphous with $FeSO_4.7H_2O$, and double salts of the formula $M^I_2Cr(SO_4)_2.6H_2O$ can be crystallised from solutions containing the alkali metal sulphates. Other hydrates include $Cr(ClO_4)_2.6H_2O$, $CrCl_2.4H_2O$, and $CrBr_2.6H_2O$.

The red crystalline acetate $Cr(OCOCH_3)_2.2H_2O$ is readily prepared in the laboratory. A chromium(II) solution is forced under hydrogen into a concentrated solution of sodium acetate. The chromium(II) acetate precipitates; it is much less readily oxidised than solutions of Cr^{II} but is nevertheless unstable in air. Figure 13.2 shows the dimeric structure of this acetate. In this structure

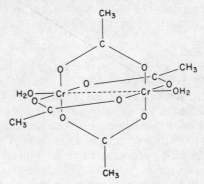

Figure 13.2   The structure of $Cr(OCOCH_3)_2.2H_2O$

each Cr atom is surrounded by an approximately square planar array of oxygen atoms from the two bridging acetate groups. The short Cr—Cr distance found (246 pm) implies a strong interaction between the two metal atoms. The compound is in fact diamagnetic, so the four metal electrons on each chromium are fully paired. Copper(II) acetate dihydrate has a similar structure although the Cu—Cu distance is larger (264 pm) and the spin moment is not completely quenched at room temperature.

Especially air-stable chromium(II) complexes are formed by hydrazine. The pale violet $CrCl_2(N_2H_4)_2$ precipitates when a chromium(II) chloride solution is passed into an excess of aqueous hydrazine. This complex has the polymeric hydrazine-bridged structure found for all $MX_2(N_2H_4)_2$ compounds of the first transition series (figure 13.3); it is paramagnetic, with $\mu_{eff} = 4.9$ B.M. The

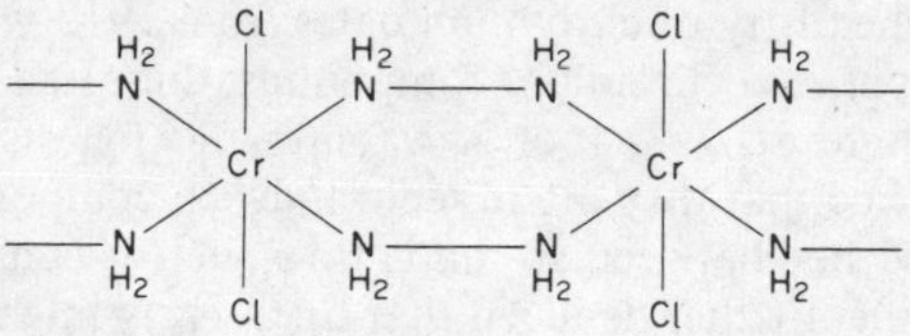

Figure 13.3   The structure of $CrCl_2(N_2H_4)_2$

$[Cr(N_2H_4)_6]^{2+}$ complex has not been prepared but 2,2'-bipyridyl forms $[Cr(bipy)_3]^{2+}$ which in alkaline solution undergoes disproportionation

$$2[Cr(bipy)_3]^{2+} \rightarrow [Cr(bipy)_3]^+ + [Cr(bipy)_3]^{3+}$$

Anionic complexes containing the ions $[Cr(NCS)_6]^{4-}$ and $[Cr(CN)_6]^{4-}$ are also known. The common features of chromium(II) complexes are that they have the preferred co-ordination number of six and are usually high-spin ($t_{2g}^3 e_g^1$); the hexacyanide complex is low-spin (t_{2g}^4) with $\mu_{eff} = 3.2$ B.M.

The anhydrous halides CrX_2 (X = F, Cl, Br) are prepared by the reactions of the metal with the hydrogen halides at around 700°. The iodide is prepared by direct combination of the elements at around 600°. With donor molecules such as pyridine and triphenylphosphine oxide, tetragonal polymeric complexes $CrX_2.2L$ are formed. Complex halides of the types $M^I_2CrCl_4$ and M^ICrCl_3 have been obtained in melts.

13.6 Compounds of Chromium in Low Oxidation States

The hexacarbonyl $Cr(CO)_6$ has the noble-gas configuration and is a white crystalline diamagnetic air-stable solid. It has the octahedral structure. A large number of derivatives have been prepared in which one or more of the carbonyl groups are replaced by ligands such as pyridine, tertiary phosphines, tertiary arsines, etc. Halide ions similarly displace carbon monoxide, for example

$$Cr(CO)_6 + Et_4NI \xrightarrow{THF} Et_4N[Cr(CO)_5I] + CO$$

The $[Cr(CO)_5I]^-$ ion can be oxidised by, for example, Fe^{3+} to the deep blue $Cr(CO)_5I$ which is one of the best-known octahedral chromium(I) compounds. Reduction of $Cr(CO)_6$ by sodium in liquid ammonia affords the trigonal bipyramidal chromium(–II) anion $[Cr(CO)_5]^{2-}$. If the reduction is carried out using sodium borohydride in boiling THF, the anion $[HCr_2(CO)_{10}]^-$ is obtained in which the hydrogen atom bridges two $Cr(CO)_5$ groups. Many olefin and arene complexes, such as $C_6H_6Cr(CO)_3$, are obtained by heating the carbonyl with the hydrocarbon.

Other ligands capable of stabilising low oxidation states on chromium are the isocyanides as in $C_1(CNPh)_6$, cyanide ion as in $K_6Cr(CN)_6$, and 2,2′-bipyridyl which gives compounds analogous to those described for titanium and vanadium.

13.7 Organometallic Compounds of Chromium

The organometallic chemistry of chromium dates from 1919 when Hein prepared what were called polyphenyl chromium compounds; these had the general formula Ph_xCr and Ph_xCrI where x = 2, 3, 4, or 5. A reinvestigation of these compounds in 1954 by Zeiss showed that they are in fact sandwich compounds of chromium with the metal sandwiched between aromatic rings such as benzene and biphenyl. At about the same time, Fischer deduced that $Cr(C_6H_6)_2$ might be stable, by analogy with the known structure of ferrocene, and he proceeded successfully to synthesise dibenzenechromium (chapter 3).

While many olefin, cyclopentadienyl, and a few π-allyl and σ-bonded organo-chromium compounds are now known, the most celebrated organochromium compound is still dibenzenechromium. In this black diamagnetic solid the chromium atom is sandwiched between two benzene rings (figure 13.4). Since the

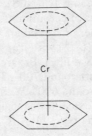

Figure 13.4   Dibenzenechromium

rings are uncharged, the compound contains chromium(0) and is isoelectronic with ferrocene. Unlike ferrocene, however, it does not undergo electrophilic substitution reactions in the aromatic rings; rather, oxidation occurs to the $(\pi\text{-}C_6H_6)_2Cr^+$ cation. Some ligands completely replace both benzene rings in their reaction with dibenzenechromium, for example

$$Cr(C_6H_6)_2 + 6PF_3 \rightarrow Cr(PF_3)_6 + 2C_6H_6$$

14 Manganese

14.1 The Element

Manganese is the third most abundant transition metal, being widely distributed in the earth's crust. It heads group VIIA of the periodic table, lying above technetium and rhenium. The maximum oxidation state +7 is shown in the permanganates, and all oxidation states from +7 to −3 are known; this is the widest range shown by a first-row transition element. Pure manganese is largely produced by electrolysis from aqueous solutions containing manganese(II) sulphate.

The metal resembles iron in appearance. Four allotropic modifications are known; the α-form, the stable form at room temperature, is hard and brittle. The largest use of manganese is in ferroalloys. Manganese is an electropositive metal, being attacked slowly by water and dissolving readily in dilute acids to form Mn^{2+}.aq with evolution of hydrogen. It differs from the earlier transition elements in the great stability of the +2 oxidation state, which for manganese corresponds to the d^5 completely half-filled shell. Thus Mn^{II} compounds are formed also when the metal combines with chlorine, bromine, iodine, and nitrogen; with fluorine, MnF_3 and MnF_4 are the major products.

14.2 Compounds of Manganese(VII) (d^0), Manganese(VI) (d^1), and Manganese(V) (d^2)

The best-known manganese(VII) species is the permanganate ion MnO_4^-. Potassium permanganate is manufactured by fusing the dioxide (pyrolusite) with potassium hydroxide in air to form the manganate

$$2MnO_2 + 4KOH + O_2 \rightarrow 2K_2MnO_4 + 2H_2O$$

The manganate is then oxidised electrolytically to permanganate

$$MnO_4^{2-} - e \rightarrow MnO_4^-$$

$KMnO_4$ forms dark purple crystals with a solubility in water at room temperature of 63 g l^{-1}. The lithium and sodium salts are considerably more soluble, while the rubidium and caesium salts are only sparingly soluble in water. The permanganate ion is a strong oxidising agent and is widely used for this purpose in the laboratory. Solutions of the ion slowly decompose on standing, with the formation of brown MnO_2

$$4MnO_4^- + 4H^+ \rightarrow 4MnO_2 + 2H_2O + 3O_2$$

This reaction is catalysed by light. In acid solution, reducing agents cause reduction to Mn^{2+}.aq

$$MnO_4^- + 8H^+ + 5e \rightarrow Mn^{2+} + 4H_2O \quad (E^\circ = +1.51 \text{ V})$$

The reaction with reducing agents is initially slow, but it is catalysed by the presence of Mn^{2+} so proceeds quickly once Mn^{2+} has been formed, or if it is deliberately added. Permanganate is widely used for the volumetric estimation of

167

other transition metal ions, for example Ti^{3+}, VO^{2+}, and Fe^{2+}, as well as for hydrogen peroxide, oxalates, formates, and nitrites. In alkaline or neutral solution reduction to MnO_2 occurs

$$MnO_4^- + 2H_2O + 3e \rightarrow MnO_2 + 4OH^- \quad (E^\circ = +0.588 \text{ V})$$

The addition of concentrated alkali to $KMnO_4$ results in the formation of the green manganate ion

$$4MnO_4^- + 4OH^- \rightarrow 4MnO_4^{2-} + 2H_2O + O_2$$

This reaction can be reversed by passing chlorine or carbon dioxide through the manganate solution

$$MnO_4^{2-} + \tfrac{1}{2}Cl_2 \rightarrow MnO_4^- + Cl^-$$

The manganate ion is the only stable representative of the manganese(VI) state.

The oxide Mn_2O_7 behaves as the anhydride of permanganic acid. It is prepared as a dark heavy oil by the action of concentrated sulphuric acid on potassium permanganate

$$2KMnO_4 + 2H_2SO_4 \rightarrow Mn_2O_7 + 2KHSO_4 + H_2O$$

This oxide decomposes explosively on heating, and in water it gives permanganic acid solution.

Permanganic acid and its dihydrate can be isolated as purple solids $HMnO_4$ and $HMnO_4 . 2H_2O$ by low-temperature evaporation of its frozen aqueous solutions. It is a violent oxidant, causing ignition of hydrocarbons on contact and decomposing (often explosively) above 3°.

No simple halides of manganese(VII) are known but two oxyhalides exist. MnO_3F is prepared by the reaction of potassium permanganate with anhydrous hydrogen fluoride, iodine pentafluoride, or fluorosulphonic acid. It is a dark green liquid (m.p. -78°), instantly hydrolysed by moisture and decomposing explosively above room temperature to MnF_2, MnO_2, and O_2. The molecule is tetrahedral in the vapour phase. MnO_3Cl is similar; it is formed together with brown MnO_2Cl_2 and green $MnOCl_3$ in the reaction of Mn_2O_7 with HSO_3Cl; these oxyhalides can be handled more safely in carbon tetrachloride solution.

Manganates(V) containing the blue MnO_4^{3-} ion can be isolated from fused or concentrated aqueous sodium hydroxide. Magnetic susceptibility measurements confirm the $+5$ (d^2) oxidation state in these compounds; $Na_3MnO_4 . 10H_2O$ is isomorphous with Na_3VO_4.

14.3 Compounds of Manganese(IV) (d^3)

Manganese occurs naturally in this oxidation state as pyrolusite MnO_2. Its stability lies partly in its insolubility in water. Solutions of manganese(IV) compounds undergo ready oxidation to manganese(VI) in alkaline solution, and reduction to manganese(II) in acid solution. Manganese(IV) therefore has not an extensive aqueous solution chemistry.

MnO_2 is a black powder when anhydrous; the aquated form prepared in aqueous solution (by reduction of MnO_4^- or oxidation of $[Mn(H_2O)_6]^{2+}$) is

brown. It evolves oxygen when heated to above 500° in air or when treated with
hot concentrated sulphuric acid; it acts as a catalyst in the decomposition of hydrogen peroxide and potassium chlorate, and in the oxidation of carbon monoxide by copper(II) oxide. In cold concentrated hydrochloric acid, the hexachloro-
manganate(IV) ion is formed

$$MnO_2 + 6HCl \rightarrow MnCl_6{}^{2-} + 2H^+ + 2H_2O$$

In hot acid, reduction to manganese(II) occurs

$$MnO_2 + 4HCl \rightarrow MnCl_2 + Cl_2 + 2H_2O$$

The only known binary halide of manganese(IV) is the fluoride MnF_4. This is a
blue hygroscopic solid which decomposes slowly at room temperature into MnF_3
and fluorine. It obeys the Curie–Weiss Law with μ_{eff} = 3.84 B.M. at room temperature. Complex halides such as K_2MnF_6 and K_2MnCl_6 are known, but the
latter salt liberates chlorine in dry air at room temperature. The hexacyano-
manganate(IV) $K_2Mn(CN)_6$ has been prepared by the oxidation of $K_3Mn(CN)_6$
with nitrosyl chloride in dimethylformamide. A series of complex iodates
$M^I{}_2Mn(IO_3)_6$ is also known.

14.4 Compounds of Manganese(III) (d⁴)

Manganese(III) is still an oxidising state and is also unstable with respect to disproportionation in aqueous solution

$$2Mn^{3+}.aq + 2H_2O \rightarrow Mn^{2+}.aq + MnO_2 + 4H^+$$

The manganese(III) state can be stabilised by using acidic solutions, increasing the
Mn^{2+} concentration, or by complex formation. It is prepared in aqueous solution
by chemical or electrolytic oxidation of Mn^{II}, or by reduction of Mn^{IV} or Mn^{VII}
The aquo-ion $[Mn(H_2O)_6]^{3+}$ occurs in the alum $CsMn(SO_4)_2.12H_2O$; it is
strongly acidic

$$[Mn(H_2O)_6]^{3+} + H_2O \rightleftharpoons [Mn(OH)(H_2O)_5]^{2+} + H_3O^+ \quad (pK_1 \approx 0.9)$$

Hydrolysis of the aquo-ion is thus considerable even at high acidity. The electronic
spectrum of this ion shows a single band with Jahn–Teller distortion at around
$21\,000$ cm^{-1} corresponding to the $^5E_g \rightarrow {}^5T_{2g}$ transition in the octahedral d⁴
ion.

Aerial oxidation of $Mn(OH)_2$ gives the brown hydrous oxide Mn_2O_3.aq
which upon drying at 100° gives $MnO(OH)$ (this also occurs naturally, as man-
ganite). Mn_2O_3 also occurs in nature (braunite); it arises in the thermal decomposition of MnO_2 above 500°, and in the ignition of manganese(II) salts
in air. Above 940°, Mn_2O_3 evolves oxygen, giving the manganese(II, IV) oxide
Mn_3O_4.

The only thermally stable halide is MnF_3. This is a red-purple solid with
μ_{eff} = 5.0 B.M. at room temperature. It is instantly hydrolysed by moisture but has
been extensively used as a fluorinating agent in organic chemistry. The crystal
structure consists of distorted ($t_{2g}{}^3 e_g{}^1$ ground state) MnF_6 octahedra which share
corners. The black $MnCl_3$ has been obtained by the reaction of manganese dioxide

with hydrogen chloride in ethanol at $-63°$ followed by precipitation with carbon tetrachloride. It decomposes above $-40°$ but with amines it forms 1 : 3 complexes which are thermally stable at room temperature but extremely sensitive to moisture. Complexes with halides are similarly more thermally stable; these include K_3MnF_6 (prepared in melts), $K_2[MnF_5(H_2O)]$ (prepared in aqueous HF), and the tris-(1,2-propanediamine)cobalt(III) salt of the $MnCl_6^{3-}$ ion.

One of the most stable Mn^{III} salts is the acetate $Mn(OCOCH_3)_3 . 2H_2O$. This is prepared by the oxidation of a solution of manganese(II) acetate in acetic acid with, for example, permanganate. Similarly, the acetylacetonate $Mn(acac)_3$ is obtained by the oxidation of manganese(II) solutions containing an excess of acetylacetone. The red sulphate $Mn_2(SO_4)_3 . H_2SO_4 . 4H_2O$ crystallises from solutions of Mn_2O_3 in cold moderately concentrated sulphuric acid.

The hexacyano-complex $K_3Mn(CN)_6$ is obtained by treatment of manganese(III) acetate with an excess of aqueous potassium cyanide followed by precipitation with ethanol. The $[Mn(CN)_6]^{3-}$ ion is also readily formed by oxidation of Mn^{II} salts in an excess of cyanide ions; it is, unlike all the other Mn^{III} complexes we have mentioned, low-spin.

14.5 Compounds of Manganese(II) (d^5)

This is the most common oxidation state shown by manganese. Because of the high-spin d^5 configuration in most of its compounds, the salts of manganese(II) are a very pale pink colour (see section 6.2.7). The aquo-ion $[Mn(H_2O)_6]^{2+}$ is barely acidic, so manganese(II) forms an extensive series of salts with common anions; the carbonate $MnCO_3$ is even found in the natural state as manganese spar. Many of the hydrated salts, for example $MnSO_4 . 7H_2O$ and $Mn(ClO_4)_2 . 6H_2O$, contain the $[Mn(H_2O)_6]^{2+}$ ion. The addition of hydroxide ions to solutions of these salts causes precipitation of the white $Mn(OH)_2$; this is a true hydroxide and is largely basic. In the presence of air, alkaline suspensions of $Mn(OH)_2$ oxidise rapidly to the brown $Mn_2O_3 . aq$. Oxidation of the Mn^{II} to permanganate can be achieved by periodate or persulphate ions in acid solution

$$2Mn^{2+} + 5S_2O_8^{2-} + 8H_2O \rightarrow 2MnO_4^- + 10SO_4^{2-} + 16H^+$$

The halides MnX_2 are pink crystalline solids which form hydrates in water and a large number of adducts with organic molecules, for example octahedral polymeric $MnCl_2 . 2py$ and tetrahedral $MnCl_2 . 2(OPPh_3)$. Complex halides are readily prepared in nonaqueous media; they are of two types. In the M^IMnX_3 type the manganese is octahedrally co-ordinated by halide ions while in the $M^I_2MnX_4$ type tetrahedral MnX_4^{2-} ions are present.

A large variety of complexes and stereochemistries are exhibited by Mn^{II}. Since these are mostly d^5, that is high-spin complexes, crystal-field stabilisation energies make no contribution to their overall thermodynamic stability. Octahedral and tetrahedral stereochemistries are common, and in $MnCl_2(DMSO)_3$ (DMSO = dimethyl sulphoxide) are found manganese ions in both stereochemistries, that is $[Mn(DMSO)_6]^{2+}[MnCl_4]^{2-}$. Ammonia forms octahedral $[Mn(NH_3)_6]^{2+}$ cations in its reactions with the anhydrous salts, while some terdentate amines such as 2,2',2''-terpyridyl (terpy) form five-co-ordinate complexes, for example

MnI_2(terpy). Dodecahedral eight-co-ordination occurs in the tetranitrato-manganate(II) anion in $[Ph_4As]_2Mn(NO_3)_4$.

Low-spin complexes are formed by cyanide ion and isonitriles, for example $[Mn(CN)_6]^{4-}$ and $[Mn(CNR)_6]^{2+}$. These are more readily oxidised than the high-spin complexes, probably because of increased CFSE in the more highly charged Mn^{III} derivatives (the CFSE in the Mn^{II} complexes is only slightly greater than the pairing energy).

14.6 Compounds of Manganese in Low Oxidation States

The manganese atom requires eleven electrons in order to achieve the stable noble-gas configuration. The carbonyl $Mn(CO)_5$ has seventeen outer electrons and exists as the dimer $Mn_2(CO)_{10}$ in which each manganese atom has the required share in eighteen electrons. In the $Mn(CO)_5^-$ anion, however, the eighteen electron rule is again upheld and this is in fact a stable species.

$Mn_2(CO)_{10}$ has a metal–metal bond joining the two $Mn(CO)_5$ fragments; it forms yellow crystals that are slowly oxidised by the air. Reduction by sodium amalgam in tetrahydrofuran gives sodium pentacarbonylmanganate(–I) $NaMn(CO)_5$ which can be used to prepare a variety of compounds by reaction with organic halides

$$NaMn(CO)_5 + RX \rightarrow RMn(CO)_5 + NaX$$

If phosphoric acid is used in place of RX, the volatile liquid hydride $HMn(CO)_5$ is obtained. The carbonyl halides $Mn(CO)_5X$ are obtained in the reactions of the halogens with $Mn_2(CO)_{10}$ or $NaMn(CO)_5$. Thermal decomposition of the $Mn(CO)_5X$ compounds at $100°$ gives the halogen-bridged binuclear carbonyl halides $(CO)_4MnX_2Mn(CO)_4$. Many substitution products of these carbonyls with ligands are known, including two volatile nitrosyl carbonyls, the red $Mn(CO)_4NO$ and the green $Mn(CO)(NO)_3$.

As well as in carbonyl halides, Mn^I occurs in the $[Mn(CN)_6]^{5-}$ ion (obtained by Al/NaOH reduction of $[Mn(CN)_6]^{4-}$) and in the $[Mn(CNR)_6]^+$ cations that arise in the reactions of MnI_2 with RNC.

14.7 Organometallic Compounds of Manganese

The pure σ-bonded alkyls and aryls of manganese are unstable in air; dimethyl-manganese is a polymeric powder which catches fire on exposure to the atmosphere. In contrast, alkyls containing π-acceptor groups are considerably more stable; the diamagnetic $CH_3Mn(CO)_5$, for example, is stable to air and water.

The unusual stability of the d^5 Mn^{2+} state results in manganese forming an ionic bis(cyclopentadienide) $Mn^{2+}(C_5H_5^-)_2$ rather than the covalent sandwich structured type shown by the other M^{2+} ions in the first transition series. Thus $Mn(C_5H_5)_2$ reacts with water to give cyclopentadiene and MnO (π-bonded cyclopentadienyls do not react with water). In the presence of other π-bonding ligands, however, manganese does form π-cyclopentadienyl complexes, for example $(\pi\text{-}C_5H_5)Mn(CO)_3$ and $(\pi\text{-}C_6H_6)Mn(\pi\text{-}C_5H_5)$.

15 Iron

15.1 The Element

Iron is the fourth most abundant element in the earth's crust; in the elemental form it occurs only rarely but in the combined state it is universally common. Iron has a very ancient history; it was probably first used by man nearly 6000 years ago (iron from meteorites), iron production beginning around 1200 B.C. Iron is the first member of the group VIII triad, lying above ruthenium and osmium. Unlike all the elements we have so far considered, iron does not show the maximum oxidation state corresponding to the removal of all (eight) of its valence electrons. The maximum oxidation state shown is +6 but states above +3 are all relatively unimportant; notice that with iron it is the +3 state that has the completely half-filled $(3d^5)$ shell. As with manganese, however, the predominance of the +2 oxidation state continues to assert itself, and in the elements to follow, that is Co, Ni, and Cu, compounds in this state assume even greater preponderance.

Pure iron finds no great industrial use; it can be prepared by thermal decomposition of iron pentacarbonyl, by hydrogen reduction of pure iron oxides, or by aqueous electrolysis of pure iron salts. Two structural types of iron occur in the solid state. The α-form, stable at room temperature, has a body-centred cubic lattice. At about 910° the α-form is transformed into the γ-form which has a cubic close-packed structure. Iron is ferromagnetic up to 768° (the Curie point); thereafter it is paramagnetic. The magnetic and other physical properties of iron are dependent on the purity of the iron and the nature of its impurities. A very large number of steels and other iron alloys are produced for specific purposes.

Iron combines with most nonmetals on heating. The finely divided metal is pyrophoric in air at room temperature, the massive metal forming Fe_2O_3 and Fe_3O_4 in air above 150°. Steam reacts above 500°, forming Fe_3O_4 and FeO with liberation of hydrogen. Iron dissolves in dilute mineral acids to form iron(II) solutions; the dissolution of impure iron in dilute sulphuric acid is accompanied by the evolution of a gas with a characteristic odour. The hydrogen evolved contains traces of the hydrides of other elements (for example C and S) present as impurities in the iron. Concentrated nitric acid has (after a momentary reaction) a passivating effect on the iron owing to the formation of an oxide film layer on the surface of the metal. Even passivated iron dissolves in reducing acids such as dilute hydrochloric acid.

Rusting. The most economically important reaction of iron is the formation of hydrated oxide in the presence of oxygen and water, that is rusting. For this corrosion to occur at room temperature, oxygen, water, and an electrolyte all appear to be essential. The mechanism of rusting is electrochemical, the rate being governed by processes occurring at the water—iron interface. At this interface oxygen is reduced by a stepwise cathodic reaction that can be summarised by the equation

$$O_2 + 2H_2O + 4e \rightarrow 4OH^-$$

Iron enters solution as Fe^{2+} in an anodic reaction that provides the four electrons
required above

$$2Fe \rightarrow 2Fe^{2+} + 4e$$

Thus Fe^{2+} and OH^- ions are present in solution, and in the presence of air the
yellow-brown hydrated iron(III) oxide (rust) precipitates.

   The Mössbauer effect. Iron compounds are particularly suitable for study by
Mössbauer spectroscopy. The nuclide ^{57}Fe, daughter of ^{57}Co, has an excited
state ($t_{1/2} = 1.0 \times 10^{-7}$ s) at 14.4 keV above the ground state. If an iron compound
is irradiated with γ-rays from a ^{57}Co source, resonant absorption of γ-rays will
occur if the iron nuclei in the compound are in an environment identical to that of
the nuclei in the source. No absorption will occur if the source and compound
nuclei are in different environments. The resonant absorption of γ-rays can be
made to occur by giving the absorbing compound a velocity relative to that of the
source. This velocity changes the energy of the incident quanta until, at a
particular velocity, the energy corresponds to that required for resonant absorption. The positions of the absorption peaks are thus usually expressed in
velocities (m s^{-1}), and shifts are related to some standard, for example sodium
pentacyanonitrosylferrate(II) or stainless steel, arbitrarily taken as zero. These
isomer shifts are directly proportional to the total s electron density at the nucleus.
Thus iron(II) compounds give larger shifts than iron(III) compounds, since the 4s
electrons in Fe^{2+} are more strongly screened by the extra 3d electron. This is
particularly true of high-spin complexes, for which isomer shifts are of the order
+1.0 to +1.8 mm s^{-1} for Fe^{2+} and +0.4 to +0.9 mm s^{-1} for Fe^{3+} ($Na_2[Fe(CN)_5(NO)]$
standard). Further information can be obtained when a splitting of the
absorption peak occurs; splittings occur when there is an asymmetric electronic
charge distribution, for example in high-spin $Fe^{2+}(d^6)$ but not in Fe^{3+} (d^5), or
when magnetic dipole interactions occur as in ferro- and antiferro-magnetic
compounds.

15.2 Compounds of Iron in High Oxidation States

Compounds of iron in +4, +5, and +6 oxidation states are known with iron in
combination with oxygen but, rather surprisingly, not with fluorine.

   The alkali metal ferrates(VI) K_2FeO_4 and Na_2FeO_4 are obtained by oxidation
of a suspension of hydrous iron(III) oxide in concentrated alkali with hypo-
chlorite

$$2Fe(OH)_3 + 3ClO^- + 4OH^- \rightarrow 2FeO_4^{2-} + 3Cl^- + 5H_2O$$

The deep-red FeO_4^{2-} ions are most stable in alkali; in neutral or acid solution,
decomposition to iron(III) is rapid

$$2FeO_4^{2-} + 10H^+ \rightarrow 2Fe^{3+} + 5H_2O + {}^3/_2O_2$$

Ferrates(VI) are very strong oxidising agents; ammonia is oxidised to nitrogen
at room temperature. They contain the discrete tetrahedral FeO_4^{2-} ion, the
potassium salt being isomorphous with K_2SO_4 and K_2CrO_4. The magnetic

moments of ferrates(VI) are in the range 2.8–3.1 B.M., in accord with magnetically dilute d^2 systems.

Potassium ferrate(V) K_3FeO_4 can be prepared by thermal decomposition of K_2FeO_4 at 700°; above this temperature it decomposes into $KFeO_2$, potassium oxide, and oxygen.

Ferrates(IV) of three types are known, that is FeO_3^{2-}, FeO_4^{4-}, and FeO_5^{6-}. The strontium and barium salts of the FeO_4^{4-} ion are prepared as fine black crystals by oxidation of a mixture of hydrous iron(III) oxide and the alkaline earth metal oxide or hydroxide with oxygen at 700–800°

$$Sr_3[Fe(OH)_6]_2 + Sr(OH)_2 + \tfrac{1}{2}O_2 \rightarrow 2Sr_2FeO_4 + 7H_2O$$

The sodium salt is obtained similarly; it undergoes immediate disproportionation in dilute alkali

$$3Na_4FeO_4 + 8H_2O \rightarrow Na_2FeO_4 + 2Fe(OH)_3 + 10NaOH$$

These ferrates(IV) do not contain discrete FeO_4^{4-} ions and are best regarded as mixed oxides. A compound FeO_2 ($\mu_{eff} = 4.91$ B.M.), which may be iron(IV) oxide, has been obtained by thermal decomposition of the reaction product obtained in the reaction of iron pentacarbonyl with dinitrogen tetroxide ($[FeNO_3]O$).

The only non-oxo compounds of iron in high oxidation states are the diarsine complexes $[Fe(diars)_2X_2]^{2+}$ (X = Cl or Br). These arise in the oxidation of the iron(III) complexes $[Fe(diars)_2X_2]^+$ with concentrated nitric acid. The magnetic moments of the iron(IV) compounds are in agreement with a low-spin d^4 configuration; Mössbauer studies support a tetragonal (D_{4h}) structure with the low-spin ground state ($d_{xy}^2, d_{xz}^1, d_{yz}^1$).

15.3 Compounds of Iron(III) (d^5)

In this oxidation state, iron forms mainly octahedral complexes. It resembles the d^5 Mn^{II} state in forming predominantly high-spin complexes; these have magnetic moments close to the spin-only value of 5.9 B.M. The few low-spin complexes, that is with CN^-, bipy, and phen as ligands, have moments in excess of the spin-only value (1.73 B.M.) because of orbital contribution arising from the $t_{2g}^5 e_g^0$ ground state. Iron(III) tends to be stabilised [relative to iron(II)] by anionic ligands, having its greatest affinity for oxygen donors such as phosphates, tartrate, citrate, oxalate, and EDTA. These complexes are usually pale in colour as a result of the $^6A_{1g}$ ground state and the occurrence of spin-forbidden bands only in the visible.

15.3.1 Aqueous Chemistry

The pale violet hexaquo-ion occurs in the solid state in, for example, iron(III) alums as well as in acidic solutions of iron(III) containing anions of low co-ordinating ability. The acidity of the $[Fe(H_2O)_6]^{3+}$ ion has been considered in detail in section 3.1.2. The very pale hydrates $Fe(ClO_4)_3.10H_2O$, $Fe(NO_3)_3.6H_2O$, and $Fe_2(SO_4)_3.9H_2O$ (as well as others) can be crystallised from aqueous solution. These salts all undergo hydrolysis in water with the formation of a brown colour;

this colour disappears if the solution is acidified with the appropriate noncomplexing acid, the hexaquo-ion being re-formed.

In the presence of co-ordinating anions, substitution at the hexaquo-ion occurs. Thus, in hydrochloric acid various chloro-complexes are formed ranging from $[FeCl(H_2O)_5]^{2+}$ to $FeCl_4{}^{2-}$. With thiocyanate the complex ion $[Fe(SCN)(H_2O)_5]^{2+}$ is produced; its well-known blood-red colour is utilised in the qualitative detection of iron(III); however, if hydrous iron(III) oxide is dissolved in aqueous thiocyanic acid containing an alkali metal cation, salts of the $[Fe(SCN)_6]^{3-}$ ion can be isolated. Cyanide ion gives the species $[Fe(CN)_5(H_2O)]^{2-}$ and $[Fe(CN)_6]^{3-}$ Salts of the latter ion, the hexacyanoferrates(III) or ferricyanides, are most usually prepared by oxidation of the $[Fe(CN)_6]^{4-}$ salts. Potassium hexacyanoferrate(III) forms red poisonous crystals having $\mu_{eff} = 2.25$ B.M. at room temperature. In hot dilute sulphuric acid, hydrogen cyanide is evolved; the concentrated acid liberates carbon monoxide.

Very stable anionic complexes are formed by oxo-anions. Thus iron(III) chloride solutions (coloured yellow because of charge-transfer absorptions) are decolorised by the addition of phosphoric acid; the stability and colourless nature of the iron(III) phosphate complexes is used in the Zimmermann–Reinhardt method for estimating iron(II) with permanganate in the presence of chloride ions. The exact nature of the complexes is a matter of some dispute; in equimolar mixtures of $FeCl_3$ and H_3PO_4 the cationic species $[Fe(HPO_4).aq]^+$ is important, whereas solutions of iron(III) phosphate in phosphoric acid contain the anions $[Fe(PO_4)_3]^{6-}$ and $[Fe(HPO_4)_3]^{3-}$. The green tris(oxalato)-complexes containing the $[Fe(C_2O_4)_3]^{3-}$ ion are readily prepared by the addition of an excess of an alkali metal oxalate to an iron(III) salt solution. The ion can be resolved into its optical isomers but these racemise in less than one hour at room temperature. Citrate and tartrate complexes are similar; their formation is often used to prevent the precipitation of iron(III) in alkaline solution. EDTA (H_4Y) gives the seven-co-ordinate complex $[FeY(H_2O)]^-$ which in the rubidium salt has the roughly pentagonal bipyramidal arrangement of the donor atoms about the metal.

Stable neutral iron(III) complexes are formed by chelating oxygen and sulphur ligands. The red acetylacetonate $Fe(acac)_3$ is prepared by the aqueous reaction of iron(III) chloride with acetylacetone in the presence of a base (for example sodium acetate). It is high-spin, a nonelectrolyte in water, and has a fairly regular octahedral arrangement of oxygen atoms about iron. The dialkyl-dithiocarbamates $Fe(S_2CNR_2)_3$ are similarly prepared from aqueous iron(III) chloride and the sodium salt of the dialkyldithiocarbamate. They have interesting magnetic properties (see section 7.3.1), with a thermal equilibrium between high and low-spin states. When treated with concentrated hydrohalic acids the black compounds $FeX(S_2CNR_2)_2$ are formed. These constitute examples of five-co-ordinate iron(III); they have the square pyramidal structure (figure 15.1) with the iron atom 63 pm above the plane of the four sulphur atoms. They have magnetic moments around 3.9–4.0 B.M. corresponding to the three unpaired electrons (and no orbital contribution) expected for this stereochemistry.

Cationic complexes containing unidentate nitrogen donors are unstable in aqueous solution. The hexammines $[Fe(NH_3)_6]X_3$ are formed in anhydrous ammonia but are decomposed in water, the liberated ammonia enhancing the hydrolysis of

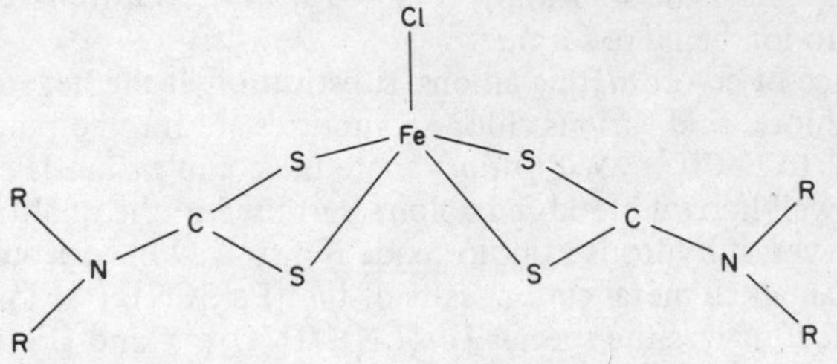

Figure 15.1   Chlorobis-(N,N-dialkyldithiocarbamato)iron(III)

the aquo-ion, so $FeO(OH)$. aq precipitates. Chelating ligands form water-stable complexes such as $[Fe(bipy)_3]^{3+}$ and $[Fe(phen)_3]^{3+}$; these strong-field ligands form low-spin complexes that are fairly inert to ligand substitution reactions. Oxygen donors form some stable cationic species, for example $Fe[OC(NH_2)_2]_6^{2+}$ and $[Fe_3O(OCOR)_6(H_2O)_3]^+$.

15.3.2   Simple Compounds

Iron(III) oxides and hydroxides. Treatment of aqueous iron(III) solutions with alkali leads to the precipitation of the red-brown hydrous iron(III) oxide. The hydroxide $Fe(OH)_3$ is not known, and this precipitate is believed to be best formulated as $FeO(OH)$. On ignition above $200°$ the red-brown α-Fe_2O_3 is formed. This oxide occurs in nature as haematite; it is used in red pigments and as a polishing agent in rouge. It is paramagnetic and has the corundum (Al_2O_3) structure. A second form of the oxide, γ-Fe_2O_3, is obtained by careful oxidation of Fe_3O_4 or by heating lepidocrocite γ-$FeO(OH)$; it is ferromagnetic and has a spinel-like structure with the iron atoms distributed randomly over the tetrahedral and octahedral sites.

The black Fe_3O_4 is a mixed iron(II, III) oxide occurring naturally as magnetite. It is strongly ferromagnetic and is oxidised to Fe_2O_3 on heating in air.

Iron(III) halides. Only FeF_3, $FeCl_3$, and $FeBr_3$ are known in the solid state; the iodine—iron reaction produces merely FeI_2. Even in aqueous solution, iron(III) oxidises iodide ions quantitatively to iodine. The brown-black $FeBr_3$ is difficult to obtain pure because it decomposes into $FeBr_2$ and bromine at temperatures only slightly in excess of those required for its preparation ($200°$). Aqueous solutions of $FeBr_3$ decompose on boiling, with evolution of bromine.

The black crystalline $FeCl_3$ has a semicovalent layer structure in which each iron atom is octahedrally surrounded by chlorines. In the vapour phase at $400°$, Fe_2Cl_6 dimers occur, with tetrahedral co-ordination around iron. When heated in a vacuum to above $500°$, chlorine is evolved and $FeCl_2$ formed. The reaction with Fe_2O_3 at $350°$ in a sealed tube produces the little-studied oxide chloride $FeOCl$. Iron(III) chloride fumes with hydrolysis in moist air and is very soluble in water. This solution is strongly acidic; crystallisation of the solution gives the yellow-brown 'hexahydrate' which has the structure *trans*-$[FeCl_2(H_2O)_4]Cl.2H_2O$. Iron(III) chloride reacts with a wide range of organic ligands to produce what appear to be addition compounds. Such compounds are formed, for example, by alcohols, ethers, aldehydes, ketones, and amines. The structures of many of these compounds are unknown; some which appear to be adducts have been shown to contain ionic species, for example $FeCl_3$ (DMSO)$_2$ and $Fe(NO_3)_3.N_2O_4$ in figure 15.2. With

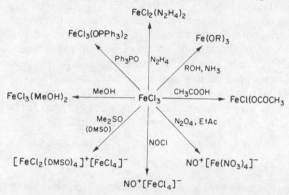

Figure 15.2   Some reactions of $FeCl_3$

phenols, iron(III) chloride usually forms charge-transfer complexes but in some instances the phenol is oxidised to a coloured derivative; for example, β-naphthol gives β-dinaphthol. As well as adduct formation, iron(III) chloride undergoes redox and substitution reactions with ligands, examples of which are to be found in figure 15.2.

Complex halides of iron(III) resemble the simple halides in their thermal stability. Thus no complex iodides are known, and the complex bromides that have been reported are thermally unstable with respect to evolution of bromine and formation of iron(II). The fluoride and chloride complexes are of three principal types: $M^I_3FeX_6$, $M^I_2FeX_5(H_2O)$, and M^IFeX_4. While it is doubtful if $FeCl_6{}^{3-}$ exists in solution, the pale yellow hexachloroferrates can be isolated from solution by using large cations having the same charge as the anion, for example $[Co(NH_3)_6]^{3+}$. The yellow tetrachloroferrates(III) are more readily obtained with a variety of cations and from a variety of solvents. The $FeCl_4{}^-$ ion is extractable from hydrochloric acid into ether, and this is of use in separations by solvent extraction. The crystal structure determination on $Ph_4As[FeCl_4]$ has confirmed that the $FeCl_4{}^-$ ion is roughly tetrahedral.

15.4 Compounds of Iron(II) (d^6)

The +2 state is in general a reducing state for iron; most iron(II) salts oxidise slowly in air although double salts (with alkali metal or ammonium salts) are more stable. The hydrated salts are pale green, these salts being known with all the common anions. They are often isomorphous with the same salts of other transition metals in the +2 oxidation state. The solubilities of iron(II) salts also resemble those of other M^{2+} ions in the first-row transition elements, that is the halides, nitrate, sulphate, and perchlorate are soluble, while the hydroxide, carbonate, phosphate, and oxalate are relatively insoluble.

15.4.1   Aqueous Chemistry

In the absence of other ligands, solutions of iron(II) in water contain the pale green $[Fe(H_2O)_6]^{2+}$ ion. This ion is present in many salt hydrates in the solid state, for example $FeSO_4.7H_2O$ and $(NH_4)_2SO_4.FeSO_4.6H_2O$. There is considerable distortion from octahedral symmetry for this ion; in iron(II) ammonium sulphate hexahydrate, for example, the tetragonal and slight rhombic distortions give $Fe-OH_2$ distances of 214, 188, and 185 pm. The aquo-ion is barely acidic, so $FeCO_3$ and $Fe(OH)_2$ can be prepared as discrete species from aqueous solution. The hydroxide is somewhat soluble in strong caustic soda solution, with the probable formation of $[Fe(OH)_4]^{2-}$ ions.

The dissolution of iron in nonoxidising acids gives $Fe^{2+}.aq$. This Fe^{2+} state is thermodynamically unstable to atmospheric oxidation

$$Fe^{3+} + e \rightarrow Fe^{2+} \quad (E^\circ = +0.771 \text{ V})$$

$$O_2 + 4H^+ + 4e \rightarrow 2H_2O \quad (E^\circ = +1.229 \text{ V})$$

This oxidation occurs only slowly in acid solution. Strong oxidising agents such as H_2O_2 readily oxidise Fe^{2+} in acid solution, the reactions with permanganate and dichromate ions being used in the volumetric estimation of iron. In alkaline solution, $Fe(OH)_2$ precipitates and the Fe^{II}/Fe^{III} potential changes dramatically

$$Fe(OH)_3 + e \rightarrow Fe(OH)_2 + OH^- \quad (E^\circ = -0.877 \text{ V})$$

The reducing power of iron(II) is thus enormously increased; one of the contributing factors to this greater ease of oxidation is the lower solubility of '$Fe(OH)_3$' compared with that of $Fe(OH)_2$. In alkaline solution therefore oxidation by air is rapid; precipitates of $Fe(OH)_2$ (white when pure) darken rapidly as they oxidise on standing in air.

As well as changes in pH, changes in the ligands have a marked effect on the redox potentials of the Fe^{II}/Fe^{III} system. The Fe^{2+} ion (d^6) is a good π-donor, and it might be expectedly stabilised in combination with π-acceptor ligands such as 2,2'-bipyridyl and 1,10-phenanthroline which have low-lying vacant π^* orbitals. The Fe^{3+} ion (d^5) is a poorer π-donor, largely because of its higher charge. The potentials for these low-spin systems are

$$[Fe(bipy)_3]^{3+} + e \rightarrow [Fe(bipy)_3]^{2+} \quad (E^\circ = +0.96 \text{ V})$$

$$[Fe(phen)_3]^{3+} + e \rightarrow [Fe(phen)_3]^{2+} \quad (E^\circ = +1.12 \text{ V})$$

The tris-(1,10-phenanthroline)iron(II) sulphate is widely used in volumetric analysis as the indicator *ferroin;* with strong oxidising agents the colour change is quite striking

$$[Fe(phen)_3]^{2+} \rightarrow [Fe(phen)_3]^{3+} + e$$

deep red   pale blue

15.4.2   Some Simple Compounds

Iron(II) *halides.* The pale iron(II) halides (FeF_2 white, $FeCl_2$ and $FeBr_2$ pale yellow, FeI_2 grey) do not melt until red heat and are deliquescent in moist air. They crystallise from aqueous solution as hydrates. The tetrahydrate $FeCl_2(H_2O)_4$ which separates from solution above 12.3°, has discrete $FeCl_2(H_2O)_4$ units, and

$FeCl_2.6H_2O$, which crystallises below $12.3°$, contains *trans*-$[FeCl_2(H_2O)_4]^+$ units. Nitrogen, phosphorous, and arsenic ligands react with iron(II) halides to give complexes of three general types; FeX_2L_2 (often tetrahedral), FeX_2L_4 (tetragonal), and $FeL_6{}^{2+}$ (octahedral). Some of these, as well as examples of those formed by other ligands, are illustrated in figure 15.3.

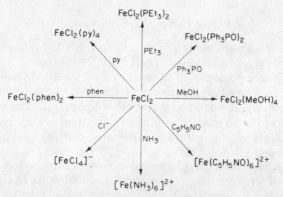

Figure 15.3   Some reactions of $FeCl_2$

Iron(II) *sulphides.* The brassy-yellow FeS_2 occurs in two forms, *pyrites* and *marcasite.* Pyrites can be regarded as having a rock-salt like arrangement of Fe^{2+} and $S_2{}^{2-}$ ions. Both sulphides are diamagnetic and thus contain spin-paired Fe^{2+} $(t_{2g}{}^6)$. The grey FeS can be prepared directly from the exothermic reaction between the elements, or by reaction of aqueous solutions of iron(II) salts with alkali metal sulphides. It is antiferromagnetic; it finds use in the laboratory preparation of crude hydrogen sulphide.

Iron(II) *sulphate.* The heptahydrate $FeSO_4.7H_2O$ forms green crystals, which are very soluble in water. The white monohydrate and anhydrous $FeSO_4$ can be obtained by thermal dehydration. At red heat decomposition occurs

$$2FeSO_4 \rightarrow Fe_2O_3 + SO_2 + SO_3$$

The double salts (Tutton salts) $M^I{}_2SO_4.FeSO_4.6H_2O$ can be crystallised from solutions containing the alkali metal sulphates. The ammonium salt, commonly known as ferrous ammonium sulphate or Mohr's salt, finds extensive use in volumetric analysis where its greater stability to aerial oxidation both in the solid state and in acid solution is advantageous over the simple salt $FeSO_4.7H_2O$.

15.4.3 Complexes

Cationic complexes. These arise in the substitution reactions of the aquo-ion and in the disproportionation reactions of iron carbonyls in inert solvents

$$[Fe(H_2O)_6]^{2+} + 3en \rightarrow [Fe(en)_3]^{2+} + 6H_2O$$

$$5Fe(CO)_5 + 6py \rightarrow [Fe(py)_6][Fe_4(CO)_{13}]$$

Most of these cations are octahedral but the bidentate ligand 1,8-naphthyridine (figure 15.4) forms the eight-co-ordinate complex $[Fe(LL)_4](ClO_4)_2$. Most of the

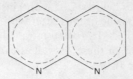

Figure 15.4   1,8-Naphthyridine

octahedral cations are high-spin ($t_{2g}^4 e_g^2$); the strong-field ligands bipy, phen, and diars give low-spin (t_{2g}^6) diamagnetic $[Fe(LL)_3]^{2+}$ complexes.

Neutral complexes. Examples of these formed by the iron(II) halides have already been given. Several crystalline complex hydrides exist, for example the octahedral *trans*-FeH$_2$(diphos)$_2$ and *trans*-FeHCl(diphos)$_2$ [diphos = o-C$_6$H$_4$(PEt$_2$)$_2$]. These are prepared by the reaction of lithium aluminium hydride on the bis(phosphine)iron(II) halides in tetrahydrofuran. They are sensitive to oxidation but are thermally stable up to their melting points (>230°).

β-Diketones readily form complexes with iron(II) in aqueous solution, especially in the presence of tertiary nitrogen bases (which neutralise the liberated protons). Acetylacetone forms Fe(acac)$_2$(py)$_2$ and Fe(acac)$_2$.1.5H$_2$O, which are believed to contain high-spin six-co-ordinate iron(II) ($\mu_{eff} \approx 5.4$ B.M.), while dipivaloylmethane (DPMH) forms tetrahedral Fe(DPM)$_2$, which is high-spin with $\mu_{eff} = 5.0$ B.M.

Perhaps the most important iron complex is haem. Haemoglobin containing the iron in blood consists of haem bound to a protein called globin. Haem (figure 15.5)

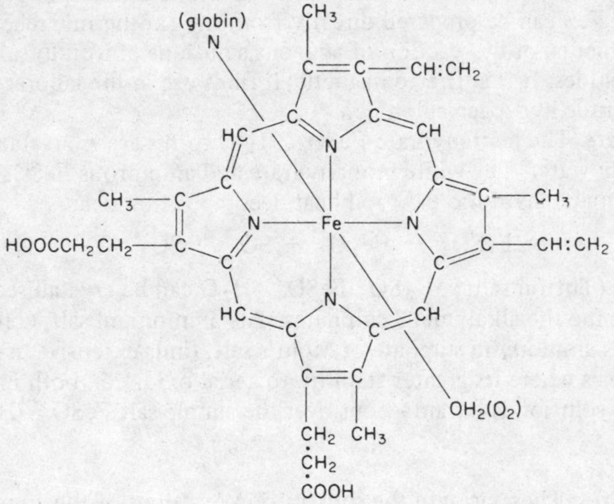

Figure 15.5   Haem

is an iron(II) complex of porphyrin in which the six-co-ordinate iron atom is bonded to four nitrogen atoms from pyrrole rings in a plane, and to a nitrogen atom in globin perpendicular to this plane. The sixth position is occupied by a water molecule. It is at this position that reversible uptake of oxygen can occur, and this enables the red blood cells to transport oxygen from one part of the body to another. Haem has an even greater affinity for carbon monoxide than for oxygen, and it is the irreversibility of the formation of carboxyhaemoglobin that makes carbon monoxide poisonous.

Anionic complexes. Salts containing the tetrahedral FeX_4^{2-} ions can readily be isolated from alcoholic mixtures of iron(II) halides and quaternary ammonium halides. The most important anionic complexes, however, are those with cyanide ions. The octahedral hexacyanoferrate(II) ion is readily formed by the addition of an excess of cyanide ions to an aqueous solution of an iron(II) salt. It is a stable nonpoisonous diamagnetic anion which forms salts with many cations. The alkali metal salts crystallise as hydrates, for example $K_4Fe(CN)_6.3H_2O$. The free acid $H_4Fe(CN)_6$ can be obtained as an etherate by treatment of the potassium salt with a strong acid such as HCl in ether; the ether can be removed in a stream of hydrogen at 80° to leave the acid as white crystals. Many substitution products of the $[Fe(CN)_6]^{4-}$ ion are known, for example $[Fe(CN)_5(H_2O)]^{3-}$ and $[Fe(CN)_5NO]^{2-}$. However, the best-known reaction of the ion is that with an iron(III) salt.

The reaction between $[Fe(CN)_6]^{4-}$ and an excess of Fe^{3+}.aq produces a blue precipitate called *Prussian blue*. Similarly, the reaction between $[Fe(CN)_6]^{3-}$ and Fe^{2+}.aq produces a blue precipitate known as *Turnbull's blue*. These compounds give the same Mössbauer spectrum and X-ray powder diffraction pattern, and are formulated as $Fe_4[Fe(CN)_6]_3$, that is iron(III) hexacyanoferrate(II). *Soluble Prussian blue* $KFe^{III}[Fe^{II}(CN)_6]$ is formed in the reaction between 1:1 molar proportions of Fe^{3+} and $[Fe(CN)_6]^{4-}$. Its electronic and Mössbauer spectra indicate that it contains high-spin iron(III) octahedrally surrounded by six nitrogen atoms and low-spin iron(II) having six carbon atoms as nearest neighbours. The intense colour of the compound is due to intravalence electron transfer (section 6.1.3).

15.5 Compounds of Iron in Low Oxidation States

Iron has a fairly extensive chemistry with π-acceptor ligands such as CO, NO, phosphines, bipyridyl, and bidentate sulphur ligands. We shall concern ourselves here with the carbonyl and nitrosyl complexes; these are the most extensive, and the other complexes in low oxidation states are similar to those discussed previously for other metals.

15.5.1 Iron Carbonyls

Iron forms three well-known carbonyls: $Fe(CO)_5$, $Fe_2(CO)_9$, and $Fe_3(CO)_{12}$. The pentacarbonyl is a yellow liquid insoluble in water but soluble in organic solvents such as benzene. It thus behaves as a covalent monomer and has the trigonal bipyramidal structure. Some of its very extensive chemistry is illustrated in figure 3.2. When solutions of the pentacarbonyl are exposed to sunlight or ultraviolet radiation, dark yellow platelets of the enneacarbonyl $Fe_2(CO)_9$ are formed. The dodecacarbonyl $Fe_3(CO)_{12}$ is obtained from the pentacarbonyl by treatment with aqueous triethylamine to form $[Et_3NH][Fe_3(CO)_{11}H]$ which is then acidified and the carbonyl extracted into petroleum ether. $Fe_3(CO)_{12}$ forms green diamagnetic crystals that decompose at 140°.

In $Fe_2(CO)_9$ (figure 15.6) the iron atoms are surrounded in an approximately

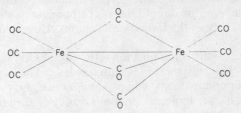

Figure 15.6   The structure of $Fe_2(CO)_9$

octahedral fashion by carbon monoxide molecules. Since the compound is diamagnetic it must contain a metal–metal bond. The Mössbauer spectrum of $Fe_3(CO)_{12}$ shows the presence of two types of iron atom; two of these atoms show quadrupole splitting and have isomer shifts similar to those in $Fe_2(CO)_9$ while the third iron atom shows little splitting, as would be expected for an octahedral environment. Such a structure (figure 15.7) is confirmed by the X-ray data. The

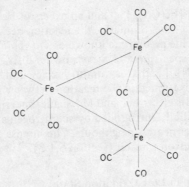

Figure 15.7   The structure of $Fe_3(CO)_{12}$

anion $[Fe_3(CO)_{11}H]^-$ has a similar structure in which one of the bridging carbonyl groups is replaced by a bridging hydrogen atom.

Carbonylate anions containing iron(–I) and iron(–II) are obtained in the reactions of bases with these carbonyls. For example, the reaction of aqueous alkali with $Fe(CO)_5$ gives yellow solutions containing the $[Fe(CO)_3H]^-$ ion; in the presence of air or oxidising agents, oxidation to the anions $[Fe_2(CO)_8]^{2-}$ and $[Fe_3(CO)_{11}]^{2-}$ occurs. The $[Fe(CO)_4]^{2-}$ ion is also formed in these reactions, as well as in disproportionation reactions of $Fe(CO)_5$ with ligands, for example

$$2Fe(CO)_5 + 3en \xrightarrow{145°} [Fe^{II}(en)_3][Fe^{-II}(CO)_4] + 6CO$$

As well as the simple carbonyls, there are a variety of carbonyl halides and hydrides. The $Fe(CO)_4X_2$ (X = Cl, Br, I) compounds are probably octahedral; they are hydrolysed by water with evolution of carbon monoxide. The unstable gaseous hydride $H_2Fe(CO)_4$ is evolved when solutions of $[HFe(CO)_4]^-$ are acidified.

Similarly, polynuclear hydrides such as $H_2 Fe(CO)_8$ and $H_2 Fe_3(CO)_{11}$ are obtained by acidification of solutions of the polynuclear carbonylate anions.

15.5.2   Nitrogen Monoxide Complexes of Iron

Iron tetranitrosyl $Fe(NO)_4$ is obtained as black crystals when the pentacarbonyl is heated to 45° with nitrogen monoxide in an autoclave. It is an involatile and very reactive substance. Its infrared spectrum shows N–O stretching frequencies at 1810, 1730, and 1140 cm^{-1}, that is in both the NO^+ and NO^- regions, so the structure $Fe(NO^+)_3 (NO^-)$ has been suggested for the tetranitrosyl. Nitrosyl halides, for example $Fe(NO)_3 Cl$, and nitrosyl carbonyls, for example $Fe(NO)_2 (CO)_2$, are also known.

Perhaps the best-known iron–nitrosyl compounds are the 'brown-ring compounds' observed in the qualitative test for the nitrate ion. Iron(II) sulphate solution absorbs nitrogen monoxide reversibly up to Fe : NO = 1 : 1. The magnetic susceptibility of these solutions indicates that the complex has three unpaired electrons, while the infrared spectra show absorptions in the 1730–1850 cm^{-1} region, that is characteristic of NO^+. The complex cation present is thus best formulated as $[Fe^I(NO) (H_2 O)_5]^{2+}$ in which NO donates three electrons.

The nitrosylpentacyanoferrate(II) or nitroprusside ion $[Fe(NO)(CN)_5]^{2-}$ is also a well-known species. It arises in the reactions of nitric acid with $[Fe(CN)_6]^{3-}$ or $[Fe(CN)_4]^{4-}$ ions, for example

$$[Fe(CN)_6]^{4-} + 4H^+ + NO_3^- \rightarrow [Fe(NO) (CN)_5]^{2-} + CO_2 + NH_4^+$$

The red sodium salt $Na_2 [Fe(NO) (CN)_5].2H_2 O$ is diamagnetic, with the N–O stretching frequency at 1944 cm^{-1}. It is therefore formulated as containing Fe^{II} and NO^+.

15.6 Organometallic Compounds of Iron

Iron has played a very important role in the development of organometallic chemistry. Following the pioneering work of Reppe in 1949, the reaction of iron pentacarbonyl with acetylenes is now known to lead to an enormous variety of organic and organometallic compounds. The field of transition-metal organometallic compounds containing π-bonded aromatic rings began, however, in 1951 when two different groups of workers somewhat accidentally discovered ferrocene, bis-(π-cyclopentadienyl)iron $Fe(\pi\text{-}C_5 H_5)_2$.

Ferrocene is an orange crystalline air-stable solid; it is insoluble in water but soluble in organic solvents. It is readily prepared by the reaction of sodium cyclopentadienide with iron(II) chloride in tetrahydrofuran

$$2NaC_5 H_5 + FeCl_2 \rightarrow Fe(C_5 H_5)_2 + 2NaCl$$

X-ray diffraction studies have established that in ferrocene the iron atom is sandwiched between two cyclopentadienyl rings, these rings being staggered relative to each other (figure 15.8). The bonding between the $C_5 H_5^-$ rings and the Fe^{2+} ion consists principally of a π bond formed by overlap of the d_{xz} and d_{yz} orbitals on the metal with the π molecular orbitals on the rings that have the correct symmetry for overlap.

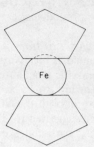

Figure 15.8   Ferrocene

The reactions of ferrocene are largely those of the aromatic rings; a very extensive organic chemistry of ferrocene is now known. Thus it undergoes Friedel—Crafts acylation and alkylation reactions but does not undergo the Diels—Alder reaction characteristically shown by conjugated dienes. Nitration and halogenation of the rings cannot be carried out because of oxidation to the ferricenium ion $Fe(C_5H_5)_2^+$.

Iron also forms arene complexes, the bis(benzene)iron(II) cation $[Fe(C_6H_6)_2]^{2+}$ being isoelectronic with ferrocene. Most interesting is the formation of cyclobutadiene complexes when the ligand itself is unstable. Much organic chemistry has been carried out on cyclobutadiene; it is stabilised in bonding, for example, to an $Fe(CO)_3$ group. Butadieneiron tricarbonyl is a pale yellow crystalline solid prepared by the reaction of cis- or trans-3,4-dichlorocyclobutene with $Fe_2(CO)_9$

16 Cobalt

16.1 The Element

Cobalt compounds have been used in coloured glass for at least 4000 years but the metal has been produced industrially only during this century. It is a widely distributed but relatively uncommon element in the earth's crust. It occurs biologically in vitamin B_{12}, which contains Co^{3+} bonded octahedrally to five nitrogen atoms (four from pyrroline rings and one from a benzimidazole ring) and the carbon atom of a CN^- group. The industrial extraction of the metal is usually an ancillary process to the extraction of other metals such as copper and lead.

Cobalt ($4s^2$ $3d^7$) stands between iron and nickel in group VIII and above rhodium and iridium. It shows high oxidation states even less willingly than iron, +3 being the highest oxidation state of any significance. Cobalt resembles iron and nickel in appearance, and like these metals it is ferromagnetic; it finds uses in a variety of steels designed to have specific magnetic properties. The massive metal is oxidised in air above $300°$ with the formation of Co_3O_4 and CoO. Steam forms CoO at red heat. Many nonmetals react when heated with cobalt; fluorine gives CoF_3, the other halogens giving CoX_2. Cobalt is more resistant than iron to attack by mineral acids, and it is not attacked by dilute alkalis.

16.2 Compounds of Cobalt(III) (d^6)

The important oxidation states of cobalt in aqueous solution are +3 and +2. The nature of the ligand bonded to cobalt has a dramatic effect on the stabilities of these oxidation states. Simple compounds of cobalt(III) are strong oxidising agents, and in water the $[Co(H_2O)_6]^{3+}$ ion is a very powerful oxidising agent; it oxidises water to oxygen

$$[Co(H_2O)_6]^{3+} + e \rightarrow [Co(H_2O)_6]^{2+} \qquad (E° = +1.84 \text{ V})$$

However, other ligands, for example ammonia, stabilise the cobalt(III) state to varying extents in aqueous solution

$$[Co(NH_3)_6]^{3+} + e \rightarrow [Co(NH_3)_6]^{2+} \qquad (E° = +0.11 \text{ V})$$

With some ligands, for example CN^-, the stabilisation of the +3 state is so great that the cobalt(II) complex reduces water to hydrogen

$$[Co(CN)_6]^{3-} + e \rightarrow [Co(CN)_5]^{3-} + CN^- \quad (E° = -0.8 \text{ V})$$

Further, the cobalt(II)–cyanide complex will even react with hydrogen to produce what is regarded as a cobalt(III) complex

$$2[Co(CN)_5]^{3-} + H_2 \rightarrow 2[Co(CN)_5H]^{3-}$$

There is thus no simple aqueous chemistry of cobalt(III) but there is an extensive range of complexes. For cobalt(II) the reverse is true; simple salts and aquo-complexes are stable in aqueous solution but the addition of other ligands makes for more facile conversion into the +3 state.

16.2.1 Simple Compounds

Cobalt(III) *fluoride.* CoF_3 is a light-brown powder which reacts violently with water with evolution of oxygen. It is obtained by the action of fluorine on the heated cobalt(II) halides CoF_2 or $CoCl_2$. It is a useful fluorinating agent; for example, it converts hydrocarbons into fluorocarbons

$$4CoF_3 + -CH_2- \rightarrow -CF_2- + 4CoF_2 + 2HF$$

With N_2O_5 the green crystalline nitrate $Co(NO_3)_3$ is formed.

A blue hydrate $CoF_3.3.5H_2O$ has been prepared by electrolytic oxidation of CoF_2 in 40 per cent HF. It oxidises water like the simple fluoride.

The other cobalt(III) halides are unknown, cobalt(III) being too strong an oxidising agent to exist solely with chloride, bromide, or iodide ions.

Cobalt(III) *oxide.* The oxidation of $Co(OH)_2$ in alkaline media gives a brown precipitate of uncertain structure; it is best written as $Co_2O_3.aq$. When this is dried at $150°$ the composition $Co_2O_3.H_2O$ is formed, which is probably $CoO(OH)$. On further heating to $300°$, oxygen begins to be evolved (as well as water) and black Co_3O_4 is formed. The dark-brown $CoO(OH)$ is formed when $Co(OH)_2$ is heated in air at $100°$; it has a layer lattice in which each cobalt is surrounded octahedrally by oxygen atoms.

Cobalt(III) *sulphate.* This is the most readily available cobalt(III) salt. The hydrate $Co_2(SO_4)_3.18H_2O$ is prepared by oxidation of cobalt(II) sulphate in $8N H_2SO_4$ either electrolytically or with ozone or fluorine. It is stable in the dry state but liberates oxygen from water. Some alums, for example $KCo(SO_4)_2.12H_2O$, can be crystallised from sulphuric acid solutions; they, like the hydrated sulphate, are believed to contain the $[Co(H_2O)_6]^{3+}$ ion. As well as being acidic (hydrolysis to $[Co(OH)(H_2O)_5]^{2+}$ occurs in 0.1M perchloric acid), this ion is a strong oxidising agent. Therefore no salts of organic anions such as formate, tartrate, and citrate can be obtained since these anions are oxidised to carbon dioxide and water by $[Co(H_2O)_6]^{3+}$. However, cobalt(III) acetate is believed to be present in dilute acetic acid solutions of cobalt(II) acetate that have been oxidised with ozone.

16.2.2 Complex Compounds

We discussed in chapter 1 the part played by cobalt(III) complexes in the historical development of co-ordination chemistry. The fact that cobalt(III) forms an enormous number of octahedral complexes having the inert t_{2g}^6 configuration has resulted in these being used extensively for rate and mechanism studies on octahedral substitution reactions (chapter 8).

Cobalt(III) has a great affinity for nitrogen donors especially ammonia, amines, nitro $-NO_2$, and $-NCS$ groups. Other ligands such as water molecules, halide, hydroxide, or carbonate ions may also be present in these complexes with nitrogen donors, and various stereoisomers can often be isolated from such mixed ligand systems. The preparation of these complexes usually involves the addition of the nitrogen donor to a cobalt(II) solution followed by oxidation with air or hydrogen peroxide. Perhaps the most famous of these compounds are the cobaltammines. The orange hexammines, containing the $[Co(NH_3)_6]^{3+}$ ion, are obtained from solutions of cobalt(II) salts in aqueous ammonia in the presence of added ammonium salts containing the required anion of the cobaltammine, for example

$$4[Co(H_2O)_6]Cl_2 + 4NH_4Cl + 20NH_3 + O_2 \rightarrow 4[Co(NH_3)_6]Cl_3 + 26 H_2O$$

The hexammines are formed preferentially in the presence of a charcoal catalyst and by using air as oxidant. When charcoal is absent and oxidation is carried out by hydrogen peroxide, the aquopentammine species predominates. Treatment of this with concentrated hydrochloric acid gives the red chloropentamminecobalt(III) chloride

$$2[Co(H_2O)_6]Cl_2 + 2NH_4Cl + 8NH_3 + H_2O_2 \rightarrow 2[Co(NH_3)_5(H_2O)]Cl_3 + 12H_2O$$

$$[Co(NH_3)_5(H_2O)]Cl_3 \rightarrow [Co(NH_3)_5Cl]Cl_2 + H_2O$$

Other cobaltammines can be prepared similarly, as well as by ligand exchange with these cobaltammines; thus, for example, the chloropentammine reacts with hot aqueous ethylenediamine to give $[Co(en)_3]Cl_3$.

In addition to these mononuclear ammines, the uptake of molecular oxygen by cobalt(II)—ammonia solutions leads to polynuclear cobaltammines containing the peroxo —O—O— bridge. Two types of these complexes exist. There are diamagnetic, red or brown complexes containing the unit $[Co^{III} - O_2 - Co^{III}]$ as in, for example, $[(NH_3)_5Co-O-O-Co(NH_3)_5]^{4+}$. There are also the paramagnetic, green complexes, for example $[(NH_3)_5Co(O_2)Co(NH_3)_5]^{5+}$, which is believed to contain two octahedrally co-ordinated cobalt(III) ions bridged by the superoxide (O_2^-) ion. These compounds arise in the one-electron oxidation of the diamagnetic species.

The most important neutral cobalt(III) complexes of the $[CoL_3]$ type are those with β-diketones. The green acetylacetonate $Co(acac)_3$ is insoluble in water but soluble, and monomeric, in organic solvents. When it is reduced by aluminium alkyls in the presence of triphenylphosphine and under a nitrogen atmosphere, the orange molecular nitrogen complex $CoH(N_2)(PPh_3)_3$ is produced. This has an approximately trigonal bipyramidal structure with the Co atom slightly above the plane of the three phosphorus atoms. One of the apical positions is occupied by the N—N group, the atoms Co—N—N being almost collinear; hydrogen occupies the other apical position.

The more important anionic cobalt(III) complexes are those with F^-, CN^-, and NO_2^- as ligands. The fluorides $M^I_3CoF_6$ are obtained as blue solids from hydrofluoric acid solutions of CoF_3 and alkali metal fluorides. They constitute rare examples of high-spin cobalt(III) complexes $(t_{2g}^4 e_g^2)$, having magnetic moments around 5.4 B.M.; their crystal structures contain octahedral CoF_6^{3-} units. The electronic spectra of these compounds show two bands around 14 500 and 11 800 cm^{-1} corresponding to the Jahn—Teller states of the $^5T_{2g} \rightarrow {}^5E_g$ transition.

The yellow cyanides $[Co(CN)_6]^{3-}$ are diamagnetic and low-spin. They give precipitates of hexacyanocobaltates(III) with heavy metals, and are generally inert to attack by reagents such as chlorine, hydrochloric acid, and alkalis. The well-known orange hexanitrocobaltate(III) ion, as found in 'sodium cobaltinitrite', is also low-spin and is used to precipitate potassium ions from aqueous solution as $K_3Co(NO_2)_6$. The sodium salt is prepared by blowing air through an aqueous mixture of a cobalt(II) salt and an excess of sodium nitrite in the presence of acetic acid

$$2[Co(H_2O)_6]^{2+} + 12NO_2^- + 2CH_3COOH + \tfrac{1}{2}O_2 \rightarrow 2[Co(NO_2)_6]^{3-} +$$
$$2CH_3COO^- + 13H_2O$$

It is precipitated by the addition of ethanol. The nitro groups are bound to cobalt via nitrogen and surround the cobalt octahedrally. In acid solution the $[Co(NO_2)_6]^{3-}$ ion is decomposed with the formation of cobalt(II) salts.

16.3 Compounds of Cobalt(II) (d^7)

16.3.1 Aqueous Chemistry

Like iron(II), cobalt(II) shows an extensive range of hydrated salts, which can be crystallised from aqueous solution. The pink aquo-ion $[Co(H_2O)_6]^{2+}$ occurs (together with small amounts of $[Co(H_2O)_4]^{2+}$) in solutions which are free of complexing anions, as well as in several solid salts, for example $Co(NO_3)_2.6H_2O$, $CoSO_4.7H_2O$, and $Co(ClO_4)_2.6H_2O$. The aquo-ion is barely acidic, and the carbonate can be precipitated as $CoCO_3.6H_2O$ by alkali metal carbonate solutions, so long as a pressure of carbon dioxide is maintained over the solution. In the absence of added ligands the aquo-ion is not a strong reducing agent like $[Fe(H_2O)_6]^{2+}$

16.3.2 Simple Salts

Cobalt(II) halides. The pink CoF_2 is conveniently prepared by heating the double salt $CoF_2.2NH_4F$; ammonium fluoride sublimes. The blue $CoCl_2$, green $CoBr_2$, and blue-black CoI_2 are obtained by direct combination of the elements or by dehydration of their hydrates.

Crystalline $CoCl_2$ contains octahedrally co-ordinated cobalt; in the pink hydrate $CoCl_2.6H_2O$ each cobalt is surrounded by four water molecules at the corners of a distorted square with two chloride ions making up the distorted octahedron. $CoCl_2$ is very soluble in water, giving a pink-red solution, and in ethanol giving a blue solution. The addition of ligands to these solutions gives complexes that either precipitate directly or can be crystallised; some of these are mentioned under cobalt(II) complexes.

Cobalt(II) hydroxide. The addition of ammonia or alkali metal hydroxides to cobalt(II) solutions results in the precipitation of either the pink or the blue form of $Co(OH)_2$. The pink form is the more stable and is formed when a suspension of the blue form is warmed or allowed to stand. $Co(OH)_2$ is amphoteric; it dissolves in an excess of alkali to form blue solutions of the $[Co(OH)_4]^{2-}$ ion. In slightly alkaline solution $Co(OH)_2$ suspensions are rapidly oxidised by air to brown $CoO(OH)$. An excess of aqueous ammonia converts $Co(OH)_2$ into cobalt(II) ammines such as $[Co(NH_3)_6]^{2+}$ which again are rapidly oxidised by air, to the cobaltammines.

Solutions of $Co(OH)_2$ in aqueous acids give solutions of cobalt(II) salts which can be crystallised as hydrates. These are usually red or pink, containing octahedrally co-ordinated cobalt, often as the hexaquo-ion.

16.3.3 Complexes

Complexes of cobalt(II) are very numerous. They fall into two general groups; there
are the pink or red basically octahedral complexes, and the intensely blue basically
tetrahedral complexes. Cobalt(II) forms more tetrahedral complexes than any other
transition metal ion (see the CFSE argument in section 5.5.5). In aqueous solution
the equilibrium

$$[Co(H_2O)_6]^{2+} \underset{H_2O}{\overset{Cl^-}{\rightleftharpoons}} [CoCl_4]^{2-}$$

             pale pink; octahedral   intensely blue; tetrahedral

has been of historical importance in our understanding of the curious problem of
the formation of pink and blue compounds of cobalt(II). Donnan and Bassett in
1902 correctly identified the blue species as anionic, and even considered $CoCl_4^{2-}$
as a likely species. The connection between colour and stereochemistry is of much
more recent origin. However, a word of warning is necessary. While the majority of
cobalt(II) complexes fall within the categories pink—octahedral and blue—tetrahedral,
there are many that have neither of these stereochemistries, and some for which the
colours are reversed. We saw, for example, that anhydrous $CoCl_2$ is blue yet contains octahedrally co-ordinated cobalt(II).

Octahedral complexes of cobalt(II) may be either high-spin $t_{2g}^5 e_g^2$ or
low-spin $t_{2g}^6 e_g^1$. In fact, rather high values of the ligand-field splitting
parameter Dq are required to cause spin pairing, so only a few low-spin octahedral
complexes occur (that is with the strongest ligand fields). Tetrahedral complexes
are high-spin with the $e^4 t_2^3$ configuration.

Cationic complexes. Anhydrous cobalt(II) salts react with ammonia to form
ammines, for example $CoX_2.6NH_3$ (X = Cl, Br, I, ClO_4, BF_4, etc), which contain
the octahedral $[Co(NH_3)_6]^{2+}$ ion. These ammines can be prepared in aqueous
ammonia but they are readily oxidised in this medium; the $[Co(NH_3)_6]^{2+}$ ion is
decomposed in pure water. Other octahedral cations are formed in the reactions
of cobalt(II) salts with bases, for example $[Co(N_2H_4)_6]^{2+}$, $[Co(en)_3]^{2+}$, and in the
disproportionation reactions of cobalt carbonyls, such as

$$3Co_2(CO)_8 + 12MeNH_2 \rightarrow 2[Co(MeNH_2)_6][Co(CO)_4]_2 + 8CO$$

Sometimes auto-complex formation occurs when a cobalt(II) salt is dissolved in a
donor solvent. Two examples of this are

$$2CoCl_2 + 6Me_2SO \rightarrow [Co(OSMe_2)_6]^{2+} + [CoCl_4]^{2-}$$

$$2CoI_2 + 6MeCN \rightarrow [Co(NCMe)_6]^{2+} + [CoI_4]^{2-}$$

Some tetrahedral cationic species are known, for example in the blue
$[Co(OAsPh_3)_4]I_2$ and $Co[SC(NH_2)_2]_4(ClO_4)_2$.

Neutral complexes. The reactions of cobalt(II) salts with ligand molecules give rise
to a very large number of complexes; the principal types can be illustrated by the
formulae CoX_2L, CoX_2L_2, and CoX_2L_4, where X is a univalent anion. Because of
the small stability difference between the octahedral and tetrahedral stereochemistries for cobalt(II), it frequently happens that a given salt plus ligand

combination gives complexes of both stereochemistries, and sometimes these exist together in equilibrium. Complexes of pyridine have been extensively studied and serve to illustrate these points.

Cobalt(II) chloride forms violet and blue forms of the complex $CoCl_2(py)_2$. The violet form is stable at room temperature; it contains octahedrally co-ordinated cobalt in a polymeric chlorine-bridged structure. The blue form is metastable at room temperature and has a monomeric tetrahedral structure. The reactions of pyridine with $CoBr_2$ and CoI_2 result in the tetrahedral $CoX_2(py)_2$ species only, while $Co(SCN)_2$ and $Co(SeCN)_2$ form only the octahedral $CoX_2(py)_2$ complexes. However, when dissolved in organic solvents, all these complexes give tetrahedral species; with added pyridine a tetrahedral—octahedral equilibrium occurs

$$CoX_2(py)_2 + 2py \rightleftharpoons CoX_2(py)_4$$

In a few cases complexes that do not have even approximately tetrahedral or octahedral stereochemistry arise. In the purple $Co(NO_3)_2(Me_3PO)_2$, for example, the nitrate ions are bidentate and the cobalt is six-co-ordinate in a highly irregular structure. High-spin five-co-ordinate complexes are formed by a few ligands, for example $NH(CH_2CH_2NH_2)_2$ (den) which gives CoX_2—(den) complexes with cobalt(II) halides. Square planar complexes are formed by several bidentate anions such as dimethylglyoximate and similar ions.

Anionic complexes. The solid complex halides $M^I_2CoX_4 (X = Cl, Br, I)$ are readily prepared from aqueous or (for organic base cations) alcoholic solutions of M^IX and CoX_2. The $[CoCl_4]^{2-}$ ion is tetrahedral with some angular distortion the extent of which varies with the cation present. Hydrochloric acid solutions of $CoCl_2$ contain $[Co(H_2O)_6]^{2+}$ and $[CoCl(H_2O)_5]^+$ ions when the acid concentration is below 3M; above 8M the principally occurring species are $[CoCl_3(H_2O)]^-$ and $[CoCl_4]^{2-}$. Complex thiocyanates resemble the complex halides. The tetrahedral $[Co(NCS)_4]^{2-}$ ion is almost quantitatively precipitated as the mercury salt $HgCo(NCS)_4$ by the addition of a solution of a mercury(II) salt to a solution containing cobalt(II) and an excess of thiocyanate ions. The blue crystalline salt is used as a calibrant in the Gouy method for the determination of magnetic susceptibility. In the tetranitratocobaltate(II) $(Ph_4As)_2[Co(NO_3)_4]$, the cobalt atom is eight-co-ordinate with bidentate nitrate groups.

When cobalt(II) cyanide dissolves in aqueous potassium cyanide, an olive-green solution is formed which is rapidly oxidised by air. The addition of ethanol precipitates the violet $K_6[Co_2(CN)_{10}].4H_2O$, which is diamagnetic and probably has the cobalt—cobalt bonded structure $[(NC)_5Co—Co(CN)_5]^{6-}$ in the anion. The green solution (which contains either $[Co(CN)_5]^{3-}$ or $[Co(CN)_5(H_2O)]^{3-}$) absorbs molecular hydrogen to form $[Co(CN)_5H]^{3-}$ ions. This activation of molecular hydrogen has been useful in a variety of reduction processes. Thus the $Co(CN)_2$—KCN solutions are homogeneous catalysts for the hydrogenation of unsaturated organic compounds; for example, styrene is reduced to ethylbenzene and benzil to benzoin. When oxygen is passed through the $K_3Co(CN)_5$ solution, the peroxo-anion $[(CN)_5Co^{III}OOCo^{III}(CN)_5]^{6-}$ is formed; this diamagnetic anion is oxidised by bromine in alkaline solution to the red paramagnetic $[(CN)_5CoOOCo(CN)_5]^{5-}$ ion. Other molecules such as SO_2, $SnCl_2$, C_2F_4, and

C_2H_2 also give insertion products of the type $[(CN)_5CoSO_2Co(CN)_5]^{6-}$ in their reactions with $K_3Co(CN)_5$.

16.4 Compounds of Cobalt in Low Oxidation States

16.4.1 Cobalt Carbonyls

The simple carbonyls of cobalt are $Co_2(CO)_8$, $Co_4(CO)_{12}$, and $Co_6(CO)_{16}$. Dicobalt octacarbonyl is a brown diamagnetic solid having two bridging CO groups and a cobalt—cobalt bond (figure 16.1). Tetracobalt dodecacarbonyl is a black solid

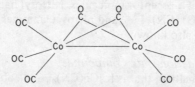

Figure 16.1   Dicobalt octacarbonyl

having the structure shown in figure 16.2. It consists of a tetrahedron of cobalt atoms three of which are bonded to two terminal CO and two bridging CO groups while the third is bonded to three terminal CO groups and the other three cobalt atoms.

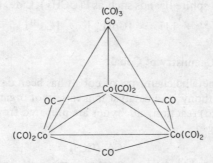

Figure 16.2   Tetracobalt dodecacarbonyl

All three carbonyls are air-sensitive. They undergo both substitution and disproportionation reactions with bases, for example

$$Co_2(CO)_8 + 2PPh_3 \rightarrow 2[Co(CO)_3(PPh_3)]_2 + 2CO$$

$$Co_2(CO)_8 + 12NH_3 \rightarrow 2[Co(NH_3)_6][Co(CO)_4]_2 + 8CO$$

In the dimeric substitution products it is the bridging CO groups that are replaced by the ligands. Carbonylate anions such as $[Co(CO)_4]^-$ are formed in aqueous alkali or by alkali metal reduction of the carbonyls in tetrahydrofuran or liquid ammonia. Acidification of solutions containing the tetracarbonylcobaltate(−I) ion yields the yellow volatile hydride $HCo(CO)_4$. This undergoes interesting reactions with unsaturated organic molecules; for example with olefins σ-bonded alkyl cobalt carbonyls result

$$Me(CH_2)_2 CH\!\!=\!\!CH_2 + HCo(CO)_4 \rightarrow Me(CH_2)_2 \underset{\underset{Co(CO)_4}{|}}{CHCH_3} + CH_3(CH_2)_3 CH_2 Co(CO)_4$$

16.4.2   Cobalt Nitrosyls

No simple nitrosyl of cobalt is known. The dark-red diamagnetic $Co(NO)(CO)_3$ is obtained in the reaction of nitrogen monoxide with $Co_2(CO)_8$ at $40°$. It is monomeric in the vapour phase and has the tetrahedral structure. In its reactions with donor molecules it is the CO groups that are replaced, illustrating that NO is one of the most powerful π-acceptor molecules. Nitrosyl halides $[Co(NO)_2 X]_2$, nitrosyl cyanides $[Co(NO)(CN)_5]^{3-}$, and nitrosyl ammines, for example $[Co(NO)(NH_3)_5]^{2+}$ are also known.

16.4.3   Other Low Oxidation-state Compounds

Complex cyanides of cobalt(0), that is $K_4 Co(CN)_4$, and cobalt(I), that is $K_3 Co(CN)_4$, are formed in the reduction of $K_3 Co(CN)_6$ with potassium in liquid ammonia. These compounds are rapidly oxidised in air; $K_4 Co(CN)_4$ liberates hydrogen from water. The crystalline cobalt(I) salts $[Co(CNR)_5]X$ are prepared by heating a cobalt(II) salt with the isocyanide in alcohol. The cation has the trigonal bipyramidal structure in $[Co(CNR)_5]ClO_4$. Five-co-ordinate cobalt(I) cations are also formed in an unusual disproportionation reaction of cobalt(II). When heated with strongly π-acidic polycyclic phosphite ligands such as $P(OCH_2)_3 CMe.$ (L), the reaction is

$$2[Co(H_2O)_6]^{2+} + 11L \rightarrow [Co^IL_5]^+ + [Co^{III}L_6]^{3+} + 12H_2O$$

16.5 Organometallic Chemistry of Cobalt

The extensive organometallic chemistry of cobalt has been developed in part from the use of dicobalt octacarbonyl as a catalyst in a variety of organic syntheses. In the hydroformylation (oxo) reaction, aldehydes are prepared from alkenes, carbon monoxide, and hydrogen

$$>\!C\!\!=\!\!C\!< + CO + H_2 \xrightarrow[100°, 100\ atm]{Co_2(CO)_8} \underset{\underset{H}{|}}{-C-}\ \underset{\underset{CHO}{|}}{C-}$$

As a further example of the catalytic use of cobalt carbonyls, we may quote the conversion of alkyl halides into esters by using tetracarbonylcobaltate(−I) ions as catalyst

$$RX + CO + [Co(CO)_4]^- \rightarrow RCOCo(CO)_4 + X^-$$

$$RCOCo(CO)_4 + R'OH \rightarrow RCOOR' + HCo(CO)_4$$

The process is catalytic in cobalt if a tertiary amine is present to regenerate the $[Co(CO)_4]^-$ ion

$$HCo(CO)_4 + R_3N \rightarrow [R_3NH]^+ [Co(CO)_4]^-$$

No simple alkyls or aryls of cobalt exist. However, they can be stabilised with

π-bonding ligands such as CO, CN^-, PR_3, or π-C_5H_5. One cobalt alkyl occurs naturally in the coenzyme of vitamin B_{12} (cobalamin). Vitamin B_{12} (cyanocobalamin) can be reduced and then alkylated to give synthetic alkyl cobalamins.

Cobalt forms many π-cyclopentadienyl complexes including the paramagnetic cobaltocene $Co(\pi$-$C_5H_5)_2$, as well as arene complexes such as that with hexamethylbenzene, $[Co(C_6Me_6)]^+$.

17 Nickel

17.1 The Element

Nickel derives its name from the ore kupfernickel which was at one time believed to be an ore of copper. While nickel is not an uncommon element in the earth's crust, there are relatively few known deposits of nickel ores that are capable of being worked economically. One such deposit is that of pentlandite $(Ni, Fe)_9 S_8$ at Sudbury, Ontario. The metal is extracted either electrolytically or by the Mond carbonyl process (section 10.2.2).

Nickel $(3d^8 4s^2)$ continues to show the trend in decreasing stability of the high oxidation states as we pass along the first transition series. Unlike cobalt, the +3 state is relatively unimportant; for nickel only the +2 state is of importance in aqueous solution. The metal is ductile and resistant to corrosion, it is used in food handling and pharmaceutical plant wherever a nonpoisonous noncorrosive metal is required. It also finds use in catalysts for industrial processes, for example in the hydrogenation of unsaturated organic compounds. Raney nickel is a catalyst prepared by dissolving away the aluminium from the alloy $NiAl_3$ with alkali, thus leaving a porous and highly active form of the metal. The massive metal is attacked by dilute mineral acids but is resistant to attack by caustic alkalis and liquid hydrogen fluoride. It finds uses therefore in plant for handling caustic soda and in valves on cylinders of gases such as the hydrogen halides. The metal is readily dissolved in dilute nitric acid. Many nonmetals combine with heated nickel; steam reacts with red-hot nickel to give NiO and hydrogen. Carbon monoxide reacts at relatively low temperatures to give the carbonyl $Ni(CO)_4$.

17.2 Compounds of Nickel in High Oxidation States

The only important simple compound of nickel in an oxidation state above +2 is the hydrous nickel(III) oxide NiO(OH). No simple halides are known in oxidation states above +2.

The oxidation of alkaline nickel(II) solutions with chlorine, bromine, persulphate ions, or electrochemically, results in the formation of a black precipitate of β-NiO(OH). This evolves oxygen and water vapour at $140°$ leaving NiO. A separate phase γ-NiO(OH) is produced when nickel is added to a melt of sodium peroxide and sodium hydroxide at $600°$ and the cooled melt is treated with ice—water. It dissolves in sulphuric acid with evolution of oxygen.

Complexes of the unknown nickel(III) halides are sometimes obtained in the oxidation of the nickel(II) complexes. Thus the complexes $NiX_3(PR_3)_2$ result from the oxidation of $NiX_2(PR_3)_2$ compounds with the corresponding nitrosyl halides. $NiBr_3(PEt_3)_2$ has a magnetic moment of 1.72 B.M. at room temperature, and is believed to have a *trans*-trigonal bipyramidal structure. The hexafluoronickelates-(III) and -(IV), containing the NiF_6^{3-} and NiF_6^{2-} ions, are obtained by fluorination of melts of KCl–$NiCl_2$ mixtures at moderate temperatures and pressures of fluorine. The violet K_3NiF_6 evolves oxygen from water and undergoes disproportionation in liquid hydrogen fluoride

$$2K_3NiF_6 \rightarrow 6K^+ + NiF_6{}^{2-} + NiF_2 + 4F^-$$

The nickel(IV) complex K_2NiF_6 is diamagnetic having the low-spin $t_{2g}{}^6$ ground state. It evolves oxygen from water but forms bright-red solutions in hydrogen fluoride. When it is heated to 350° in a vacuum, fluorine is evolved.

$$3K_2NiF_6 \rightarrow 2K_3NiF_6 + NiF_2 + F_2$$

17.3 Compounds of Nickel(II) (d^8)

17.3.1 Aqueous Chemistry

The hexaquo-ion $[Ni(H_2O)_6]^{2+}$ is green, and this colour is shown by most hydrated simple salts and their aqueous solutions. Many salts crystallise with this ion, for example $Ni(NO_3)_2.6H_2O$, $NiSO_4.7H_2O$, and $NiSO_4.6H_2O$. The hexaquo-ion is barely acidic; no hydroxo- or oxo-nickelates(II) are known. The carbonate $NiCO_3.6H_2O$ is precipitated by sodium bicarbonate solution. The aquo-ion is also labile; a large number of complexes can be obtained by substitution reactions. Because of the great stability of the +2 oxidation state, redox reactions are uncommon in aqueous nickel(II) solutions.

17.3.2 Simple Compounds

Nickel(II) halides. Nickel is unique among the elements of the first transition series in forming simple halides in the +2 oxidation state only. The yellow NiF_2, $NiCl_2$, and $NiBr_2$, and black NiI_2 are high-melting solids. They give green solutions in water from which hydrates can be crystallised. The hexahydrate $NiCl_2.6H_2O$ contains *trans*-$[NiCl_2(H_2O)_4]$ units while the dihydrate $NiBr_2.2H_2O$ has nickel atoms in an environment of 4Br and $2H_2O$.

Nickel(II) hydroxide and oxide. The hydroxide $Ni(OH)_2$ is precipitated as a finely divided green powder by the addition of an alkali metal hydroxide solution to an aqueous solution of a nickel(II) salt. It is notoriously difficult to filter but becomes more crystalline on standing. Basic salts, for example $NiCl_2 . Ni(OH)_2$, may precipitate if very strong solutions of nickel(II) salts are used. The oxide NiO is produced in the thermal decomposition of the hydroxide, carbonate, or hydrated nitrate; it has the rock-salt structure.

Nickel(II) cyanide. The pale blue hydrated cyanide is precipitated by the addition of potassium cyanide solution to aqueous nickel(II) salt solutions. The structure of the hydrate may be $[Ni(H_2O)_x]^{2+}$ $[Ni(CN)_4]^{2-}$; it can be dehydrated at 140° to the yellow anhydrous $Ni(CN)_2$. A very interesting reaction of nickel cyanide is that with aqueous ammonia in the presence of benzene (or a similar molecule). Pale-violet clathrate compounds of the type $Ni(CN)_2.NH_3.C_6H_6$ are precipitated. In this compound, the benzene is trapped in the crystal structure as shown in figure 17.1*a*. As can be seen by studying figure 17.1*b*, the Ni and CN groups in this compound form layers with ammonia molecules bonded above and below the planes of the layers on alternate nickel atoms. Half the nickel atoms are thus octahedrally co-ordinated to nitrogen, while the other half are bonded to carbon atoms in a square plane. The average magnetic moment per nickel atom is 2.2 B.M. in accord

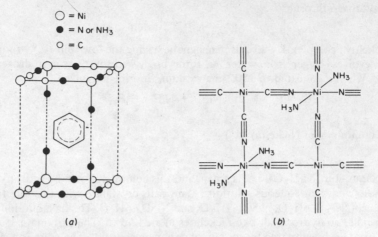

Figure 17.1   The benzene ammine nickel(II) cyanide clathrate

with the mixture of diamagnetic square nickel(II) and paramagnetic octahedral nickel(II) ions. Many other organic molecules, including pyridine, aniline, pyrrole, and thiophen, may be similarly trapped between the layers of the Ni—CN groups. The infrared spectra of the clathrates show absorptions similar to those of the free organic molecules, indicating that no bonding occurs between the aromatic molecule and the nickel atom.

17.3.3 Complexes

These are exceedingly numerous. The maximum co-ordination number shown is six in the octahedral and tetragonal complexes. Nickel(II) also forms many five-co-ordinate as well as square and tetrahedral four-co-ordinate complexes. However, it should be pointed out that tetrahedral complexes of nickel(II) are considerably less common than those of cobalt(II); this is to be expected in terms of the CFSE argument (section 5.5). A characteristic feature of nickel(II) chemistry is the low energy difference between these stereochemistries, which often results in equilibria between various structural types occurring in solution, as well as in the crystallisation of complexes containing nickel in two different stereochemistries. These various equilibria and 'anomalous' nickel(II) complexes have been discussed in section 7.3.2. We shall consider here those complexes that occur predominantly in one stereochemical form.

Cationic complexes. The substitution of water molecules in $[Ni(H_2O)_6]^{2+}$ by nitrogen donor molecules usually results in a colour change from green to violet as a result of the new ligand field present. In strong ammonia solutions the violet and octahedral hexammine $[Ni(NH_3)_6]^{2+}$ is formed, and salts of this ion can be precipitated with anions such as Br^- and I^-. Ethylenediamine forms octahedral chelated complexes containing the $[Ni(en)_3]^{2+}$ ion. There are also many octahedral cations containing a mixture of ligands, for example $[Ni(en)_2(NO_2)]BF_4$ which contains a polymeric distorted octahedral cation having bridging NO_2 groups. The monomeric cation $[Ni(diars)_2Cl_2]^+$, which has the four arsenic atoms in the square plane around nickel, is probably a nickel(II) complex containing a diarsine radical cation (diars$^{\ddot{+}}$).

Co-ordination numbers other than six are rare in cationic complexes. There are several five-co-ordinate cations such as $[\text{NiLCl}]^+$ in which L is a tetradentate ligand, but tetrahedral cations are rare and occur only with bulky ligands such as hexamethylphosphoramide which forms $[\text{Ni(HMPA)}_4](\text{ClO}_4)_2$.

Neutral complexes. The reactions between nickel(II) salts and donor molecules, particularly those containing nitrogen or phosphorus as the donor atom, give rise to a large number of neutral complexes (in addition to the cationic complexes illustrated above). Most commonly these are of the type NiX_2L_2 (which may be octahedral, square planar, or tetrahedral), but NiX_2L_4 and NiX_2L (L = unidentate ligand, X = univalent anion) also occur commonly. These complexes can be prepared by direct reaction between the salt and the ligand, or by thermal decomposition of $[\text{NiL}_6]\text{X}_2$ complexes which frequently proceeds via the stages NiL_4X_2, NiL_2X_2, and NiLX_2. The pyridine complexes will serve to illustrate these compound types. The direct reaction between nickel(II) chloride and pyridine produces the blue $\text{Ni(py)}_4\text{Cl}_2$ which has the *trans*-octahedral structure. On thermal decomposition above $110°$, pyridine is evolved and the yellow-green $\text{Ni(py)}_2\text{Cl}_2$ is formed; at $170°$ this decomposes into pyridine and Ni(py)Cl_2. Both this and the bis(pyridine)-complex contain six-co-ordinate nickel. The hexa(pyridine)nickel(II) cations are formed only in the presence of weakly co-ordinating anions; for example, nickel(II) nitrate gives $[\text{Ni(py)}_6](\text{NO}_3)_2$.

Phosphorus ligands frequently give four-co-ordinate complexes. Trialkylphosphines give square planar $\text{NiX}_2(\text{PR}_3)_2$ compounds while the aryl phosphines give tetrahedral $\text{NiX}_2(\text{PAr}_3)_2$ complexes. The unusual $\text{NiBr}_2(\text{PhCH}_2\text{PPh}_2)_2$ complexes have been discussed in section 7.3.2.

Numerous neutral nickel(II) chelates are formed with chelating anionic ligands; the β-keto-enolates, salicylaldimine, and *vic*-dioxime complexes (section 7.3.2) are examples of this type. The best-known chelate is the bright-red bis(dimethylglyoximato)nickel(II) (figure 17.2). This precipitates quantitatively

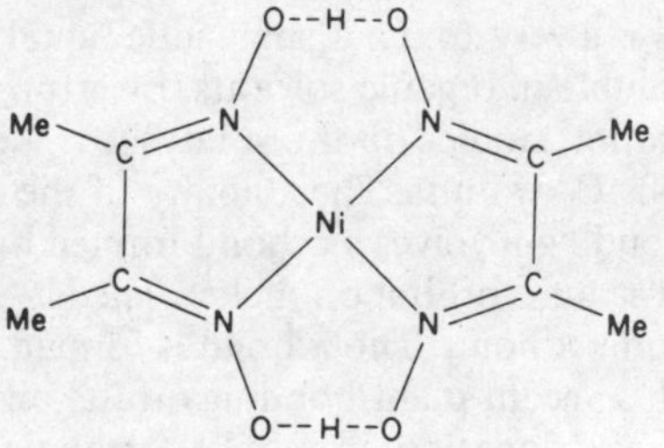

Figure 17.2   Bis(dimethylglyoximato)nickel(II)

when a neutral or ammoniacal solution of a nickel(II) salt is treated with an alcoholic solution of dimethylglyoxime; the reaction is used in the gravimetric estimation of nickel. The structure has some interesting features. The nickel atom is at the centre of a square plane of nitrogen atoms. These planes are stacked one above the other in the crystal, alternate molecules being twisted through $90°$, so the nickel

atoms form chains with a Ni—Ni distance of 325 pm. There are strong intra-
molecular hydrogen-bonds within each molecule. The diamagnetic complex may
thus be regarded as containing either square planar or tetragonally co-ordinated
nickel (with metal—metal bonding).

Anionic complexes. The tetrahedral $M^I_2 NiX_4$ (X = Cl, Br, I) complexes are readily
prepared from the nickel(II) halide and a quaternary ammonium or phosphonium
salt in nitromethane or ethanol; they are not easily obtained from aqueous solution.
Substituted tetrahalonickelates(II) such as $[NiBr_3(PPh_3)]^-$ are similarly prepared
in the presence of the added ligand; they have a distorted tetrahedral structure.
These complexes have the intense blue colour and the high magnetic moments
(3.5—4.1 B.M.) expected of tetrahedral d^6 complexes; the electronic spectrum of
the $[NiCl_4]^{2-}$ ion is shown in figure 6.10. The tetrafluoronickelates(II) $[NiF_4]^{2-}$
and the trihalonickelates(II) $M^I NiX_3$ (X = F, Cl, Br) contain octahedrally
co-ordinated nickel.

The tetracyanonickelate(II) ion $[Ni(CN)_4]^{2-}$ is formed when nickel(II) cyanide
is dissolved in an excess of aqueous potassium cyanide; the potassium salt crystallises from the orange-red solution as the orange $K_2 Ni(CN)_4 . H_2 O$. This salt is
diamagnetic and contains square planar $[Ni(CN)_4]^{2-}$ ions. These ions are extremely
stable with an overall stability constant β around 10^{30}; even chelates such as
bis(dimethylglyoximato)nickel(II) dissolve in potassium cyanide solution (compare
the stability constants for $Ni^{2+} - NH_3$ and $Ni^{2+} - $ EDTA complexes in tables 3.1 and
3.2). There is some evidence for the formation of diamagnetic $[Ni(CN)_5]^{3-}$ ions in
the presence of an excess of cyanide ions; no evidence has been found for
$[Ni(CN)_6]^{4-}$ species.

17.4 Compounds of Nickel in Low Oxidation States

17.4.1 Nickel Carbonyl

The tetracarbonyl $Ni(CO)_4$ was the first metal carbonyl to be discovered (Mond,
1890). It is produced industrially as the intermediate in nickel refining (see section
10.2.2). Nickel carbonyl is a very toxic, diamagnetic liquid (b.p. $42°$); it is
insoluble in water but soluble in organic solvents. Electron diffraction studies on the
vapour, and X-ray diffraction studies on the solid, show the molecule to be
tetrahedral with linear Ni—C—O units. The stability of the molecule cannot be
understood if the sole bonding involves a σ bond formed by overlap of a filled
carbon σ orbital with a vacant σ orbital on nickel. The Ni—C bond is thus believed
to consist of a σ bond and a π bond. The π bond is formed by overlap of a filled
d_π orbital on nickel with a vacant p antibonding orbital on carbon monoxide. Such
bonding is synergic since the donation of metal electron density into the carbon
monoxide orbitals will enhance the σ-donor power of the latter while at the same
time increasing the acceptor power of the nickel atom along the σ bond.

Nickel carbonyl is inflammable in air; its solutions oxidise slowly in air. Reduction by alkali metals in tetrahydrofuran gives carbonylate anions, for example

$$4Ni(CO)_4 + 2Na \rightarrow Na_2 [Ni_4 (CO)_9] + 7CO$$

In liquid ammonia the thermally unstable carbonyl hydride can be isolated as a tetra-ammoniate

$$2Ni(CO)_4 + 2Na + 6NH_3 \rightarrow [NiH(CO)_3]_2.4NH_3 + 2CO + 2NaNH_2$$

With donor molecules, disproportionation reactions are common, especially with nitrogen donors

$$3Ni(CO)_4 + 6py \rightarrow [Ni(py)_6][Ni_2(CO)_6] + 6CO$$

Phosphorus, arsenic, and antimony donors, as well as isonitriles and unsaturated organic molecules, give substitution products that are stable with respect to disproportionation, for example

$$Ni(CO)_4 \xrightarrow{\text{PPh}_3} Ni(CO)_3(PPh_3) \xrightarrow{\text{PPh}_3} Ni(CO)_2(PPh_3)_2$$

17.4.2 Nickel Nitrosyls

No simple nitrosyl of nickel is known, the reaction of the carbonyl with NO in the absence of a donor solvent proceeding according to

$$Ni(CO)_4 + 4NO \rightarrow Ni(NO)(NO_2) + 4CO + N_2O$$

The blue air-sensitive $Ni(NO)NO_2$ ignites on contact with water; it is believed to be polymeric, resembling the nitrosyl halides $Ni(NO)X$. Substituted nickel carbonyls with NO give nitrosyl complexes such as $Ni(NO)_2(PPh_3)_2$. Two series of nitrosyl halides are known. When nickel(II) halides react with NO in the presence of a halogen acceptor such as zinc, the nitrosyl monohalides $Ni(NO)X$ are formed. The dihalides $Ni(NO)X_2$ arise in the reactions of nickel carbonyl with nitrosyl halides. The structures of these presumably polymeric compounds are unknown; both types show infrared absorptions in the NO^+ region.

17.4.3 Other Low Oxidation-state Compounds

The fairly extensive chemistry of nickel(0) includes compounds in which nickel is bonded to CN^-, acetylide ion, phosphines, aryl isocyanides, and acrylonitrile, as well as to those ligands such as 2,2'-bipyridyl and 1,10-phenanthroline which stabilise the zero oxidation states of all the first-row elements that we have so far considered.

The reduction of $K_2Ni(CN)_4$ in aqueous solution with, for example, sodium amalgam gives Bellucci's salt $K_4[Ni_2(CN)_6]$ containing nickel(I). This compound is diamagnetic, the anion being binuclear with a nickel–nickel bond. With potassium in liquid ammonia the reduction of $[Ni(CN)_4]^{2-}$ proceeds via this red nickel(I) complex to the yellow precipitate of $K_4Ni(CN)_4$ containing nickel(0). The $[Ni(CN)_4]^{4-}$ anion is isoelectronic with $Ni(CO)_4$ and is believed to have the tetrahedral structure. Other nickel(0) compounds can be obtained by substitution reactions on this cyano-complex

$$Ni(SbPh_3)_4 \xleftarrow{\text{SbPh}_3} K_4Ni(CN)_4 \xrightarrow{\text{CO}} K_2[Ni(CO)_2(CN)_2]$$

Nickel reacts with many phosphines directly, for example

$$Ni + 4PF_3 \xrightarrow[\text{35 MPa}]{100°} Ni(PF_3)_4$$

$Ni(PCl_2Me)_4$ and $Ni(Ph_2PCH_2CH_2PPh_2)_4$ can be prepared under less extreme conditions; all these phosphine–nickel(0) compounds are believed to have the tetrahedral structure.

17.5 Organometallic Compounds of Nickel

Simple alkyls and aryls of nickel are usually too unstable to be isolated on their own. However, ligand-stabilised aryls can be made from square $(R_3P)_2NiX_2$ complexes and aryl Grignard reagents. Such compounds as $(R_3P)_2NiAr_2$ are yellow or brown and diamagnetic with the *trans*-square planar structure.

Nickel forms olefin and acetylene complexes as well as cyclobutadienyl and cyclopentadienyl derivatives. Perhaps the most important of the π-bonded organometallic compounds of nickel are the π-allyls. Bis-(π-allyl)nickel is obtained by the Grignard route

$$NiBr_2 + 2CH_2=CHCH_2MgBr \rightarrow Ni + 2MgBr_2$$

This sandwich-structured compound is pyrophoric in air but its ethereal solutions are stable to deoxygenated water. It is a very active catalyst for the cyclotrimerisation of butadiene to cyclododeca-1,5,9-triene

18 Copper

18.1 The Element

Copper has an ancient history, being used extensively in the Bronze Age. Its compounds are widely distributed in the earth's crust, and occasionally it occurs naturally as the element. It is the last member of the first transition series; the configuration $3d^{10} 4s^1$ gives rise to an extensive chemistry of the +1 oxidation state. In this respect it differs from all the other members of the first transition series. The next element, zinc with the $3d^{10} 4s^2$ configuration, shows the +2 state almost exclusively and does not have the transition-metal properties associated with the variable oxidation states.

Copper is a soft reddish metal noted for its high termal and electrical conductivities. The pure metal is used extensively in, for example, electrical equipment, as well as in alloys such as brass (Cu–Zn), bronze (Cu–Sn), and monel metal (Cu–Ni). The metal oxidises in air at red heat to CuO; at higher temperatures Cu_2O is formed. In dry atmospheres at room temperature little corrosion occurs; in moist atmospheres a green film of basic carbonate coats the metal. Copper is readily attacked by the halogens and by sulphur at relatively low temperatures. It dissolves in oxidising acids such as HNO_3, concentrated H_2SO_4, or dilute H_2SO_4 in the presence of air but is resistant to reducing acids such as dilute HCl, in accord with its noble position in the electrochemical series.

$$Cu^{2+} + 2e \rightarrow Cu \qquad (E^\circ = +0.34 \text{ V})$$

18.2 Compounds of Copper(III) (d^8)

Copper(III) is isoelectronic with nickel(II) but only a few compounds of copper in this oxidation state have been characterised (no compounds of copper in oxidation states greater than +3 are known). There are no simple halides CuX_3 but fluorination of a 3 : 1 mixture of KCl and $CuCl_2$ in a flow system at 250° gives the pale green K_3CuF_6. This reacts violently with water and has a magnetic moment of 2.8 B.M. as expected of a $t_{2g}^6 e_g^2$ configuration.

Oxidation of $Cu(OH)_2$ in alkaline solution with chlorine gives a strongly oxidising solution which evolves oxygen on acidification. The solid cuprates(III) can be obtained by heating mixtures of CuO and alkali metal superoxides in oxygen. The steel-blue $KCuO_2$ is diamagnetic and may therefore have a square planar structure in the anion. A few other complex salts such as the diamagnetic $Na_7[Cu(IO_6)_2].12H_2O$ and $Na_9[Cu(TeO_6)_2].16H_2O$ can be obtained by oxidation of copper(II) salts in alkaline solutions of the alkali metal periodate or tellurate.

18.3 Compounds of Copper(II) (d^9)

Copper(II) is the most stable state in aqueous solution. Its d^9 configuration gives rise to Jahn–Teller distortions (see sections 5.5.6 and 6.2.4) both in its simple compounds and in complexes. A wide range of stereochemistries are exhibited by copper(II) compounds, with four, five, and six-co-ordination predominating; in each structure variations from idealised geometries occur through bond length and bond angle distortions. The distorted octahedral compounds have magnetic moments in excess of the spin-only moment, being usually around 1.9 B.M. (section 7.2.1).

18.3.1   Aqueous Chemistry

The blue aquo-ion $[Cu(H_2O)_6]^{2+}$ is formed when copper(II) salts are dissolved in an excess of water. It is tetragonally distorted, with two water molecules further away from copper than the other four, which are in the square plane around the metal. The mild acidity of the ion results in the frequent precipitation of basic salts from aqueous solution. Thus the addition of sodium carbonate solution causes precipitation of the green basic carbonate $CuCO_3.Cu(OH)_2.aq$

$$[Cu(H_2O)_6]^{2+} + H_2O \rightleftharpoons [Cu(OH)(H_2O)_5]^+ + H_3O^+$$

$$2[Cu(OH)(H_2O)_5]^+ + CO_3{}^{2-} \rightarrow [Cu(OH)(H_2O)_5]_2CO_3$$

$$\text{that is } CuCO_3.Cu(OH)_2.aq$$

In neutral or acid solution, the copper(II) ion is a mild oxidant

$$Cu^{2+} + e \rightarrow Cu^+ \qquad (E^\circ = +0.15 \text{ V})$$

The oxidation of iodide ions to iodine is used in the volumetric estimation of copper(II) with thiosulphate. This oxidation of iodide ions by Cu^{2+} proceeds despite the potential

$$\tfrac{1}{2}I_2 + e \rightarrow I^- \qquad (E^\circ = +0.54 \text{ V})$$

because of the insolubility of the copper(I) product, CuI. The potential for the system

$$Cu^{2+} + I^- + e \rightarrow CuI \qquad (E^\circ = +0.86 \text{ V})$$

thus shows that the oxidation of iodide ions according to the equation

$$Cu^{2+} + 2I^- \rightarrow CuI + \tfrac{1}{2}I_2$$

is thermodynamically favourable.

Substitution reactions of $[Cu(H_2O)_6]^{2+}$ with added ligands give rise to a large number of co-ordination compounds. With concentrated aqueous ammonia, the substitution proceeds as far as $[Cu(NH_3)_4(H_2O)_2]^{2+}$ (sections 3.1.1. and 3.1.2). Similarly, bidentate ligands such as ethylenediamine readily form the intensely blue complexes of the type $[Cu(en)_2(H_2O)_2]^{2+}$, with completely substituted compounds such as $[Cu(en)_3]^{2+}$ formed only in the highest concentrations of ethylenediamine. Cations such as $[Cu(NH_3)_6]^{2+}$ are readily formed under anhydrous conditions. Some neutral ligands such as hydrazine and methylhydrazine cause reduction of Cu^{II} in aqueous solution.

With chloride and bromide ions the blue $[Cu(H_2O)_6]^{2+}$ solution becomes yellow and brown respectively, and from such solutions a variety of halo-complexes have been isolated with different cations. The yellow $[CuCl_4]^{2-}$ ion is planar in $(NH_4)_2[CuCl_4]$, but with larger cations, such as $Cs_2[CuCl_4]$, the anion adopts a flattened tetrahedral structure. The red trichlorocuprates(II) $[CuCl_3]^-$ also show two distinct structural types. The potassium and ammonium salts contain planar binuclear anions $[Cu_2Cl_6]^{2-}$, which are stacked in the crystal lattice in such a manner that there is an elongated tetrahedral co-ordination about each copper atom. In $CsCuCl_3$ there are square planar $CuCl_4$ units linked by a single bridging atom; two terminal chlorines complete the distorted octahedral array around each copper atom. Cyanide ions cause precipitation of brownish-yellow $Cu(CN)_2$ which decomposes at room temperature with evolution of cyanogen (compare the reaction of Cu^{2+} with I^-). In an excess of cyanide the colourless $[Cu(CN)_4]^{2-}$ and $[Cu(CN)_4]^{3-}$ species are formed.

18.3.2 Simple Compounds

Copper(II) oxide and hydroxide. The black CuO results from thermal decomposition of the basic carbonate, hydroxide, or nitrate. It is readily reduced to the metal by hydrogen or carbon monoxide at around 300°, and is used as an oxidant in organic microanalysis for carbon and hydrogen.

The blue hydroxide $Cu(OH)_2$ is obtained (often contaminated with basic salts) on the addition of an alkali metal hydroxide solution to an aqueous copper(II) solution. It darkens on standing, becoming black above 30° as dehydration to the oxide takes place. In concentrated alkalis $Cu(OH)_2$ shows amphoteric character, forming deep-blue solutions of hydroxocuprates(II); in ammonia the soluble deep-blue $[Cu(NH_3)_4(H_2O)_2]^{2+}$ ion is formed.

Copper(II) halides. The fluoride CuF_2 is a white hygroscopic solid, the chloride is yellow, and the bromide black; CuI_2 cannot be prepared because of the reduction of Cu^{2+} by iodide ions. The crystal structure of $CuCl_2$ consists of infinite chains of planar $CuCl_4$ groups (Cu−Cl = 230 pm), with two chlorine atoms from other chains completing (Cu−Cl = 295 pm) the tetragonally elongated octahedral environment around each copper atom. In the dihydrate $CuCl_2.2H_2O$, the square plane around copper is composed of two Cl at 228 pm and two H_2O at 193 pm, with the two further chlorine atoms completing the distorted octahedral structure at Cu−Cl = 295 pm.

Many complexes of copper(II) halides have been prepared. Nitrogen donors frequently form tetragonal CuX_2L_2 complexes, for example $CuCl_2(py)_2$, but some tertiary amines, phosphines, and arsines cause reduction to copper(I) complexes. With phosphine oxide ligands, tetrahedral complexes are formed as in $CuCl_2(Ph_3PO)_2$. 2,2′-Bipyridyl forms $CuX_2(bipy)_2$ adducts which contain the trigonal bipyramidal cations $[Cu(bipy)_2X]^+$.

Copper(II) sulphate. The blue pentahydrate $CuSO_4.5H_2O$ has copper co-ordinated to four water molecules in a plane with the tetragonal structure completed by two oxygen atoms from sulphate ions. The fifth water molecule is hydrogen-bonded as shown diagrammatically in the representation of part of the giant structure in figure 18.1. The pentahydrate is readily dehydrated thermally to the white $CuSO_4$; the d−d transition in the anhydrous salt has moved into the infrared as a result of the weaker (spectrochemical sense) ligand field.

From solutions of copper(II) sulphate saturated with ammonia, the dark blue ammine $CuSO_4.4NH_3.H_2O$ can be crystallised. This compound has a square pyramidal structure around copper, with the metal atom lying 20 pm out of the

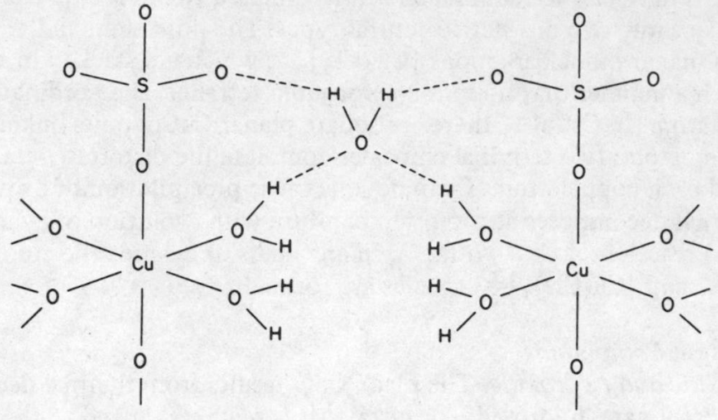

Figure 18.1   Representation of part of the structure of $CuSO_4 \cdot 5H_2O$

plane of the four nitrogen atoms. With gaseous ammonia, $CuSO_4$ forms the violet pentammine $CuSO_4 \cdot 5NH_3$.

Copper(II) nitrate. The anhydrous $Cu(NO_3)_2$ (section 10.3.2) exists in two crystal forms; these have complex structures in which copper ions are linked in infinite array by nitrate ions. The nitrate can be sublimed in a vacuum without decomposition. Discrete $Cu(NO_3)_2$ molecules occur in the vapour, in which the copper atoms are four-co-ordinated by oxygen atoms from bidentate nitrate groups. The deep-blue hydrate $Cu(NO_3)_2 \cdot 3H_2O$ crystallises from aqueous solutions; it is very soluble in water.

Copper(II) acetate. Copper(II) carboxylates frequently have dimeric structures giving rise to anomalous magnetic properties. The acetate $[Cu_2(OCOCH_3)_4] \cdot 2H_2O$ has the same structure as the chromium(II) analogue (figure 13.2). The Cu–Cu interaction gives rise to incomplete quenching of the spin moments of the ions, so low magnetic moments are observed in this type of carboxylate.

18.4 Compounds of Copper(I) (d^{10})

Copper(I) has the closed-shell $3d^{10}$ configuration, so its compounds are diamagnetic and colourless (except for those salts having coloured anions or in which charge-transfer transitions are of low energy). In solid compounds, copper(I) is often the thermodynamically stable state at moderate temperatures; for example, CuO and $CuBr_2$ give Cu_2O and CuBr on heating.

18.4.1 Aqueous Chemistry
If the potentials

$$Cu^+ + e \rightarrow Cu \qquad (E^\circ = +0.52 \text{ V})$$
$$Cu^{2+} + e \rightarrow Cu^+ \qquad (E^\circ = +0.15 \text{ V})$$

are compared we see that the Cu^+ ion (in the first equation) is capable of oxidising the Cu^+ ion (in the second equation) according to

$$Cu^+ + Cu^+ \rightarrow Cu + Cu^{2+} \quad (E^\circ = +0.37 \text{ V})$$

That is, disproportionation of copper(I) aquo-ions occurs. Of course, in the presence of added ligands, the potentials are considerably affected, so the relative stabilities of Cu^I and Cu^{II} do depend on the nature of such ligands. The surprising instability of Cu^I in aqueous solution is probably connected with the low hydration energy of the Cu^+ ion (which may form $[Cu(H_2O)_2]^+$) compared with that of the Cu^{2+} ion which has the higher charge and co-ordination number.

Only the insoluble salts of copper(I) are stable in water, that is salts such as CuCl, CuI, and CuCN. Those that dissolve undergo disproportionation; thus Cu_2SO_4, prepared for example from Cu_2O and dimethyl sulphate, reacts immediately with water

$$Cu_2SO_4 + 6H_2O \rightarrow Cu + [Cu(H_2O)_6]^{2+} + SO_4^{2-}$$

Several water-soluble copper(I) complex species are stable with respect to disproportionation in aqueous solution. The complexes of copper(I) resemble those of silver(I) in having a preferred co-ordination number of two; however, three and four-co-ordination also occur. In aqueous ammonia, CuCl dissolves to form the colourless $[Cu(NH_3)_2]^+$ ion (compare Ag^I). This is readily oxidised in air to the blue $[Cu(NH_3)_4(H_2O)_2]^{2+}$ species; however, it is stable to disproportionation, the reaction

$$[Cu(NH_3)_4(H_2O)_2]^{2+} + Cu \rightarrow 2[Cu(NH_3)_2]^+ + 2NH_3 + 2H_2O$$

proceeding from left to right in the absence of air. Chloro-species of copper(I) are similarly prepared from copper(II) chloride solutions in hydrochloric acid

$$[CuCl_4]^{2-} + Cu \rightarrow 2[CuCl_2]^-$$

These two-co-ordinate complexes are linear like their silver(I) analogues. Complexes such as $[Cu(CN)_4]^{3-}$ and $[CuCl_3]^{2-}$ (in solid K_2CuCl_3) contain tetrahedrally co-ordinated copper(I) while planar three-co-ordination occurs in $K[Cu(CN)_2]$ (section 4.2.1).

18.4.2 Simple Compounds

Copper(I) oxide. The red Cu_2O occurs naturally in *cuprite;* it is precipitated as a finely divided orange form when alkaline solutions of copper(II) are treated with mild reducing agents such as glucose. In Fehling's solution the $[Cu(H_2O)_6]^{2+}$ ion is converted into a tartrate complex so that no $Cu(OH)_2$ is precipitated when alkali is added. In dilute sulphuric acid Cu_2O undergoes disproportionation

$$Cu_2O + 2H_3O^+ + 3H_2O \rightarrow Cu + [Cu(H_2O)_6]^{2+}$$

However, in solutions of co-ordinating agents, for example concentrated hydrochloric acid or concentrated aqueous ammonia, it dissolves to give solutions of copper(I) complexes.

Copper(I) halides. The fluoride is not known as a stable species at room temperature. The white CuCl, CuBr, and CuI are prepared by aqueous methods. The chloride is most easily prepared by reduction of a copper(II) salt solution in hydrochloric acid with copper metal or sulphur dioxide. The resulting solution of chloro-complexes such as $CuCl_2^-$ is poured into a large volume of water, and the CuCl precipitates. The iodide CuI precipitates when iodide ions are added to copper(II) salt solutions. In the solid state these halides have copper ions tetrahedrally surrounded by halide ions; in the vapour, CuCl forms principally a trimeric species, which is believed to have a puckered cyclic structure.

These halides are very insoluble in water but dissolve readily in solutions of ligands such as Cl^-, CN^-, $S_2O_3^{2-}$, and NH_3. Solutions of CuCl in hydrochloric acid contain the various $CuCl_2^-$, $CuCl_3^{2-}$, and $CuCl_4^{3-}$ ions depending on the chloride ion concentration.

Complexes of copper(I) halides with neutral donor molecules can be prepared by direct reaction or, often more conveniently, by reduction of copper(II) halides with ligands. Such complexes are often polymeric. In the azomethane complex $(CuCl)_2 MeN{=}NMe$, the copper atoms are tetrahedrally co-ordinated by two chloride ions and two nitrogen atoms from the bridging *trans*-azomethane molecules. Complexes of CuCl with azo- and diazo-compounds have found use in organic chemistry. In the Sandmeyer reaction an $-NH_2$ group attached to an aromatic ring is replaced by a $-Cl$ group when the amine is diazotised and treated with copper(I) chloride; the intermediate CuCl–diazonium salt complex decomposes evolving nitrogen. Azobenzene is converted into benzidine by treatment with CuCl in hydrochloric acid.

The iodide gives 1 : 1 adducts with phosphines and arsines CuI.L. Molecular weight measurements indicate some of these to be tetrameric, and the crystal structure determination on $CuI(AsMe_3)$ shows the presence of a tetrahedral array of copper atoms linked together by bridging iodine atoms. Each copper is tetrahedrally co-ordinated to one terminal arsenic atom and three shared iodine atoms.

18.4.3   *Carbonyl and Organometallic Compounds*

Solutions of copper(I) halides in hydrochloric acid or ammonia absorb carbon monoxide to form colourless complexes, the halogen-bridged dimer $[CuCl(CO)]_2$ can be isolated as colourless crystals.

Copper(II) forms no σ-bonded alkyls or aryls, Grignard reactions causing reduction to copper(I). The copper(I) alkyls and aryls, CuR, are thermally unstable and probably have polymeric structures. The only organometallic compounds of copper of any significance are those with acetylenes and olefins. Solutions of CuCl in hydrochloric acid absorb acetylene in forming complexes such as $CuCl.C_2H_2$ and $[CuCl_2(C_2H_2)]^-$. The explosive orange and diamagnetic acetylides $CuC{\equiv}CR$ and Cu_2C_2 are formed by using ammoniacal solutions of CuCl or with potassium acetylide in liquid ammonia. The insolubility of these acetylides suggests that they are polymeric, the copper atoms forming both σ and π bonds to the acetylide groups.

Copper(I)–olefin complexes are prepared by the reduction of alcoholic solutions of copper(II) halides with sulphur dioxide in the presence of the olefin. The most stable complexes are formed by chelating diolefins such as cyclo-octa-1,5-diene and norbornadiene. In the complex of cyclo-octa-1,5-diene $[C_8H_{12}CuCl]_2$, each copper atom is tetrahedrally surrounded by two bridging chlorine atoms and two olefinic double bonds.

Bibliography

Chapter 1

1. F. Basolo and R. Johnson, *Co-ordination Chemistry,* Benjamin, New York (1964)

Chapter 2

1. R.G. Pearson, *J. chem. Educ.,* **45** (1968), 581–7 and 643–8
2. S. Ahrland, J. Chatt and N.R. Davies, *Q. Rev. chem. Soc.,* **12** (1958), 265–276

Chapter 3

1. J.P. Hunt, *Metal Ions in Aqueous Solution,* Benjamin, New York (1963)
2. *Stability Constants* and *Stability Constants Supplement No. 1,* Special Publication (Nos 17 and 25 of the Chemical Society (London) 1964 and 1971)

Chapters 4 and 5

1. R.J. Gillespie, *Molecular Geometry,* Van Nostrand (1972)
2. F.A. Cotton and G. Wilkinson, *Advanced Inorganic Chemistry,* Interscience, London (3rd edn 1972)
3. S.F.A. Kettle, *Co-ordination Compounds,* Nelson, London (1969)
4. E. Cartmell and G.W.A. Fowles, *Valency and Molecular Structure,* Butterworths, London (3rd edn 1966)

Chapter 6

1. D.S. Urch, *Orbitals and Symmetry,* Penguin, Harmondsworth (1970)
2. A.B.P. Lever, *Inorganic Electronic Spectroscopy,* Elsevier, Amsterdam (1968)
3. D. Sutton, *Electronic Spectra of Transition Metal Complexes,* McGraw-Hill, London (1968)
4. B.N. Figgis, *Introduction to Ligand Fields,* Wiley, New York (1966)
5. H.H. Jaffé and M. Orchin, *Theory and Applications of Ultraviolet Spectroscopy,* Wiley, New York (1962)

Chapter 7

1. B.N. Figgis, *Introduction to Ligand Fields,* Wiley, New York (1966)
2. B.N. Figgis and J. Lewis in *Techniques of Inorganic Chemistry,* vol. 4, Interscience, New York (1965), p. 137

3. B.N. Figgis and J. Lewis, *Progress in Inorganic Chemistry*, vol. 6, Interscience, New York (1964) p. 37

4. A. Earnshaw, *Introduction to Magnetochemistry*, Academic Press, London (1968)

Chapter 8

1. F. Basolo and R.G. Pearson, *Mechanisms of Inorganic Reactions*, Wiley, New York (2nd edn 1967)

2. D. Benson, *Mechanisms of Inorganic Reactions in Solution*, McGraw-Hill, London (1968)

3. C.H. Langford and H.B. Gray, *Ligand Substitution Processes*, Benjamin, New York (1965)

4. A.G. Sykes, *Kinetics of Inorganic Reactions*, Pergamon, Oxford (1966)

Chapters 9 and 10

1. H. Remy, *Treatise on Inorganic Chemistry*, Elsevier, Amsterdam (1956)

2. D.J.G. Ives, *Principles of the Extraction of Metals*, Royal Institute of Chemistry Monographs for Teachers No. 3, London (1960)

3. R. Colton and J.H. Canterford, *Halides of the First Row Transition Metals*, Wiley, London (1969)—metal halides

4. C.C. Addison and N. Logan, *Preparative Inorganic Reactions*, vol. 1, Wiley, New York (1964), p. 1—metal nitrates

Chapters 11–18

For inorganic factual material, the most comprehensive reference text is *Gmelins Handbuch der Anorganischen Chemie*, Verlag Chemie. This is a continuing series so that some sections contain more recent information than others. The best sourcebook in the English Language is *Comprehensive Inorganic Chemistry*, Pergamon, Oxford (1973); volume 3 contains the factual chemistry of the transition elements.

The following standard texts will be found useful in supplementing the data in chapters 11–18.

1. A.F. Wells, *Structural Inorganic Chemistry*, Oxford (3rd edn 1962)—solid-state crystal structures

2. G.E. Coates, M.L.H. Green and K. Wade, *Organometallic Compounds*, Methuen, London (1968)—vol. II contains the transition-metal organometallic compounds

3. R.J.H. Clark, *The Chemistry of Titanium and Vanadium*, Elsevier, Amsterdam (1968)

4. H. Remy, *Treatise on Inorganic Chemistry*, Elsevier, Amsterdam (1956)—vol. II contains the transition elements

5. R. Colton and J.H. Canterford, *Halides of the First Row Transition Metals*, Wiley, London (1969)

6. F.A. Cotton and G. Wilkinson, *Advanced Inorganic Chemistry*, Interscience, London (1972, 3rd edn)

7. D.L. Kepert, *The Early Transition Metals*, Academic Press, London (1972)

Index

Contents

Foreword

Reading this important new book has helped me note how far afield and adrift I have gone from practising as a person-centred therapist who once held complete trust in the core principles as being truly necessary and sufficient in themselves for therapy to be successful. As a result it helped me to return 'home' again with more trust in myself and the client.

Let's see what a 'cornucopia' of offerings are provided in this book, since this important exploration of Carl Rogers' work is about so much more than empathy, covering in fact nearly all the major constructs and concepts of Rogers' work during his lifetime.

Throughout this book, to support his subtle analyses of the principle of empathy in particular, Steve Vincent uses original source material faithfully, including impeccable research going as far back as 1937, to seek Carl's earliest conceptualisations. Steve has also minutely delved the depths of the major 1959 theoretical and scientific analysis by Carl of his own theory, using quotes from it throughout the work, thus both enlarging upon and deepening our understanding of the principles. If the reader were never to read these source works of Carl's in their entirety, then this book would provide beautifully clear explications of them.

The quotes and extracts from Rogers are always to the point, and explorations of the concepts rich and original, each amplifying, yet not changing, Carl's meanings. Steve also reminds us of some of Rogers' own seminal thinking that has perhaps been lost over the years; for example, citing an early quote of Carl's that the 'process of change' is rather an existential state of 'changingness', a unique paradox that in itself helps both the practitioner and the client be comfortable with an ongoing state of evolving and emergent ways of being.

While the audience for this book might best be considered to be those in training as therapists, or students using the book as a university text, it will also be most helpful for practitioners who want to review and renew a deeper understanding of Rogers' approach after practising for so long that they may have forgotten or put aside the core principles. Potential clients, in seeking a safe haven for their deep explorations, may also profit greatly from this book as a guide in their search.

How does this book advance our thinking?

The depth of Steve's intellect is shown in his creative new contributions to the advancement of our thinking and the refinement of our practices. As a contribution to the literature, the author minutely dissects the complexity of the principle and practice of empathy to a level rarely achieved elsewhere, and of other concepts as well. In the latter case, for example, he makes very clear for the learner/reader/practitioner the distinction between the actualisation tendency and self-actualisation, and between actualisation and the formative tendency, constructs that so often get misused, confused and misunderstood.

He also makes an important distinction between, and provides a precise definition of, resistance and avoidance, arguing that these concepts are inconsistent with the person-centred approach (PCA). In addition, he makes it clear that Carl was adamant about the disservice to relationship emanating from the concept of transference. Steve also objects to such a concept because it separates, not connects, people rather than creating what I term a 'reciprocal multi-directional empathy' that emerges in all directions as therapy evolves, without interference from such conceptual frames of separation.

There are many non-traditional insights, emphases and ideas throughout this book, such as the fresh notion that unconditional positive regard is the highest level of 'confrontation' and congruence. The examples the author uses to support his points are moving and precise, such as the young boy cutting himself, who moves towards self-understanding with his therapist by respecting and accepting all perspectives of his inner 'configurations'. In the writing about the 'as if' quality of empathy, there is a fine example of what the best of being objective means, and its relationship to being empathic. There is a creative acronym (VALID) to help us remember the state of clients when they enter therapy, and subtle analyses of each 'condition' needed for successful therapy, be there 3, 6 or 12! Enumerating the five early premises with which Carl began to articulate his approach from 1961 onwards is very helpful. The author makes a brilliant 'circular' reverse summation, intertwined and non-linear in nature, that shows the complete tapestry of person-centred principles as they weave together as a whole cloth.

I appreciate Steve's idea of the community being defined as various configurations' of self/selves, expressed through different voices. His clarification of the use of that term by Dave Mearns, and sourced originally by Carl himself, is quite eye-opening. In each chapter the author invites us to ask ourselves intellectually profound questions with no simple answers, helping us to dig deeper into even more fertile soil for thinking through our values and beliefs. This book is thus a profusion of possibilities for us all.

I am particularly interested in language, and in those parts of the book which focus on use of language the reader discovers very original contributions and additions to the PCA dialogue. Steve defines words such as perception, awareness, symbolisation, etc., with great precision, as Carl seemed to use them.

I may disagree somewhat with the author in his critique of there being too much emphasis on the importance of language in skills training. He seems to contradict himself on occasion, acknowledging the importance of the right word for a client, and thus the therapist, yet in some instances discounting the importance of words, thus presenting an internal contradiction for the reader to resolve for themselves, or to live with as a dual truth and find their own path – very person-centred!

How is this book different from others on the same subject?

This book has a unique format and style, merging tradition with innovation and whimsy. In the middle of the book, we are suddenly presented with a personal dialogue with Steve and an interviewer, which changes the pace and rhythm of the book entirely. In fact this interview encapsulates what the book is all about.

The interview is both intellectually stimulating and very personal, including the 'I' voice so derided in academia, and which Carl himself initiated for the first time in academic writing. It is a very thorough interview, engaging and stylistically different from the rest of the book. Steve reminds me here of Brian Thorne and Carl himself in how very honest and forthcoming he is about himself. This gives a further order of credibility to the rest of book.

I was delighted with the wit, humour, play on words, catchy headings, and the like which Steve uses throughout the text, whether in his personal interview segment, or within the theoretical analyses, making the latter sparkle, and more human and understandable for the reader. Steve uses these lighter devices while at the same time asking the reader to be intellectually stretched by the demands of the book. For example, he has fun with clever yet profound sub-titles, such as 'Thou shalt not skill', and at the same time addresses the deeper issue of: once I am 'de-skilled', who am I as a therapist? This is a very important idea, as if only through emptying my self can I discover myself.

No matter what the writing style or rhythm, the nature of Steve's enquiry is evocative and thorough, while he also takes strong and clear stands, such as objecting to some current misunderstandings of the person-centred approach and certain aspects of eclecticism popular today.

What might have been Carl's view of this book?

While none of us can speak for Carl, my guess is that he would be delighted to see how profoundly Steve has dissected, in a loving, respectful and scientific way, Carl's lifetime of evolutionary thinking, honouring his own scientist in that sense. I trust that he would be particularly pleased with how Steve has minutely parsed Carl's major 1959 work, of which he himself often said he was most proud as a mathematical and scientific conceptualisation. Steve shows both what a profound thinker Carl was, as well as the profound thinker that Steve is. When compared with the reductionistic, stereotypic depiction of Rogers' work in so many previous texts outside the PCA community, this book is a breath of fresh air.

The future

I would like to end this foreword with a particular emphasis on the values of love, spirituality and peace building, which, while not elaborated upon in particular by Steve in this text, seem to permeate this entire work and make it clear that, in fact, Carl's approach has always expressed these values. Although through no intention of the author, this book has thus deepened my own belief that this theory and its core concepts, particularly empathy, are readily transferable to other arenas beyond individual therapy, such as the larger social conflicts and crises wounding our suffering planet. Unfortunately some 43 years later, with a world in turmoil, Carl's 1961 quote below is still applicable. Perhaps it is our role to be the current catalysts to such realisation of these principles both inside the therapy office and for the larger public good.

> *Our civilization does not yet have enough faith in the social sciences to utilize their findings.*

> (Rogers 1961: p. 335)

In the closing years of his life, Carl made even more explicit his sense of the importance of love in our work, and his ever-present concern for the need for balance between the individual's self-concerns with an acknowledgement of our inevitable connectedness to each other. In the poignant quote below, published in 2002, he belies those critics who had long discounted or misunderstood the place of relationship in his philosophy by labelling it as narcissistic and hyper-individualistic.

> *in becoming more aware of self and others, in becoming more acceptant of self and others, and in becoming more understanding of self and others, people become more able to meaningfully live their lives, more . . . able to love, able to receive love.*
>
> (Rogers and Russell 2002: p. 258)

In this re-discovery of the richness of Rogers' approach that Steve has created for us by writing this book, he has guided us with elegance, insight, wisdom and compassion towards a deeper understanding of the genius and profundity of Carl Rogers' work and principles. For that reason, I look forward with enthusiasm and anticipation to his next books on the other core conditions, genuineness and unconditional positive regard.

Gay Leah Barfield PhD, MFT
Former Co-Director
Carl Rogers Institute for Peace
December 2004

Foreword

A step towards a better world

Every human being strives to live a life that brings them satisfaction and happiness. We search for ways to live such a life in an outer world, and that is not an easy enterprise.

Each race imbues in its members the ways and means of being that have been culturally and historically tested by the experience of previous generations. This take place through childbearing in families, through the process of education, through peer exchange of experiences, and in many other ways.

This process of providing culturally tested ways of being in the world is of great value and provides the essence of each person to become Human. Yet, it often imposes on a person ways and means of being (norms, values, ideas and ways of acting) that are inconsistent with the person's deepest and authentic strivings and intentions. The person starts to live an imposed life. This may be convenient, yet it does not bring deep satisfaction and happiness to that person, and the World (society, culture) is denied the full potential of that person's input, energy and creativity. Both the person and the World lose out.

In this book Steve Vincent deeply and carefully explores Empathy as a way of facilitating personal growth and being. It considers the way a person is able to become fully Themself in the World, bringing deep satisfaction and happiness to that person, and their full potential of energy, creativity and productivity to the World. (I am far from believing that the way is not without troubles and difficulties, but I do believe that this is the way for achieving dynamic harmony of the Person, the World and the Person in the World!)

At a personal and professional level, Steve Vincent is a writer to trust on the subject of empathy. Empathy for Steve, as I know him, is not just a technique for professional psychological work. Empathy is for him a way of being with people. This personal and professional commitment to empathy makes Steve Vincent's exploration of empathy really deep and comprehensive. In a very personal yet accurate way he presents the history of the development of the concept and practice of empathy in psychotherapy. What is evident is the deep impression and influence that Carl Rogers, the founder of the Person-centered Approach and one of the most famous and influential psychologists of the 20th century, had on Steve.

Throughout this book, the author – as scientist, professional psychotherapist and inquiring person – carefully and accurately presents his view on empathy in psychotherapy and interpersonal relationships. In Chapter 4, I value highly the systematic and convincing analysis made by Steve of the role and place of empathy in the process of facilitating personal growth. And I find very sincere and touching the self-Inter: View Steve makes in Chapter 3 as a way of exploring and expressing his personal notions of empathy.

The book is intended for professional psychologists and psychotherapists who look for ways of deeper understanding of, and ways of improving, their work with clients. The book will be invaluable for psychology students and interns who work on the understanding of human nature and on the development of their counseling style. But what is most valuable and inspiring for me is the way the book brings all of us closer to being ourselves, to enable one to be oneself with persons close to us – relatives, friends or clients – and to make the World we live in better!

Thank you, Steve, for your exquisite work!

Veniamin Kolpachnikov
Assistant Professor
Department of Psychological Counselling
Moscow City Psychology–Pedagogical University
Russia
December 2004

About the author

Steve Vincent, born in 1954, completed a sociology degree at the University of Sheffield, during which he was Chair and Trainer for Student Nightline. This was followed by a one-year full-time postgraduate Diploma in Counselling at the University of Aston in Birmingham. In 1999, at the University of East Anglia, he completed his Masters Degree, *On Becoming Person-Centred?* (Vincent 1999a) which involved researching the person-centred training of client-centred therapists.

Steve has worked as a youth counsellor in Liverpool (1980–82) and as a student counsellor in Southampton (1982–95), where he founded a person-centred training unit within a college of further education. After 21 years in further education, he resigned in an attempt to get a more person-centred life! He has been Membership Secretary, Chair and Newsletter Editor for BAPCA (British Association for the Person-Centred Approach) and is currently a BACP (British Association for Counselling and Psychotherapy) accredited practitioner as well as a Board member of CSP (Carl Rogers Center for Studies of the Person, La Jolla, California).

He now devotes most of his time to offering client-centred therapy in several GP practices, being a partner in a family and, from September 2002, a father for the first time. Carl Rogers, in reply to a question about the best way to make an organisation person-centred, said 'Start one' – so that's exactly what Steve did, and Client-Centred Therapy, Person-Centred Approach Services (CCTPCAS) was formed. It is now a project of CSP, and through it Steve and others offer a variety of person-centred events and activities in the UK.

You can contact Steve Vincent at cctpcas@aol.com or visit the CCTPCAS website at www.cctpcas.com, and you can find out more about CSP at www.centerfortheperson.org.

Introduction: the shape of this book

Prelude, substance and interlude

The primary purpose of this book is to provide a substantial text on empathic understanding – a kind of 'All you ever wanted to know about empathy, but were afraid you wouldn't find in one place' volume. As I reflected upon the development of my own understanding of, capacity for, experience and communication of empathy, two main factors stood out, namely learning about client-centred therapy (with reference to Carl Rogers in particular), and my own experiencing (which includes reflecting upon that experiencing). I wanted to somehow combine these two elements as both, it seemed to me, had validity and meaning. Furthermore, they have cross-fertilised and nourished each other over the years.

However, it seems that writing in the first person and from one's own experience is unacceptable to many publishers, who maintain that there are 'academic standards to uphold, reputations to consider', and so on. Yet mindful of Carl Rogers saying that what is most personal is most meaningful, I was delighted when researching some of the background for this book to come across the following:

> In the science of man, we are going to try to include emotions and affect, etc. Yet we operate on the absolute, unquestioned assumption that the best way of doing so will be somehow to put that in cognitive and intellectual terms . . . I can't but help wonder whether there may not some day be a new conception of science which would involve communication of more than intellectual symbols. . . . There may be ways of getting at knowledge and communicating knowledge which question the whole damn underlying assumption that science consists of nothing but symbols, all of which must necessarily be intellectual.
>
> (Rogers 1989: p. 164)

So, at one time, I tried to blend the academic and the experiential into the ongoing text. Up to a point this seemed to work well enough, yet beyond that point it began to feel a little fuzzy, a little unclear. Next I decided to leave all of my personal musings out altogether, yet this felt like unwelcome censorship, whether undertaken by myself or by a would-be publisher. Then, finally, I hit on the idea of 'interludes' – interspersing the core journey, derived overwhelmingly from the work of Carl Rogers, with sightseeing trips of my own, derived from my experience. Several more personal (and hence meaningful, at least to me) articles were earmarked for inclusion as interludes, among them *On Becoming More Whole? Necking in Glasgow* (Vincent 1999b), *A Training Trip Through Empathy* (Vincent 1999c), *From Whole to Hole to Whole* (Vincent 2000a), *Failing or Fortunate? Striving for Revising* (Vincent 2000b), *Setting Aside or Erupting?* (Vincent 2001), *Client-Centred Therapy: Conditions, Process and Theory* (Vincent 2002a), *Empathy: an inter-view* (Vincent 2002b) and a few poems. It transpired that the inclusion of these as interludes would have taken the word count of this book way beyond the optimum publishing size, so while *Empathising in action: an inter-view* has been retained as Chapter 3 of this book (in the belief that it serves to bring together many of the emergent threads up to that point and to pave the way for some of what follows, all in a less formal fashion), all of the others are omitted (the omitted articles are available via the CCTPCAS website at www.cctpcas.com).

I have tried to use gender-free terminology in my own writing. Quotes from Carl Rogers (of which there are many) tend to be couched in masculine form, and these are given verbatim. In his later years, Carl apologised for this, writing, for example:

> I regret that in this paper, written some years ago, I used 'he' as the generic term. I hope my feminine readers will forgive me. I have learned better since.
>
> (Rogers 1986: pp. 2–3, 23)

Although some readers may care to skip my introductory comments, a 'prelude' is nevertheless provided in an attempt to offer a very basic 'level playing field' – a shared understanding from which the rest can follow.

I hope that it works for you.

Rogers; Carl; Carl Rogers; Rogers, Carl

In writing this book, I struggled somewhat in determining how to refer to Carl Ransom Rogers (1902–1987). In various drafts I tried only using *Carl*, only using *Rogers*, and using only *Carl Rogers*. Feedback from my 'critical friends' was divided. Some liked the use of *Carl*, as it had a kind of familiar, friendly, warm feeling to it. Others disliked this – for precisely the same reason (together with an opinion that it didn't come across as sufficiently 'academic'). This group tended to prefer the use of *Rogers*, yet the first group felt this to be impersonal and cold, even 'needlessly academically conventional'. The use of *Carl Rogers* (or even *Carl R Rogers*) all the time seemed even more formal, put the word count up, and frankly became somewhat tedious, even boring, yet clearly some critical friends preferred it this way. It seemed that whatever form of address I chose, one-third would be happy and two-thirds unhappy. Funnily enough, in all draft versions all 'proper' quoted *references* were to Rogers, Carl – no problem there! (And most of the references in this book *are* to Carl Rogers, Carl or Rogers – this is a book about client-centred empathy as a Carl Rogers companion.)

And what of my own feelings and thoughts? Well, I settled on a mix of all three forms of address. Often *Carl* somehow just 'felt right' and sometimes *Rogers* seemed more apt, while at other times *Carl Rogers* seemed to fit best. There were other times when there was little sense of 'rightness', so my choice was either based on varying the text or was random. 'Maybe', I ended up thinking, 'even if I certainly couldn't please *all* of the readers *all* of the time, I could at least please *some* of the readers *some* of the time'.

And you know I really don't think that Carl would have minded much one way or the other.

Client-centred therapy/person-centred approach

References throughout this book are, in the main, to *client-centred therapy*. The term *person-centred* is less frequently used. However, the words *counselling* and *therapy* are used interchangeably (consistent with Carl Rogers who also treated these words as being synonymous). Carl wrote and spoke many times of the changes in terminology that were introduced over the years – for example, as applications drawn from therapy became more and more diverse:

> As client-centered therapy became applied in the field of education, it wasn't therapy, so what was it? Student-centered education? Gradually it became quite evident that we weren't talking about therapy in these other fields, so what was it? That was when we began to adopt the term *person-centered*, which still implies a lot of reliance on the internal frame of reference, but it's not therapy.
>
> (Rogers and Russell 2002: p. 253)

Debates over what is client-centred and what is person-centred (or indeed whether the two are one and the same thing) abound, it seems. Although I have my own views, I'd like Carl to have the last (or even the definitive) word on this. Just prior to his death in February 1987, he wrote that wrangling over the differences between these two terms was futile, as he, Carl Rogers, was the *same person*, with access to the same personal resources, whether he was meeting a client for one-to-one therapy or facilitating a large group. Historically, the term *person-centred approach* arose from a need to encompass the many diverse applications of the philosophy, theory and practices that grew out of the learning gathered from *client-centred*, one-to-one therapy – *activities* such as trainee-centred learning, encounter groups, relationships (including parenting), conflict transformation, and so on. Rogers conceded that a label for this diversity could well be more open to loose use, misuse and misunderstanding, yet he stated that nevertheless he 'would like a term to describe what I do when I am endeavoring to be facilitative in a group of persons who are *not* my clients', and that term was *person-centred* (Rogers 1987a: p. 12).

To me, this couldn't be more clear. One-to-one counselling is client-centred therapy, whereas non-client-work activities based on the principles that arose from client-centred therapy constitute a person-centred approach. The facilitative and/or therapeutic *person* may well be the same (or very similar) in both, but the *activity* is different. Nonetheless, as the bulk of Carl's writing about empathy is in the context of client-centred therapy, that term takes precedence in this book.

Client-centred therapy: core/necessary and sufficient conditions

What a horrible word *conditions* is in some respects! The word 'respects' in this context brings to mind the word *respect* in another context – that is, linked with *unconditional* positive regard. How strange it seems to have *unconditional* as part of a *condition*!

Over the years, Carl Rogers penned several versions of 'conditions' – for therapy, for personality change, for trainee-centred learning, for facilitative group leadership, and so on. In 1942, for example, there were four conditions, namely *objectivity, respect for the individual, understanding of the self* and *psychological knowledge* (*see* Chapter 1). At other times, there have been four quite different *necessary and sufficient* conditions, or five, or six – or there are what have become known as the three *core* or *therapist* conditions, namely congruence, unconditional positive regard and empathic understanding, although even the wording of these changed over time. For instance, what was first known as *acceptance* in the early days of client-centred therapy later came to be known as *unconditional positive regard*. Some people erroneously believe that Rogers planned to introduce a *seventh* condition of intuition, or an *eighth* condition of spirituality. Their belief is erroneous because Carl saw intuition as part of empathic understanding, and a 'spiritual' (although he was wary of this word) or 'transcendental' connection as part of both unconditional positive regard and empathy.

And so on. No doubt the debates will continue.

For the purposes of this book, the central version referred to is Carl's *Conditions of the Therapeutic Process* (Rogers 1959), otherwise known as the six necessary and sufficient conditions for therapeutic personality change, as described below.

For therapy to occur it is necessary that the following conditions exist.

1 Two persons are in *contact*.
2 The first person, whom we shall term the client, is in a state of *incongruence*, being *vulnerable* or *anxious*.
3 The second person, whom we shall term the therapist, is *congruent* in the *relationship*.
4 The therapist is *experiencing unconditional positive regard* toward the client.
5 The therapist is *experiencing* an *empathic* understanding of the client's *internal frame of reference*.
6 The client *perceives*, at least to a minimal degree, conditions 4 and 5, the *unconditional positive regard* of the therapist for him, and the *empathic* understanding of the therapist (Rogers 1959: p. 213).

At times, Carl referred to condition 1 as a *pre*-condition, leaving five 'main' conditions. At other times, he wrote that conditions 1 and 2 could be considered as pre-conditions, leaving four 'main' conditions, three of which had been termed 'core' conditions, and so on. Academically or historically interesting as it may be to wrangle over just how many 'conditions' there are (or indeed over which is the 'correct' or 'definitive' version), what I would personally prefer is to distil and preserve the *essence* of the spirit of client-centred principles, attitudes and beliefs as I believe Carl Rogers intended them.

Before the beginning . . .

First and foremost, let us not forget throughout this exploration that empathy, or empathic understanding, is but one of the conditions for therapeutic personality change. When Carl Rogers and his colleagues researched what seemed to be effective and meaningful in therapeutic encounters, certain threads were identified as integral parts of the whole experience. While it is possible (and perhaps necessary) to examine each strand of this rich weave as a discrete aspect of client-centred therapy, it is helpful to remember that each necessary and sufficient strand remains a *part* of a *whole* – to remember to put back together the entire fabric once we have examined a weft, weave or thread. Alternatively, while it might be desirable to carve up the meal that is client-centred therapy into bite-sized chunks, let us not lose sight of the entire dish. In illustration of this point, Carl's daughter Natalie wrote:

> I remember one discussion group, however, that has stayed in my mind and heart. A large group of person-centered folks (including Carl) were sitting in a circle. The topic turned to 'Is it client-centered to do this, or client-centered to do that?'. There were disagreements, and case samples focusing on the minutiae of what a therapist said or did, with criticisms of this or that wording or behavior. My guts were tight, my breathing tense. I looked across the circle at Carl and his shoulders were slumped, his eyes focused on the floor. My unspoken thoughts were: 'Is the person-centered approach going to self-destruct by nit-picking every word a therapist says and losing sight of the essence of the I–thou relationship? Are we focusing on the precise verbalisations while losing the vision of being in tune with the client? Do we really know how to be a companion with the client on her road to self-discovery?'. I was angry and disappointed at the level of the discussion. Then Carl spoke. He heaved a big sigh and said 'Boy, am I glad that when I discovered this process I didn't have to figure out "Is it client-centered or not?" I was really just looking to see what worked for the client.' My spirit rose and in my heart I said 'Right on, Dad!'.
>
> (Rogers N 1997: p. 271)

So perhaps we can keep in mind the essence of all that constitutes a client-centred whole as we explore that portion of it which is empathic understanding.

Prelude

This brief overview is included for two main reasons – in part to define some key concepts for readers new to client-centred therapy, and in part because even the understanding of those familiar with client-centred therapy often seems to vary (for example, those who have stated that they are unaware of any difference between the *actualising tendency* and *self-actualising*, or those who see no contradiction in delivering skills training).

Client-centred therapy: an overview

According to Carl Rogers, the most well known of the founders of client-centred therapy, this approach to counselling could be deemed to have begun in 1940:

> On December 11, 1940, before an invited audience at the University of Minnesota, he delivered a lecture entitled 'Newer Concepts in Psychotherapy', and he subsequently came to consider the date of this event as the birthday of client-centered therapy.
>
> (Thorne, cited in Rogers and Russell 2002: p. 11)

Not long before he died, Carl Rogers (writing in partnership with Nat Raskin) stated that client-centred therapy (and the person-centred approach) was:

> a clearly stated theory, accompanied by the introduction of verbatim transcriptions of psychotherapy, [that] stimulated a vast amount of research on a revolutionary hypothesis: that a self-directed growth process would follow the provision and reception of a particular kind of relationship characterised by genuineness, non-judgmental caring, and empathy.
>
> (Rogers and Raskin 1989: p. 155)

Rogers linked this self-directed growth process with the idea of an organic *actualising tendency*, believing that in parallel with a *formative tendency*, based on *syntropy* (an innate universal drive to grow and become more complex, as opposed to *entropy*, the decaying of matter), there is a fundamental and integral characteristic of humans, *given the right conditions*, to be constructive, cooperative, social, responsible beings. Talking about his 1942 publication *Counseling and Psychotherapy*, Rogers said:

> The central theme was the notion that the potential for better health resided in the client . . . So the counsellor endeavored to create an interpersonal situation in which material could come into awareness. [By 1950] There was a great deal of stress on both the philosophical and attitudinal characteristics of the therapist, and a definite move away from techniques with perhaps a bit of groping toward a close personal relationship.
>
> (Rogers and Evans 1975: p. 24)

Rogers wrote and spoke of his growing disillusionment with both his 'traditional' working practices and his theoretical training. At one stage he wrote:

> It is the client who knows what hurts, what directions to go, what problems are crucial, what problems have been deeply buried. It began to occur to me that unless I had a need to demonstrate my own cleverness and learning, I would do better to rely on the client for the direction of movement in the process.
>
> (Rogers 1961: p. 12)

A consequence of this dawning belief in the client being at the heart of therapy (the client's *internal frame of reference* – how the client sees their own world) was 'out with the old and in with the new' ways of being. Counsellors aligned with such beliefs began to refrain from advising, manipulating, directing, interpreting or teaching clients, as such directive practices were increasingly seen as either at best irrelevant, or at worst damaging with regard to clients accessing and utilising their own capacities. The primary aim became as follows:

> That the client should grow toward *himself*, not toward Freud, Jung, or Rogers, is the fundamental understanding of nondirectiveness. To advise him how to behave, what to do, what to think, or how to act is a violation of the priority of his own phenomenal world.
>
> (Barton 1974: p. 180)

Thus in its origins what became known as client-centred therapy began as a reaction against traditional therapist attitudes and behaviours:

> There is no doubt that the term *nondirective* was in itself accurate, but it was a term of rebellion. It was saying what we were against. We were opposed to the highly directive kind of procedures being used. Then, as we became more secure, I think, in the fact that this was a new way, a legitimate way of proceeding, we began to recognize that it was centered in the client's frame of reference – that what we were endeavoring to do was to get within the client and release what was there. It was centered in the client in a way that was quite specific and quite deep. That term is often misunderstood. A lot of people say, 'I'm centered in the client', meaning 'I pay a lot of attention to my client'. We meant something deeper than that. Really this was built around the client's internal frame of reference, his inner world.
>
> (Rogers and Russell 2002: p. 253)

Carl Rogers was a leading pioneer in developing a more 'humanistic' approach to psychotherapy. However, learning derived from therapy began to be applied in increasingly diverse areas, and likewise the terminology that was used to describe what was going on came to have increasingly diverse interpretations, too. Rogers wrote:

> I smile as I think of the various labels I have given to this theme during the course of my career – nondirective counselling, client-centred therapy, student-centred teaching, group-centred leadership. Because the fields of application have grown in number and variety, the label 'person-centred approach' seems the most descriptive.
>
> (Rogers 1981: pp. 114–15)

Patterson (and Watkins) wrote that:

> Rogers uses two terms: *client-centered therapy* and the *person-centered approach*. Later, in an article entitled 'Client-Centered? Person-Centered?' Rogers states he would like 'a term to describe what I do when I am endeavoring to be facilitative in a group of persons who are not my clients.' That term is *person-centered*. *Client-centered* is the term used with clients, in therapy.
>
> (Patterson and Watkins 1996: p. xv)

Perhaps the most comprehensive presentation of what client-centred therapy is all about (certainly according to Carl Rogers himself) was published in 1959. Carl said:

> I worked harder on this theoretical formulation than on anything I have written before or since. It is, in my estimation, the most rigorously stated theory of the process of change in personality and behavior which has yet been produced . . . I suspect these volumes are in fact very little used. Certainly my chapter 'A Theory of Therapy, Personality, and Interpersonal Relationships, as Developed in the Client-Centered Framework' is the most thoroughly ignored of anything I have written. This does not particularly distress me, because I believe theories too often become dogma, but it has, over the years, perplexed me.
>
> (Rogers and Evans 1975: pp. 135–6)

Drawing from this and other sources, we can try to achieve a brief overview of client-centred attitudes, beliefs, processes and outcomes. What is it all about? Rogers wrote:

> Psychotherapy deals primarily with the organization and the functioning of the self.
>
> (Rogers 1951: p. 40)

If 'dealing' with the self is a fundamental aspect of therapy, perhaps it might be helpful to know a little of what is meant by this. It is interesting that originally Rogers was not too concerned with the 'self' as:

> I certainly didn't start psychological work being interested in anything as vague as the self. To me, that seemed like old-fashioned introspectionism. I was really forced to examine self and forced to define it for myself, because my clients in therapy kept using that term in all kinds of significant ways.
>
> (Rogers and Evans 1975: p. 15)

In the client-centred personality theory that developed as a consequence of researching and exploring the notion of 'self,' it was stated that the forming of a *self-concept* (through which experience is perceived) follows basic human needs in infancy for *positive regard* from others (a need for love and affection, which when experienced result in feelings of warmth, liking, respect, sympathy and acceptance to and for the other person). This in turn generates needs for *positive self-regard* (a good feeling towards the self that is no longer dependent on the attitudes of others) – and all too often this results in the actualising tendency becoming thwarted by these *conditions of worth* (the positive or negative valuing of experiencing based on the values of others rather than whether the experiencing enhances or fails to enhance the person). Rogers had thought about and researched how clients entering therapy seemed to be, finally settling on the term *incongruence* as a description:

> It seemed to me that psychological maladjustment is the degree of discrepancy between the self as perceived and the experiences that are going on within the organism. In psychological adjustment there is a greater congruence between these two.
>
> (Rogers and Russell 2002: p. 251)

In other words, incongruence is a psychological process whereby experiences that are not consistent with a person's self-concept, or self-image, may be *denied* to awareness altogether – or the perception of such an experience may be 'twisted' to such an extent that the experience can be admitted to consciousness, but only in a *distorted* fashion. Feelings or sensations accompany these psychological processes:

> When the individual has no awareness of such incongruence in himself, then he is merely vulnerable to the possibility of anxiety and disorganization. . . . If the individual dimly perceives such an incongruence in himself, then a tension state occurs which is known as anxiety.
>
> (Rogers 1957a: p. 97)

Clients are likely to be *defensive* if their self-concept is *threatened* by experiencing, and they may feel *low self-esteem*. Thus they bring both their *incongruence* (including vulnerability, anxiety, tension, low self-esteem and psychological defensiveness) and their *potential* to therapy, where they come into contact with a person who holds genuineness, non-judgemental caring and empathy as fundamental 'core' conditions of being a counsellor. These therapist attitudes were renamed and modified over time, until eventually the terms *congruence, unconditional positive regard* and *empathic understanding* became the most generally accepted terminology. It is these core conditions that Rogers believed were both *necessary* and *sufficient* to 'interface' with the client's conditions of worth and to facilitate constructive change:

> The goal of therapy is fostered by the therapist through the providing of three conditions. . . . If they are sufficient, then no other conditions are necessary. The only means of influencing the client are through these conditions. Any other active intervention by the therapist is inconsistent with the basic assumption of the existence of the drive toward self-actualization. The three conditions offered by the therapist frees the operation of this drive in the client. The perception of these conditions by the client results in client self-disclosure, self-exploration, self-directed and self-discovery learning leading to changes in client perceptions and attitudes that result in changes in behavior. These changes are elements of the self-actualizing process. They are unique for each client (though there are some common elements).
>
> (Patterson 2000: p. 183)

Thus six client-centred conditions, considered as both necessary and sufficient, emerged as two people being in contact, the client being incongruent, the therapist being congruently respectful and empathic, and the client perceiving the therapist's unconditional positive regard and empathic understanding (*see* Introduction, page 4).

Although it is not known for sure, it seems likely that either Carl Whitaker or a colleague of his coined the term congruence, for in an interview in 1958, Rogers said 'I think I essentially picked up this idea from the Atlanta group that's working with Carl Whitaker' (Kirschenbaum 1979: p. 196). Carl Rogers defined *congruence*, or genuineness, as meaning:

> the psychotherapist is what he *is*, when in the relationship with his client he is genuine and without 'front' or facade, openly being the feelings and attitudes which at that moment are flowing in him . . . the feelings the therapist is experiencing are available to him, available to his awareness, and he is able to live these feelings, be them, and able to communicate them if appropriate.
>
> (Rogers 1961: p. 61)

A client-centred therapist attempts to experience unconditional positive regard for clients. Stanley Standal (a student studying with Rogers) was somewhat dissatisfied with the term 'acceptance' as being too open to different interpretations, so he coined the term *unconditional positive regard* as meaning a deep, respectful and genuine *unconditional* caring or 'prizing' of clients – 'unconditional' in the sense that the caring is not contaminated by evaluation or judgement of the client's feelings, thoughts and behaviour as good or bad (Standal 1954). Therapists value and warmly accept the client without placing stipulations on that acceptance.

> The therapist expresses this quality of genuine regard through *empathy*. Being empathic reflects an attitude of profound interest in the client's world of meanings and feelings as the client is willing to share this world. The therapist receives these communications and conveys appreciation and understanding, with the effect of encouraging the client to go further or deeper. An interaction occurs in which [the therapist] is a warm, sensitive, respectful companion in the typically difficult exploration of another's emotional world.
>
> (Rogers and Raskin 1989: p. 157)

Client-centred therapy: a summary

The acronym 'VALID' can be used to characterise clients entering therapy, in that they are likely to be:

Vulnerable and
Anxious, feeling
Low self-esteem or worth, being in a relative state of
Incongruence and
Defensive.

Ideally, it is hoped that upon commencing therapy, clients will engage with a therapist who will *be* congruent and from whom they will receive unconditional positive regard *for* them and empathic understanding *of* them. These *therapist conditions* interact with the *client's conditions* (of worth), thereby generating a therapeutic *process* within which change and growth can occur, primarily through clients 'rediscovering' the potential of their actualising tendency. However, as has so often been commented upon:

> Rogers has elaborated on most of these points many times. Yet there appear to be many who call themselves client-centered who seem to be unaware of their implications for practice. Such therapists appear to have little faith in the actualizing tendency in their clients.
>
> (Patterson 2000: p. 183)

Client-centredness is 'borne on the therapist's shoulders' (Barton 1974: p. 185), for if a therapeutic encounter is to be client-centred in nature, it is up to the *therapist* to initiate and maintain it:

> If I can provide a certain type of relationship, the other person will discover within himself the capacity to use that relationship for growth, and change and personal development will occur.
>
> (Rogers 1961: p. 33)

This process of growth flows through many stages or phases:

> With all his ambivalences, the client wants to grow, wants to mature, wants to face his problems and work them through. Accept and clarify his initial expressions of feeling, and a fuller, deeper expression of feelings will follow. Accept and clarify these and insight will spontaneously occur. Accept and clarify these insights, and the client will begin to take positive actions in his life, based on this insight. Accept and clarify the meaning the client sees in his positive actions, and at some point, when he feels enough self-acceptance, self-understanding, and confidence in his ability to continue to deal with his own problems, he will end the relationship.
>
> (Kirschenbaum 1979: p. 128)

Client-centred theory is, as Rogers put it, of the *'if–then'* variety, the first *if* being the therapist core conditions. *If* a therapist holds a belief and trust in the inner capacities of clients and is authentically respectful and understanding in contact with incongruent clients, *then* a therapeutic process will naturally follow. *If* a therapeutic relationship exists, characterised by a genuine counsellor accepting and empathically understanding clients as they unfold through this process, *then* predictable outcomes follow:

> The conditions we have chosen to establish predict such behavioral consequences as these: that the client will become self-directing, less rigid, more open to the evidence of his senses, better organized and integrated, more similar to the ideal which he has chosen for himself. . . . We have set the conditions which predict various classes of behavior – self-directing behaviors, sensitivity to realities within and without, flexible adaptiveness – which are by their very nature

unpredictable in their specifics . . . scientific progress has been made in identifying those conditions in an interpersonal relationship which, if they exist in *B*, are followed in *A* by greater maturity in behavior, less dependence on others, an increase in expressiveness as a person, an increase in variability, flexibility, and effectiveness of adaptation, an increase in self-responsibility and self-direction . . .

(Rogers and Evans 1975: pp. lxxvi–vii)

In other words, while the outcomes of therapy are *predictable*, what can be predicted is the *unpredictable*, because clients grow towards actualising the unique and rich, complex and ever-changing individual potential that resides within all human beings.

Chapter 1

Empathic understanding: origins and developments

In the beginning, there was the word

Dictionary corner . . .

. . . and the word was *empathy*. The *Concise Oxford Dictionary* defines empathy as the:

> Power of identifying oneself mentally with (and so fully comprehending) a person or object of contemplation.

Derived initially from the Greek 'empiateia' (meaning 'passionate affection') and later from the German language, current usage of the word *empathy* comes from the German words *ein* meaning 'in' and *fuhlung* meaning 'feeling' (from the Greek *pathos*, literally meaning 'suffering', also related to the Greek *penthos*, meaning 'grief'). Thus we could elaborate by saying that empathy means the power of projecting your own personality into, fully comprehending and being *in-feeling* with another – possibly with their suffering or grief. The *Concise Oxford Dictionary* also refers us to the word *sympathy*, which is defined as a '(capacity for) being simultaneously affected with the same feeling as another; tendency to share or state of sharing another person's emotion or sensation or condition; mental participation *with* another in his trouble or *with* another's trouble; compassion or approval (*for*); agreement (*with*) in opinion or desire', and the word *sympathetic* is defined as 'touches the feelings by association, etc.'. The meaning of the prefix *sym* is derived from 'alike' or 'as'.

Comparing *empathy* and *sympathy* might be helpful. Despite the italicising of 'with' in the definitions of sympathy, most common usage of the word would seem to be about feeling sympathy *for* or *towards* another, rather than 'projecting' *into* the feelings of another. Empathy might be likened to *insight* into the feelings of another ('This is how you feel . . .'), as compared with a sympathetic *associating with* or *feeling alike* ('I know just how you feel . . .'), or indeed questioning ('How do you feel . . .?'). In the definitions of *sympathy* appear the words '*approval for*' and '*agreement with*', which again might help us to differentiate. *Empathy* is not to do with agreement or approval, but rather it is to do with insight into the feelings of another – *without judgement*.

Then there was Carl Rogers

Just one of several possible ways into an exploration of empathy is to chart the history of this therapeutic condition with reference to developments in client-centred thinking and practice. To this end, 'snapshots' have been selected from near the beginning, somewhere around the middle, and near the end of the career of Carl Ransom Rogers (1902–1987). So what follows is selective, for while researching this chapter it became abundantly clear that charting the historical development of empathy in the writings of Rogers could not be undertaken adequately without significant reference to other con-textual elements – and especially to what were to become known as the *core conditions* (of congruence, unconditional positive regard and empathic under-standing), and indeed to all six of the necessary and sufficient conditions for therapeutic personality change.   Nevertheless, the snapshot references that follow were chosen with a primary focus on empathy – after all, this is a book about empathy!

Later in his life (in 1986), Carl Rogers was asked (by David Russell) to recall his first ever experience of empathy. Carl had vivid memories of being interviewed for a sexual issues survey by Dr Gilbert Van Tassel Hamilton. Rogers said that up until that moment in the 1930s:

> I think I had learned how to go more deeply into social interchange with friends, but as far as being really, thoroughly, deeply listened to, I don't know that that had been a part of my experience.
>
> (Rogers and Russell 2002: pp. 97–8)

Through being interviewed by Dr Hamilton, Carl Rogers realised all kinds of things about the self-expression of 'hidden' subject matter and how risky it can be to both say things and be heard. He also learned, through 'being on the receiving end', something about how someone might really listen to another person.

Way back in 1942, Carl Rogers wrote in his book *Counseling and Psychotherapy* that the *four* necessary qualifications to practise counselling were *objectivity*, *respect for the individual*, an *understanding of the self* and *psychological knowledge* (Rogers 1942: pp. 253–6). *Objectivity*? Students of client-centred therapy, especially those more familiar with modern thinking and practice, might feel somewhat surprised that the very first condition identified by Carl Rogers was *objectivity*, as it has a somewhat cold, dispassionate, disengaged feel to it. However, even in the early days of Carl's work we can see the origins of what was to come. For instance, he wrote:

> There are certain personal qualities which should be present if the individual is to develop into a good counselor, but there is no reason to believe that these are any more rare than the qualities which would be necessary for a good artist or a first-class aviation pilot. This is definitely true if we are talking of the client-centered type of counseling.
>
> (Rogers 1942: p. 253)

Here, then, is a message of hope for the would-be therapist. Even in 1942 Carl Rogers was saying that *being a person* was a primary requisite for counsellors – that therapists are not a 'class apart' from others. However, Rogers did write somewhat scathingly of 'directive' counsellors, stating that even in recorded interviews directive counsellors opined with some assurance on issues as diverse as the philosophy of life, racial prejudice, history, parents, marriage, vocational choice, discipline and:

> in fact, all the puzzling personal questions which a perplexed indi-
> vidual can face. The directive counselor, to be sure, has need of more
> omnipotent qualities . . . obviously a generous portion of supernatural
> wisdom is required of the individual who takes such an attitude
> toward counseling.
>
> (Rogers 1942: p. 253)

could be applied to ? T approach

Rogers proposed a more modest aim for client-centred counselling, namely, to free individuals in order that they might utilise their own capacities and resolve issues for themselves. Within this context, he believed that a prerequisite for counsellors was that they would be 'sensitive to human relationships', and he wrote of awareness and 'social sensitivity' – sensing and being alert to subtle differences – and of these qualities combining to form a solid and natural foundation for the would-be therapist.

Objectivity

Having presented social sensitivity as a *pre*-condition, Rogers then identified *objectivity* as the first condition:

> It is generally conceded that to be helpful as a therapist the clinical
> worker needs to have an objective attitude. This has been variously
> described as 'controlled identification', as 'constructive composure',
> and as 'an emotionally detached attitude'.
>
> (Rogers 1939: p. 281)

This, as stated above, was back in 1939 – and it would seem to bear little relation to what we now understand as empathic understanding. Yet the seeds of what was to grow into the fifth necessary and sufficient condition for client-centred therapy were evident. Perhaps just as Carl remarked upon the actualising tendency being akin to tuber tendrils seeking the light in a darkened environment, so an analogy might be drawn from the words that followed, when he quoted himself:

> The term as used in clinical practice is defined somewhat differently. . . .
> There is included in the concept a capacity for sympathy which will
> not be overdone, a genuinely receptive and interested attitude, a deep
> understanding which will find it impossible to pass moral judgments or
> be shocked and horrified.
>
> (Rogers 1939: p. 283)

So in a sense, although Rogers used the word 'objectivity', he seemed to be somehow refining or redefining its everyday definition, stressing that this objectivity was neither a 'cold and impersonal detachment' nor 'deeply sympath-etic and sentimental'. He also stated that the client-centred counsellor does not become so immersed in the world of the client as to be unable to enable that client. There is, he wrote:

> a degree of sympathetic 'identification' . . . sufficient to bring about an understanding of the feelings and problems . . . but an identification which is 'controlled', because understood, by the therapist.
>
> (Rogers 1939: p. 283)

Might we be able to see herein the origins of what Rogers later described as the 'as if' quality of empathic understanding – entering a client's world of feelings and meanings as if it were your own, yet without ever losing the *as if* quality? Just as empathic understanding is now seen as the fifth of the six necessary and sufficient conditions, so in 1942 *objectivity* was set within the context of other conditions, which were *respect for the individual, self-understanding* and *psychological knowledge*.

Respect for the individual

Carl wrote that *respect for the individual* meant a 'deep-seated respect' for the integrity of the client. The zealous, reforming therapist, who seeks to mould clients in the image of the counsellor, would be unable to establish a growth-promoting relationship with clients. *Acceptance of clients as they are*, and permitting clients the freedom to discover their own personal solutions, were seen as the key factors in respecting the individual. Here we can see the beginnings of what would later become known as *unconditional positive regard*.

Self-understanding

An *understanding of the self* included self-knowledge in terms of 'emotional patterns, and . . . limitations and shortcomings', for:

> Thoroughly to understand and be objective . . . the therapist must have some insight into his own personality.
>
> (Rogers 1939: p. 283)

Carl Rogers wrote of the necessity of personal insight in order to achieve an awareness of inner warps and biases due to personal feelings and prejudices, stressing the importance of supervision and counselling for therapists (the therapy being provided by the therapist's supervisor if necessary!). So here we can see the origins of what became known as *congruence*. It is perhaps interesting to note at this point that in 1942 Rogers placed objectivity *first*, whereas in later years empathic understanding became the *last* of the three therapist conditions. Yet even in these early days, there was a sense in which respect and self-understanding (later to become unconditional positive regard and congruence) were seen as necessary precursors to objectivity.

Psychological knowledge

Finally, Rogers wrote of *psychological knowledge*, stating that:

the therapist can scarcely expect to do satisfactory work without a thorough basis of knowledge of human behavior and of its physical, social and psychological determinants.

(Rogers 1939: p. 284)

Carl believed that many might expect psychological knowledge to be the *first* condition – yet stunning intellect was no guarantor of effective therapy:

The essential qualifications of the psychotherapist lie primarily . . . in the realm of attitudes, emotions and insight, rather than in the realm of intellectual equipment.

(Rogers 1939: p. 284)

Even earlier than this (in *Family* magazine in 1937), Carl had written of non-forcing, non-critical, accepting 'workers' who gave clients the fullest opportunity to grow.

Just four years on from the publication of his book entitled *Counseling and Psychotherapy*, marked developments could be found. In 1946 in his article entitled 'Significant aspects of client-centered therapy', Carl Rogers was still writing of 'non-directive' therapy and 'counselor techniques and procedures', and although he did acknowledge the influences of Freud, Rank, Taft and Allen, he also dismissed eclectic approaches as being 'not so fruitful'. He was able to state that, derived from practical clinical experience, there had been for him a movement *away from* guiding and directing and *towards* new ways of being (often seen as 'heretic' by other approaches to therapy) that seemed more helpful in terms of bringing about a meaningful therapeutic process. He identified three distinctive elements of what was beginning to be termed client-centred counselling, namely a *predictable process*, a *trust in the capacity of clients*, and *the therapist/client relationship*.

Predictable process

Carl Rogers stated that *if* a relationship founded on certain characteristics is initiated, *then* a predictable chain of events unfolds. At this time, he identified six elements (or 'conditions') necessary to bring about an experience within which the growth forces within clients are most likely to be tapped. These were that the counsellor:

1 holds a belief that individuals are responsible for themselves and is willing for people to keep self-responsibility
2 trusts in the growth forces within individuals, rather than in counsellor expertise and knowledge
3 creates a warm and permissive atmosphere within which clients are free to explore (or withhold) anything they wish to
4 sets limits only on behaviours, not on feelings or attitudes (deemed mostly relevant to working with children)
5 uses only procedures and techniques that convey deep understanding (such as 'sensitive reflection and clarification'), and does not express approval or disapproval
6 remains committed to working within the above five principles, in that the

counsellor does not question, probe, blame, interpret, advise, suggest, persuade or reassure.

Rogers maintained that whether the therapeutic process is relatively shallow or relatively deep, client-centred characteristics are fundamentally the same.

The capacity of the client

What made the therapeutic process predictable for Rogers was the discovery that:

> within the client reside constructive forces whose strength and uniformity have been either entirely unrecognized or grossly underestimated.
>
> (Rogers 1946: p. 417)

Carl had come to believe that clients, due to their tendencies to actualise, are *capable* of many things – such as beneficial emotional release at their own time and pace, effective self-exploration (even of denied feelings and attitudes), looking after their psychological selves, exploring the connections between self and non-self, and being strong and resourceful enough to come up with steps towards a positive journey through life. *If* the counsellor initiates a client-centred relationship, *then* all of these things become more likely. In stating that the more deeply the therapist trusts in the constructive forces within individuals the more deeply these tendencies are connected with, Rogers placed this emerging client-centred approach at an 'opposite pole' – both theoretically and practically – to the 'omnipotent counsellor with supernatural wisdom'. Rogers said that he now knew:

> that the client knows the areas of concern which he is ready to explore; that the client is the best judge as to the most desirable frequency of interviews; that the client can lead the way more efficiently than the therapist into deeper concerns; that the client will protect himself from panic by ceasing to explore an area which is becoming too painful; that the client can and will uncover all of the repressed elements which it is necessary to uncover in order to build a comfortable adjustment; that the client can achieve for himself far truer and more sensitive and accurate insights than can possibly be given to him; that the client is capable of translating these insights into constructive behavior which weigh his own needs and desires realistically against the demands of society; that the client knows when therapy is completed and he is ready to cope with life independently. Only one condition is necessary for all these forces to be released, and that is the proper psychological atmosphere between client and therapist.
>
> (Rogers 1946: pp. 418–19)

How different 'the client will protect himself from panic by ceasing to explore an area which is becoming too painful' feels to identifying *resistance* that must be overcome by the therapist, or *avoidance* that the analyst must circumvent in some way! The therapist can empathise, understand, accept, be with and, as a consequence, *move through* such phases *alongside* the client, rather than challenging, overcoming or circumventing. Another issue arising from this trust in the capacity of the client might be to question how 'the client knows when therapy is completed' fits with the introduction of *time-limited* therapy, which many believe to be driven by economic rather than therapeutic considerations. (Incidentally, figures from several sources indicate that where therapy is *not* time-limited, the average number of sessions per client is no more than six. This raises the question of why many service providers insist that client time must be managed, when philosophically, therapeutically and economically there would seem to be evidence that clients do indeed know best what is beneficial.) (e.g. Budman and Gurman 1988; Tolman 1990; Mellor-Clark *et al.* 2001.)

Client-centred therapy is so named because of the trust in and primary reliance upon the internal resources of the client, and Rogers believed that:

> There is no other aspect of client-centered therapy which comes under such vigorous attack. It seems to be genuinely disturbing to many professional people to entertain the thought that this client upon whom they have been exercising their professional skill actually knows more about his inner psychological self than they can possibly know, and that he possesses constructive strengths which make the constructive push by the therapist seem puny indeed by comparison. The willingness fully to accept this strength of the client, with all the re-orientation of therapeutic procedure which it implies, is one of the ways in which client-centered therapy differs most sharply from other therapeutic approaches.
>
> (Rogers 1946: p. 419)

Client-centred therapeutic relationships

Unlike therapies in which the counsellor uses skills, techniques and expertise upon clients, in client-centred work the fundamental starting point is the creation of an atmosphere within which clients may choose to grow. Carl identified some of the characteristics of such a relationship. For instance, clients would experience it as safe, warm, understanding and accepting. He also stated that 'the sense of communication' was a very important factor, in that not only could clients communicate anything they wished to, but also such communications would be understood rather than evaluated in any way. *The therapy session really does belong to the client.* Rogers quoted a Minister in counselling training as saying that one of the difficulties (trainee) therapists may face is when clients believe that the counsellor ought to be able to offer guidance – and the therapist values client-centred theory yet wonders whether it is enough, whether it goes far enough in actual practice. How can one leave a person struggling and helpless when one could help them by showing them the way out? The Minister posed some fundamental questions:

> Do you believe that all people truly have a creative potential in them? That each person is a unique individual and that he alone can work out his own individuality? Or do you really believe that some persons are of 'negative value' and others are weak and must be led and taught by 'wiser, stronger' people?

> In other methods you can shape tools, pick them up for use when you will. But when genuine acceptance and permissiveness are your tools it requires nothing less than the whole complete personality. And to grow oneself is the most demanding of all.
>
> (Rogers, 1946: p. 420)

Rogers wrote that the use of the term 'client-centred' arose through therapy being more effective the more the therapist strives to understand clients as clients experience themselves.

By 1946, Carl Rogers had realised that traces of 'subtle directiveness' were still to be found, and thus there was an increasing emphasis on the therapist discarding the need to diagnose (together with a personal need to display 'diagnostic shrewdness'), professional evaluations and prognoses, and the need to guide the client in order to be helpful. Rather, the client-centred therapist would strive to *provide understanding of the way the client seems to himself at this moment*, and the client would have the capacity to 'do the rest'. Rogers wrote:

> Client-centered counseling, if it is to be effective, cannot be a trick or a tool. It is not a subtle way of guiding the client while pretending to let him guide himself. To be *effective*, it must be genuine.
>
> (Rogers 1946: p. 420)

An authentically client-centred therapeutic relationship creates a 'psychological atmosphere which releases, frees, and utilises deep strengths in the client'.

Somewhere in between

Sigmund Koch invited Carl to contribute to his series of books entitled *Psychology: a study of a science*, and Carl had been wondering whether his developmental learning and growth might be drawn together into a consistent and understandable framework. We have already heard that he put a great deal of effort into this piece of work (*see* Prelude), and recalled:

> As one young psychologist with a background in mathematics said to me recently, 'It is so precise! I could restate it in mathematical terms.' I must confess this is close to my opinion.
>
> (Rogers and Evans 1981: pp. 135–6)

Yet Carl was puzzled and it *is* puzzling: his most rigorous piece of hard work being the most thoroughly ignored. Let us ignore it no longer!

Rogers gave us many important and fundamental definitions within his 1959 chapter, several of which follow. Given that the very term client-centred arose from the primary focus on the internal frame of reference of the person seeking help, let us begin there.

Internal frame of reference

> This is all of the realm of experience which is available to the awareness of the individual at a given moment. It includes the full range of sensations, perceptions, meanings, and memories which are

available to consciousness. The internal frame of reference is the subjective world of the individual. Only he knows it fully. It can never be known to another except through empathic inference, and then can never be perfectly known.

(Rogers 1959: p. 210)

The only way that I can *internally* validate my own experience is to focus further on my experiencing – to question, probe, examine, explore and reflect upon myself. If I attempt to check out my feelings with others, I observe myself as an object and thus am viewing myself from *external* reference points. As we shall see by and by, Rogers was to define the *process* of empathising as 'perception-checking', and the origins of this can be seen here:

> Knowledge which has any 'certainty', in the social sense, involves the use of empathic inference as a means of checking, but the direction of that empathy differs. When the experience of empathic understanding is used as a source of knowledge, one checks one's empathic inferences with the subject, thus verifying or disproving the inferences and hypotheses implicit in such empathy. It is this way of knowing which we have found so fruitful in therapy.
>
> (Rogers 1959: pp. 211–12)

External frame of reference

To perceive another person (or object) *without* empathising is to operate from an external frame of reference. Viewing objects in this way is common. Rarely, for instance, do we experience empathy for the chairs we sit upon, or the clothes we wear, or the pavements we walk along, or for the pen or word-processing equipment we use to write with! It could be said that none of these objects have any awareness of their own experiencing with which it would be possible to empathise. However, if *people* are viewed in the same way as objects (as, sadly, they so often seem to be, both from within and from outside the 'caring professions'), then human beings too become like inanimate objects.

An example comes to mind. A trainee counsellor on a client-centred course who was on placement with a bereavement service had been taught by that agency that there were five stages of grieving which clients needed to go through (and within a limited time frame, too). Counsellors were encouraged to identify the stage a client was in and attempt to move them on to the next stage, then the next, and so on. No attention at all was paid to fully receiving clients in the moment, and little attention was given to the inner world of client experiencing – little respect was shown for the capacity of clients to creatively discover their own constructive processes. Clients became objectified (in effect, a group of people had been pathologised) – they had to fit a model and both the therapist and the bereaved client needed to work in a way and at a pace dictated by that model. Furthermore, the model was somewhat white and Eurocentric, and if the bereaved client was not 'sorted' by having moved (or more accurately, by having been moved by the counsellor) through all five stages within the set time limit, their grieving was deemed to be 'abnormal' and subject to possible psychiatric referral.

Empathy

> The state of empathy, or being empathic, is to perceive the internal
> frame of reference of another with accuracy, and with the emotional
> components and meanings which pertain thereto, as if one were the
> other person, but without ever losing the 'as if' condition. Thus it
> means to sense the hurt or the pleasure of another as he senses it, and
> to perceive the causes thereof as he perceives them, but without ever
> losing the recognition that it is *as if* I were hurt or pleased, etc. If this 'as
> if' quality is lost, then the state is one of identification.
>
> (Rogers 1959: pp. 210–11)

A theory of therapy: if–then

Carl Rogers described the theory of client-centred therapy as being of the *if–then*
variety, even 'reverting' to theoretical logic:

$$B = (f)A \qquad C = (f)A \qquad B + C = (f)A \qquad C = (f)B$$

In the above formula, (f) represents *is a function of*, A denotes *conditions of the
therapeutic process*, B denotes *the process of therapy* and C denotes *outcomes in
personality and behaviour*. In other words, *if* (A) two persons are in contact, one
(the client) being incongruent and the other (the therapist) being congruent,
respectful and empathic, and the client perceives the therapist's unconditional
positive regard and empathic understanding at least to a minimal degree, *then* as a
consequence there will be (B), a therapeutic process. Also, *if* (A) *then* there will as
another consequence be predictable outcomes in terms of personal growth. Thus
both therapeutic process (B) and personal growth (C) are *functions of* the six
necessary and sufficient conditions. Likewise, predictable outcomes in terms
of personality change (C) are a consequence of (A), the necessary and sufficient
conditions. Rogers restated this by saying that the necessary and sufficient
conditions are independent variables, and client-centred therapeutic process
depends upon their existence. Thus (B) is a dependent variable. Once (B)
(therapeutic process) exists, then it becomes the independent variable, and (C)
(outcomes) are dependent variables. We can note, too, the idea that the greater
the degree of the existence of the necessary and sufficient conditions, the more
profound the process and outcomes are likely to be.

Necessary and sufficient

Carl Rogers believed that client-centred therapy differed markedly from conventional therapies, in that not only had *necessary* conditions been identified, but also
these conditions were deemed to be *sufficient*. He wrote that irrespective of the
particular characteristics of clients:

> It has been our experience to date that although the therapeutic
> relationship is used differently by different clients, it is not necessary
> or helpful to manipulate the relationship in specific ways for specific
> kinds of clients. To do this damages, it seems to us, the most helpful

and significant aspect of the experience, that it is a genuine relation-ship between two persons, each of whom is endeavoring, to the best of his ability, to be himself in the interaction.

(Rogers 1959: pp. 213–14)

Having defined conditions that were both necessary and sufficient, Carl then wondered *which* of the therapist conditions was the more important. He comes to the conclusion that:

For therapy to occur the wholeness of the therapist in the relationship is primary, but a part of the congruence of the therapist must be the experience of unconditional positive regard and the experience of empathic understanding.

(Rogers 1959: p. 215)

This cannot be over-emphasised. It is simply not enough for the therapist to *be* congruent. Client-centred congruence has little meaning or value unless uncon-ditional positive regard and empathic understanding are fundamentally embedded aspects that permeate the therapist's way of being, and that are therefore authentically experienced by the therapist and may be perceived as genuine by clients. However, to live by such profound attitudinal values towards others in a sustained way might not be possible or even desirable (the therapist has needs, wishes and rights, too). So Rogers also considered it important to note that client-centred theory focuses upon conditions *within the therapeutic relation-ship*, pointing out that if it were to be necessary for the client-centred therapist to be congruent at all times, there would be no client-centred therapists!

Thus it is that imperfect human beings can be of therapeutic assistance to other imperfect human beings.

(Rogers 1959: p. 215)

The actualising tendency

Carl Rogers also referred to the fundamental place in client-centred therapy of the *actualising tendency*, which he defined as:

the inherent tendency of the organism to develop all its capacities in ways which serve to maintain or enhance the organism.

(Rogers 1959: p. 196)

Carl believed that certain therapeutic consequences of a belief in the actualising tendency were inescapable, as:

It means that psychotherapy is the releasing of an already existing capacity in a potentially competent individual, not the expert manip-ulation of a more or less passive personality. Philosophically it means that the individual has the capacity to guide, regulate and control himself, providing only that certain definable conditions exist.

(Rogers 1959: p. 221)

The belief that human beings are, by their inner natures, constructive and trustworthy was passionately held by Carl, who stated that:

> For me this is an inescapable conclusion from a quarter-century of
> experience in psychotherapy.
>
> (Rogers, 1957b: p. 299)

Rogers believed that there was no need to devise therapies based on control or
socialisation, for when free from defensiveness, humans can be trusted to be
positive, forward-moving, constructive, social beings who display deep needs for
affiliation and communication with others. Humans have innate needs to like,
love, respect and care for others beneath their sometimes more superficial hostile
or aggressive behaviours. Indeed, such 'negative' actions are often the result of a
squashing or thwarting of our fundamental natures. Carl wrote:

> I have little sympathy with the rather prevalent concept that man is
> basically irrational, and that his impulses, if not controlled, will lead to
> destruction of others and self. Man's behavior is exquisitely rational,
> moving with subtle and ordered complexity toward the goals his
> organism is endeavoring to achieve. The tragedy for most of us is
> that our defenses keep us from being aware of this rationality, so that
> consciously we are moving in one direction, while organismically we
> are moving in another.
>
> (Rogers 1957b: p. 300)

In other words, the internal barriers created by our incongruence serve to thwart
our ability to be fully functioning individuals.

By 1967, the research of Carl Rogers and his colleagues (in this instance,
Eugene Gendlin, Donald Kiesler and Charles Truax) was focused on such
questions as 'What do we as therapists do that actually leads to constructive
change in our clients or patients?'. It had already been ascertained that only a
small fraction of the events occurring in counselling sessions actually had any
really significant therapeutic impact, so what were these key factors? Likewise, it
had also become clear that therapeutic relationships differ from therapist to
therapist, *and* that an individual therapist will be different with different clients,
and that an individual therapist will differ over time with the same client. Thus a
therapist might adopt different language styles (from 'restricted' to 'elaborate')
with different clients, and might move from gentle, tentative communications in
early sessions with a client to more confident, understanding communications in
later sessions with that same client. This would seem to give strong support to the
client-centred belief that attitudes rather than behaviours are most relevant to the
process of therapy.

Rogers pointed out that many writers from many orientations (psychoanalysts
Ferenczi, Alexander, Schafer, Halpern and Lesser, and eclectic writers Strunk,
Raush and Bordin, Strupp, Hobbs, Fox and Gordin) as well as client-centred
authors (Dymond, Jourard, Truax, and Rogers himself) had stressed the importance of:

> the therapist's ability to understand sensitively and accurately the
> inner experiences of the client or patient, . . . the maturity of the
> therapist and his integration or genuineness within the relationship . . .
> [and] warmth and his acceptance of the individual with whom [the

therapist] is working . . . Cutting across parochial viewpoints, they can be considered as elements common to a wide variety of therapies.

(Rogers 1967: p. 98)

Thus we were again reminded of what became known as the three 'core' or therapist conditions for client-centred therapy. Rogers emphasised that all three of these common elements were 'logically intertwined'. For instance, while empathic understanding is essential, it is a prior requirement that acceptance and prizing be present if a therapist is to be 'deeply sensitive to the moment-to-moment "being" of another person'. While unconditional positive regard is necessary for empathy, neither have much meaning unless they are authentic.

Congruence

Rogers pointed out that we are unlikely to reveal our inner selves to people who 'put up a front' or play a role (e.g. a professional façade) or feign understanding – in other words, we are not likely to go too far or too deep with incongruent (inauthentic or phony) people. We are more likely to share with therapists who are *being*, not *denying* (or distorting) who they are, and where a striving to be respectful and empathic is deeply embedded within that way of being. In an attempt to circumvent misunderstanding, Carl also clarified some of the things that being congruent does *not* mean. For instance, it does not mean that therapists overtly express all of their feelings (thereby over-burdening clients), nor does it mean an impulsive blurting out of feelings without thought, nor does it mean total therapist self-disclosure.

Congruence *does* mean that the therapist does not *deny* any experienced feelings, and that there is a willingness to *transparently be* any persistent feelings that arise within the relationship, and to *communicate* them within the relationship *if appropriate*. When might it be appropriate to share such feelings? For Carl, this would seem to be either if the feeling persists over time or if the feeling gets in the way of deeply hearing a client. When it *is* appropriate to share such feelings, Rogers stated that it was important for the therapist to *own* them (rather than 'project' them on to the client). The client-centred therapist being in touch with inner experiencing is ongoing, and this can be communicated, too. Thus the discomfort in or anxiety about sharing the feeling can accompany the feeling itself. This, stated Rogers, enables the relationship to grow and become more real.

Thus if a therapist feels boredom and communicates this to a client (owning the feeling and their apprehension about sharing the feeling), it emerges that the therapist experiences distance from the client and would like to connect more deeply. The boredom – a barrier – dissipates and a new sensitivity emerges. It may be that the therapist becomes more able to deeply hear the client again – even to hear the client's shock or hurt at what has been said. Both client and therapist will be more genuine as a consequence of the therapist daring to be real. Carl Rogers said that there had been a gradual shift between 1950 and the mid to late 1960s, and that client-centred counsellors had come to feel increasingly free about being 'whole persons' in the therapeutic encounter. Carl said that his 'trial ground' for being more fully himself (for reasons that he did not fully understand) was working with groups, within which:

I express anger, and affection, and annoyance and all kinds of things, as well as being very responsive to hurt. Hurt arouses in me feelings of really wanting to be empathic so that a lot of the therapeutic attitudes that I've stressed, I think, are very real parts of me, and of many therapists, and so they need expression as well, but other feelings, too, have equal validity.

(Rogers 1969, in Frick 1989: p. 102)

It may be that therapists struggle more with the communication of feelings that are experienced, anticipated or heard as being *negative*. Ideally, it might possibly be said that the therapist would only experience positive attitudes – yet if this is not the case, Rogers believed that it is harmful to clients if such feelings are hidden. He considered that every therapist has negative attitudes from time to time, but that it is preferable for them to be expressed, and for the counsellor to be real, not false.

In my relationships with persons I have found that it does not help, in the long run, to act as though I were something that I am not. It does not help to act calm and pleasant when actually I am angry and critical. It does not help to act as though I know the answers when I do not. It does not help to act as though I were a loving person if actually, at the moment, I am hostile. It does not help for me to act as though I were full of assurance, if actually I am frightened and unsure.

(Rogers 1961: p. 16)

Why might it be appropriate or reasonable for therapists to share persistent feelings that arise within the therapeutic relationship with their clients? According to Carl Rogers, it is because:

It is not an easy thing for the client, or for any human being, to trust his most deeply shrouded feelings to another person. It is even more difficult for a disturbed person to share his deepest and most troubling feelings with a therapist. The genuineness, or congruence, of the therapist is one of the elements in the relationship which makes this risk of sharing easier and less fraught with dangers.

(Rogers 1967: p. 102)

Carl Rogers gave some possible indicators of therapist *incongruence*. They included defensiveness (which might be apparent through inconsistencies between therapist words and paraverbal or non-verbal communicating), or making 'textbook responses' that are intended to sound good rather than being expressions of authentic, respectful empathy. Rogers identified voice qualities and manner of expression as the most usual indicators of congruence, saying that it was fortunate that we are all pretty adept at distinguishing between the authentic and the phony. Carl did become increasingly concerned about the frequent misunderstanding of the notion of therapist self-expression within the concept of congruence, writing that:

it does not mean that the therapist burdens his client with overt expression of all his feelings. Nor does it mean that the therapist discloses his total self to his client. It does mean, however, that the therapist denies to himself none of the feelings he is experiencing and that he is willing to experience transparently any *persistent* feelings that exist in the relationship and to let these be known to his client. It means avoiding the temptation to present a façade or hide behind a mask of professionalism, or to assume a confessional–professional attitude.

(Rogers 1989: pp. 11–13)

Unconditional positive regard

The attitude of high regard held by the client-centred counsellor for the dignity and worth of individuals is a fundamental differentiation between client-centred therapy and a great many other approaches. This had come to mean that, *without reservation*, the therapist *experienced and communicated* a non-judgemental, acceptant warmth, a deep and genuine caring, and a non-evaluative 'pure' ('uncontaminated') prizing of client feelings, thoughts or behaviours. This deep respect for clients included a trust in both their capacity and their right to self-direction (underpinned by a trust in the actualising tendency).

Rogers stated that, as therapists, we might even find painful and negative client expressions easier to accept than good or positive ones, yet *unconditional* positive regard means the acceptance of *all* expressions – no 'ifs' or 'buts' – it is a *non-possessive caring for the client as a separate person*. Yet so often, it seems, we hear phrases like 'I accept the person but not their behaviour', though as Carl said, 'it is doubtful if this is an adequate or true comment', in that although a client behaviour may feel 'bad' or socially unacceptable to a therapist, there is a kind of 'there but for the grace of god' element in acceptance: the behaviour is accepted 'as a *natural consequence* of the circumstances, experiences, and feelings of this client'. By providing a non-threatening climate within which clients are able to choose to explore their inner selves (rather than approving or disapproving), the therapist becomes a companion searching with the client for *client* meanings and values.

Carl Rogers believed that many people who are striving to become client-centred therapists do so because:

> [their] philosophical orientation has tended to move in the direction of greater respect for the individual.

(Rogers 1951: p. 21)

Becoming a client-centred therapist offers an opportunity to put this philosophical belief into practice. Such a trainee counsellor:

> soon learns that the development of the way of looking upon people which underlies this therapy is a continuing process, closely related to the therapist's own struggle for personal growth and integration. He can be only as 'nondirective' as he has achieved respect for others in his own personality organization.

(Rogers 1951: p. 21)

Thus a link emerges between congruence and unconditional positive regard, for as Carl stated time and time again, respect for clients needs to be a deeply integrated, fundamental and integral aspect of the therapist's personality. On the one hand, experiencing deep respect for another can be deeply enriching and energising for the therapist. On the other hand, it can be a strenuous and problematic struggle:

> *I have found it highly rewarding when I can accept another person.* I have found that truly to accept another person and his feelings is by no means an easy thing, any more than is understanding. Can I really permit another person to feel hostile toward me? Can I accept his anger as a real and legitimate part of himself? Can I accept him when he views life and its problems in a way quite different from mine? Can I accept him when he feels very positively toward me, admiring me and wanting to model himself after me? All this is involved in acceptance, and it does not come easy.
>
> (Rogers 1961: pp. 20–21)

The more acceptance the therapist experiences and offers, the more a useful helping relationship exists:

> By acceptance I mean a warm regard for him as a person of unconditional self-worth – of value no matter what his condition, his behavior, or his feelings. It means a respect and liking for him as a separate person, a willingness for him to possess his own feelings in his own way. It means an acceptance of and regard for his attitudes of the moment, no matter how negative or positive, no matter how much they may contradict other attitudes he has held in the past. This acceptance of each fluctuating aspect of this other person makes it for him a relationship of warmth and safety, and the safety of being liked and prized as a person seems a highly important element in a helping relationship.
>
> (Rogers 1961: p. 34)

Accurate empathic understanding

Empathic understanding also needs to be deeply embedded within the personality of the counsellor:

> In client-centered therapy the client finds in the counselor a genuine alter ego in an operational and technical sense – a self which has temporarily divested itself (so far as possible) of its own selfhood, except for the one quality of endeavoring to understand.
>
> (Rogers 1951: p. 40)

Carl Rogers offered us a challenge, and through trying this we could test the degree of our empathic understanding:

> The next time you get into an argument with your wife, or your friend, or with a small group of friends, just stop the discussion for a moment and for an experiment, institute this rule. 'Each person can speak up for himself only *after* he has first restated the ideas and feelings of the previous speaker accurately, and to that speaker's satisfaction.' You see

what this would mean. It would simply mean that before presenting your own point of view, it would be necessary for you to really achieve the other speaker's frame of reference – to understand his thoughts and feelings so well that you could summarize them for him. Sounds simple, doesn't it? But if you try it you will discover it is one of the most difficult things you have ever tried to do.

(Rogers 1961: p. 332)

Once therapist congruence and unconditional positive regard *are* present:

The ability of the therapist accurately and sensitively to understand experiences and feelings *and their meaning to the client* during the moment-to-moment encounter of psychotherapy constitutes what can perhaps be described as the 'work' of the therapist

(Rogers 1967: p. 104)

Rogers identified two major aspects of empathic understanding, namely *immediacy* and *communication*. Over the years it seems as though there have been many interpretations of the term 'immediacy', including the therapist spontaneously blurting out every passing feeling (which Carl so much disliked). What Carl Rogers meant by this term was that *immediacy in experiencing empathy* was a necessity. After all, it is of little use if the therapist only achieves insight or feels completely at home in the universe of the client some time after the counselling session has ended (for example, in the car on the way home!). The capacity to sensitively communicate client inner meanings back to the client in such a way that they *belong* to the client – and in a language *attuned* to the client – was the other crucial factor for effective client-centred therapy. Such communicating serves to clarify and expand client awareness of feelings or experiences, and includes the 'sensitive play of voice qualities which reflect the seriousness, the intentness, and the depth of feeling' as well as the actual words used.

Rogers again differentiated between *internal* and *external* frames of reference. At that time, the most common form of understanding in therapies other than client-centred was an external, even impersonal, evaluative diagnostic formulation that implied 'I know what is wrong with you' (an understanding from the outside looking in), as compared with 'an accurate and sensitive grasp of events and experiences and their *personal meaning to the client*' (in other words, an understanding from within the world of client feelings and meanings). Carl offered us questions that he had found useful when exploring the degree to which he could permit himself to experience empathy. They included the following:

Can I let myself enter fully into the world of his feelings and personal meanings and see these as he does? Can I step into his private world so completely that I lose all desire to evaluate or judge it? Can I enter it so sensitively that I can move about in it freely, without trampling on meanings which are precious to him? Can I sense it so accurately that I can catch not only the meanings of his experience which are obvious to him, but those meanings which are only implicit, which he sees only dimly or as confusion? Can I extend this understanding without limit?

(Rogers 1961: pp. 53–4)

Rogers also pointed out that the communication of *striving* to understand can in itself be therapeutic – if nothing else, it both lets clients know that they are *worth* the effort, and encourages them to communicate further. However, a consistent inability to do more than strive might not be so helpful. Indeed, if the therapist does not *achieve* accurate empathic understanding over a significant period, a consequence may be that the client feels ever more helpless and despairing.

Carl also gave some pointers to factors characteristic of low, intermediate, high and very high levels of empathy. At the *extremely negative* end of the scale, the therapist shows no comprehension at all of the feelings and meanings that the client is trying to communicate, and *fairly negative* elements included the therapist being perceived as hostile, superior, and even disgusted. Clues to *low levels* of empathy included the therapist going off at a tangent, communicating only the client's most obvious feelings, misinterpreting, being intellectually preoccupied, attending to content rather than what the client is experiencing in each moment, evaluating, giving advice, using inappropriate language or timing, or reflecting upon therapist experiencing rather than the client's experiencing.

Intermediate levels of empathy are indicated by the therapist accurately sensing and communicating the client's most obvious feelings, yet only occasionally sensing and communicating less apparent ones. In so doing, the counsellor 'may anticipate feelings that are not current or may misinterpret the present feelings'.

High levels of empathy are indicated by the therapist communicating deeper yet slightly inaccurate awareness of feelings and experiences – 'pointing' to more hidden feelings yet 'unable to grasp their meaning':

> At a *very high level of empathic understanding* [my italics] the therapist's responses move, with sensitivity and accuracy, into feelings and experiences that are only hinted at by the client. At this level, underlying feelings or experiences are not only pointed to, but they are specifically identified so that the content that comes to light may be new but it is not alien. At this high level the therapist is sensitive to his own tentative errors and quickly alters or changes his responses in midstream, indicating a clear but fluid responsiveness to what is being sought after in the patient's own explorations. The therapist's words reflect a togetherness with the patient and a tentative trial-and-error exploration, while his voice tone reflects the seriousness and depth of his empathic grasp.
>
> (Rogers 1967: p. 106)

Rogers was interested in the findings of two research programmes undertaken by Fiedler, who found that therapists who were able to totally engage with client communications were most characteristic of a good therapeutic relationship, followed by accurate therapist comments, co-operative therapists, power equality, good therapist understanding, consistent striving to understand, accurate tracking of the client, and a sharing tone of voice (Rogers 1951: pp. 53–4). The theme of levels of empathy is explored more fully in Chapter 4.

Empathic understanding enriches both therapist and client, for as Rogers wrote:

> When I can permit myself to understand . . . it is mutually rewarding. And with clients in therapy, I am often impressed with the fact that even a minimal amount of empathic understanding – a bumbling and faulty attempt to catch the confused complexity of the client's meaning – is helpful, though there is no doubt that it is most helpful when I can see and formulate clearly the meanings in his experiencing which for him have been unclear and tangled.
>
> (Rogers 1961: p. 54)

Towards the end

Carl Rogers stated the obvious – that all approaches to counselling and psychotherapy have a focus on clients. However, he also stated that in client-centred therapy there is a seldom explicated technical connotation, in that while in most approaches achieving understanding of clients is seen as a step towards eliciting a case history for future reference and analysis, or behaviour modification or re-education or whatever (the 'real work' of such therapists):

> Instead, the client-centered therapist aims to remain with this phenomenal universe throughout the entire course of therapy and holds that stepping outside it – to offer external interpretations, to give advice, to suggest, to judge – only retards therapeutic gain.
>
> (Rogers 1989: p. 21)

In 1980, Carl wrote that he was so shocked and appalled at the way empathy had become known as a set of techniques (like 'reflecting' or 'mirroring'), and had become such an insulting caricature of what was in fact 'one of the most delicate and powerful ways we have of using ourselves', that for many years he had written or said little on the subject. Then he wrote *In Retrospect: Forty-Six Years* because he was:

> getting tired of the misunderstanding of the concept of empathy, of superficial notions such as saying back to the person what they said and things like that.
>
> (Rogers and Russell 2002: p. 211)

In reminiscing, Carl remembered how in his earlier days he had learned that 'simply listening' to a client was helpful:

> Yet listening, of this very special kind, is one of the most potent forces for change that I know.
>
> (Rogers 1980: p. 116)

However, the phrase 'simply listening' might be a little misleading:

> I think that one of the elements most likely to be misunderstood is the intensity of empathic listening. People feel, you know, that listening is a sort of passive experience. I feel that's one thing that's badly

misunderstood . . . One of the complexities of the person-centered approach is that in many ways it is a very disciplined approach . . . And there's a discipline involved in helping to make that shift [into the client's world]; there's a discipline involved in being deeply empathic. It means you really do shut things out, and you are focused. Because the essence of it is simple, people forget that there is also disciplined learning and disciplined action in the interaction.

(Rogers and Russell 2002: p. 284)

Then, when Rogers moved to Teachers College, Columbia University, he learned (from Elizabeth Davies or Leta Hollingworth – he quoted Elizabeth in Rogers and Raskin (1989) and Leta in Rogers (1985) – both Rankian social workers) that listening for feelings and emotions was effective, and that these feelings could be 'reflected back' to the client. Although he expressed gratitude for this as it improved his work at that time, he also wrote of:

'Reflect' becoming in time a word that made me cringe.

(Rogers 1980: p. 138)

Rogers then moved to Ohio State University, and the recording of counselling sessions began. Although he believed throughout his life that the recording and analysis of interviews is one of the best routes to self-improvement as a therapist, and although a great deal of client-centred theory was derived from such 'microscopic' studies, there were deeply regrettable consequences, especially in terms of a focus on skills and techniques, as compared with 'a *way of being* that is rarely seen in full bloom in a relationship'.

So for many years Rogers spoke and wrote only of empathic attitude rather than of how empathic attitudes might be communicated, choosing to focus instead on congruence and unconditional positive regard because, he said, although they too were 'often misunderstood . . . at least they were not caricatured'. Only in 1980 did he return to commenting on how an empathic attitude might be *implemented*. He stated that in part this was due to his view that an increasingly technological society was breeding approaches to therapy (citing gestalt, psychodrama, primal therapy, bioenergetics, rational–emotive therapy and transactional analysis as examples) in which the *therapist* was very much the expert and which involved clients being actively manipulated, whereas he believed empathy to be 'possibly the most potent factor in bringing about change and learning', and that an empathic way of being empowered rather than disempowered clients.

Experiencing

Carl Rogers referred to Eugene Gendlin, who formulated the view that at all times human beings can refer to their ongoing inner experiencing. The client-centred therapist accurately gathers this 'felt meaning', and the perceiving of this by clients helps them to focus further on such meanings. Thus clients can access this 'ongoing psycho-physiological flow within' and use it as an inner referent (hence 'internal frame of reference'). Carl said that this psychophysiological flow is very real:

It's the kind of process that enables the therapist to be really in touch with the correct words that are going on in the client's mind . . . One word will not fit; another one that seems quite similar may absolutely fit. It's as though there's some physiological click, and you can see it in the client's face and manner of expression: 'No, that's not quite it; yes, that's it'. Their response indicates whether it is a click or not.

(Rogers and Russell 2002: p. 254)

Imagine, if you will, the therapist venturing an empathic communication: 'You seem angry . . .'. The client pauses a moment, and says, 'Kind of . . . Maybe more *frustrated* than angry . . .'. What is the client *doing* during the pause? When *I* pause in this way, I am taking a few moments to sort of 'stop' myself, or 'freeze-frame' – and I kind of 'look inside myself' to see what is 'really' going on. Take a moment now – do it yourself and *see*.

Thus the client-centered therapist aims to concentrate on the *immediate phenomenal world* of the client. For he believes that it is in confusions or contradictions within this world that the client's difficulties lie. This exclusive focus in therapy on the present phenomenal experience of the client is the meaning of the term 'client-centered'.

(Rogers 1989: p. 22)

However, there is a difference of opinion among client-centred practitioners with regard to quite how this focus on the client's phenomenal world is implemented:

Some aim to convey an understanding of just what the client wishes to communicate. For Rogers, it has felt right not only to clarify meanings of which the client is aware, but also those just below the level of awareness.

(Rogers and Raskin 1989: p. 171)

Also:

This is not to say, however, that the client-centered therapist responds only to the obvious in the phenomenal world of his client. If that were so, it is doubtful that any movement would ensue in therapy. Indeed, there would be no therapy. Instead, the client-centered therapist aims to dip from the pool of implicit meanings just at the edge of the client's awareness.

(Rogers 1989: p. 21)

From such statements we can glean something of different levels of empathy within a therapeutic process. When speaking of the therapist being a companion to clients, Carl said:

I think that the therapist's function is that of being a companion to the client in the client's search for the innermost aspects of self and experience. It is much easier to face a denied experience if someone is right with you.

(Rogers and Russell 2002: pp. 281–2)

It is far better to journey with a *companion* than to travel alone . . .

In striving to 'track' a client, the therapist is sometimes a little way behind yet striving to be alongside, sometimes there is an equal reciprocity, and sometimes the therapist is aware of feelings and meanings only vaguely experienced or perceived by the client. We saw that being too far behind (being unempathic) is not at all helpful – being too far ahead can be hurtful, too. Carl Rogers was mindful of the work of radical psychoanalyst Heinz Kohut, and said:

> Similarly, way back in my early years of training young counselors, some of the most sensitive of them were so empathic that they understood material the client was not yet aware of, was quite frightened of. And when it was verbalized, the client vanished. We began to call them 'blitz therapists'. They had the makings of very good counselors, but they had to learn that you don't verbalize an under-standing that is deeper than the client needs, or wants, or can deal with consciously.
>
> (Rogers and Teich 1992: p. 56)

From state to process

One development that was clear by 1980 was that Rogers had changed from writing about empathy as a *state of being* to defining empathy as a *way of being*, a *process*:

> An empathic way of being with another person has several facets. It means entering the private perceptual world of the other and becom-ing thoroughly at home in it. It involves being sensitive, moment by moment, to the changing felt meanings which flow in this other person, to the fear or rage or tenderness or confusion or whatever he or she is experiencing. It means temporarily living in the other's life, moving about in it delicately without making judgments; it means sensing meanings of which he or she is scarcely aware, but not trying to uncover totally unconscious feelings, since this would be too threatening. It includes communicating your sensings of the person's world as you look with fresh and unfrightened eyes at elements of which he or she is fearful. It means frequently checking with the person as to the accuracy of your sensings, and being guided by the responses you receive. You are a confident companion to the person in his or her inner world. By pointing to the possible meanings in the flow of another person's experiencing, you help the other to focus on this useful type of referent, to experience the meanings more fully, and to move forward in the experiencing.
>
> (Rogers 1980: p. 142)

Carl Rogers added:

> One thing that seems very true to me is that there is no such thing as a perception without a meaning. That is, the human organism immedi-ately attaches a meaning to whatever is perceived . . . The world of reality for the individual is his own field of perception, with the meanings he has attached to those various aspects . . . None of us

knows for sure what constitutes objective reality and we live our whole lives in the reality as perceived.

(Rogers and Evans 1981: pp. 9–10)

Here then, yet again, we see a focus on both feelings *and meanings* as integral aspects of empathic understanding. We see, too, a more dynamic *process* of an *ongoing perception checking* of our empathic understanding than would be the case, say, with simply paraphrasing, or summarising, or asking 'open' questions, or reflecting (feelings or otherwise) in response to what a client has just said. Indeed, it might be helpful if trainee client-centred therapists were to eliminate the word 'response' from their therapeutic vocabulary – it could well be more fruitful to think in terms of striving to respectfully *communicate* empathic understanding than *respond* to what has just been said. It might be, too, that focusing on each individual counsellor 'response' carries with it the risk of empathy being seen as a *state* (any individual therapist communication being rated on a scale from empathic to non-empathic – a kind of 'in that moment it either is or is not empathic' approach), thereby losing sight of the flow, the *process*, the essence of an empathic *way of being* and a therapeutic, reciprocal relationship.

If would-be therapists simply learn to reflect, paraphrase, summarise and the like, there is a tragic loss of both unique individuality and potential. Research concludes that there is a great variety of counsellor communications, influenced by both the life history of a counsellor and their personal style. Individual therapists must *choose* each communication that they utter:

> Each therapist has his favorite themes, specific sensitivities, and degrees of high attunement. With each special attunement, there are characteristic areas of low emphasis in which each therapist is fairly opaque, insensitive, and even neglectful.
>
> (Barton 1974: p. 201)

Perhaps a couple of personal examples might be helpful at this point. As a humanist I do not believe in any God or gods, and I became aware that I was less sensitive and open to religious beliefs than I would ideally like to be; my experiences of the church and other religious establishments had also generated bias and prejudice. I have therefore had to work on my acceptance in these areas. On the other hand, being a male who once experienced a sexual assault (when I was 12 years old) has, I believe, served to sensitise me to clients who have been violated (although I have to be constantly vigilant in differentiating between empathy and identification). A third example might be worthy of mention. Often in training students decide to explore empathic communications (*not* reflecting or paraphrasing) in a group setting, and several different communications emerge in response to one client disclosure. Yet examine these varied communications, and many of them will be positively therapeutic – celebrate diversity!

To clarify yet further the range of empathic expressions available to individual therapists, we can turn to Peggy Natiello, who took a close look at a 30-minute session that a client called Mary had with Carl Rogers:

> Even in the half-hour relationship presented here, there is a subtle change and expansion of Rogers' style of response. Most of his responses are geared toward empathic understanding, but there is a

good deal of variety in the form they take. He asks direct questions to further his understanding; repeats what he thinks he is hearing, sometimes literally, sometimes metaphorically; links ideas from earlier in the session to later disclosures; completes the client's sentences; helps her find the right words; speaks in the first person as though he were her; and even jokes a little.

(Farber *et al.* 1996: p. 126)

Some trainees and practitioners struggle with the notion of therapists making choices each time they communicate to clients, for might this not imply that client-centred therapy is not so 'non-directive' as they had thought it to be? Maria Villa-Boas Bowen tackled Rogers on this very issue. He replied (in a personal communication, dated 16 January 1983):

When I try to respond to the feeling that is most important to the client, my choice of what is most important is certainly influenced by my own personality, past history, and so forth. I agree that that is most assuredly true and I don't think I have ever said [otherwise]. . . . If it is openly recognized that such responses are partly shaped by the therapist's *perception* and that this perception is shaped both by the client's expression and by the therapist's personality, we may be able to avoid imitative 'modeling'.

(Farber *et al.* 1996: pp. 90–91)

As we explore the various aspects involved in empathic understanding, we see once again the importance of the therapist being sensitive, non-judging and unafraid, for:

To be with another in this way means that for the time being, you lay aside your own views and values in order to enter another's world without prejudice. In some sense it means that you lay aside your self; this can only be done by persons who are secure enough in themselves that they know they will not get lost in what may turn out to be the strange or bizarre world of the other, and that they can comfortably return to their own world when they wish . . . being empathic is a complex, demanding, and strong – yet also a subtle and gentle – way of being.

(Rogers 1980: p. 143)

Here we might again make links with both *congruence* and *unconditional positive regard*, for only if this intense and respectful focus on the inner world of the client is a consequence of deeply held and integrated beliefs can it be deemed to be authentic – that is, congruent. I need to be *genuinely* interested in the world of feelings and meanings of the other, not simply playing a role or somehow pretending to be absorbed by the other. This suggests most strongly that if I am to become a client-centred therapist, then developing my own personal growth and awareness far outweighs the learning and implementation of 'non-directive' skills and techniques.

Gay (Swenson) Barfield quotes some of Rogers' 1980 definition when exploring empathy with him in a filmed exchange of views in 1985, in which Carl acknowledged the benefits of empathy in a variety of settings, such as the classroom, intimate relationships (such as marriage and parent/child relation-ships), hostile groups and international relations. Carl said:

> I know that many tensions exist because of failures in communication, and each group has arrived at the pattern of feeling: 'I'm right and you're wrong. I'm good and you're bad'. And the trouble is that the other group feels exactly the same way: 'I'm right and you're wrong. I'm good and you're bad'. This destroys communication; it is also a good example of an absolute lack of empathy.
>
> (Rogers and Teich 1992: p. 58)

However, Carl went on to mention his unhappiness at the misuse and abuse of empathy that has taken place, quoting counselling training programmes as an example of this – courses wherein 'wooden, mechanical' techniques of 'reflecting feelings' and paraphrasing are taught and learned, and there is a lack of sensitivity and affect. Indeed, Rogers recalled one trainer saying to him after a demonstra-tion session that only three or four of Carl's empathic 'responses' would have been acceptable on his course! Of empathy as a technique or skill, Rogers said:

> I regard that as not only a distortion but really a gross misunderstand-ing of the whole concept, that often turns out people who are not helpful at all as counselors.
>
> (Rogers and Barfield 1985)

Empathy, said Rogers, involves the whole person of the therapist intensely yet delicately (because of the possibility of hurt) focusing on the inner world of the client – so much so that an empathic way of being at its most intense was 'a slightly altered state of consciousness' in which the therapist is being a companion to clients as they explore, sometimes fearfully, their inner selves in search of new personal meanings and feelings. The inner world of another person is clearly 'not a place to move around like a bull in a china shop', and the therapist needs to be a *confident* companion in order not to be shocked or judging. Stating that truly being understood and sensing that you are *worth* understanding is a 'very precious feeling to have', Carl recalled a demonstration session with a therapist from a different approach, who said afterwards 'You know, I've heard all the jokes about client-centred therapy, but it feels *so good* to be *heard*'.

If we are not mechanically reflecting or paraphrasing or whatever, what are we doing when we empathise? Rogers described the process of empathy as checking one's perceptions *of* the client *with* the client, using different forms of speech and styles as appropriate. Indeed, Rogers often found himself speaking *for* the client as if he were in the client's inner world, stating that clients would soon let him know if he was inaccurate.

Intuition

It could be said that just as Carl Rogers implicitly trusted the innate resources and potential of clients, so he grew to trust his own capacities, too:

> As a therapist, I find that when I am closest to my inner, intuitive self, when I am somehow in touch with the unknown in me, when perhaps

> I am in a slightly altered state of consciousness in the relationship, then whatever I do seems to be full of healing.
>
> (Rogers *et al.* 1986: p. 130)

It is regrettable (in my view at least, although I know that I am not alone) that there are some who have argued that Rogers intended intuition to become an additional therapist condition, whereas he himself clearly placed his intuitive experiencing within the realm of empathic understanding. It is even more regrettable (again in my view, and again I am not alone) that some people have abused segments like 'the unknown in me' and 'slightly altered state of consciousness' in their attempts to argue that Carl intended to add spirituality as an additional therapist condition. As we shall see – hogwash!

Reciprocity

Carl and Gay also spoke of the *reciprocity* of empathy, saying that although in other relationships the experiencing and offering of empathy could be somewhat one-sided (and therefore unhealthy), in therapy, although it is the therapist who mostly offers empathy, the empathic process is not *entirely* one-directional. In essence, this means that the therapist relies on clients to empathise with and understand his or her empathic understanding of them – there is a reciprocal process. Thus 'two persons being in contact' and 'the client receives, at least to a minimal degree, the unconditional positive regard and empathic understanding of the therapist' (the first and sixth necessary and sufficient conditions) can be linked with the notion of reciprocity in empathic process. Another illustration of 'reciprocal empathy' is when the therapist chooses to express an apparently non-therapeutic or 'negative' feeling that has persisted in the counselling relationship, such as anger or boredom. Carl stated that in such moments the therapist relies on the empathy and understanding of the client to get through such phases.

Gender

Although it is difficult to find many references to gender issues in the writings of Carl Rogers, it might be of interest to note that in this interview with Gay Barfield, Rogers did comment on a trend, namely males becoming more acquainted with and accustomed to their 'softer' or more 'feminine' side, rather than simply being dominating and 'macho' men. It may be that this greater willingness of men to strive to be empathic and understanding is linked with a move away from more directive approaches to psychotherapy.

It is abundantly clear when watching the video *Empathy: an Exploration* that in his later years Rogers believed passionately that empathy was a compassionate, delicate and sensitive way to *be with* people, not some technique or skill to *use on* them.

Summary

What have we learned of empathic understanding by trawling through its origins and developments? Here are six headings:

1 Context
2 Clients are active, not passive
3 Being a person
4 Does not, is not
5 Empathic understanding is . . .
6 Predictable process and consequences.

Context

Empathic understanding occurs within the context of other necessary and sufficient conditions, namely being in meaningful contact with another person (the client), the other person being relatively incongruent, the therapist being congruent and experiencing unconditional positive regard and empathic understanding, and the client perceiving the unconditional positive regard and empathic understanding of the therapist. These conditions are both necessary and sufficient – client-centred therapy is not eclectic. Although it is true that therapeutic relationships can include broader applications of client-centred therapy, known as the person-centred approach, it is also a fact that Carl Rogers went so far as to state that the use of different techniques for different people in one-to-one therapy:

> goes a long way towards destroying the possibility of a growthful experience for clients.
>
> (Rogers and Sanford 1985)

Clients are active, not passive

At the heart of client-centred theory lies the *actualising tendency*, a belief that individuals have their own *inner* resources (which have traditionally been underestimated or even ignored). Clients are capable of setting their own time and pace in therapy (for beneficial emotional disclosure, for instance) and of self-exploring at self-controlled depth. Clients may be self-protecting (withdrawing when appropriate, as compared with resisting or avoiding) and, given the right conditions, are strong, resourceful, positive and constructive. Clients are capable of achieving insight, which in turn may lead to behavioural choices. Clients also know best when to end therapy. People have access to their experiencing and their inner psychophysiological flow, and thus have a capacity for organismic valuing. The *client* does the probing, examining, exploring and reflecting. The constructive capacity of clients is immense, and the efforts of the therapist are puny by comparison. The client-centred therapist goes to work *with* a client – not *on* one!

Being a person

The prime requisite for becoming a client-centred therapist is being a person – an imperfect, whole person is good enough. This person strives in a disciplined way to be delicately sensitive, true, honest, trustworthy, compassionate, secure, transparent, real, non-defensive, non-possessive, non-threatening, self-aware (of inner warps, biases and prejudices), aware of others and their social/cultural contexts, genuinely receptive and interested, caring, deeply understanding, warm and permissive, and engages only in appropriate self-disclosure. Unconditional positive regard (characterised by a 'there but for the grace of some deity go I' attitude) and empathic understanding *must* be integrated and owned parts of the congruence of the therapist, as must a respect for the sovereignty and autonomy of individuals and their potential capacities and resources.

Does not, is not

Such a person finds it impossible to moralise, be prejudicial, judge, be horrified, blame, interpret, probe, question (open or otherwise), advise, suggest, paraphrase, summarise, persuade, reassure, diagnose, evaluate, deliver prognoses, guide, manipulate, go off at a tangent, reflect on therapist (their own) experiencing, use inappropriate language, mirror, reflect, become lost or over-immersed in the other's inner world, project a professional façade or front, pretend or be otherwise phony, direct, criticise, approve or disapprove, praise, lose the essence of the I–thou relationship by focusing on a frozen state of empathy (individual 'responses', for example), be the expert or blurt out every passing feeling.

Technological society has tended to produce experts who use skills on people (going to work *on* clients, not *with* them). As a result, a regrettable caricature of empathic understanding arose. However, in client-centred therapy empathy is not at all to do with skills, techniques, tools, tricks, expertise or wooden, mechanical, textbook counsellor responses.

Tricks? The therapist *using tricks on people* conjures up the image of the conjurer. *Con* suggests manipulative, and a *juror* decides, or is judgemental. Why, expertly and professionally, pull this *technique* out of the hat, or that *skill* from up your sleeve, or suddenly wave your wand to make that *tool* appear as if from nowhere? Authentic unconditional positive regard and empathic understanding *are* the magic. So:

> *Why be the magician*
> *when you can be the magic?*

Empathic understanding is . . .

Being empathic is demanding – the therapist needs to be strong yet also gentle. To be empathic means to enter the world of another as if it were one's own, yet without ever losing the 'as if' quality. The therapist is flexible, altering, changing, fluid – engaging in a sensitive togetherness as a delicate, confident companion. The therapist may at first 'simply' listen and really hear, and then, through ongoing perception checking, enter what is almost an altered state of consciousness – a reciprocal process for both therapist (from tentative to more confident) and client. The essence of the I–thou relationship is an authentic and respectful striving (and accuracy) to do with sensing and communicating with immediacy (not in one's car on the way home!) one's being attuned to client feelings and meanings, including the unspoken and underlying feelings and meanings of which the client is only dimly aware. The therapist strives with an intense focus to understand clients as they experience themselves to be, from the client's own internal frame of reference (their sensations, perceptions, meanings and memories).

For those who say that client-centred therapy is all about feelings, or too heavily focused on feelings, Rogers stressed the importance of feelings *and* *meanings* throughout his entire career.

Predictable process and consequences

If the necessary and sufficient client-centred conditions are present, then the power of the actualising tendency will be released. This is an empowering relationship that taps the growth forces within the client, enabling them to be more in touch with their inner psychophysiological flow. It can be said to have an if–then quality. *If* certain conditions are present, *then* a predictable process will follow; *if* the process ensues, *then* predictable outcomes accrue.

Carl Rogers can provide our final summary of the origins and developments of empathic understanding:

> I think at first I realized that it paid to listen carefully, and I gradually realized that I was listening to understand the feelings and the personal meanings. Sometimes people skip that second part – not only listening to the feelings but the personal meaning that experience has for the individual. Then I realized that I was entering into the private, inner world of the individual, trying to sense the exact nature of that inner world and to move around in it freely and maybe sense it from the inside. I realized that empathy also meant being a companion to the person as they search their own experience. Then I began increasingly to trust my own intuitive understanding; sometimes I would want to say things which had no relation to what the client had just said, and yet it seemed to me important to say.
>
> Gradually my understanding of empathy extended to an intuitive capacity for empathy, where I would find something rising in myself that wanted to be said. It might be bizarre. It might be out of context. But I found that if I voiced it, it often rang a bell with the person and opened up all kinds of areas that had been dimly sensed by the client but not really experienced. I don't really understand the workings of intuition. Do I simply pick up nonverbal cues? I don't think that's sufficient to explain it. Somehow there is a way in which the inner core of me relates to the inner core of the other person, and I understand better than my mind understands, better than my brain understands – I'll put it that way. Mind is greater than brain and, somehow, my nonconscious mind understands more than my conscious mind understands, so I'm able to respond to something in this other person that I didn't know I was responding to.
>
> (Rogers and Russell 2002: pp. 284–5)

Chapter 2

Empathic understanding is not a technique

Thou shalt not skill!

Having taken a selective journey through some of the origins and developments of empathic understanding in Chapter 1, we now take a reverse angle view in this section. Very often it seems that people (such as Carl Rogers, trainers, trainees, correspondents and authors) have defined empathic understanding, at least in part, through a process of eliminating what it is *not* as much as by defining and celebrating what empathic understanding *is*. Although this may very well be understandable, it has perhaps also been unfortunate in some respects.

This tendency to define empathy by stating what it is not may be understandable for two reasons. The first is the 'everyday habit' of many human beings of describing how they *are* by saying how they *are not*, as in such phrases as 'I'm not feeling so good', 'I don't feel very happy about that' and, when asked how they are, 'Not too bad'. The second reason is more specific to client-centred therapy, in that to some degree the approach was generated through Carl Rogers and others discovering that traditional counselling methods were not satisfactory either to the therapist or to the clients with whom they worked, and so there was a rebellion against 'directive' therapy. We can readily see, for instance, that the early term '*non*-directive' described, in effect, what this type of counselling was not – simply put, it was not directive. Indeed, throughout the development of client-centred therapy (and the person-centred approach), frequent references can be found that reject *doing to* people in favour of *being with* them.

The regrettable element of this is that defining what something *is not* has a rather defensive feel to it when compared with celebrating fully what something *is*. For instance, so many client-centred counselling (or counselling skills) training courses begin with looking at what client-centred therapy is *not* and what the client-centred therapist *does not do* – for example, the therapist does not show shock or surprise, does not praise, support or encourage, does not diagnose or interpret, and so on.

One feature of client-centred therapy is that it is an opportunity for clients to explore and examine their conditions of worth, including the values of others that have been internalised due to a need to be loved, accepted or valued, rather than because they 'feel right' (or are in harmony with 'organismic valuing'). An example of this might be the woman who has come to realise that first she was someone's daughter, then someone's girlfriend, then someone's employee, then someone's wife, then someone's mother, and so on. Eventually her children become more independent, and she realises that her whole life has been spent meeting the needs of or pleasing others. Those were her roles ('keeping everyone happy'), and those roles carried with them values, guidelines and rules about how they should be fulfilled – 'scripts', if you will. There comes a time when although the *roles* may have been fulfilled, the *person* does *not* feel fulfilled, and she wonders, 'What about me?' – usually accompanied by feelings of guilt, shame and embarrassment about even experiencing such a 'selfish' thought. Yet, as if from deep within, there is *something* (often described as a 'voice') that persists with 'What about me?', and that just cannot successfully or permanently be ignored, suppressed or denied. If this message is heard to the extent of actually acting upon it, often another phase is a feeling of liberation – perhaps the client makes what seems to her at the time to be her own free choices, feeling a greater sense of self-assertion. Yet after a while this sense of new-found freedom loses its depth of meaning, because she realises (sometimes gradually, sometimes as a sudden shock) that what she is now doing is the exact opposite of the previous values, guidelines and rules, and if a construction of self is determined in exact opposition to something, then that something retains a central and powerful position. Thus if he used to say 'Buy blue', she now says 'I'll buy red', and somehow *his blue* is still the determinant. In client-centred theoretical terms, the locus of evaluation is still external ('his red', not 'her blue'). Often another phase is that the woman effectively asks herself 'So who on earth am I, really?', or alternatively she states 'I don't know who I am any more'.

Now turning to counselling training, a colleague described facilitating the first day of a counselling diploma course, when trainees had decided to pool their current knowledge. A line was drawn down the middle of a board under the heading 'Client-Centred Counselling', and one column was marked 'The counsellor does' and the other 'The counsellor does not'. The facilitator was writing up what the students were calling out: 'does not praise, does not rescue, does not reassure', and so on (interestingly, the 'does not' column was the starting point). After writing a few entries in the 'does not' column, the facilitator turned to the group and said, 'Why? Why doesn't the client-centred practitioner do these things?'. Following some debate and exploration, the only explanation or reason that trainees could come up with was that this is what they had been *told* (or had learned) on their previous skills course. Returning to the analogy of our female client, if traditional therapies were to be seen as 'his red' it would be most regrettable if client-centred therapy were to be the equivalent of 'her blue' for no other reason than a dislike of or rebellion against 'his red' – yet maybe something akin to this was an understandable developmental phase in the generation of client-centred therapy. It may well be, too, that counsellor trainees who begin to feel *de*-skilled as a consequence of discovering (or being told) that the techniques that they had learned (and had certificated) are no longer praiseworthy might well enter a phase similar to asking 'So who am I, really?' or 'I don't know who I am or how I'm meant to be any more'. Well, one of the central aims of this book is that moving towards an authentically empathic way of being – a way of being that is *you through and through* – is eminently preferable to learning a (relatively impersonal) plethora of skills and techniques.

There is, perhaps, a vital difference between the analogy of the female client and the person who is hoping to become a client-centred therapist. For the client-centred counsellor trainee, it might go something like this: 'My previous skills course tutors taught and told me *red*. Now these diploma facilitators are saying no, not red, *blue* – and furthermore not only is this a higher-level course, but also they can back up *blue* by quoting Carl Rogers, so they must be right'. Yet to jettison or otherwise discard something for which a certificate has been achieved might not come easy. There is a strong argument for neither red nor blue *if they are simply or only the introjected values of others* (be they authors, facilitators or even Carl Rogers), rather than being based on an internal locus of evaluation – *experiencing* first hand what seems to work for you (preferably both as therapist and client). Better a chosen red experienced at first hand than a second-hand introjected (internalised, the value absorbed or taken on from another) blue! Reading a book may indeed be a part of the entire experiencing of a trainee counsellor, yet the message has to be *try it and see*, or, better still, *be it and see*.

While rejecting 'traditional red' was undeniably a significant feature in the development of client-centred therapy, there was also the equally important element of discovery – of looking for and at what seemed to be more meaningfully therapeutic. It is tragic to witness the way of being that was so dear to Carl Rogers reduced to nothing more than the 'blind' introjection of client-centred beliefs, values and technique-oriented practices – not based on personal discovery, experiencing, feeling and meaning, but on the teachings of trainers (that is, an external frame of reference to learners). What might be the qualitative impact if the starting point for client-centred counsellor training were to be the celebration of what client-centred therapy *is* – taking pride and rejoicing in the journey ahead towards becoming a sensitive companion to clients?

You don't know what you've got until you lose it

It was stated earlier that *to some degree* the development of client-centred therapy was a consequence of discarding many traditional beliefs and methods. However, Rogers also built his research activities around looking at what *did* seem to work and what *was* effective (as described in Chapter 1 and Chapter 3 onwards). Let us nevertheless spend time exploring some of the therapist ways of being with clients that did *not* seem to be effective, or which were deemed to be philosophically dehumanising. Can you, for instance, imagine a therapist saying some or all of the following?

> You poor thing. It is perfectly understandable why you feel the way that you do. Still, you can confide in me, your secrets are safe with me. You know that I'll back you all the way, for I'm on your side. You're not alone. Well done! *(You will be rewarded with oodles of attention and compassion if you focus on your feelings.)* Yes, you can do it – it will be okay. You have been badly treated and, you know, another client of mine had a similar experience and . . . And yet I'm glad you're beginning to sort yourself out, it must be really hard. You are most certainly justified in feeling as you do. Don't you think that a sensible course of action might be to . . .? Yes, you could do this; we could work in achievable stages towards . . . Your tears represent feelings of loss at your unresolved grief and you lack the ego strength to fully deal with this right now, so it will clearly take some time before you do have the inner resources – it is as if I were your parent and as if you were my child. If I were you I'd . . . because this is clearly a case of . . . I understand that when you say this you mean that, don't you? When this happened to me I . . .

No, you cannot imagine a therapist saying all of those things, yet if a bet were to be placed on you recognising and being able to imagine *some* counsellors saying *some* of those things *some* of the time, the person placing the wager would not get good odds! Perhaps the above example can be used to further this process of eliminating what empathic understanding is not.

The *intent* of empathic understanding is not the same as *sympathy* – feeling sorry for a client ('you poor thing') – nor is it an attempt at *rescuing* or *normalising* ('it is perfectly understandable and natural that you feel the way you do'). Empathic understanding is not akin to *befriending* ('you can confide in me, your secrets are safe with me'), nor is it about *supporting* ('I'll back you all the way, I'm on your side'), or *meeting client needs* (whatever they may be). Nor is empathy about *praise* ('well done', or an implied 'you'll be rewarded with oodles of attention and compassion if you focus on your feelings' – the latter hardly conveys *unconditional* positive regard), *encouragement* ('you can do it') or *reassurance* ('it will be okay'). Empathic understanding is not the same as *agreement* ('yes, you have been badly treated'), nor is it about making *comparisons* ('another client of mine had a similar experience and . . .'), nor is it *patronising* ('I'm glad you're beginning to sort yourself out, it must be really hard'). Empathy is not about making *value judgements* ('you are most certainly justified in feeling as you do'), nor does it involve giving *advice* ('a sensible course of action might be to . . .') or *answers* (which are more often than not disempowering). Empathic understanding has nothing to do with *problem solving* ('you could do this . . .'), *goal setting* ('we could work in achievable stages towards . . .') or *having an agenda* or preconceived ideas (although the therapist *will* have preconceived *beliefs* – a philosophy that underpins the therapeutic approach). Empathic understanding is not about *interpreting* ('your tears represent feelings of loss at your unresolved grief . . .'), *diagnosing* ('and you lack the ego strength to fully deal with this right now'), or giving *prognoses* ('it will clearly take some time before you do have the inner resources'), nor is it based on theories of *transference* ('it is as if I were your parent') and/or *counter-transference* ('and it is as if you were my child'). Empathy is not *confrontational* (in the everyday meaning of 'confrontational' as typified by something like 'I understand that when you say this you mean that, don't you?'), and does not include *inappropriate self-disclosure* ('when this happened to me I . . .'), nor is it about the counsellor *knowing best* ('if I were you I'd . . .' or 'this is clearly a case of . . .').

Nor is empathic understanding *all about feelings*.

And nor is empathic understanding concerned with understanding *about*.

On client-centred training programmes, the issues that cause the most initial concern among trainees are usually praise, encouragement, reassurance, support, rescuing and, above all else, *questioning*. There has frequently been an additional confusion between acceptance and agreement, often linked with confrontation as a 'style' (skill or technique) within counselling. Some research evidence (Vincent 1999a) in support of this comes from having modified, expanded and administered (to counselling and counselling skills training groups), over a period of some 18 years, the *Counsellor Attitude Scale* developed by Richard Nelson-Jones and Dr Cecil H Patterson (Nelson-Jones and Patterson 1975). It can be stated through this and other research (see below) that, *inevitably*, acceptance, agreement, challenging, confrontation, diagnosis, encouragement, evaluation, explaining, focusing, interpretation, leading, mirroring, normalising, paraphrasing, praise, probing, prognosis, questioning, reassurance, reflecting, rescuing, responsibility for clients, suggestion, summarising, support, therapist self-disclosure, transference and counter-transference, 'uh-huh', and 'understanding about' have featured as key issues.

When reviewing studies by Julius Seeman (Seeman 1949) and W Snyder (Snyder 1943) of the early days of client-centred counselling, Carl Rogers noted that:

> counsellors used a number of responses involving questioning, inter- preting, reassuring, encouraging, suggesting. Such responses, though always forming a small proportion of the total, would seem to indicate on the counsellor's part a limited confidence in the capacity of the client to understand and cope with his difficulties. The counsellor still felt it necessary at times to take the lead, to explain the client to himself, to be supportive, and to point out what to the counsellor were desirable courses of action.
>
> (Rogers 1951: pp. 30–31)

It is important to stress that tendencies to praise, support and encourage (and so on) are usually well meant, and indeed may well form part of a helping relationship. Thus there is no need for people who experience such impulses to feel 'bad' about themselves. Nevertheless, it is also true to say that the person who *intends* to praise, support and encourage (and so on) has not fully understood or integrated a *client-centred* way of being.

The 'client-centred counsellor does not' list above contains 30 items in all. It might prove helpful to look at these characteristics, now grouped under 13 headings, in an attempt to grasp more clearly what empathy is not.

Acceptance and agreement

Acceptance

> There's . . . a misunderstanding of unconditional positive regard: To care for the client, to prize the client, doesn't mean that you approve of everything the client does; it means a real caring for the essence of the client.
>
> (Rogers and Russell 2002: p. 284)

As we have seen, Stanley Standal, a student working with Carl Rogers in the early days, came to the conclusion that 'acceptance' was a term which was too easily misunderstood. For instance, acceptance is not the same as *agreement* (see below), nor is it the same as *belief* – yet the term *could* be held to mean either (Standal 1954). For example, if person A believes that the earth is flat, person B can *accept* that this is A's *belief*. Person B does not have to *agree* with A, or *accept* that the earth is, in actual fact, flat. In this instance, if person B were to *agree* with person A, or to *accept* person A's *belief* as *fact*, person B would be in a state of incongruence, because they do *not* believe that the earth is flat, and maybe their *agreement* with person A derives from their own needs for positive regard from person A. In this case there is a strong argument for person B engaging in self-awareness and personal growth work if they wish to become a client-centred therapist! Person B's *agreement* would also be based on their own frame of reference – after all, it is person B who is doing the agreeing.

Incidentally, a more socialist critique of 'acceptance' might be that it can be ,seen as akin to 'accepting your lot' – that acceptance along the lines of 'that's just the way it is' dissipates energy for action (individual or group) through accepting the status quo.

Agreement

Agreement would often seem to be confused with acceptance (as described above). For instance, how often in a counsellor training group have two people been in some form of dispute where person A *disagrees* with person B and is then *accused* by person B of lacking *acceptance*?

In more basic counselling terms, therapist agreement or disagreement would seem to imply a locus of evaluation that is *external* to the client and *internal* to the therapist (or the theory held by that therapist). Very often counsellor trainees seem to think that it is okay (or 'nice' or 'warm' or 'compassionate') to *agree*, yet not okay (or 'nasty' or 'cold' or 'dispassionate') to *disagree*. However, in terms of the power dynamics in the relationship, does not the right to agree also imply the right (or authority) to disagree? Often agreement is verbalised while disagreement is not. Perhaps we can deduce from this that it feels okay (or acceptable) to be 'nice' and 'warm' and 'compassionate', but not to be 'nasty' or 'cold' or 'dispassionate' – despite the power dynamic being exactly the same.

Clearly agreement *and* disagreement involve *value judgements*, usually from either a therapist or a theoretical frame of reference (or both), and as such both agreement and disagreement are directive, potentially manipulative, and generate a power imbalance in the therapeutic relationship.

Agreement, disagreement, non-agreement and collusion

Trainees and supervisees often become concerned that their acceptance will be received by clients as somehow *condoning* or *colluding* with actions or behaviours that the *counsellor*, in private at least, finds unacceptable or even appalling, such as the client who admits aggressive behaviour towards or abuse of a partner, or self-aggressive or abusive acts. This raises a number of issues, not least of which are the congruence of the therapist and the potential consequences in terms of social responsibility and conscience.

Ideally, it could be argued that a wholly non-judgemental therapist will not face such issues. Yet few counsellors, it seems, have escaped grappling with such feelings, thoughts and meanings. If a therapist *does* experience difficulty in 'reality', what then? If the therapist somehow feigns or puts on a show of acceptance when 'really' he or she is anxious about being perceived as colluding with a hurtful behaviour, the therapist is hardly in a state of congruence. Furthermore, many believe that if a therapist 'privately' agrees or disagrees with a client, then the client will sense that approval or disapproval *at some level*, for clients too have a capacity to be empathic ('reciprocal empathy, being in contact'). Some then go on to say that if clients perceive therapist discomfort at some level, then it is probably best for the therapist to self-disclose, as *unspecified* client sensing could potentially be more harmful than transparency and openness. (This theme is explored more fully in Chapter 3 when looking at empathic understanding in relation to congruence.)

Some practitioners and theorists maintain that the therapist accepts *the client* but *not* her or his *behaviour*, which as we have seen in Chapter 1 is not a wholly satisfactory answer. For instance, if being client-centred includes a belief in engaging with *whole* persons, how can one aspect of a person (their behaviour) be separated out as if it does not matter? Surely behaviour is an integral part of the whole person, and a significant part at that? We have also seen in Chapter 1 that acceptance can be seen as a 'there but for the grace of some deity go I' kind of attitude – if a person has *this* experience, then maybe *that* behaviour is a natural and understandable consequence.

Is it, though, possible for a client to sense *non*-agreement, as compared with *agreement* or *disagreement*? Can a therapist 'suspend' *all* judgement, both positive and negative? Might it even be important for a client to actually *sense* non-agreement? For if a client senses therapist agreement, then feelings of collusion or condoning might well be an outcome, and if a client senses therapist disagreement, devalued and worthless feelings might be the result.

Perhaps this can be brought more to life through an illustration. A young boy is referred for counselling because he has begun to cut himself at school. Each time he does so, those around him (mostly the adults, but peers, too) are very concerned, and their caring leads them to try their utmost to find a way of stopping or preventing such self-harming activities. They can 'accept' the young boy as a valuable person for whom they feel compassion, but they find it very difficult indeed to accept (or have compassion for) him physically cutting himself. The therapist is a human being, too, and *of course* does not like to see self-inflicted scars on such a young person – and would most certainly feel uncomfortable at the notion of therapy somehow condoning or colluding with such behaviour. However, the therapist takes a genuine interest in what is going on for the boy. The therapist is *not* trying, overtly or covertly, to *stop* the behaviour. Rather, the therapist is authentically respectful of the behaviour, and is striving to sense the feelings and meanings associated with this self-harm. Striving to engage with the boy as a *whole* person, it becomes clear that within the boy's self-structure are at least three significant aspects, namely a 'perpetrator' (the boy who is cutting himself), a 'victim' (the boy who is being harmed) and some kind of 'observer/ mediator' who watches over and tries to make sense of what is going on.

If the therapist was to be anxious about *colluding with* or *condoning* the self-harm, this would seem to be to do with 'siding with' the perpetrator construct, and maybe with a failure to rescue or protect the victim, too. This is hardly *unconditional* positive regard, for the perpetrator is 'bad' and the victim is worthy of protection (and the observer/mediator would seem to have been somewhat neglected in this equation, too). The boy at first seems surprised (even shocked) that here is someone who is not trying to *prevent* or *correct* an aspect of his behaviour. Rather, here is someone who actually seems to be taking a 'no strings attached' interest in anything and everything that the boy *is* (including what he *does*) – and he seems to be valued as no more or less worthy *whatever* he decides to disclose. The boy even doubts whether this can be true. However, in part as a consequence of this *unconditional* positive regard, the boy gradually becomes free of any need to be defensive, and thus shares more and more of his inner feelings and meanings, slowly moving towards a greater trust in the therapist's acceptance of his *whole being* as genuine. The boy discovers that his inner 'perpetrator' is actually doing its best to keep him safe – for when the boy feels hurt by *others* he experiences not being able to 'handle it', yet he had also discovered that if he inflicted the pain *on himself by himself*, he *could* manage it. Suddenly, it seems, it becomes possible for him to value his inner perpetrator as a part of him that is well meaning, creative, protective and constructive. Had the therapist neglected, ignored, lacked compassion for or disapproved of the boy's inner perpetrator, it is far less likely that *client* self-acceptance would have emerged.

It is so easy – and it seems so tempting – to be 'seduced into naturally' extending all of our compassion to the *victim*, with the consequence that the perpetrator and the observer/mediator are neglected. When this happens, our positive regard is not unconditional, and opportunities for therapeutic movement may well be thwarted. The young boy in this illustration, through a process of being able to own, value, understand and respect himself more fully, changed his own behaviour. No one had to *do* anything – the therapist 'simply' had to strive to *be*, as fully as was possible, the client-centred core conditions.

The boy did the rest.

Challenging and confrontation

Confrontation often appears as a *technique* that is either taught or learned on many counselling skills courses (and, regrettably, on many counselling pro- grammes, too). Indeed, at the time of writing 'confrontation and/or challenging skills' is on the syllabus of at least three major counselling course validating bodies in the UK. At its 'worst', confrontation would seem to assume a hectoring quality, in that the *implementer of the skill* is seemingly determined to get the client to see things as the counsellor does. At its 'best', the user of counselling skills perceives client incongruence, for example, and is determined to confront the client with this observation in order to ensure that the client sees it, too.

Yet is there really anything more 'confrontational' than authentic unconditional positive regard and genuine (and accurate) empathic understanding? For instance, if clients have low self-esteem as a result of their conditions of worth, introjected values, and needs for positive regard (and so on), and the therapist offers unconditional acceptance and empathic understanding of them, then their very self concept is effectively 'challenged' or 'confronted'. However, the crucial differentiating factor here is to do with therapist *intent*, for in genuinely striving to experience and offer acceptance and empathy, the therapist's intent is *simply and only that* – to strive to the best of their ability to experience and offer the fourth and fifth necessary and sufficient conditions from a position of congruence (the third condition).

Study the writings of Carl Rogers and you will find very few references indeed to challenging or confrontation. *Find* a reference and it will be linked with congruence, not empathy. For example, writing about relationships other than therapy (such as partners, sex partners, colleagues, friends, employers and employees), Rogers states that *congruence* is probably the most important of the three core conditions, and that it:

> may involve confrontation and the straightforward expression of personally owned feelings.

> (Rogers 1980: p. 160)

It is important to note that Rogers *excluded* counsellor–client relationships from this, and included 'confrontation' within the client-centred core condition of *congruence, not* within empathic understanding.

When it was stated earlier in this chapter that empathic understanding is not about having an agenda or preconceived ideas, yet it might involve preconceived *beliefs*, remember that one fundamental client-centred belief is in the potential for (client) autonomy and self-direction. Client-centred therapists trust in the re-sources of clients to 'make what they will' of their therapist's authentic striving to experience and offer unconditional positive regard and empathic understanding. However, this is far from a 'couldn't care less' or 'pass the buck' attitude – if counsellors genuinely intend to offer the core conditions, then they are providing an opportunity for therapeutic growth. This is as far as it is possible to be from *setting out with an intent to* 'confront' or 'challenge'.

During some 20 years spent as a trainer of counsellors, I was intrigued as to why trainees would abandon client-centred principles and resort to confronting or challenging their clients, and two main reasons emerged. Either trainee therapists felt that their clients, the process, or they themselves had become 'stuck', or trainees had been influenced by their supervisor (many of whom described themselves as 'person-centred') into importing ways of being that are contrary to client-centred attitudes. Carl Rogers recognised this himself when he said:

> Sometimes people feel that client-centered therapy is good for going only so far, and when you really strike difficult problems you should probably be more confronting or more this or more that. I think – and I feel quite strongly from my experience – that that is really a mistaken line of thought. I think that when the situation is most difficult, that's when a client-centered approach is most needed, and what is needed there is a deepening of the conditions we've talked about – not trying something more technique oriented. It might be worth saying what I've probably said before: in this technological age, an approach that simply lives a philosophy and puts its trust in the capacities of the client seems lacking in glamour, compared to things that try all sorts of tricks and techniques and do different things and try different procedures. It interests me that people who use that sort of an approach often burn out because they just come to the end of their rope; whereas when you get nourishment from the client by seeing the client grow, that's very rewarding.
>
> (Rogers and Russell 2002: pp. 258–9)

Diagnosis, evaluation and prognosis

> The whole diagnostic look at an individual is in sharp contrast to the I–Thou kind of relationship.
>
> (Rogers and Evans: 1975, p. 25)

Therapist diagnosis, evaluation and prognosis clearly do not respect the inner resources of *clients* and their potential and capacity for self-direction, as there is an obvious implication that actually the therapist, not the client, knows best. Carl Rogers wrote that:

If we can provide understanding of the way the client seems to himself at this moment, he can do the rest. The therapist must lay aside his preoccupation with diagnosis and his diagnostic shrewdness, must discard his tendency to make professional evaluations, must cease his endeavors to formulate an accurate prognosis, must give up the temptation subtly to guide the individual, and must concentrate on one purpose only; that of providing deep understanding and acceptance of the attitudes consciously held at this moment by the client as he explores step by step into the dangerous areas which he has been denying to consciousness.

(Rogers 1946: p. 420)

Writing of client-centred principles applied to a learning context, Carl differentiated between empathic understanding and evaluative ways of being, stating of empathy that:

This kind of understanding is sharply different from the usual evaluative understanding, which follows the pattern of 'I understand what is wrong with you'. When there is a sensitive empathy, however, the reaction in the learner follows something of this pattern: 'At last someone understands how it feels and seems to be *me*, without wanting to analyze or judge me. Now I can blossom and grow and learn.'

This attitude of standing in the students' shoes, of viewing the world through their eyes, is almost unheard of in the classroom. But when the teacher responds in a way that makes the students feel *understood* – not judged or evaluated – this has a tremendous impact.

(Rogers 1980: pp. 272–3)

Diagnosis, evaluation and prognosis necessarily involve *judgements*, and being judgemental is not in accord with congruently experiencing and offering both unconditional positive regard and empathic understanding:

Yet it would appear to be true that whether the theme is evaluative or self-concerned, there is slightly less of full respect for the other person than in . . . thoroughly empathic understandings.

(Rogers 1951: p. 45)

If therapists are absorbed with their own processing, or are considering courses of action, then without doubt there is a decreased focus on the world of the client. Even accuracy of evaluation is irrelevant here, because any 'objective' judgement means less respect for client self-direction:

On the other hand, to enter deeply with this man into his confused struggle for selfhood is perhaps the best implementation we now know for indicating the meaning of our basic hypothesis that the individual represents a process which is deeply worthy of respect, both as he is and with regard to his potentialities.

(Rogers 1951: p. 45)

Returning to this theme in 1980, Carl Rogers was able to categorically state that:

Brilliance and diagnostic perceptiveness are unrelated to empathy. It is important to know that the degree to which therapists create an empathic climate is not related to their academic performance or intellectual competence (Bergin and Jasper 1969; Bergin and Solomon 1970). Neither is it related to the accuracy of their perception of individuals or their diagnostic competence. In fact, it may be negatively related to the latter (Fiedler 1953). This is a most important finding. If neither academic brilliance nor diagnostic skill is significant, then clearly an empathic quality belongs in a different realm of discourse from most clinical thinking – psychological and psychiatric. I believe that therapists are reluctant to accept the implications.

(Rogers 1980: pp. 149–50)

Much of Carl's earlier writing to do with diagnostic evaluation stemmed from his reactions against the psychoanalytic approaches that were so prevalent at the time. For instance, Rogers believed that the diagnostic testing of patients objectifies people rather than prizing them, and thus was 'worse than a waste of time. It is destructive . . .' (Rogers and Evans 1975: p. 98). While Carl Rogers did believe that childhood experience (such as the introjection of values from significant others, culminating in our conditions of worth) is a force to be reckoned with, and therefore that behaviour can be influenced by the past, he added:

I'm objecting to putting hard-and-fast labels on these processes. I don't like the pigeonholes that Freudian and other theories have promulgated. I think people handicap their thinking when they think in terms of so many labels.

(Rogers and Evans 1975: p. 8)

As we have seen, Rogers went further than simply objecting to labelling when he described targeting individuals in this way as destructive. He added of practitioners who do label:

I feel that they are destructive, primarily because they make pigeonholed objects out of human beings. If I say you're a paranoid, then you're quite naturally placed. I don't even have to think about you as a person. You belong in that pigeonhole. I've been very interested to see this at work among psychiatrists. If one psychiatrist doesn't like another psychiatrist, then he begins to label him, calls him a manic, or says that he's got paranoid tendencies, or that he's essentially a schizophrenic. They're quite deceptive in that they're really finding a different way of saying they don't like that guy. They have to admit there's some truth in that. Now I think it's a pseudo-thing; I hate pseudo-things. They make it sound as though 'I know what illness you've got'. Physicians overuse labeling, even in physiological diseases; but there are real diseases which they can identify and which deserve a label. To my way of thinking, in psychological states, there are certain patterns. Sure, there are people who are suspicious, and I would call them that instead of giving them a high-falutin label like paranoid. At any rate, labels give a pseudoscientific sound to what is actually a very loose and unfounded categorization.

(Rogers and Evans 1975: pp. 96–7)

A frequent consequence of diagnostic evaluation was (and often still is) medical prescription, of which Rogers wrote:

Medication is seen so often as a means of dealing with psychotic disorders. Perhaps sometimes that's justified; I don't want to try and step into the field of medicine, but it is most certainly overdone to a great degree. Often patients are medicated in order to be of help to the people caring for them rather than for the welfare of the patient. I think most psychological disorders are psychological in their origin. Their treatment should be dealt with in that fashion.

(Rogers and Evans 1975: pp. 290–91)

Carl believed that much diagnostic testing was *institutionalised*, yet, to further evidence what a useless waste of time such activities are, he pointed out that once a person *leaves* working for 'the establishment' in order to work privately, that practitioner almost never repeats:

[the bureaucratic] rigmarole of diagnoses . . . which shows that he doesn't believe at all in a lot of this folderol he's been sponsoring in the institution. As I see it, rigid classification has no logical or helpful basis. It has just become established as the custom or the tradition, and I wish tradition weren't so hard to change.

(Rogers 1975: pp. 98–9)

Can the client-centred therapist completely do away with diagnosis and evaluation (especially, perhaps, if he or she is working in a medical setting)? This is a vexing question, for even Rogers acknowledged that:

There are difficult border lines, and the border line between what is physical, physiological, medical and illness and what is psychological is sometimes difficult to determine.

(Rogers and Evans 1975: p. 291)

This strongly demonstrates the importance of both supervision and access to medical consultancy.

Carl believed that client *creativity* is generated when people experience freedom from evaluation and measurement from an external locus of evaluation, and wrote that:

Evaluation is always a threat, always creates a need for defensiveness, always means that some portion of experience must be denied to awareness.

(Rogers 1961: p. 357)

If, for example, some aspect of a client is evaluated as 'good' by a therapist, then that makes it all the more difficult for that client to 'admit' (to the therapist or to themselves) that actually they *dislike* this aspect of themselves. Conversely, if a counsellor evaluates some client characteristics as bad, then it makes it all the more difficult for clients to openly express and be those characteristics. Client creativity is fostered through clients moving towards an increasingly inner locus of evaluation.

> It is impossible to be accurately perceptive of another's inner world if you have formed an evaluative opinion of that person. If you doubt this statement, choose someone you know with whom you deeply disagree and who is, in your judgment, definitely wrong or mistaken. Now try to state that individual's views, beliefs and feelings so accurately that he or she will agree that you have sensitively and correctly described his or her stance. I predict that nine times out of ten you will fail, because your judgment of the person's views creeps into your description of them.
>
> (Rogers 1980: p. 154)

Consequently, true empathy is always free of any evaluative or diagnostic quality. The recipient perceives this with some surprise: 'If I am not being judged, perhaps I am not so evil or abnormal as I have thought. Perhaps I don't have to judge myself so harshly.' Thus the possibility of self-acceptance is gradually increased.

None of this means that the therapist becomes some kind of 'cold mirror' – on the contrary, for Carl Rogers differentiated between having a *reaction* and making an *evaluation*. For instance, there is a real contrast between your saying to me 'I am not enjoying reading this book' and 'You are a bad writer'! Your stating to me that you are not enjoying this book leaves me relatively free to keep my own evaluation of my writing, whereas telling me that I am a lousy author condemns me. Note that the same effect would be true of 'I am enjoying reading this book' and 'You are a great writer' – for with the former I remain free to evaluate my own writing, and with the latter the locus of evaluation is effectively taken away from me.

> When the counsellor perceives and accepts the client as he is, when he lays aside all evaluation and enters into the perceptual frame of reference of the client, he frees the client to explore his life and experience anew, frees him to perceive in that experience new meanings and new goals. But is the therapist willing to give the client full freedom as to outcomes? Is he genuinely willing for the client to organize and direct his life? Is he willing for him to choose goals that are social or antisocial, moral or immoral? If not, it seems doubtful that therapy will be a profound experience for the client.
>
> (Rogers 1951: p. 48)

There may well be what might appear to be a conundrum, in that the more accepting the therapist can be of client choice, autonomy and self-direction, the more likely it is that the client will choose healthy, positive directions – the actualising tendency is trustworthy. Carl stated that it could be even more difficult for the therapist to be truly accepting of clients who appear to be 'regressing' rather than growing, to be moving in a direction the therapist believes to be counter to mental well-being – even to be choosing death over life. Yet Rogers wrote that:

> only as the therapist is completely willing that any outcome, any direction, may be chosen – only then does he realize the vital strength

of the capacity and potentiality of the individual for constructive action.

(Rogers 1951: p. 48)

We began this section with a quote from Carl Rogers about diagnosis and evaluation being 'in sharp contrast to the I–Thou kind of relationship'. Let us end this section on a similar, comparative note:

> It was just incredible to have that kind of contact . . . He was a troubled person who was not able to cope with society, that was certain. The one thing that brought him out of it, I feel, was that we were able to form a close person-to-person relationship, and the fact that his feeling about that relationship has continued through eight years is really astonishing. It's even more interesting because, until the very last portion of our contacts together, he would never have openly and consciously admitted that our relationship meant anything to him. It was only by his behavior of continuing to come for interviews that he showed it had a meaning for him. So it's things like that make me feel that if you can reach the psychotic in a real relationship, you've got a chance of bringing him out of his psychosis. I don't underrate the difficulty of reaching such a person as him because, for most individuals who have turned to psychosis as an escape, life has hurt them so much and so many times, and they've been so disappointed in all their personal relationships that they don't believe you. They don't believe you care; they don't believe you're interested; they won't let you reach them because you may hurt them again. We did learn in our research how terribly hard it is to reach some of these people. I still feel that a close human relationship is far more significant than anything doctors with superior knowledge of a patient who is ill can think of doing. I just feel the medical model is ridiculous.
>
> (Rogers and Evans 1975: pp. 95–6)

Explaining, interpreting and normalising

Carl Rogers was keen to distinguish between attempts to *understand* feelings and meanings, and attempts to *explain* or *interpret* them. When writing about group facilitation, for instance, he referred to those facilitators who try to interpret (apparently on behalf of group learning and awareness) that which might not be in conscious awareness, saying that in his experience:

> where leaders have used the technique of interpreting 'unconscious' meanings, such interpretations are usually not facilitating and are frequently disrupting . . . Certainly it is not even justified to say that 'reflecting meanings and intents' is not interpretation in a sense. Nevertheless, there does seem to be a valid distinction between the two, at least in operational terms.
>
> (Rogers 1951: p. 353)

What are these 'operational terms'? Although Rogers grew to dislike the term 'reflection', he differentiated between an attempt to *reflect* feelings and meanings as they appear in the here-and-now awareness of clients, and interpretation, which implies trying to bring 'unconscious' (Carl's quotation marks) experiencing into conscious awareness. He wrote:

> This difference is probably essentially the same as the one that appears to exist between interpretation as used by some psychoanalytic therapists and the method of 'adopting the client's frame of reference' used by client-centred therapists.
>
> (Rogers 1951: p. 354)

Any interpretation of the client by the therapist diminishes the dignity of that client. Carl's reaction against this was a passionate one:

> It made me realize that if I wanted to look like a smart psychologist, I could go ahead and diagnose and advise and interpret. But if I wanted to be effective in working with people, then I might just as well recognize that this person has the capacity to deal with his own problems if I could create a climate where he could do it. Since that time, I would say my whole effort has been focused on the kind of psychological climate that helps the individual to resolve his problems, to develop and to grow.
>
> (Rogers and Evans 1975: p. 27)

'Normalising' is a term that arose in a recent supervision session. The supervisee related how *the counsellor* had *explained* to (or instructed) the client that it was perfectly normal to have the feelings that the client had disclosed. The supervisee seemed somewhat surprised – perhaps *affronted* would be more accurate – that 'normalising' was even worth exploring in supervision, for it was something that had been learned on a client-centred counselling diploma course (and therefore 'normalising' must be 'right'). Yet Carl Rogers quoted the findings of Bergman:

> Evaluation-based and interpretive responses or 'structuring' responses are followed, more often than would be expected by chance, by abandonment of self-exploration.
>
> (Bergman 1951: 216–24)

So how can the *therapist* who has a specific goal *on behalf of* clients, and who thus *explains* something like 'normalcy' to a client in order to achieve that goal, be considered even remotely *client*-centred? Once again, the notion of therapist *intent* is so crucial – there is a huge difference between a client feeling heard, accepted and understood and being *told* that it is normal to have particular feelings.

To the incongruent client and the incongruent therapist, feeling 'stuck' might be accompanied by feelings of discomfort, and the implementation of techniques may well be a defensive reaction aimed at easing such tension. Yet Rogers believed that:

> No approach which relies upon knowledge, upon training, upon the acceptance of something that is *taught*, is of any use. These approaches seem so tempting and direct that I have, in the past, tried a great many of them. It is possible to explain a person to himself, to prescribe steps which should lead him forward, to train him in knowledge about a

more satisfying mode of life. But such methods are, in my experience, futile and inconsequential. The most they can accomplish is some temporary change, which soon disappears, leaving the individual more than ever convinced of his inadequacy.

(Rogers 1961: pp. 32–3)

This reminds me of a client who had received counselling from a predecessor at a medical practice for which I was working. The client began by saying that she had sought help over a bereavement, and that her previous counsellor had been 'fantastic' in the way that she applied techniques and methods (such as writing letters to the deceased, placing them in a box, and then taking them to the park – accompanied by the therapist – to perform a ritual burning). At the time, the client felt truly grateful to her counsellor. Yet some months later (when she arrived to see me), her gratitude had turned to anger and resentment, for she felt as though her previous counsellor had not only taken something away from her (her natural grieving process), but had also left a 'gaping dark space' where her organismic coping methods would have been. In other words, little respect was shown by my predecessor for the potential of this client – her innate tendency to actualise in a constructive way was squashed. What a contrast this is to Rogers saying that:

> I think that we do accept the individual *and* his potentiality. I think it's a real question whether we could accept the individual as he is, because often he is in pretty sad shape, if it were not for the fact that we also in some sense realize and recognize his potentiality. I guess I feel, too, that acceptance of the most complete sort, acceptance of this person as he is, is the strongest factor making for change that I know. In other words, I think that it does release change or release potentiality to find that as I am, exactly as I am, I am fully accepted – then I can't help but change. Because then I feel there is no longer any need for defensive barriers, so then what takes over are the forward-moving processes of life itself, I think.
>
> (Rogers 1989: p. 61)

Carl maintained that striving to see a client's world *as that client sees it* is very precious and rare. Yet all too often we instead:

> offer another type of understanding which is very different. 'I understand what is wrong with you'; 'I understand what makes you act that way'; or 'I too have experienced your trouble and I reacted very differently'; these are the types of understanding which we usually offer and receive, an evaluative understanding from the outside. But when someone understands how it feels and seems to be *me*, without wanting to analyse me or judge me, then I can blossom and grow in that climate. And research bears out this common observation. When the therapist can grasp the moment-to-moment experiencing which occurs in the inner world of the client as the client sees it and feels it, without losing the separateness of his own identity in this empathic process, then change is likely to occur.
>
> (Rogers 1961: pp. 62–3)

Talking about a demonstration session he had 'performed', Carl spoke about how thoughts about the client occurred to him *after* the session had ended. Such thoughts can always, of course, be the subject of self-reflecting and supervision. *At the time*, though:

> Yesterday's client – I don't remember her exact words – but she said something about how close she felt, and I said we were almost one in this. That's often the feeling that exists, because I'm not trying to think about [how to respond] or interpret, lead, persuade or advise. None of those things are present; I'm just with that client at this moment in exactly the way that he or she is, even though I may not understand what's going on.
>
> (Rogers and Russell 2002: p. 280)

Given the notion (referred to in Chapter 1 and again towards the end of this book) that sociologically, economically and politically there has been a trend, perhaps driven by values more akin to technology than to human beings, to be mechanically problem solving, thereby requiring experts to deliver strategies, tactics, action plans and so on to us, how does this impact upon a client-centred way of being therapeutic with clients? Working in GP surgeries, where there are underlying assumptions based on a medical model of being broken (*ill*) and therefore in need of fixing (a *cure*), patients very often expect their counsellor to be an expert at mending them. Here is a light-hearted caricature:

Patient:	*'My problem is that I am ill, feeling very depressed.'*
Therapist:	*'Perhaps you could tell me more about that . . .'*
Patient:	*'Well, I discovered my partner being unfaithful, I've just been made redundant, I've had test results saying I have an incurable illness, the car packed up the day after the warranty expired, my house burned down . . . oh, and the dog was run over.'*

We might laugh – yet I have seen patient after patient who regard their *problem* as being depression, when in fact they really would be 'nuts' if they were *not* feeling incredibly low (and very often there is a catalogue of tragedies, not just one). Yet they are convinced that they *must* be ill, for they have a prescription from their GP for antidepressants, and surely their doctor would not be prescribing drug treatment if there was no illness to cure. Worse still, there is an expectation that in being referred to see me, they will meet an expert who can help to fix their ailment.

So the temptation to 'normalise' or explain to patients that they are not actually ill can be huge, yet if I do so, I am putting myself in the role of expert, thereby diminishing any trust in the personal resources of clients, and possibly colluding with values internalised into the client's self-concept about being inadequate (for instance, 'I should be able to cope'). My intent in such situations is to be as close to the therapist conditions as I can be, relying on the client to receive these conditions and move with them. If, for instance, a client meets with congruent therapist unconditional positive regard (perhaps a calm acceptance, without judgement, of who and how they are) and empathic understanding, it has been my experience that they themselves will very quickly come to the conclusion that they have every right to feel depressed, and that their feelings and thoughts are worth the effort of understanding and accepting, rather than failings to be fought.

Encouragement, rescuing, support and responsibility

Encouragement

You may recall from earlier on that *encouraging*:

> would seem to indicate on the counsellor's part a limited confidence in the capacity of the client to understand and cope with his difficulties.
> (Rogers 1951: pp. 30–31)

At a fairly mundane level, encouraging a client to pursue a particular course of action (maybe typified by such statements as 'you could have much to gain by approaching this person and talking it through with them' or maybe the more indirectly implied 'have you talked with . . .?') can easily be demonstrated as not being in harmony with client-centred beliefs. The client is being encouraged, directly or indirectly, to take a course of action that the *therapist* thinks is desirable – and there is little respect shown for the inner resources of the client.

Perhaps there is something, too, about actions that seem on the face of it to be *positive*, and those that seem *negative*. A great many trainee counsellors, when asked whether or not it is acceptable to *discourage* a client, would instantly reply that it is blatantly *not* acceptable to discourage. Yet the same trainee may feel far less strongly about *encouraging*. However, just as with *agreement* and *disagreement*, surely the right to do one implies the right to do the other? The power imbalance in the relationship is skewed (that is, primarily with the therapist, not the client) whether encouragement or discouragement is the factor. The trainee counsellor who encourages clients yet baulks at the idea of discouraging them may well *withhold* encouragement at times, rather than actively discouraging – the relationship still remains conditional and imbalanced. Clearly both encouragement and discouragement involve value judgements, and judgements of this kind clearly do not communicate unconditional positive regard and empathic understanding.

Having stated earlier that empathy is *not all about feelings*, all too often, it seems, in the early days of learning the clients of trainees appear to be encouraged to focus almost solely on their feelings, often at the expense of content, thought, behaviours, meanings or memories (as they influence the here and now). Indeed, being 'heady' or cognitive often seems to be frowned upon – and as we have seen, specious arguments about accepting a person yet not that person's behaviour are often utilised in an attempt to justify this. At this perhaps more subtle level, if the therapist consistently responds to disclosures of feelings and not to thoughts, meanings, behaviour or content, then clients are effectively being *encouraged* (or *manipulated* or *directed*) to 'operate' at a feeling level if they are to gain the attention and respect of, and feel valued by, their therapist. Such *therapist* direction cannot be described as *client*-centred.

Rescuing

Rescuing, as with many of the qualities in this section, may also come from an honourable, humane and understandable place although, as with other such characteristics, that 'place' is likely to reside within the *therapist*, not the client. Once again a *judgement* is involved – that the client is worthy of rescuing or needs to be rescued. In this instance there is a lack of respect for the client's own inner resources, and in addition the therapist might be encouraging, creating or promoting dependency. In rescuing a client, the therapist may in effect be colluding with client denials and/or distortions, colluding with the perpetuation of a self-concept that includes the personal construct of being in need of rescue.

At a fundamental level, the counsellor may be colluding with the thwarting of a client's actualising tendency.

Support

Support is not something that the client-centred therapist *intends* to offer to clients. Carl Rogers stated of client-centred therapy:

> It is experienced as basically supporting, but it is in no way supportive. The client does not feel that someone is behind him, that someone approves of him.
>
> (Rogers 1951: p. 209)

People often respond with not a little scepticism to the statement that client-centred therapists do not intend to be supportive, for how could a therapist be so cold and uncaring yet describe their approach as client-centred? Yet if the therapist sets out to support clients, a judgement has been made – no matter how humane or caring. This judgement is that the client is worthy of (and maybe in need of) the support of the therapist. The therapist has the power to *withdraw* or *withhold* support, and may even be in danger of fostering dependency. Is offering support in keeping with beliefs about client autonomy and self-direction? The answer is no. Is offering support in keeping with beliefs about the inner strengths and resources derived from the actualising tendency? Again the answer is no.

So client-centred therapists do not *intend* to offer support to clients, but rather their *intent* is to strive to both experience and offer authentic unconditional positive regard and empathic understanding. Clients may well experience, at times at least, this genuine offering of respect and empathy *as supportive*. However, that is up to *clients* to experience and decide. The therapist *intent* remains fundamental – to offer certain conditions which it is believed are not only necessary but also sufficient to create a relationship within which there is the potential for therapeutic growth.

Responsible to and not for

The notion that therapists are responsible *to* clients but not *for* them frequently arises in counsellor training programmes, and links with what Carl Rogers meant when he wrote of the 'basic struggle' of the counsellor:

> only when the counsellor, through one means or another, has settled within himself the hypothesis upon which he will act, can he be of

maximum aid to the individual. It has also been my experience that the more deeply he relies upon the strength and potentiality of the client, the more deeply does he discover that strength.

(Rogers 1951: p. 48)

In much the same way as John Shlien wrote that 'undeserved denigration of the technique (of reflecting) leads to fatuous alternatives in the name of congruence' (Shlien 1986), so too can the idea of responsible *to* but not *for* be misunderstood or even abused. Imagine two people, A and B, in relationship. Throughout the relationship A is critical of B, constantly and incessantly knowing best, putting B down, saying that B is wrong, is no good, and so on. B eventually discovers the resources to voice a dislike of A's 'superior' way of being, only to hear from A that *B* is wholly responsible for what B feels, for 'no one can *make* B feel anything'. This is an utter cop-out! Sometimes even an utter cop-out, an attempt at the abdication of responsibility such as this, can mistakenly be 'justified' in the name of congruence. Why 'mistakenly'? Because in this instance, unconditional positive regard and empathic understanding are clearly not integrated and deeply held beliefs that guide A's way of being. At the same time, if therapists are without influence, why be a therapist? Perhaps it would be more accurate to say that client-centred therapists need to be responsible *to and for themselves*, and responsible *to* but *not for* their clients. And therapists being responsible *to and for* themselves *includes* integrated and deeply held beliefs and attitudes that underpin a disciplined client-centred way of being:

> It would appear that for me, as counsellor, to focus my whole attention and effort upon understanding and perceiving as the client perceives and understands, is a striking operational demonstration of the belief I have in the worth and the significance of this individual client. Clearly the most important value which I hold is, as indicated by my attitudes and my verbal behaviour, the client himself. Also the fact that I permit the outcome to rest upon this deep understanding is probably the most vital operational evidence which could be given that I have confidence in the potentiality of the individual for constructive change and development in the direction of a more full and satisfying life. As a seriously disturbed client wrestles with his utter inability to make any choice, or another client struggles with his strong urges to commit suicide, the fact that I enter with deep understanding into the desperate feelings that exist but do not attempt to take over responsibility, is a most meaningful expression of basic confidence in the forward-moving tendencies in the human organism.
>
> (Rogers 1951: pp. 35–6)

When might a therapist struggle with the notion of not being responsible *for* a client? Other than with malpractice, the most difficult moments would seem to be when there is a considerable risk or danger, either to or from the client. Carl said that therapists need to ask themselves just how far their respect for clients actually goes. Can the therapist, for instance, maintain a deep respect for the capacity of individuals when life is literally at stake? If therapists are reaching or have reached the end of their capacity to experience and offer unconditional acceptance and respect, what are the consequences then? Rogers wrote that:

One would be the hypothesis that 'I can be successfully responsible for the life of another'. Still another is the hypothesis 'I can be temporarily responsible for the life of another without damaging the capacity for self-determination'. Still another is 'The individual cannot be responsible for himself, nor can I be responsible for him, but it is possible to find someone who can be responsible for him'.

(Rogers 1951: p. 47)

Thus one scenario might be that there are those therapists who believe that they *can* be (or even *are*) responsible for the lives of others, namely their clients. Another example might be therapists who believe that neither they nor their clients are *able to* (or *should*) take responsibility, in which case the client might either be referred to someone from the first extreme, or the client might somehow end up being cast adrift. Yet a third possibility is the therapist who tries to *temporarily* set aside client-centred beliefs in the fundamental trustworthiness of the actualising tendency, believing that client autonomy can be restored to them once this 'blip' has passed. Carl Rogers questioned what the crucial factor might be in all of this – is it communication from a frame of reference external to the client (such as therapists voicing that they are willing to continue seeing the client, or communicating a belief in the client's ability to choose, or referring the client to another resource), or is it holding the client in deep respect, whether indicated from the internal or external frame of reference? Rogers entered the moral and ethical debate as to whether the counsellor has the authority to permit clients to seriously consider suicide or psychosis without making a positive effort to prevent or circumvent such choices, asking whether not making such an effort is an abdication of our general social responsibility. He concluded that:

these are deep issues, which strike to the very core of therapy. They are not issues which one person can decide for another. Different therapeutic orientations have acted upon different hypotheses. All that one person can do is to describe his own experience and the evidence which grows out of that experience.

(Rogers 1951: p. 48)

And that is precisely what Carl Rogers did. In his formulation of the necessary and sufficient conditions, he placed unconditional positive regard *before* empathic understanding, believing that empathy is the primary vehicle for communicating our deep respect for the actualising tendency:

We might say, then, that for many therapists functioning from a client-centred orientation, the sincere aim of getting 'within' the attitudes of the client, of entering the client's internal frame of reference, is the most complete implementation which has thus far been formulated for the central hypothesis of respect for and reliance upon the capacity of the person.

(Rogers 1951: p. 36)

It would seem, then, that it is vital for client-centred therapists to explore the limits to which they are prepared to go within this philosophy, to determine just how strong their belief in the right of clients to self-determination actually is. It could even be posited that the more exceptions there are to unconditional positive regard, the less client-centred and therapeutic the relationship will be. And this feels really, really tough – for having established that it is in no way client-centred to *rescue*, nor could it be described as compassionate to stand idly by watching someone drown. Taking responsibility for the thrashing client might involve diving in the water and dragging them to the safety of dry land – yet what if the client somehow *needed* to flail prior to treading water? Perhaps worse still, what if the client really, truly, deeply wanted to drown? Could the therapist sleep at night knowing that the client could have been saved? Abdication of responsibility might involve referring to a lifeguard – the therapist is unable to help, clients cannot help themselves, so there is no option other than to refer to a third party who can help. Then again, the sinking client could be thrown a lifeline by the therapist – the client can learn to swim at a later date, once they have been saved. These are real issues – you cannot counsel a cadaver. When writing of person-centred group work, Rogers stated that one obligation of the facilitator was to define the limits within which unconditional positive regard could be offered. Perhaps the same obligation could be said to hold true for the client-centred therapist.

Clearly there are some deep and, to many, troublesome issues surrounding the extent to which beliefs in unconditional positive regard and the right of clients to self-determination can be lived out – matters of personal and social conscience. What a difference there can be between a philosophical classroom debate and losing a client through suicide. Yet let us not forget the paradox, too – the more the actualising tendency is trusted, the more experience of its trustworthiness there tends to be.

Focusing, leading and suggestion

Focusing

Some fine distinctions need to be made when looking at the issues that surround focusing. If it is the *therapist* who decides what is meaningful to focus upon, then clearly the *therapist* is leading (see below), and this could not be deemed to be client-centred. However, most therapist communications involve a *choice* on the part of the therapist, as client disclosures very often involve many feelings and meanings and/or potential sensations and meanings, and rarely can the therapist effectively capture all of them in one verbalisation. To attempt to communicate *all* feelings and meanings (and all *potential* feelings and meanings) can become somewhat clumsy or cumbersome, or the therapist might be trying too hard to 'get it right' (maybe connected with their own conditions of worth), and so on. So what does guide therapists with regard to communicating their sensing of different aspects of client disclosures? Perhaps a fairly basic paraverbal communication exercise might illustrate this ('paraverbal' means the *way* we speak, rather than *what* is spoken; it includes volume, speed, pitch, and so on). Consider the following statement:

I never said he stole that money.

Imagine saying the above sentence placing the stress on the word 'I', and the meaning becomes *I didn't say anything, but someone else did*. Place the stress on the word 'never' and it becomes *a protestation of innocence*. Place the stress on 'said' and there is *a belief but not a verbalisation*. Stress 'he' and *the accusation is levelled at a different person*. Stress 'stole' and the meaning might become *borrowed*. Stress 'that' and *it was different money that he stole*. Stress 'money' and *the theft was of something different*. Thus this sentence, consisting of just seven words, can have at least seven *different* meanings depending upon the paraverbal nature of the way it is spoken. The therapist who communicates a sense of righteous indignation because the word 'never' was stressed by the client is closer to empathic understanding than the therapist who picks up on theft because his or her own house was burgled last week! In stressing the importance of paraverbal communication, Rogers wrote that:

> even the transcripts of our recorded cases may give to the reader a totally erroneous notion of the sort of relationship which existed. By persistently reading the counselor responses with the wrong inflection, it is possible to distort the whole picture.
>
> (Rogers 1951: pp. 27–8)

The same kind of deliberation also applies to *non-verbal* communication – the client might make significant eye contact when speaking a particular word, or grimace, or tense a part of their body. If 'focusing' in this way is based on what seems significant *to* the client, *from* the client, then it could well be seen as part of being client-centred. If focusing is based on anything other than what seems significant to the client (that is, it is significant to the therapist, or psychological theory, for instance), then it is not client-centred.

Incidentally, a further development has been that some theorists and practitioners regard 'focusing' as meaning that therapists focus primarily upon themselves. Carl Rogers clearly wished to distance himself from this view, in that while it is true that therapist congruence is a core condition for client-centred therapy, it is fundamentally to do with embodying authentic unconditional positive regard and empathic understanding – and the more that therapists become absorbed with themselves (*'therapist*-centred therapy'!), the less attention is available for clients. For therapists to internally monitor themselves along the lines of 'How am I experiencing myself in relation to and in relationship with this client?' seems fair enough in terms of striving to be congruent, and especially if the aim of such self-awareness is to access, maintain and enhance the degree of unconditional positive regard and empathic understanding that the therapist is able to give to the therapeutic relationship. However, if the focus of the therapy was to shift too far away from the client because the therapist is so self-absorbed and self-monitoring, then that relationship would lose client-centred therapeutic integrity.

Leading and suggestion

Issues about leading and suggestion bear many similarities to the reasoning outlined above for *non-acceptance, evaluation* and *rescuing*. Something that Carl Rogers wrote about person-centred group facilitation can be reapplied to client-centred therapy:

> What the facilitator does *not* do is perhaps as important as what he or she does do. A facilitator committed to a person-centered way of being does not guide, push or prod. He or she is content to go with the

process of the group, whether rapid or slow. He or she does not persuade or take sides. The facilitator does not offer solutions. His or her basic trust is in the wisdom which, from experience, he knows resides in the group, if he can only create the psychological climate in which it can emerge.

(Rogers and Ryback 1984: p. 6)

Put simply and briefly, client-centred therapy is about the therapist either following or 'tracking' clients, or being alongside clients (as a respectful and sensitive companion) in their journeys of self-discovery. *If* the *therapist* leads (other than occasionally voicing a perception of which the client is as yet only dimly aware), *then* the relationship *cannot* be client-centred. If any suggestion comes from a frame of reference that is *external* to the client, the therapy cannot be client-centred.

A brief journey into power

Most procedures in psychotherapy may be placed on a scale having to do with power and control. At one end of the scale stand orthodox Freudians and orthodox behaviorists, believing in a politics of authoritarian or elitist control of persons 'for their own good', either to produce better adjustment to the status quo or happiness or contentment or productivity or all of these. In the middle are most of the contemporary schools of psychotherapy, confused, ambiguous or paternalistic in the politics of their relationships (though they may be very clear regarding their therapeutic strategies). At the other end of the scale is the client-centered, experiential, person-centered approach, consistently stressing the capacity and autonomy of the person, her right to choose the directions she will move in her behavior, and her ultimate responsibility for herself in the therapeutic relationship, with the therapist's person playing a real but primarily catalytic part in that relationship.

(Rogers 1978: pp. 20–1)

Let us temporarily step out of *'Thou Shalt Not Skill'* to look at the notion of power and control, for many of the attitudes and practices rejected by Carl Rogers link with beliefs about the location of power and control in psychotherapy. Carl said that elements we might explore include:

Who will be controlled? Who will exercise control? What type of control will be exercised? Most important of all, toward what end or what purpose, or in the pursuit of what value will control be exercised?

(Rogers and Evans 1975: p. lxii)

Therapists, said Rogers, first need to think about both their general and their specific goals when practising therapy, to settle within themselves their purpose in being therapists in the first place. Although he agreed with the prevailing scientific view that research can help us to determine the means of achieving the goals that we have settled upon, Rogers also thought that the tradition had been for an individual or group (such as a therapeutic approach) to gain the power to use their methods. Individuals are then exposed to these established conditions and there is a high probability that outcomes will be in line with them, leading to an ongoing, socialised organisation that continues to produce the desired behaviour in patients or clients. He believed, with regard to the first point, that:

> It is possible for us to choose to value man as a self-actualizing process of becoming; to value creativity, and the process by which knowledge becomes self-transcending.
>
> (Rogers and Evans 1975: p. lxxviii)

Client-centred therapist conditions can be set with a *minimum* of power or control, and client outcomes are to do with self-actualising, a process whereby clients move towards greater self-responsibility:

> Thus such an initial choice would inaugurate the beginnings of a social system or subsystem in which values, knowledge, adaptive skills, and even the concept of science would be continually changing and self-transcending. The emphasis would be upon man as a process of becoming.
>
> (Rogers and Evans 1975: p. lxxix)

Not only do therapists establish the conditions in which they work, but also clients give power to therapists:

> The client, a somewhat lost, confused, not too well-ordered person, has typically never met anyone with such a powerfully directed focus and lived belief. As already indicated, the client's position as a failed person hoping to find his way also lends substance to the therapist's power.
>
> (Barton 1974: p. 191)

And therapist power has a value base that the therapist cannot but help communicate:

> The methods the therapist uses . . . represent values. They com-municate the essential therapeutic values . . . or the lack of them. The therapist's methods are not simply objective techniques but are part of the therapist as a person. The therapist as a person relates to the client as a person. The responses of the therapist reveal what the therapist values. These responses also reveal whether the therapist considers himself or herself an expert by leading, questioning, inter-preting, guiding, suggesting, advising, or whether the therapist places the responsibility on the client by . . .
>
> (Patterson 2000: p. 121)

Many trainee therapists have said that their main reason for wanting to become a client-centred therapist is their *desire to be of help* (wanting either to 'give something back' to repay valued help that they themselves received, or to offer to others the assistance they themselves never received) – yet Carl Rogers believed that even this ideal or goal was somewhat secondary, for:

in the most real moments of therapy I don't believe that this intention to help is any more than a substratum on my part . . . Surely I wouldn't be doing this work if it wasn't part of my intention. And when I first see the client that's what I hope I will be able to do, is to be able to help him. And yet in the interchange of the moment, I don't think my mind is filled with the thought of 'now I want to help you'. It is much more 'I want to understand you. What person are you behind that paranoid screen, or behind all these schizophrenic confusions, or behind all these masks that you wear? In your real life, who are you?'. It seems to me that is a desire to meet a person, not 'now I want to help'. It seems to me that I've learned through my experience that when we can meet, then help does occur, but that's a by-product.

(Rogers 1989: p. 55)

Let us now return to the '*Shalt Nots*' of meeting with a valued client.

Mirroring, paraphrasing, reflecting and summarising

Let one thing be absolutely clear:

There is no such thing as a client-centred skill.

Likewise, Carl Rogers wrote of therapist communicating that:

If it is simply reflection, that's no good. That's just a technique. It must be a desire to understand empathically, to really stand in the client's shoes and to see the world from his vantage point.

(Rogers and Evans 1975: p. 29)

Many people react with astonishment to the statement that *there is no such thing as a client-centred skill*, yet we have seen in Chapter 1 how Carl was *appalled* at the way in which client-centred therapy (and empathic understanding in particular) had become a caricature of itself, and he was utterly dismayed at the way in which in many counsellor training programmes the learning or teaching of 'wooden techniques' had somehow replaced *becoming a person*, and *developing a client-centred way of being*.

It is clear that, in part, techniques such as mirroring, paraphrasing, reflecting and summarising became associated with client-centred therapy because they are relatively 'non-directive' when compared with more overtly manipulative practices (to give an extreme example, paraphrasing a client's words is less directive than referring a client for electroconvulsive treatment). It is also true (and by his own admission), that *in the early days* of developing client-centred therapy, Carl Rogers and his colleagues *did* pay considerable attention to therapist techniques. It may be, too, that those who desire to be of help to others readily grasp skills and techniques that are relatively easily taught or learned.

Approaching this from a different angle, surely *a skill is a skill is a skill* whatever the orientation of the therapist? If, for instance, a transactional analyst mirrors or reflects something that a client has said, does it mean that for that moment only the therapist is suddenly 'being client-centred'? If a psychodynamic practitioner paraphrases a client disclosure, has that analyst suddenly ceased to become psychodynamic and miraculously converted to being client-centred? If a cognitive–behavioural therapist summarises what the client has just shared, has that therapist somehow 'integrated' client-centredness into cognitive–behavioural therapy? The answer to all of these questions is *of course not*. In part we can say 'of course not' because Rogers stated that the six client-centred conditions of contact, client incongruence, therapist congruence, unconditional positive regard and empathic understanding, and client perception of the therapist's respect and empathy were both necessary *and sufficient* – and that the 'importing' of elements from other therapies goes a long way towards *destroying* the possibilities for a therapeutic encounter. And in part we can assert 'of course not' because a reflection is a reflection *whatever* the therapeutic approach of the therapist (indeed, such communication skills have even been adopted to enhance sales techniques and other forms of 'people management'). So let it be repeated, *there is no such thing as a specifically client-centred skill.*

To *paraphrase* or *summarise* what a client has just said, the practitioner needs to develop a fairly good memory that includes the ability to absorb, store and recall information, have a reasonable vocabulary and understanding of the language being spoken (including synonyms), and have the ability to re-order words (usually into a more succinct form) without losing their essential meaning. While this may, then, involve a degree of cleverness (and indeed may form the basis of a helpful relationship), it does not necessarily have to involve the demanding task of truly and delicately entering the world of another and striving to respectfully experience and understand it as the other person experiences it, including their feelings, meanings and memories. Paraphrasing or summarising hardly represents an attempt to communicate the fruits of a counsellor having strived to sensitively and delicately enter the world of another and to respectfully experience and understand it as the other person experiences and understands it.

Another way of seeing how client-centred therapy is more to do with an empathic way of being than with the implementation of skills might be to look at the technique called 'reflecting' – and 'reflecting' or 'reflecting feelings' appear on the curriculum of the majority of counselling skills and even (allegedly) client-centred counsellor training programmes. Yet Carl Rogers wrote of the influence of a Rankian trained practitioner:

> The next social worker that I employed in Rochester was Elizabeth Davis . . . and one of the things she taught me was to focus on the feelings that were being expressed and to respond to those feelings. It was this that led to this whole idea of reflection of feelings, which is much misunderstood.
>
> (Rogers and Russell 2002: p. 113)

Although Carl expressed gratitude for this having improved his work early in his career, he later wrote of '*reflect* becoming in time a word that made me cringe' (Rogers 1980: p. 138). It is possible to reach the conclusion from Carl's words that the learning of reflecting skills (or the technique of 'reflecting feelings') *could* therefore be a useful step on the way towards developing an empathic way of being. That may be – yet imagine the counselling *skills* trainee who has received positive feedback and indeed certification for both open-ended questioning and reflecting discovering later (when training to become a client-centred *therapist*) that not only is there nothing specifically client-centred about the generic skill of reflecting, but also Rogers 'cringed' at the notion. Worse still, if the implementer of the skill of reflecting primarily focuses on *feelings*, the relationship comes to have a narrow focus, as the therapist is not engaging with a *whole* person – it becomes *directive* (the *client* is not leading) and *judgemental* – and the client's world of personal meanings, thoughts, behaviours and memories may be neglected.

To illustrate this point, if another person is speaking, it is not too difficult for listeners to train themselves to listen for key words or phrases and to 'reflect them back' to the speaker. However, this is largely an exercise in discerning listening, memory and recall – the listener does not need to invest their whole being in this exercise, as it is primarily their intellect that is utilised. For instance, it is possible to reflect a client's words with little or no compassion or caring. Indeed, one could listen for and reflect the key words or phrases of someone speaking on television, or from a passage in a book. There is no necessity for many of the necessary or sufficient conditions – there is no absolute need for 'psychological contact' or a quality of presence, no essential requirement for the listener to be congruent, no necessity for great respect for the inner world of client feelings and meanings (if engaged in a primarily intellectual exercise), and no real demonstration of empathy or understanding. Instead, this could be a demonstration of competency in hearing and recall. Hence Rogers wrote that he had become more and more 'allergic' to the use of the term 'reflecting'.

However, Carl Rogers did acknowledge that viewing or listening to his demonstration counselling interviews could easily result in the idea that he did regularly 'reflect'. It was a friend and former colleague of his, Dr John Shlien, who wrote to Rogers that 'reflection is unfairly damned', for while Shlien agreed with him that reflecting purely as a technique or skill lacked quality of presence, Shlien also maintained that reflecting can be 'an instrument of artistic virtuosity in the hands of a sincere, intelligent, empathic listener' and that 'it made possible the development of client-centred therapy, when the philosophy alone could not have. Undeserved denigration of the technique leads to fatuous alternatives in the name of congruence' (Rogers 1986: pp. 375–7).

This correspondence with John Shlien enabled Carl to further clarify his process of communicating empathy. Rogers was *not* trying to 'reflect feelings' but *was* engaged in a process more akin to *'perception checking'* – by which he meant a cautious (or tentative) offering of his sensing of what the client was experiencing from moment to moment, and being constantly prepared to amend his sensing in response to client feedback. At the same time, Rogers acknowledged that there was a sense in which the therapist is 'holding up a mirror' to the client, a 'mirror' that enables clients to see and experience themselves more clearly. Carl stated that Sylvia Slack expressed this 'beautifully' when she wrote:

> It was like Dr Rogers was a magical mirror. The process involved my sending rays toward that mirror. I looked into the mirror to get a glimpse of the reality that I am. If I had sensed the mirror was affected by the rays being received, the reflection would have seemed distorted and not to be trusted. Although I was aware of sending rays, their nature was not truly discernible until they were reflected and clarified by the mirror. There was a curiosity about the rays and what they revealed about me. This experience allowed me an opportunity to get a

view of myself that was untainted by the perceptions of outside viewers. This inner knowledge of myself enabled me to make choices more suited to the person who lives within me.

(Slack 1985: pp. 35–42)

Thus therapist communications need to be undistorted, and that means not 'polluting' the internal frame of reference of the client.

Watching videos of Carl Rogers in action has been experienced as confusing by some viewers. In one session, for instance, Rogers seemed almost to repeat, word for word, what the client had just said. However, in the question-and-answer session that followed the demonstration, he explained that in *that* moment, with *that* particular client, using the client's own words and speaking in the first person (as if he were that client) seemed the best way to try to enter her frame of reference, her world of feelings and personal meanings. In other demonstration therapy sessions, Carl was perceived quite differently, so we can see that striving to empathically understand another is far from being a 'wooden', 'cold' or systematic technique. Rather, empathy is a fluid yet consistent way of being with another, that will vary from client to client and with the same client over time (from, for instance, tentative to more confident, as we saw in Chapter 1). In an interview given just months prior to his death (transcription published posthumously in 2002), Carl Rogers said:

It is interesting that the term *reflection of feelings* still is a current issue . . . When one looks at it from an external point of view, and studies a transcript of an interview, for example, it is quite clear that the therapist very often is giving a restatement, a fresh verbal organization of the feelings and personal meanings the client has expressed. In that sense, reflection of feelings seems like a fairly accurate term. But from the point of view of the therapist, when good therapy is going on, the therapist is not trying to reflect feelings. He is trying to check understandings or test perceptions. He's trying to say, 'Is this the way it is in you? Am I getting the flow of your feeling? Am I understanding the nuances of what's going on in you? Are you saying that you're angry with your father?' Although those questions as I'm now stating them are usually not verbalized, that's always the intent.

(Rogers and Russell 2002: pp. 253–4)

Praise and reassurance

Praise

Many people over the years have expressed how, without praise, their learning would have been impaired in some way, or that if only they had received more positive feedback they would have done better. Praise is often viewed as a human need, and as a very good thing indeed.

Once again, though, praising involves the *judgement* that something *merits* positive feedback. Does the counsellor feel the same way about voicing criticism or negative feedback to clients? If the counsellor does *not* feel the same way, how can it be argued that therapist positive regard is *unconditional*? For sure, if therapists grant themselves the power and authority to praise and encourage and support and reassure, then by implication they also grant themselves the power and authority to chastise or criticise, or at the very least to withhold or withdraw praise, encouragement, reassurance and other apparently 'positive' feedback.

Empathic understanding is about entering the *internal* frame of reference of the other. Praise indicates that the *therapist* believes that the other person is deserving of positive feedback – from a locus of evaluation *external* to the client. Lest it again be thought that the client-centred therapist lacks warmth, if a client clearly feels good about themselves or something they have achieved, and the therapist empathically understands this *from within the client's frame of reference* – and the client perceives this appreciation – then that relationship will not lack warmth. Rather, being in tune with the client's internal locus of evaluation, it will have deeper and more personal meanings for that client.

Reassurance

Again, recall the words of Carl Rogers:

> reassuring . . . would seem to indicate on the counsellor's part a limited confidence in the capacity of the client to understand and cope with his difficulties. The counsellor still felt it necessary at times to take the lead . . .
>
> (Rogers 1951: pp. 30–1)

Reassurance, at least in the early days of counsellor training, is often seen as another of those 'nice' qualities. Without intending any disrespect for the word 'nice', can we nevertheless examine the concept of 'nice' in a client-centred therapist and client context?

First, let us question where the *need* to reassure comes from. If it is the *therapist's* need to reassure, then no matter how honourable or nice a place that need comes from, it is clearly the *therapist's* need, and *not* the need of the client, and so it can hardly be described as *client*-centred. Alternatively, the therapist might sense a need for reassurance coming from the client, and there are those who see counselling as being about responding to or meeting such needs. At first glance, this might even seem to make some sense, yet it does not stand up to more serious scrutiny.

The statement that client-centred therapy is not at all to do with meeting client needs often meets with surprise. Yet ask yourself 'If a client expresses the need for me to provide all the answers and solutions, is to do so client-centred?' The consensus of opinion would surely be that client-centred therapy is *not* about the therapist providing solutions, so in this instance responding to an expressed client need by simply going along with it would clearly be inappropriate. The client-centred therapist might well strive to respect and empathically understand a client's needs, but that is not the same as *judging* the need as okay or not okay (from a frame of reference *external* to that client) and responding accordingly. Another question could be asked: 'What if it is a client need to seriously abuse the therapist in some way?'. Does the therapist creating a 'permissive' atmosphere mean that counsellors respond to client needs by becoming their victim? Of course not! If it is clear that there are some client needs that a therapist would certainly *not* meet, who determines which needs it might be okay for the therapist to meet? If the *therapist* determines what is and is not acceptable, then 'say goodbye' to *unconditional* positive regard. If the *client* determines what needs the therapist will meet, 'say goodbye' to *therapist* positive self-regard! Client-centred practice is about authentically and unconditionally *respecting* all client needs – the therapist 'responds' to such needs not by trying to meet them, but by striving to sense and understand the feelings and personal meanings that generate such needs.

Thus responding to client needs has no place whatsoever in client-centred therapy. Whatever the context, *reassuring* clients is not in the spirit of being client-centred, for reassurance does not respect client autonomy and self-direction, is judgemental, and often involves a frame of reference external to the client.

Secondly, and on a quite different level, consider the situation where a therapist reassures a client that 'it will be okay' or 'you are not alone', for instance. Suppose that it will *not* be okay and that despite the therapist's assurances the client still *does* feel alone? The therapist in this example could hardly be said to be respecting the client's world of experiencing and perceptions.

As with so many of these skills and techniques, there would appear to be an anxiety that *not* to use them would result in the therapist being perceived as heartless. *It is really very important to understand and live the difference between an outcome being predictable and striving to achieve that outcome*. If a therapist strives to empathise with and understand a client's world of feelings and meanings, a predictable outcome is that the client will experience being someone worthy of being understood – and the client may well feel reassured by this. However, it is *not* the therapist's *intent* to reassure the client, nor are client feelings of reassurance a goal of client-centred therapy. As Rogers said, if the therapist can *be* the core conditions, the client can *do* the rest.

Probing and questioning

Probing

At one extreme, it is not difficult to see that client-centred therapy has nothing to do with the therapist taking a crowbar to the client in order to jemmy out their feelings or meanings – client-centred therapy is about prizing, not prising! However, clients reflecting upon the process of their therapy often describe it as having been 'like peeling back the layers of onion' until they discovered the 'core' of their personality or being. What part does the counsellor play in this process? After all, the core conditions for client-centred therapy do indicate that the therapist needs, necessarily, to be genuinely interested in (and respectful of) the client's world of feelings and personal meanings. Might *taking a genuine interest* in someone relate to probing in an effort to understand?

Returning to the theme of therapist *intent* might be helpful here. It is true that if the therapist communicates, with some degree of accuracy, authentic and respectful empathic understanding, then the client is likely to respond by taking further steps on their journey of self-discovery. However, the *intent* of the therapist is to strive to sensitively empathise, *not* to somehow 'move a client on'. In other words, a *predictable consequence* of the client perceiving authentic unconditional positive regard and empathic understanding from the therapist is that the client will continue on their journey, going deeper and further if they choose to do so. This is the *if–then* quality of client-centred therapy that was described in Chapter 1 – *if* the client receives nurturing conditions, *then* the client may grow. Just as the phrase *'Why be the magician when you can be the magic?'* emerged:

> *Why be the gardener when you can be the garden?*

Carl Rogers wrote of the 'soil of the theory' out of which client-centred therapy grew. Similarly, maybe the counsellor can provide the soil out of which clients may grow. It might be that counsellor *supervisors* can in part tend the garden, and it might be that therapist personal and professional development activities are important in the process of striving to ensure that conditions in the garden are all that they can be. However, the therapy garden is all that is necessary for the plant to grow – there is no need to prise open or go poking around, whether clumsily or smoothly, in the petals.

Questioning

> During one post-workshop discussion, a participant said to Rogers, 'I noticed that you asked questions of the client. But just last night a lecturer told us we must never do that'. Rogers responded, 'Well, I'm in the fortunate position of not having to be a Rogerian'.
>
> (Brink, personal communication, 1990,
> quoted in Farber *et al.* 1996: p. 11)

Carl Rogers and his colleagues discovered through their research activities that a distinguishing feature of client-centred counselling was that the therapist frames far fewer communications as questions than is the case with most (if not all) other approaches to psychotherapy. Likewise, at the commencement of client-centred therapy training courses (or at the end of counselling skills programmes), it is not uncommon for something in excess of 90% of trainee counsellor communications to their clients to be phrased as questions. On skills courses that purport to be 'client-centred' (or 'person-centred' or 'humanistic' or 'non-directive') most of these questions will be open-ended – yet a *question*, whether 'open' or 'open-ended' or otherwise, *cannot and does not convey empathy or understanding*.

If, for instance, therapists have empathically understood clients, why would they then need to question them? Just how respectful of client feelings and meanings is it when, after a client has just described the most harrowing and dreadful experiences, the therapist says 'How do you feel about that?'. Further-more, there are repercussions to asking questions in *process* terms, for *if* the counsellor asks questions, the onus is *then* upon the client to reply to those questions – in other words, the *counsellor* is leading the session, not the *client*. As a consequence of this therapist question–client answer–therapist question–client answer pattern, the client could be forgiven for inferring that if all of the therapist's questions are adequately answered, a solution will be given to the client by the therapist, or if not a solution, at least an accurate prognosis or diagnosis, or whatever. Yet another consequence is that if the therapist's capacities are, to whatever extent, engaged with formulating their next question to the client, how fully is the client being heard? How fully is the therapist striving to respect, empathise and understand?

To illustrate this point, many trainee therapists have, over the years, commented that when they were tape recording counselling practice sessions, they felt a need to 'perform' and to 'get it right', and they usually believed that the tape-recorded evidence was not accurately reflecting their usual practice. In other words, their capacities were largely devoted to *formulating responses*, rather than really listening, really hearing and really striving to respectfully enter their clients' worlds of feelings and meanings. However, by the end of successful client-centred therapy training, *less than 5%* of therapist communications are questions. Instead, the therapist *communicates* (albeit it tentatively and cautiously at times) empathic understanding rather than *responding* to clients with open-ended questions, or reflections, or by paraphrasing or summarising what the client has just said.

Please note that this is not to say that the client-centred therapist *never* asks questions. A general guideline is that a client-centred way of being is only to ask a question when no empathy is being experienced, and the question asked is designed to *generate* empathic understanding. This guideline can be refined somewhat, too, for Carl Rogers wrote about empathic *guessing* and empathic *inference*. That is, it is not as simple as an instant 'no empathy – therefore ask' – there is no 'if–then' formula here (as in *if* there is no empathy being experienced, *then* ask a question in order to gain some). Rather it is a case of striving to gather and absorb *all* client cues (verbal, paraverbal and non-verbal) in an attempt to enter their frame of reference. A tentative guess or inference is preferable to a non-empathic question (no matter how open-ended).

Therapist self-disclosure

'Letting it all hang out?'

It is true to say that during the course of his career Carl Rogers did change his views somewhat with regard to the therapist's *presence*. In the early days, Carl wrote on many occasions of the need for therapists to 'set their selves aside' when with clients, as typified when he quoted Nat Raskin:

> the counselor makes a maximum effort to get under the skin of the person with whom he is communicating, he tries to get *within* and to live the attitudes expressed instead of observing them, to catch every nuance of their changing nature; in a word, to absorb himself completely in the attitudes of the other. And in struggling to do this, there is simply no room for any other type of counsellor activity or attitude.
> (Raskin, quoted in Rogers 1951: p. 29)

In 1951, Rogers himself wrote:

> the relationship is experienced as a one-way affair in a very unique sense. The whole relationship is composed of the self of the client, the counselor being depersonalized for the purposes of therapy into being 'the client's other self'. It is this warm willingness on the part of the counselor to lay his own self temporarily aside, in order to enter into the experience of the client, which makes the relationship a com-

pletely unique one, unlike anything in the client's previous experience.

(Rogers 1951: p. 208)

However, within just a few years Rogers came to realise that the quality of therapist presence was a very significant factor in counselling, and by the end of his career he was speaking with Gay Barfield of *reciprocal* empathy (as we saw in Chapter 1) rather than 'a one-way affair', and in 1985 he was able to say:

> I think congruence means that one is aware of what one is experiencing inside, and consequently one is able to express it if it seems appropriate in the situation. The most important part is that one is aware of what one is experiencing. And experiencing, as we have said, has both content and feeling in it.
>
> (Rogers and Teich 1992: p. 62)

We might again refer to John Shlien's phrase 'fatuous alternatives in the name of congruence' when looking at appropriate and inappropriate therapist self-disclosure within a client-centred framework, for regrettably there are those who seem to view 'being appropriately congruent' as meaning that it is perfectly in order (or 'appropriate') for the client-centred therapist to 'let it all hang out' (a kind of 'anything goes' approach to therapy). Such people, it seems, can try to justify just about *anything* by saying 'I was just *being* congruent' or, worse still, 'I was just *communicating* my congruence'! Well, maybe it is true that such a person *is being* congruent, yet it is also true to say that this congruence hardly deeply encapsulates unconditional positive regard and empathic understanding. As Rogers stated:

> For therapy to occur the wholeness of the therapist in the relationship is primary, but a part of the congruence of the therapist must be the experience of unconditional positive regard and the experience of empathic understanding.
>
> (Rogers 1959: p. 215)

'I was just communicating my congruence' is worse still because it demonstrates a complete lack of understanding about what congruence in a client-centred framework *means* and *is*. Carl Rogers maintained that much of our experiencing is *available* to awareness (given the right conditions – more of this in Chapter 4), and congruence is a lack of *denying* significant experiencing to awareness, or not having to *distort* our experiencing in order to allow it into our awareness. To 'communicate congruence' to a client, a therapist would need to be saying something like 'Right now, I am denying none of my significant experiencing to my awareness, nor am I distorting it in order to permit it into my awareness, albeit in a skewed fashion' – this is not very likely or appropriate, nor is it very *client*-centred! Nor was therapist communication or client perception of therapist congruence initially one of the six conditions for therapeutic personality change. It was only deemed necessary for the client to receive the therapist's unconditional positive regard and empathic understanding (the fourth and fifth conditions), because at the time Rogers was less certain about the place of therapist disclosure. The notion of therapist congruence will be more fully explored later, but for now let it be stated that *there is a crucial difference between communicating congruence* and *congruent communication*. Thus while disclosing that there is no denial or distortion of significant experiencing of the therapist by the therapist is both highly unlikely and undesirable, it *is* highly desirable that the client experiences the therapist's respect and empathy as *authentic* because it *is* authentic – that is, congruent.

When speaking of the personal growth journey that is an ingredient (and probably the main ingredient) on the way to becoming client-centred, Carl did say that it was regrettable that some people never moved beyond the *self–self–self* phase (Berwick and Rogers 1983). Some members of the Center for Studies of the Person came up with the idea of a 'narcissistic paradigm' (*this is what I want, so to hell with you!*). Perhaps a 'masochistic paradigm' is worthy of mention, too (*this is what you want, so to hell with me!*). Quite frequently, internalised conditions of worth about pleasing others lead to the thwarting of the actualising tendency – the person lives their life as a server, a carer, a tender of others, hence the *masochistic paradigm* (what I want, need or desire is of meagre merit compared with catering to the needs of others). One of the predictable process phases in client-centred therapy is a movement *away from* values introjected from others, a movement away from pleasing others and *towards* pleasing the self, hence the *narcissistic paradigm* (my wants, needs or desires clearly take precedence over those of others). It is crucial to realise that *this phase does not constitute the entire journey*, and, as Rogers conceded, there are those who emerge from training and other person-centred experiences not having moved beyond *self–self–self*. Just as it is conceivable that the learning of counselling skills *might* represent steps along the way towards becoming empathic as a way of being, so entering a narcissistic phase might be a process movement towards (client-centred) congruence. Nevertheless, let us be clear that the counsellor 'letting it all hang out' or believing that 'anything goes' is not what Rogers meant client-centred therapy to be. Such attitudes are more akin to Shlien's 'fatuous alternatives'.

However, this is not to say that the client-centred therapist is a 'closed book' or a 'blank screen'. As already stated, it has been extremely unfortunate that some people seem to take words like 'transparency' and 'genuineness' or phrases like 'being real' to also mean 'let it all hang out' or 'anything goes'. *Being transparently real is only client-centred if the therapist has deeply integrated the core conditions as a way of therapeutic being.* This is not the same as stating that the client-centred therapist *never* self-discloses. When asked by John Masterson about the danger of personal growth journeys resulting in everything else having to stop this instant (perhaps 'a fatuous alternative in the name of immediacy') until personal feelings are worked through, Rogers said:

> I've seen that happen, yes. . . . The risk runs two ways. Sometimes what people learn is to be more expressive of their feelings – period. Sometimes what they learn is to be more willing to listen, and to hear – and then perhaps sit on their own feelings. Either is a risk and unfortunate. I think it has to be a balance: of being willing to express what I deeply feel, but also a willingness to hear what you deeply feel – and that I think is the basis for a stronger, richer relationship, when there is that balance.
>
> (Rogers 1985)

Appropriate therapist self-disclosure

Carl Rogers gave us useful guidelines on when it might be appropriate for a therapist to self-disclose, the primary reason being when *not* to do so would serve to block the therapist's capacity to absorb, empathise and understand – when the therapist is prevented from meaningfully entering a client's world of feelings or meanings due to some personal experiencing acting as a barrier across the gateway to the experiencing and meanings of another. However, Rogers also believed that if a therapist feeling is experienced as *persistent* and *prolonged* in the relationship, then it probably merits being voiced *in* that relationship. He quoted the experience of a colleague who, near the end of his working day, received a call saying that his son had sustained minor injuries in a road traffic accident. With only one client to see, and knowing that his son was in no dire danger, the therapist believed that he could set aside his own feelings and thoughts until later. However, during the therapy session the counsellor's feelings and thoughts *did* veer towards his injured child so, after some internal mulling over, he thought it best to 'come clean' with the client – who responded by expressing relief, because they had perceived the therapist's distraction as meaning that it was something 'unattractive' about them that had elicited a negative reaction from the therapist.

To *refrain* from self-disclosure might be anti-therapeutic:

> From the beginning, and with increasing emphasis as time has passed, Rogers has stressed the centrality of sincerity, honesty, and genuineness in the therapist's responses. It is only his genuinely felt understanding and honest positive regard that can be deeply helpful and curative; technique alone is not enough and may even be irrelevant or harmful at certain moments in psychotherapy. If, for instance, the therapist uses his nondirectiveness as a way of suppressing his own annoyance because the poor weak client couldn't possibly take it, an attitude of fundamental disrespect, annoyance, and lack of real regard for the client's powers will be communicated. Even with the best of intentions, such communication of lack of understanding and regard is damaging because the sneaky obscurity of it tends to poison the atmosphere and inhibit that flow of honest feeling that is at the heart of client-centered help.
>
> (Barton 1974: p. 195)

For a therapist to *feign* empathic understanding (perhaps by reverting to skills and techniques) or to offer only a *semblance* of unconditional positive regard would not be *authentic* self-expression – the therapist is not a whole person, the therapist is *incongruent*. Carl Rogers said that respect for the capacity of individuals was a central tenet that permeated his whole being, so when being a whole person in the counselling relationship:

> I still feel that the person who should guide the client's life is the client. My whole philosophy and whole approach is to try to strengthen him in that way of being, that he's in charge of his own life and nothing that I say is intended to take that capacity or that opportunity away from him. It is changed in this respect, that I would try to be aware of my own feelings and express them as my feelings without imposing

them on him. I would express even negative feelings. I might tell a client, 'I'm really bored by what you're saying'. It doesn't guide him. It does provide him with some rather jolting data which he must handle in some way or another. But I'm not telling him what he should do to avoid boring me. Maybe he would just as soon bore me; that's up to him . . . I was quite fascinated by the discovery that people did have much more capacity to guide themselves than I'd given them credit for. I became sort of a tourist in that, thinking 'Let's not let any of me into the situation except just an understanding of the client's feelings'. I gradually realized over the years that the tourist approach shakes the client out of what might be a very close interpersonal relationship which is much more rewarding.

(Rogers and Evans 1975: pp. 26–7)

Rogers also wrote of experiencing yet being unaware of annoyance:

When I am experiencing an attitude of annoyance toward another person but am unaware of it, then my communication contains contradictory messages. My words are giving one message, but I am also in subtle ways communicating the annoyance I feel, and this confuses the other person and makes him distrustful, though he too may be unaware of what is causing the difficulty. When . . . I fail to listen to what is going on in me, fail because of my own defensiveness to sense my own feelings, then this kind of failure seems to result. It has made it seem to me that the most basic learning for anyone who hopes to establish any kind of helping relationship is that it is safe to be transparently real.

(Rogers 1961: p. 51)

Rogers followed this by writing that, strange as it may seem, it is helpful if a therapist can be his or her own therapist. We shall see that one outcome of therapy is that clients become more self-accepting, and an outcome of that outcome is that they become more accepting towards others. The person who has achieved unconditional positive self-regard in a responsible and socialised way through a process of being sensitively aware of, respectful of and under-standing towards their own inner feelings has moved significantly towards being the necessary and sufficient conditions for being a client-centred therapist. Rogers continued:

Now, acceptantly to be what I am, in this sense, and to permit this to show through to the other person, is the most difficult task I know and one I never fully achieve. But to realize that this *is* my task has been most rewarding because it has helped me to find what has gone wrong with interpersonal relationships which have become snarled and to put them on a constructive track again. It has meant that if I am to facilitate the personal growth of others in relation to me, then I must grow, and while this is often painful, it is also enriching.

(Rogers 1961: p. 51)

In other words, if a feeling (or attitude) is persistent enough or strong enough in a relationship, it will probably be sensed by clients anyway (this could be viewed as an example of the 'reciprocal empathy' referred to in Chapter 1), so honesty and openness are preferable to the alarm or insecurity that the therapist being a 'closed book' might generate. However, such honesty springs from the integrity of the therapist, and that integrity includes the deep embodiment of a client-centred way of being.

Let us close this section with three final quotes about Carl Rogers that hopefully capture the essence of congruent therapist communication. The first illustrates his total commitment to entering and being with clients in their own personal worlds of experiencing:

> It is interesting to note that, after a demonstration interview, Rogers would occasionally tell his audience that his attention was so totally focussed on his client that the rest of the world simply disappeared for him.
>
> (Brink, cited in Farber *et al.* 1996: p. 30)

Next comes a quote that resonates with my own experience and with comments made by several associates. In it we can see how self-disclosures made by Rogers were very much located in and relevant to the client's world of feelings and meanings:

> One of our recorded clients in our research said at the last interview 'I really don't know you at all, and I've never known anyone better in my life'. And both things could be true. She didn't know anything about me as a person outside of therapy, but in therapy she knew me as someone who was understanding and real.
>
> (Rogers and Russell 2002: p. 286)

Finally, Rogers spoke of his regret at not emphasising the rigorous and disciplined nature of being client-centred:

> Being real does not involve us doing anything we want to do; it means a disciplined approach. That's one thing that I realize I have not stressed enough, and consequently it has been overlooked. I'm quite a disciplined person myself, and it comes naturally to me to think that everyone else is, too, but that's not so.
>
> (Rogers and Russell 2002: p. 286)

Transference and counter-transference

The view of Carl Rogers with regard to the theories of transference and counter-transference was unequivocally clear – they were unfounded and had outlived their time. Carl deplored them, believing them to be seductively disgusting and socially reprehensible. *There is no place at all for such thinking or practice in client-centred therapy.*

> I have a very strong conviction that psychoanalysis is a speculative, outdated theory which still hangs on, which has an incredible amount of undue importance in the training of psychiatrists and social workers. I think it is absurd that it hasn't been displaced.
>
> (Rogers and Russell 2002: p. 243)

If psychoanalysis, including such theories of 'transference' and 'counter-transference' (and 'resistance', too) are so repugnant, why are they still with us? Carl's opinion with regard to their continuing influence was as follows:

> I think it is because it appeals to a certain type of mind. Some people feel something simple must obviously be untrue or low grade, but if you have a complicated theory that nobody can quite understand, and it's really mysterious, and it takes an expert to even spell it out, and the way it applies to an individual is so complicated that it would take years and years to really understand it – 'Ah, that must be true'.
>
> (Rogers and Russell 2002: p. 243)

While some people might think that it 'must be true' because it is complex and long-winded in its learning, Carl Rogers believed there to be a distinct lack of meaningful research data to back up the alleged importance of so-called transference in therapy, and indeed a reluctance on the part of practitioners for whom 'transference' is a fundamental aspect of their counselling to open their data to public scrutiny. Carl's opinion was that:

> To deal with transference feelings as a very special part of therapy, making their handling the very core of therapy, is to my mind a grave mistake. Such an approach fosters dependency and lengthens therapy. It creates a whole new problem, the only purpose of which appears to be the intellectual satisfaction of the therapist – showing the elaborateness of his or her expertise. I deplore it.
>
> (Rogers 1987b: p. 183)

Looking at notions of 'transference' and 'counter-transference', Carl differentiated between two main types of client feelings or emotions towards the therapist, namely *feelings that are an understandable client response to the attitudes and behaviours of the therapist*, and *feelings that have little or no relationship with or to the therapist*. Thus a client might understandably feel anger or resentment at the therapist offering (premature) interpretations, or projecting an aura of expertness or superiority, or a client might understandably feel warmth or love when experiencing a deep understanding from the counsellor. Such feelings are *understandable* client responses to a client-centred therapeutic relationship. On the other hand, feelings such as love, sexual desire, adoration, hatred, contempt, fear and mistrust are projections *from* the client, they may be positive or negative, and they are usually 'transferred' from a 'significant other', or indeed they may be aspects of the client's self that the client cannot as yet accept. Furthermore, in psychoanalysis, client attitudes that are 'transferred' to the current therapeutic relationship are usually deemed to be infantile in nature and origin, and hence link with Freudian theories about psychosexual development – theories that play no part whatsoever in the client-centred framework.

In addition, Carl also stated that even if the phenomena described as transference and counter-transference become discernible, 'In client-centered therapy, however, this involved and persistent dependent transference relationship does not tend to develop' (Rogers 1951: p. 201), indicating that so-called transference phenomena might be more to do with the practitioner and the relationships that practitioners create than with some 'universal fact' of human relations. Carl proposed that *it is not necessary* for the therapist to determine whether client feelings and emotions are counsellor-caused or projections, stating that while such deliberations may be of theoretical interest, they are not problematic in client-centred therapy theoretically, for:

> A true transference relationship is perhaps most likely to occur when the client experiences another as having a more effective under-standing of his own self than he himself possesses.
>
> (Rogers 1951: p. 218)

Carl also stated that there was *absolutely no need* to make 'transferred' attitudes 'special', or to permit the dependence that features so prominently in other (especially psychoanalytic) approaches. Instead, the client-centred therapist can *accept* dependent feelings, thereby not colluding with client attempts to change the counsellor's role. In other words, accurate understanding and genuine acceptance, together with non-evaluation, non-judging and non-interpretation, result in a tendency for 'transference' attitudes to dissolve, with the feelings becoming expressed towards their true object, thus resulting in greater client awareness of and new insights into the meanings of experience.

Linking a belief in and a therapeutic focus on so-called transference with the psychoanalytic notion of 'resistance', Carl also differentiated between an *under-standable* resistance to the often painful process of admitting to awareness previously denied and distorted experience (as happens in client-centred ther-apy), and resistance *to* the therapist created *by* the therapist (as occurs in more Freudian-based therapies). Transference and resistance are interpreted *by* the therapist, usually serving no really useful purpose other than to satisfy the intellectual curiosity of the therapist (and maybe make the therapist appear very clever). Focusing on such phenomena as 'transference' and 'counter-transference' (if they even exist), interpreting them and dealing with the resistance created by the therapist's interpretations, therapeutically seems to have one main function – to needlessly prolong the counselling.

Why then might notions of so-called transference and counter-transference seem to keep re-emerging as somehow desirable, even among so-called client-centred practitioners? Why is it that every so often we hear yet another speaker, or read yet another article, which suggests that transference and counter-transference could be incorporated into client-centred therapy with no loss of integrity? Could it be that Carl was right when he said that maybe some therapists *do* experience a need to be perceived as exceptionally clever and smart by either their clients or their peers, as knowing far more about client psychological processes and experiencing than clients themselves can possibly know or dis-cover? Might there be something in the links made by Carl Rogers and Gay Barfield to the problem-solving, mechanically and technologically orientated nature of the so-called 'developed countries'? Perhaps we can let Carl have the last word on transference:

> I feel there's something about its complexity that is very seductive and disgusting. I get really upset – perhaps that word should not be used. But it is socially reprehensible that the professions have not moved beyond psychoanalysis.
>
> (Rogers and Russell 2002: p. 244)

Rogers passionately believed that the 'special virtue' of client-centred therapy is that *the client has no need to resist the therapist,* as the *client* is free to deal with any resistances found within the self.

Uh-huh

One of the more gross caricatures of client-centred therapy is that the therapist has a bouncy rubber spring for a neck, and simply sits there nodding and bobbing sagely, interjecting the occasional (but well-timed) 'hm-mm' and 'uh-huh'! It gets worse – despite all that we have learned from Carl Rogers about *thou shalt not skill,* and how *non*-client-centred it is to direct or manipulate clients, many counselling skills and, regrettably, counsellor training courses teach students 'minimal encouragers' or therapist *noises* as *techniques* designed to *encourage* the client to talk.

It may well be that if you are listening to someone very attentively, closely tracking their every word, gesture and movement, and striving to sense, absorb and understand the feelings and personal meanings that 'lie behind' client words from within their frame of reference, then you may well utter the odd 'mm' or 'uh-huh' or whatever noise comes naturally for you, but what a far cry this is from using noises like 'uh-huh' and 'hm-mm' as *techniques designed to get the client to do something*!

Here is another minimal noise:

Sigh . . .

Understanding about

It may seem somewhat surprising that *understanding* is included in a 'The client-centred counsellor does not' section of a book on empathic understanding! Carl Rogers believed that therapist communications from an external frame of reference were not necessarily 'wrong', 'bad' or to be decried, stressing that such communications are usually sympathetic – coming, as it were, from a good, understandable place. Indeed, he wrote that often such sympathetic communications are:

> even attempts to 'understand', in the sense of 'understanding about', rather than 'understanding with'. The locus of perceiving is, however, outside of the client.
>
> (Rogers 1951: p. 33)

Understanding about is perhaps best characterised by the '*You (feel) – because – (reason)*' exercise that is sometimes employed in counselling *skills* training. For instance, the person utilising this skill might say something like 'You *feel* angry *because* it seems so unfair'. The learning of such techniques may or may not represent a step on the way towards becoming empathic (perhaps for some trainees it does and for others it does not). However, this type of *response* demonstrates that the helper understands *why it is that the other person feels the way that they do*. It does *not* communicate that the listener has absorbed what the sensations of anger feel like to the client, or what personal meanings lie behind the notion of justice for that client. For instance, 'You *feel* sad *because* your partner has just died' states the obvious – it does not communicate an *as if* feeling of grief or a sensitive understanding of what this *means* to the client. With client-centred empathy:

> the client himself has been the source of understanding; he, *from his own point of view*, can gradually articulate the sense and meaning of his difficulties. Hence, there is no need for an elaborate developmental theory or complex theory of neurosis for an effective understanding of the client. The client can be helped to explain himself, to give an account of his own world, and it is *his* account, *his* description, *his* developing consciousness, and *his* expression of feeling that will make most ample sense of his life.
>
> (Barton 1974: p. 179)

Moving beyond *understanding about* and towards an *empathic way of being* is essential for the client-centred therapist.

Quack of all trades, master of none?

If we are to be *client*-centred, perhaps it might be as well to listen to our *clients*. In personal correspondence (later made public), Professor Dave Mearns wrote:

> I am concerned and frustrated at the bad name which is attaching to person-centred practice in Britain. An example is the fact that complaints against 'person-centred' members of BAC [the British Association for Counselling] far outweigh those of any other discipline, even taking into account the fact that more people ascribe to the person-centred approach than any other core model. Particularly frustrating is the fact that when you scratch beneath the surface of these complaints you almost always find that the practitioner has virtually no training in the approach . . .
>
> (Dave Mearns, quoted in Vincent 1998: p. 1)

In the UK at least, there has been a major increase in counsellor training programmes that describe themselves as 'eclectic' or 'humanistic' or, most commonly, as 'integrative'. The vast majority of them aim at just above the minimum hours requirement for validation or accreditation. So in what usually amounts to a course of around 450 hours in total, it is maintained that trainees can come to grips with, thoroughly understand and competently practise two or often three 'mainstream' approaches to therapy – that's either 225 hours or 150 hours per approach. *Why*? Why, oh why, oh why?

Do trainers who offer such programmes genuinely, honestly and deeply believe that the best interests of clients are served by 150 to 225 hours of training in each approach? Do trainers maintain this belief (and their integrity) even in the light of research evidence which demonstrates that 'hotchpotch' counsellors (many of whom call themselves 'person-centred' because it seems to be the most embracing label) attract a disproportionate number of client complaints? Or might this be down to market forces ('buy two, get one free' and 'it's more attractive to punters if they think they're getting more')? Might this even be due to there being some trainers who enjoy 'strutting their stuff', and the more stuff there is to strut the better? Whatever the reason or reasons, it is clearly bad news both for clients and for authentic client-centred therapists.

Some further evidence can be cited. A local counselling directory (Hampshire Association for Counselling 2001) contains the details of 91 counsellors, of whom just under one-third offer therapy rooted in a single approach, and of these, three counsellors are identified as client-centred and three as person-centred. The average number of approaches offered by the rest of the counsellors is slightly over three. Around 30% of the counsellors offered two different approaches, over 15% offered three different approaches, 18% offered four different approaches, and 4% offered five different approaches. Over 40% of these 'eclectic' counsellors included 'client-centred', 'person-centred' or 'Rogerian' (a term Rogers disliked) in their descriptions. Yet Carl Rogers believed that the use of different techniques and styles with the same or with different therapeutic styles actually destroys the potential for a truly therapeutic relationship with clients (Rogers and Sanford 1985), so how with any integrity can *client-centred* or *person-centred* be included in the description of an 'integrative' counsellor? We know, too, that Rogers gave us necessary *and sufficient* conditions to work within.

Carl identified two further drawbacks, the first being the trainee counsellor who becomes 'mesmerised' by or idolises a trainer, therapist or even an approach to counselling. Such people, said Carl, tend not to think for and *be* themselves:

> The other – and this is an equal danger, I feel – is when people say 'Well, I'm an eclectic therapist; I do whatever seems necessary'. That tends to mean they have no theoretical basis by which they're operating, so they have no way of determining whether they're really doing the right things or not. They sort of fly by the seat of their pants, which has certain good qualities, but also very grave deficiencies . . . The need is to think about theory, expand theory, expand hypotheses, but to do it in a thoughtful and coherent way.
>
> (Rogers and Russell 2002: p. 275)

It seems, then, that there are external threats to the integrity of client-centred therapy – for instance, from people who 'use' the core conditions some of the time or with some of their clients, or who see being client-centred as a 'good basis for the relationship' but not *sufficient* to resolve all client concerns. Sadly, there are threats from within, too. We have already looked at issues surrounding those who seem to think that being congruent means 'anything goes', or who claim to use client-centred 'reflecting' skills. When Carl Rogers was asked about 'using anything and everything', he said:

> I think that in many ways it can and, in some respects, has gone completely wild, and I regret that. In its more solid aspects, it's one of the most significant social inventions of this century because it is a way of eliminating alienation and loneliness, of getting people into better communication with one another, of helping them develop fresh insights into themselves, and helping them get feedback from others so that they perceive how they are received by others. It serves a great many useful purposes.
>
> (Rogers and Evans 1975: p. 32)

Rogers went on to say that the reputation of the person-centred approach (he was talking at the time about broader applications, not just therapy) could be damaged in the public eye.

It may well be that our reputations *have* been damaged and *continue* to be damaged by these internal and external threats to client-centred integrity. However, we can live by our own beliefs – rather than be defensive, let us celebrate what and who and how we are.

> Anyway, I think that because I'm not trying to do all these fancy things that other therapists do, I have a more complete freedom to really enter the world of the client.
>
> (Rogers and Russell 2002: p. 281)

Summary

The importance of not seeing counselling or therapy as a skills-based activity is clear from the fact that when, near the end of his life, Carl Rogers was asked whether he had any regrets, he replied:

> I regret that in early writings about therapy we were so fascinated with the enormous learnings from recording interviews that we became too much focussed on techniques, and that has misled a great many people in the counseling field. I realize that.
>
> (Rogers 1989: p. 198)

Furthermore, to take a principled stance *against* skills and techniques places the therapist in a minority, for:

> Contrary to the opinion of a great many psychotherapists, I have long held that it is not the technical skill or training of the therapist that determines his success. Instead, I believe it is the presence of certain attitudes in the therapist, which are communicated to, and perceived by, his client, that effect success in psychotherapy.
>
> (Rogers 1989: p. 10)

There is clearly an essential difference between *being skilled* and the *implementation of skills*. Being skilled is something that someone *is*, whereas implementing skills is something that someone *does* to another person. Witness Carl Rogers in action and you will most likely experience someone who is extremely skilled not only at *being* authentically respectful and understanding, but also at *communicating* genuine acceptance and empathy. The therapist who has *got* (or acquired) the skills and techniques to *use* on or apply to clients tends to *respond* by utilising tools that are designed to *do* or achieve some desired result.

There is an anecdote about an audience watching Rogers demonstrate client-centred therapy. One viewer turns to another and says 'Boy! That Carl Rogers! He sure has got it!'. After a reflective pause comes the reply 'No . . . He hasn't *got* it – he *is* it'.

Chapter 3

Empathising in action: an inter-view

Background to this inter-view

During a diploma group, a counsellor trainee asked me to describe, 'off the cuff' so to speak, my *process* of empathising. The questioner spoke of having learned all about definitions, academic understanding and the theory of empathy (such as the place of empathic understanding within client-centred therapeutic process), and had experienced both the giving and receiving of empathy in personal therapy, in practice sessions and on a client work placement. Yet this trainee was really curious about *what actually happens within a therapist when that therapist is engaged in empathising*. Although I recall feeling somewhat 'put on the spot', there was also a simultaneous experiencing of curiosity as my mind engaged with the quality of the question, followed by the thought that it deserved a full and honest answer. So . . . I took a deep breath and spoke. A dynamic dialogue ensued, in which questioners were encouraged to reflect upon their own experiencing, with further questions being put by group members, which in turn triggered further exploration, clarification and ideas.

A couple of days later I started trying to remember and type up what I had said. I recall feeling somewhat abashed in that I was quite interested in my own answers! What had emerged from the dialogue felt useful – a pulling together of some of the strands of 'empathising in action', if you will. As the session felt a bit like an interview at the time, I thought 'Who can I get to interview me?'. As I started typing, I framed the first sentence as the initial question that had been asked in the training group. Then I thought 'Why not interview myself?'. This idea seemed a little crazy at first, but I nevertheless persevered, and the process turned out to feel quite dynamic. A college manager reminded me, when I happened to mention my feeling of silliness, of 'FAQs' ('Frequently Asked Questions') computerised systems, and I became more aware that it might actually be helpful to remember and collate questions about empathic under-standing that trainees have posed over the years, and to attempt to respond to them fully and honestly.

So what follows is in part an interview with Steve Vincent, in which he responds to questions put to him by . . . Steve Vincent! Many of the questions are ones I recollect trainees and other peers having asked over the years; some of the questions are my own. This process of exploring *what actually happens within the therapist when empathising* has since been repeated in other groups and in further dialogues, and some of these sessions were tape recorded and transcribed. Here, then, is an edited version of responses to questions about empathy in action – an interview drawn from a variety of sources. I hope that it has meaning and relevance.

Empathy: an inter-view

I Would you be prepared to describe for us your process of empathising?

Well . . . I realise that I might sound a little like a stereotyped politician, but first of all I'd like to amend the question somewhat! I think that it makes more sense for me to try and describe my process of *striving* to empathise . . . Yes, this has more meaning, for it seems to me that what is most therapeutic is an *authentic striving* to empathise, rather than some kind of 'effortless and accurate' empathy *all of the time*. If we think in terms of being a compassionate companion to the client, for example, I will be journeying *alongside* rather than knowing or showing the way . . . something like that. Maybe there is something, too, about empathy being very hard work, and a place we strive to reach rather than a place we instantly find ourselves at. Maybe the journey to get there is as important as the arrival.

Anyway, I'm with a client. I think one of the primary and . . . No, let me step back a minute . . . I'm *not* with a client. Maybe there is something about preparing myself *before* I meet with the client in person. I am reminded of some of the clinical psychology research into receptiveness. In one experiment a volunteer group was divided into two equal halves, one half being told that pain would be administered at a set time, the other half being kept in the dark. When the pain was administered, those who had received prior warning experienced the pain more keenly – both subjectively and physiologically (through scientific measurement) – so the discovery was that our brains can be 'primed' to be ready to receive stimuli and sensations. So maybe something similar could be true for empathising. Maybe if I prepare myself to receive then I will perceive more keenly. This is one reason why I am all in favour of the 50-minute counselling hour! I use the space between meeting with clients to calm myself in different ways – mentally and physically. I remember now, I think Professor Susan Greenfield called this 'mental priming process' *modulation* (Greenfield 2000). For me, regrettably, this often means slipping out of the back for a cigarette and a pace around, and during that time I usually kind of mentally freewheel. It feels, when I look back at it, rather as if I am emptying myself of the last client (maybe a kind of storing, rather than a forgetting) in order to be as open as I can be to receiving the next one. I do experience myself after this short break as somehow freshened up and ready to meet with whatever comes through the door next. So I'm not smoking, I'm modulating!

Not only is there something about my own experience, but also it would not surprise me in the least if this kind of receptivity is physiologically or neurologically (or whatever) measurable. There may well be reliable clinical back-up research data to support the view that 'priming' could be a significant factor in preparing to *empathise*. It occurs to me that this view could be backed up by looking at its opposite, too. Just a few days ago I joined a group an hour or so after it had started. Although my late arrival was expected (it had been negotiated and agreed well in advance), I nevertheless felt relatively disconnected from the group for quite some time. It occurs to me, too, that I simply drove up, parked, and rushed in. As I speak, I feel sure that other factors were also relevant here (factors that I shall not go into right now as they do not feel appropriate), yet it seems to me that part of my feeling of disconnection was due to my not being adequately primed. I think that I tend to trust my experience in this – I sense that there is something very real and true in the significance of modulating my mind to receive empathically.

So where was I? Now I *am* primed and meet with a client. I think one of the primary and initial inner processes is that I am uncluttered and receptive, and that enables me to strive to *absorb* all that is passing between us, and primarily all that a client is communicating. I will strive to be watchful, in that I think it really helps to look at the person who is communicating! I strive to be attentive to the eyes, the mouth, the face, the head, the body, the hands and arms, the feet . . . And this is not some 'detached observing' that takes place. Rather I am trying to absorb all that the client is experiencing from moment to moment. I recall watching a Carl Rogers demonstration session, in which he began a communication with 'I gather . . .', and I liked that – the therapist strives to *gather* all that a client is giving off and giving out. Gathering includes absorbing *sounds* as well as vision and words. For instance, how is a client speaking? What is the quality of a client's silence? There is a *whole universe* of communication to absorb . . .

2 Can you clarify some of this by giving an example now?

Of what? I'm sorry, I see that some of you are looking quizzical or puzzled. Yet there are so many aspects to this universe of communication! I shall try to bring in examples as we go along.

Okay, I recall a demonstration session last year where I was one of a couple being counselled. The therapist missed the 'discreet' tears of my partner. Those tears were communicating something and the therapist was unaware of that communication, let alone what the crying might feel like or what personal meanings might underlie this weeping. Come to think of it, the therapist totally missed what it felt like or what it meant to me, too! I recall another instance where a client was saying nothing out loud, but her fingers were kind of trammelling (the movement that a cat's paws and claws often make) the arm of the chair. I not only noticed this, I actually (unobtrusively) made the same movement myself, in an attempt to gain some kind of insight into what she might be experiencing. I think I often *gaze* at clients when we first meet. I certainly don't *stare* in some kind of *piercing* way – I like the idea of a *gentle gaze*. I hope that this gentle gaze conveys that I am interested, curious and wanting to absorb. At least, I hope that some clients sense something of this, if they look! As I think about it, I feel that in initial meetings many clients are too absorbed in what is going on for them to notice very much of my intentions. Maybe as therapy progresses clients sense more of my countenance and what it means. Anyway, my hope is that if clients do look at me, my face (and perhaps especially my eyes) will let them accurately discern *something* of my intentions. (I'm not sure where this comes from, but I remember someone writing about how utterly weird and bizarre it is that two lumps of fatty gristle could communicate so much – yet the eyes do very often seem to say a great deal.)

So I *see* things, I do not just *hear* things. Sometimes what I see might not be what is being voiced. For instance, a client might look ashamed or embarrassed as they are speaking, yet they do not use the words shame or embarrassment. If I am striving to absorb what the client is *experiencing* in the moment, then their embarrassment or shame is as much a part of their being as their words.

3 Doesn't what you have said involve a degree of interpretation? How truly client-centred is that?

I suppose that depends upon your interpretation of interpretation! I guess that there are different degrees of interpretation. It doesn't *feel* like interpretation to me, although it does mean an absorbing and a trying to *sense the sense of* whatever is going on in clients, trying to tune into their sensations and grasp something of what their experiencing means to them. I have come across the phrase 'empathic inference' somewhere, and Rogers wrote of 'empathic guessing' at some point (in 'The Silent Young Man', Chapter 17, Rogers 1967). For me, there is something around . . . If a client *looks*, say, embarrassed, then I think that this is a pretty unequivocal communication. This is not some kind of arrogance about interpretation or some highly intellectualised theory about 'body language', and nor will any communication of empathic inference or guessing be assertive. You don't look convinced . . . okay, let us dwell on this a little longer.

I recall a television programme on the human brain that I saw recently, and one of the statements made was that one of the largest single functions of the brain is given over to face recognition. I don't want to get into pedantic semantics, but I wonder if, at one level at least, there is any kind of subtle difference between *recognition* and *interpretation*. In the Nineteen Propositions (Rogers 1951: pp. 481–533), Carl wrote about internal and external frames of reference, yet he also wrote about a *common frame of reference*, saying that there are *counterparts* of experience in each of us. Many argue that without a common frame of reference, there would be little or no communication at all between us! I am reminded of Maurice Merleau-Ponti, an existentialist philosopher who added *being-within-the-world* to the categories of *being-in-itself* and *being-for-itself*. In other words, there are inanimate objects that have no awareness whatsoever of their own existence (and hence nothing with which to empathise), and there are entities that do have awareness of their own being, yet there is also something about conscious entities (organisms that have awareness of self, others and the world about them) being born into a world where feelings and meanings pre-exist and are absorbed and internalised as we grow into that world within which we share reciprocal relationships. I am also reminded of the idea of the 'I–thou' relationship. There is the 'I', there is the 'thou' and there is a hyphen. Let us not forget the hyphen!

So do I *interpret* a look as meaning 'embarrassed' or do I *recognise* a look as meaning embarrassed? This for me would all seem to link with Carl describing the process of empathising as *perception-checking* – it certainly doesn't feel like a detached observation of client 'body language' coupled with some theoretical belief about what it signifies! And my experience has been that when I have communicated my sensing of client embarrassment to that client – perhaps somewhat tentatively if I feel somewhat uncertain – it is usually accurate or at least *nearly* accurate. And I feel fine with nearly accurate!

4 I'm a little surprised that you have referred to clinical psychology a couple of times already – that doesn't sound like you! Can you say more about that?

It grieves me that so often our focus is so narrow. It grieves me even more that there are those among us who try to develop client-centred therapy by incorporating *old stuff* from other approaches. And by 'old stuff' I mean things like transference, for that seems to rear its ugly head every so often, despite the fact that Rogers *despised* the whole notion of transference!

Yet there is so much that we could be learning and that is compatible with client-centred theory and beliefs! I'm excited, for instance, by the work of Professor Susan Greenfield. Look at Freud. He believed that it was only a matter of time before the physical structures of the id, ego and superego would be found in the human brain. We can say with some certainty that this has turned out to be utter rubbish! Yet what has been discovered, at the 'cutting edge' of neuroscience, if you will, seems to me to bode very well for client-centred therapy – for instance, some of the findings about perception, and how we create the perceptual world around us. Professor Greenfield talks of how nonsensical it is to tie down the development of the self to ages and stages. She writes of what she calls a 'transient self' that is a process of becoming – just like Rogers. I think that it is very important that we pay attention to neuroscientific discoveries and are open to learning from them. I also think that it is *fantastic* that so many of these discoveries are in harmony with client-centred thinking, and that we can develop within our approach without damaging or diluting our integrity.

5 What did you mean by feeling fine with 'nearly accurate' empathy?

Can I step back a moment? It occurs to me that when I said 'You don't look convinced' earlier, that was a live example of what I mean. You didn't *look* convinced, and I voiced my understanding of what your facial expression meant. Subsequent non-verbal signals led me to believe that on this occasion my recognition of your signals was pretty accurate.

Anyway, a few things come to mind about feeling fine with 'nearly accurate' empathy. I am aware of feeling absorbed by the concept of absorbing, so I would like to get back to that before long, too! I remember writing an article called *In Praise Of Being Off-Beam*, the gist of which was that accurate empathy, like a *zap* right between the eyes, is maybe not so therapeutic – or at least persistent and prolonged *zaps* might not be. My experience of offering therapy is of a striving to sensitively enter the world of my client. I venture, sometimes very tentatively, sometimes more confidently, my grasp of what they are experiencing, a grasp that includes, ideally, feelings, thoughts and meanings (and personal meanings may well link with memories). I hope that my empathy and understanding are not *wildly* off-beam, for if my client experiences me as being 'on another planet', then I suspect that the encounter will not be therapeutic for them. So there is something about conditions five and six – that I am actually experiencing a degree of accurate empathic understanding and that the client is receiving this. Therefore constant zaps between the eyes might be inappropriate – too scary, or over-threatening, or whatever. And missing by miles is not too good, either, but if I can get pretty close pretty much of the time – maybe with an occasional *zap* – then that feels good, to me. It's as if the client *knows* that I am striving to sensitively enter their world, and so will be co-operative in that joint venture. If I don't 'get it quite right' then they will clarify for me. And who knows? Maybe in clarifying for me, things will become clearer for them, too.

6 You did a couple of demonstration sessions, and in one I saw you as
being confident with your empathy, and in the other as really tentative.
What is that all about?

Well, that's just how it was. I remember the first session well, and I really did feel
reasonably confident that my empathising was fairly accurate, so part of being
authentic was neither to be over-confident nor to be too diffident either. In the
second session I had a heavy cold, and it felt as if I was absorbing through a layer
of cotton wool, so I definitely felt less confident (there was that word 'absorbing'
again! I would like to return to that soon . . .). Yet it occurs to me that even when I
felt more confident I remained kind of . . . humble? That doesn't feel quite right –
maybe a kind of *humility* is incorporated in my way of being, a kind of self-
acceptance that I feel confident about the accuracy of my empathy, yet also
experience a real willingness to be guided by the client. Rogers wrote about
constantly being prepared to change horses in midstream, and Dave Mearns
wrote about the 'dance of psychotherapy'. They both capture some of what I
mean, I think.

Can I just go back to the idea of near enough accurate empathy being good
enough for me for a moment? I am reminded of Carl's adage 'what I am is good
enough', and on one of the CSP/CCTPCAS videos he says more about this. It goes
something like:

> What I am is good enough. And what I am is an imperfect human
> being. Yet imperfect beings can achieve astonishing things. Especially
> when they are aware that they are imperfect beings.
> (Rogers and Sanford 1985)

I like that. And it seems to me that the degree of accuracy in my empathic
communications almost feels less important than my striving to be empathic – at
times at least. The outcome may be imperfect, yet if the striving is there, that is
kind of perfect! It's almost as though if I am authentically striving, I can't 'get it
wrong'. If the striving is there, the rest will follow.

You know, it occurs to me that there is also something about *timing* in this
confident or tentative thing. If I reflect on my work with clients, then I think that
it is fair to say that I am less likely to be very tentative the more I have come to get
inside a client's world of feelings and meanings. I will still, I hope, maintain a
degree of humility, but yes – I think that's true. The *less* I know (or the less I have
absorbed or gathered), the more tentative I am likely to be. Of course, if an
ongoing client discloses a completely new aspect, then the tentativeness might
return. So maybe increased confidence and reduced tentativeness is a therapeutic
trend, but not a straight line on a graph, if you see what I mean. Anyway, can I
return to absorbing?

I started off talking about priming myself to be alert, attentive and watchful. I
want to say something more about absorbing *sounds*. By this I mean that I think it
is unfortunate that, over the years, there has perhaps been too much attention
paid to *words*, at the expense of *sounds*. I remember a recent demonstration session
where at one point I voiced something of the client's sense of 'righteous
indignation'. In the audience feedback session that followed, considerable
surprise was expressed because the client had *said nothing* about indignation –

righteous or otherwise – so surely this *couldn't* have been client-centred and *must* have been interpretative or analytic or something. Well, for me the process of 'absorbing' a client includes hearing the *sound* of their communications, and not just the *words* that they speak. When I listened to the *sound* of this client's voice, almost a kind of 'plaintive wailing', the communication of righteous indignation was a strong current that ran through a significant section of the session, and the client later confirmed that this was a very accurate sensing. So it seems to me extremely evident that communication is about far, far more than the words that are being spoken.

Another example occurs to me. At an international gathering of person-centred folk, the idea emerged of delegates initially speaking in their first language and only *afterwards* having their words translated into English (rather than phrase by phrase). People spoke in several languages. With most delegates *I didn't understand a word that they said*, and for a few I understood at best the meaning of only a smattering of words. Yet I sure as hell *could* sense what they were *feeling* – and pretty accurately, too. Interestingly enough, for me the translations that followed tended to lack the emotive content which was expressed by the original speaker. Maybe you could try watching the television or a video with the sound muted. You might not hear a single word, yet my guess is that you will accurately perceive many of the emotions being expressed – perhaps largely through facial expression, gesture, bodily signals, and so on – so maybe these aspects of communication could be considered as the nature of 'non-verbal' signals. Or you could try listening to a tape and just focus on the *words* or, better still, look at a transcript of a tape and then listen to the tape, focusing on the sounds. Quite often, in my experience, the sounds impart information that the words do not, and this might illustrate the nature of 'paraverbal' communication.

And it does not stop there! A client who is extremely anxious might speak very rapidly. He or she might not use the word 'anxious', yet the rate of their speech might be communicating anxiety. Loudness of speech could be communicating anger, quietness might indicate feeling low or exhausted, and so on, and so on.

Thus my process of absorbing is in part about a seeing and hearing beyond the words that are spoken.

7 I'm confused. Are you saying that the client-centred therapist is somehow detached, somehow objectively absorbing and analysing – or processing, if you prefer – all of these sights and sounds?

No, that's not what I'm trying to communicate. It's interesting to me, because for many years I would experience a heated reaction to words like 'objective' in the belief that 'objective' symbolises 'cold', 'dispassionate', 'clinical' or whatever. To me, client-centred therapy is *in*volved and *compassionate* and *subjective* and *warm* and . . . It is *in*-volved because I am really trying to be 'in there' with the client. Yet I realise as I speak that I am feeling some surprise at my lack of a strong reaction to the word 'objective' that I once would have felt. I think this is because the words 'subjective' and 'objective' have come to have less and less meaning and relevance for me. They both feel like the extremes of a linear scale, and my experience is that life ain't like that. Perhaps just as Rogers felt pretty passionately about the artificiality of separating *above the neck* and *below the neck*, believing rather that we are (potentially at least) whole persons, so I feel that this apparent need we have to be purely one thing or another is nonsense. Perhaps being a *whole person* doesn't mean being *wholly this* or *wholly that* – it could mean that the whole is comprised of many gradational aspects. So maybe I don't need to ferret out whether I am being *wholly objective* or *wholly subjective*. Maybe I can be a composite of many, many different attributes.

I know I have used the word *absorbing* a lot. Maybe another word to describe my process could be *immersing*. I feel as if I am immersing myself in my client's world of feelings and meanings. Although, to quote Rogers, I keep a sense of self about me (not losing the 'as if' quality of empathy), this hardly equates, for me, to the notion of objectivity. I want to say, too, and with some passion, that nor is this a *passive* process. Not only am I *trying really hard* to absorb and immerse myself in the world of my client, but also I will strive to communicate something of what I have sensed. It feels very *active* to me, far from passive.

8 You said that you regretted the focus on words . . .

Yes. And by this I mean that, for instance, I find it bizarre that 'the powers that be' consider it desirable – or even a requirement – for someone embarking on counselling diploma training to have completed counselling skills training. Now don't get me wrong, I think that counselling skills courses can be invaluable – there are many people for whom the implementation of counselling skills is their end aim, and that's great in my view. It is also very worthwhile, in my opinion, for people to have the opportunity to explore the mainstream therapeutic approaches prior to making as informed a choice as is possible of which diploma course to pursue. Yet over and over again my experience on diploma courses is that many trainees have spent at least the first term or two *unlearning* skills – and a process of *unlearning* the old alongside *learning* the new can be extremely problematic for many.

Oh gosh – this self-referencing is getting a bit embarrassing! I wrote another article called *How Do You Do* (I was really pleased with that title!), the gist of which was that in 'non-directive' counselling skills training, a kind of stock response would be 'How do you feel about that?'. Now it seems to me that 'How do you feel about that?' (hence the *How Do* . . . part of the title I like) is a perfectly good, non-directive (up to a point – the focus is on feelings), open-ended *response*. Yet if I am *communicating* empathy, my communication is more likely to be characterised by 'You feel sad' or 'You're really angry', or whatever (hence the *You Do* . . . part of the title). Note, too, the differentiation between a *response* (to a client's *words*, more often than not) and a *communication* (of sensing a client's moment-to-moment experiencing).

This focus on *words* is perhaps even more stark when it comes to *paraphrasing*, *summarising* and, Carl Rogers forbid, *reflecting*. It seems to me that the *process* is that many counselling skills trainees *have to focus on words* in order to be able to log and recall them for the purpose of paraphrasing, summarising, reflecting or whatever. And in the process of having to concentrate on the spoken word in order to be able to demonstrate specific skills and techniques, they very often miss out on the full depth and richness of what is passing between two people. So while I think that the implementation of skills certainly has its place, I am not convinced that such training is particularly helpful when it comes to a process of becoming client-centred.

So in general it would seem that most of *my* words thus far have not been about words – and for me this is significant. Perhaps now we can turn to the words themselves.

Of course the words that a client speaks are important. *Of course* it is important that I absorb and am able to remember a good proportion of the words that have been spoken. My point thus far has been that there is far, far more to communication than words alone. It almost seems helpful to think in terms of *essence* – and perhaps *essential*, too. Am I able somehow to distil the *essence* of what a client is communicating? What are the *essential* ingredients of a client's experiencing? As I said earlier, I remember watching a video of Carl Rogers in action, and he commenced one communication with 'I gather . . .', which I really liked. It triggered for me the idea that my process when I am striving to empathise is one of *gathering*. I am a *gatherer*, an *absorber*, of *all* of the elements of client communication. And if there is a whole universe of communication to absorb, maybe the words only represent one solar system. Within that system there may well be a very significant sun that casts light, or there may be significant planets, yet that solar system is only a part of the whole universe.

I might mention, too, that Rogers was interested in the process of finding the *right* words. Choose a word and look it up in a thesaurus. You now have a list of several words, all with very similar meanings. Yet when clients describe something, only *one* – just one – of those words seems to fit exactly. And you can *see* clients very obviously *referencing themselves* in order to sense, feel and think which word is the precise symbol of what they mean. Hit on the right word first time and hey – accurate empathy! Hit on one of the others, in a humble, flexible way, and clients experience your striving to empathically understand – and they will soon put you right if the relationship is one in which they feel free to do so.

9 You seemed to suggest that 'How do you feel about that?' wasn't entirely non-directive, because it focused on feelings. Could you say more about that?

I'd be delighted to! I have also spoken (and written) a few times about feelings *and meanings*, and this links with the word thing, too. Carl Rogers wrote that the *instant* we become aware of or perceive something, we attach a *meaning* to that awareness or perception. So if I just focus on the *words*, I might miss the *feelings and meanings* that lie 'behind' the words, or 'underneath' the words, or *inform* the words, or whatever. In other words (hah!), if I focus only on what is voiced (in order to paraphrase, reflect, or whatever), I run the very real risk of not sensing 'underlying' feelings or meanings.

Part of my agenda here is to try to dispel the myth that seems to have arisen about client-centred therapy (or the person-centred approach) being all to do with *feelings – because that is simply not true*. Read Rogers and you will quickly discover that he was consistent throughout his life, both in his writing and in his practice, about linking feelings *and meanings*. In his book *A Way of Being*, Carl wrote passionately about the artificial distinction that we make between 'head' and 'gut' (above the neck and below the neck, as he put it). If we are to be *whole* persons – that is, if I am to engage as one whole person with another whole person – then let us move away from yet another example of linear extremes –

head and gut! Feelings and personal meanings interweave; there is an incredibly rich flow between and among feelings and meanings.

10 What are your views on internal and external frames of reference?

I think that they may be useful concepts on the journey towards an empathic way of being when with clients, clarifying the difference between my world view (including my view of myself) and yours. And I guess that for me, when I am with a client what is internal and what is external is *usually* fairly clear – I think that I am relatively aware of what is 'my stuff' and what is 'the client's stuff'. The idea of a common frame of reference interests me, too, as I have already said.

11 You mean that there is a third frame of reference?

Oh yes! If you picture the 'Two Circles' diagram that Rogers used to illustrate and summarise the Nineteen Propositions (*Client-Centred Therapy* 1951, front cover and pp. 526–7) . . . you are seeing two circles, side by side – yes? So, using this format, the client's *internal* frame of reference would be in the left-hand-side segment, the therapist's frame of reference (that is, a frame of reference *external* to the client) would be in the right-hand-side segment, and the overlap would represent their *common* frame of reference. Thinking *only* in terms of internal and external frames of reference feels too rigid to me, too extreme and too linear. There is how you see the world, how I see the world, how we see the world, a pre-existing world of meanings (and feelings), the world that is becoming (new meanings, meanings in a process of being formed), and so on.

12 Did Carl Rogers write or say anything about this, or is this all your own thinking?

Some of both, I guess. There is Merleau-Ponti who I mentioned earlier, too. In fact there are many, many resources. As for Carl Rogers, when he wrote Proposition Seven in *Client-Centered Therapy* he mentioned that the experience and perceptions of others have counterparts in our own experience, and that these counterparts can be used as tentative guides. Thus when a client speaks of anger, for instance, I too have experienced anger, and although I need to be constantly aware that the client's feelings and meanings are 'unique' to that individual, it would also be naive, it seems to me, not to acknowledge that without some form of shared frame of reference there would be little if any communication.

So I can, perhaps, use my own frame of reference as a tentative guide into your frame of reference. I can also use our common frame of reference as a tentative guide into your frame of reference. On a simplistic level, we would probably all agree that a particular object is what can be called a chair, so 'chair' falls within our common frame of reference. However, we need to be aware that the chair has many properties that constitute its 'chairness', and that we shall not perceive the chair identically. For instance, one person might perceive shape first, another colour, another the material of which it is made, or whatever. I'm not sure that picking on a chair is necessarily the best example, but I hope that you are catching some of the sense of what I mean. And what the chair *means* to each of us might vary. For most of us, it probably goes no further than something to sit upon. Yet what if I was bound to a similar chair during a horrific assault, or it reminds me of my mother's chair, and my mother died recently, or something like that? So there is something about humility and remaining tentative. On the one hand, it may be reasonable to assume a probability that we do understand each other when we say 'chair', but on the other hand, we need to continually perception-check if we are striving towards empathic understanding.

So if as therapist I identify something as clearly 'my own stuff' – that is, 'stuff' from a frame of reference that is *external* to the client – then it follows that it is unlikely to have any place in the counselling room. If what emerges clearly belongs to the client – that is, from a frame of reference that is *internal* to the client – then it clearly has every significance to therapy. Sometimes, though, things seem to emerge from the space between us – from the hyphen of the I–thou relationship, if you will. I have come to trust what emerges in that space more and more, but then I also do a lot of exploring of such issues in my writing, my personal reflecting, and in supervision.

13 Did you mention memory as part of the internal frame of reference of the client? Doesn't that seem more like psychoanalytic thinking?

I want to be accurate here – can I just quickly look something up? This is the 1959 paper that Carl wrote, in which he gives us lots of definitions. Ah, here it is. I quote:

> *Internal frame of reference.* This is all of the realm of experience which is available to the awareness of the individual at a given moment. It includes the full range of sensations, perceptions, meanings, and memories, which are available to consciousness.
>
> (Rogers 1959: p. 210)

So the instant we perceive something we attach a meaning to it. When we speak of the *here and now* or of *immediacy*, our *current* experiencing includes any *past* experiencing (feelings and meanings) that have relevance to the here and now. So while the client-centred therapist does not *direct* the client to focus on past experience (or to then *analyse* that past experience), it is of course true that feelings and meanings gathered along the way may well pertain to current experiencing – and if the therapist is to grasp the intricacies and depth of current client experiencing, this understanding may well include the history that clients bring into *now*.

14 I've had two or three clients now, all of them male, and I just can't seem to get through their cognition and into their feelings. What can you say about that?

Okay . . . I think that I have a few thoughts about that. Let me see . . . I feel a reaction inside of me so that I know I have an agenda again, something I feel quite passionate about. Let me try to access this. It is that I want to try to dispel what seems to be a fairly popular myth that client-centred therapy is all about *feelings*. It isn't, as I have already said. Rogers wrote, for instance (I'm sorry that I'm repeating myself – it feels important), that every time we perceive something we instantly attach a meaning to it. And the meanings have importance, as do our perceptions, feelings and memories. There is something about *wholeness*. It reminds me of those who say that 'So long as I can be client-centred, why should I need to know theory?'. These same people will often talk about being whole persons or meeting clients as whole persons, or whatever. Yet *theory is an integral part of the wholeness of client-centred therapy*, and any reading of Rogers makes this crystal clear. How can I claim to be a whole person engaging with another whole person while at the same time dismissing an integral part of the whole of the approach that I claim to belong to?

Anyway, to get back to the point, I would want to be cautious about labelling a client as 'cognitive' or 'all head and no feeling', or whatever. Another issue that Carl Rogers was passionate about was this artificial differentiation that we make between feeling and thought. Read the part of *A Way of Being* where Rogers refers to Vincent Hanna and you will see what I mean. I feel passionate about this, too – it is yet another example, as I see it, of extreme and linear categorising, and it simply doesn't fit my experiencing of myself or others. So often people (including me) say 'I feel . . .' when what follows seems more like a thought, for instance. Well, that might just be language and pedantic, yet it does seem to me that there has been a trend in some client-centred circles to place more emphasis and value on feelings, to the detriment of meanings. And I think that this is a most regrettable trend, and one that Carl would have been hesitant to follow. In any case, how can I claim to regard you *unconditionally* if I value your feelings above your thoughts or, as I prefer, your personal meanings?

So if I am faced with an apparently 'cognitive' or 'heady' client, then maybe one of the first things I will want to explore is my own value system, to check out the authenticity of my unconditional positive regard. Thus there are real links with both congruence and unconditional positive regard here, it seems to me. I am also reminded (with, I confess, not a little embarrassment) of a client of mine from a few years back. A bit like the questioner, I felt frustrated that I could not get past all of his 'head stuff' and into his feelings. How I flinch now when I hear myself saying that! As if somehow I've got to apply this crowbar to get into a client's feelings! And even if my crowbar is wrapped in fluffy cotton wool, it is hardly client-centred! *Prizing* (with a 'z') *not prising* (with an 's') is a catchphrase I've come up with. Anyway, I took a tape of this client to supervision. To my eternal chagrin, after about ten minutes my supervisor stopped the tape and said something like 'Well, to me that was full of feeling'. This was an extremely salutary lesson for me to learn, and it is one that has stayed with me. What I had spectacularly failed to recognise was that my client was talking about his *feelings* in a *cognitive* way.

Another example comes to mind. Early in counsellor training (or in counselling skills development), a client begins – in a flat, steady, toneless voice and with an expressionless face and body – to say something like:

> Well, I've just discovered I am terminally ill, my partner has been laid off and our house is to be repossessed, and the cat was run over . . .

And the counsellor replies:

> How do you feel about that?

You can laugh, but it happens! For the sake of the point I am trying to make, we can say that there were no verbal, non-verbal or paraverbal clues to emotional content. Yet the client expects us to *infer* feeling. That is why people laugh –

because it seems absurd that the counsellor would not *infer feelings*. So maybe here is another clue about how to experience empathy. Perhaps this links to what Rogers meant when he wrote about empathic inference.

It seems to me that there is a tendency in client-centred circles to veer away from such discussion and exploration, because 'inference' gets equated with *analysis*, or *interpretation*, or *diagnosis*, or whatever – so best to avoid them. There is another possible example of our value base. I feel sure that words such as 'intuition' are somehow much more acceptable than 'inference', and I really think that this is a great shame, because sometimes I think that we miss out on exploring our processes as a consequence of such p-c p-c (person-centred political correctness!). Yet if I am to engage as a whole person, and if I have a capacity for inference, might not that capacity have a part to play? And if I can remain humble – that is, retain a sense of tentative humility – if I continually engage in a process of perception checking, might not this be another strand in the process of empathising?

To get back to our 'heady' client, then – first of all, can I experience acceptance? Can I value 'headiness' just as I value emotions, perhaps especially given an awareness that, at times at least, 'head' can be 'gut in disguise'? That doesn't feel right – it might be true, yet maybe I can value 'headiness' for precisely what it is (intellect, or brains, or whatever). Do I have to relegate someone's intellect to lower divisions, being interested only in the premier league of feelings, yet kid myself (and therefore con clients) into believing that I am being unconditional? Can I remain client-centred, thereby not utilising my crowbar to get at feelings? Can I gather whether my client might actually be talking about or alluding to feelings, yet in a cognitive way?

It seems to me, too, that other clients whom I have experienced as 'heady' have been using their intellect as a *defence* against their feelings. That's okay! If a client needs to protect himself (or herself, although I note the gender implications in the question) then that is their process in the moment. Rather than trying to prise my way into feelings, why not prize the individual? I may or may not communicate something of my sensing of their defensiveness, and that may or may not loosen up the expression of feelings. Yet I do not have any aim in mind other than to come as close as I can to being the core conditions of client-centred therapy.

So with a 'heady' client, it seems to me that if I can strive to authentically experience unconditional positive regard and empathic understanding, and the client is able to perceive this from me, then many things are both possible and, indeed, predictable. To stray from *being* client-centred makes client-centred *process* an impossibility, and would result in something that I predict would be very different in many respects.

15 How does all of this relate to what I've read about 'cognitive' and 'affective' empathy?

Well, I wouldn't want to dismiss those terms out of hand, as I guess that there are contexts within which they could be helpful. For example, if I am going through a transcript of a session with a client, I might want to make use of certain 'tools' or strategies that help me to explore how I was in that session. Many trainers say that 'cognitive empathy' is about understanding *meanings*, and affective empathy is about resonating with *feelings*. Well, okay, although I have already mentioned the artificiality of such harsh and linear distinctions. Nevertheless, it might prove helpful in terms of me exploring whether I could identify any predominant trends within me within the session. For instance, if I discover that I was relating mostly to meanings at the expense of engaging with feelings (or vice versa), why was that?

I suppose that I have to say, too, that in my experience as a trainer the idea of 'cognitive empathy' has not proved too useful. I can recall several instances where this kind of understanding has really been 'from the outside' – almost as if the trainee therapist is observing the client and understanding that they are feeling something and maybe why, too. Yet it somehow lacks the rich profundity of really getting into the client's world of feelings and meanings. Does 'inside out' as compared with 'outside in' make sense? Maybe this links with Carl Rogers differentiating between understanding *about* and understanding from *within* a client's frame of reference.

16 It reminds me of the 'You – because –' exercise we did on our skills course. Is this what you mean?

Yes, I think so – nice example (he said in a non-evaluative way!). This is often presented as 'You feel – because – ', as in 'You feel angry because this situation is unfair'. The 'You feel angry' part is presented as affective empathy and the 'because this situation is unfair' as cognitive empathy, or understanding. Put the two together and hey presto – empathic understanding! Yet this feels somewhat shallow to me. Again, I don't mean to be dismissive as I'm sure that this kind of empathic understanding has its place, either in itself or as part of a journey towards other ways of being empathic. But it somehow feels 'clever', 'dispassionate' or 'detached' when compared with the incredible experience of almost becoming the other person – without losing the 'as if' quality, of course! There is a title in that somewhere – 'On Becoming Your Client' perhaps! So it occurs to me that the 'You – because – 'type of empathic understanding, albeit somewhat formulaic, has a richness that simply reflecting, paraphrasing or posing open-ended, non-directive questions does not have, yet at the same time it lacks the richness that I hope I have described. And remember that Carl did distinguish between understanding *about* and 'truly' understanding from *within* the client's frame of reference.

17 Wasn't there talk of Rogers going to include intuition as a seventh condition, or a fourth core condition?

Well, if there wasn't, it would seem as if there is right now! My understanding is that there *has* been some talk of intuition being added as a seventh condition, or a fourth 'core' condition, but not by Carl Rogers! It is true that, especially later in his life, Carl became very interested in the intuitive aspects of his empathy, although my understanding is that Rogers looked at intuition under the umbrella of empathy, so to speak.

18 So do you have views about intuition?

Of course! Intuition fascinates me. I have increasingly come to trust my intuition, too, although yet again I strive to maintain a kind of tentative humility. I have not to date encountered an explanation of this intuitive capacity that I have come to trust, so I tried to formulate my own.

I recall meeting with a client for the first time. He described his experience of a one-night stand and how rejected and worthless he felt on being spurned at college the next day. At one point I said something like 'And this feels so much more raw because this person was another guy', and I remember feeling *astonished* – it was as if the words just kind of 'popped into my head' and it felt as if I wasn't at all in control of my own process. It felt scary, almost, and certainly very bewildering. Yet after a pause, the client said 'You've just saved several weeks of therapy – but how on earth did you know?'.

I listened to a tape of this session many times, and began to formulate a theory around my own intuitive process. I think that I should perhaps stress at this point that to me this feels like an extreme example of my intuition, yet it was significant in shaping my thinking. What I discovered through listening to the tape was that the client had been very careful about using gender-free terminology – for example, several times he said things like 'the other person', or 'they', or 'them', or whatever. Not once had he used a gender-specific word. At the time I was not *consciously* aware of this, yet my belief now is that at some level just 'behind', or 'beneath' or 'at the edge of' full awareness, my brain had somehow 'clocked' this careful choosing of words – not just once but in a sequence. In retrospect, it feels as if as this sequence accumulated it eventually became strong enough to break through into awareness, and *that* was the intuitive moment. For me, this explanation of this particular instance feels right. I am not convinced that such an explanation applies to *all* intuitive moments, though. As I speak I am wondering if there might be links between what I was saying about empathic inference and empathic guessing and intuition, too.

I want to return more directly to trying to explain my process of empathising. It seems to me that another vital aspect is the sixth necessary and sufficient condition – that the client perceives, at least to a minimal degree, the unconditional positive regard and empathic understanding of the therapist. What a tragic waste if I utilise my senses to try to sense what the client is experiencing moment by moment, yet then do not communicate anything of the outcomes of my efforts! You know, again it seems to me that this is a critical point in differentiating between the implementation of counselling skills and being an empathic therapist. When being at my best empathically, I think that *I use very little of my capacity in thinking about what I am going to say next.* This feels very, very different to me from trying to identify, store and recall words for the purpose of paraphrasing, summarising, reflecting, or whatever. It also feels much more deeply immersed in the *client's* world of perceptions, feelings and meanings than offering an open-ended, non-directive question. Rather, the words seem to flow.

In really trying to *gather* and *absorb* the client's feelings and meanings, and then offering the fruits of my harvesting, I need not be too concerned about the quality of storage and recall, formulation and so on. Rather, I am striving to communicate my striving to empathise and understand. I can honestly say that thoughts about *implementing skills* (such as 'reflecting') just never, ever enter my head. It is hard to explain properly. At my best, I am just so immersed in trying to experience the feelings and meanings of the other person that when I speak it is as if I *were* that other person – or at least as if I am some close approximation to that other person. It is *incredibly* unlikely that this other person is thinking about reflecting my words back to me, or paraphrasing my words, or whatever – so it is equally unlikely that I will be! Even if a client feels stuck, or numb, then my striving is to experience what their numbness feels like and what feeling stuck means to them, and whatever will flow, will flow – or not, as the case may be. I am striving to be *in there with them*, not intending to pull them out, or move them on, or whatever. I think that sometimes counsellors do not enter the numbness, 'stuckness' or stasis of clients because not to try to pull them out or move them on somehow feels like abandoning them to their hopelessness, but far from it. I cannot be *really in there* with someone if I am hell-bent on trying to get them moving.

19 I'm sorry to disagree with the last questioner, but I was told that Rogers was going to include spirituality as the fourth core condition. Is this true?

Again, to the best of my knowledge, the answer is no, this is not true. I feel concerned and somewhat suspicious about how such rumours arise. I suspect that they say more about the people who would like to see spirituality incorporated into client-centred therapy or the person-centred approach than they do about Carl Rogers. I have no particular quibble with the therapist who is spiritual – and I guess it's a very personal, individual thing anyway. Perhaps of greater importance is therapist *congruence* – if I *am* spiritual I need to in some way *be* spiritual, and if I'm not, I won't be. I do feel concerned, though, when people say '*Carl* was going to make spirituality a condition', without backing it up. Show me where Carl said or wrote that – yes?

Carl expressed caution about even using the word 'spirituality', and this would seem to include a couple of things. First, he was eager to avoid religious connotations and connections, for if he *did* use the word, he meant it in a non-religious way. Second, he commented that the word 'spirituality' meant so many different things to so many different people that there was a lack of clarity in meaning. Carl did clarify his own meaning, though. He spoke and wrote of how, when being deeply empathic, it almost felt like an altered state of consciousness, and that he would experience a kind of connectedness between his own 'spirit' or 'essence' and the 'spirit' or 'essence' of the client. He talks with Gay Barfield about this on the *Empathy: an Exploration* video from 1985, and it is very clear to me that even if couched in terms of 'transcending the everyday' or 'altered states of consciousness', his thinking nevertheless falls very definitely within the realms of empathic understanding – an *existing* therapist condition.

So no, I know of no evidence that Carl was thinking of any new conditions, spiritual or intuitive or otherwise.

Incidentally, Carl also said in his 'A Way of Meeting Life' interview (Rogers, 1984: p. 2) that he originally had no interest in a notion so vague as a 'self' but that he researched it because client after client referred to it. If he had another lifetime and clients perpetually referred to spirit, then that's the way he would go – and he recognised a trend towards the more spiritual and was certainly open to it.

20 What exactly does 'the client receives, at least to a minimal degree' actually mean? You mentioned this just now

Now there's a question! One of my worries about training in client-centred therapy is that we don't really have any locally, nationally or internationally agreed measure of what 'minimal' means.

I think that it could be helpful to link the first and sixth necessary and sufficient conditions here – 'two persons are in contact' (or 'psychological contact' if you prefer) and 'the client receives the unconditional positive regard and empathic understanding of the therapist' (at least, as you said, to a minimal degree). With regard to being in contact, Rogers wrote about 'contact' as meaning that the therapist makes a significant difference in the perceptual (or phenomenal) field of the client. So from this we might infer (or intuit!) that 'receiving to a minimal degree' is to do with being heard, maybe to do with a degree of understanding, and so on. In the video I have just mentioned where Carl Rogers and Gay Barfield discuss empathy, and in another video where Carl takes a question from the audience on empathy, what emerges is the idea of *reciprocal empathy* – and that makes sense to me. In other words (and the video audience seemed a little surprised by this idea when Carl voiced it), *we rely upon our clients to empathise with and understand our attempts to empathise with and understand them*. This might go some way towards clarifying not only the notion of 'contact', but also the notion of the client receiving, at least to a minimal degree, the empathy of the therapist.

It also reminds me of a question in the CORE (Clinical Outcomes in Routine Evaluation) forms, which are widely used in the UK within the NHS to try to measure the effectiveness of counselling. The therapist is asked to rate the 'psychological mindedness' of clients. I guess that one of the ways in which I might think about a client's 'psychological mindedness' is the degree to which he or she is able to receive my striving to empathise with and understand him or her – although I'm also aware as I speak that an exceptionally rigid self-structure might be linked, for me, with rating psychological mindedness at the lower end of the scale.

As this goes on, I realise that two perhaps separate issues are emerging for me. One is the nature of contact, and reciprocal empathy ('the client receives . . .'), and I feel okay about that. The other is this notion of 'minimal', and I don't feel okay about that. If, say, a trainee produces a transcript of a counselling session in which there are 25 counsellor communications, three of which demonstrate empathy and 22 of which do not, are we to say that 12% adequately evidences a 'minimal degree'? Probably not, yet in terms of simple arithmetic, my guess is that 12% could accurately be described as minimal. So can someone who is communicating empathy 12% of the time accurately describe themselves as client-centred? Or can we say that if *all* of a counsellor's communications involve a *minimal* degree of empathy – yet also major degrees of some other quality that falls outside the core conditions – they are client-centred? Or is the *client* the sole determinant of receiving to a minimal degree? Is it all about the client's 'psychological mindedness' or reciprocal empathy?

I'd like to quickly throw in another couple of ideas. I found it really liberating to hear Carl say (in the Radio Telefis Eireann Dublin interview) that rather than placing a condition upon oneself (he was actually talking about unconditional positive regard) – which leaves the self wide open to feelings of failure and inadequacy if the condition is not met – we can instead think in terms of being fortunate if the condition is met to some degree. I think that I am more likely to succeed with empathy if I view achieving it as a fortunate bonus, rather than striving to avoid failure. I also like the idea of some being better than none, and more being better than some. Again, I think that holding such a view is liberating, whereas striving to meet ideal conditions is almost bound to lead to failure. Too much empathy can be a bad thing, too. Carl wrote of counsellors who tried so hard to be empathic (in order to impress themselves or others) that they became insensitive to clients.

It seems to me that if the third, fourth and fifth necessary and sufficient conditions (congruence, unconditional positive regard and empathic understanding on the part of the therapist) are met, then defining the *receiving* of these with reference to the *client* would seem to make some sense. At the same time, the second necessary and sufficient condition is that the client is incongruent, so we might ask what degree of denial and/or distortion of our empathic communications is acceptable!

I wonder if this links in any way with Carl's changes to the sixth necessary and sufficient condition. Sometimes Carl wrote of the therapist *communicating* unconditional positive regard and empathy, and at other times he wrote about the client *receiving* unconditional positive regard and empathy. My understanding is that Carl preferred the client *receives* version on the grounds that the therapist *communicating* empathic understanding could represent a regrettable pathway towards the teaching of skills, something he was most keen to avoid. So maybe there is no link there. However, we could make one if we so choose.

21 What are your views on therapist self-disclosure and the process of empathy, or empathising? And earlier you mentioned material that arises from the hyphen in the I–thou relationship . . .

I guess that it will come as no surprise that my views on therapist self-disclosure are much the same as those that Carl Rogers expressed on many occasions. Let me try the kind of 'glib, by the book' response first.

Rogers said that if a therapist experienced a personal feeling or thought *persistently*, then it probably belonged in the counselling relationship. Maybe we can break this down somewhat, though. What does 'persistent' mean? Two or three times in one session, or over a period of sessions, or what? Rogers was very clear about congruence not meaning that the therapist 'impulsively blurts out every passing feeling', so there is another link with the idea of a feeling or thought being *persistent* or occurring over a prolonged period of time. So it is my view that if *over a period of time* (and for me that would be perhaps two or three sessions) a feeling, thought or meaning is *persistent*, then I have plenty of time to reflect and explore along the way. Now let us assume that I have decided, either as a result of self-reflecting or as an outcome of supervision, that this persistent feeling or thought *does* belong in the therapeutic relationship. This does not mean that I will begin the next session with it! Nor will I be 'biding my time' until it comes up! If I introduce the feeling or thought, then the therapy is no longer client-centred and directed by the client. If I am simply 'treading water', eager for my opportunity, then I am not as fully present for the client as I could be. For me, this process is more about being open to sharing this persistent feeling or thought *should it occur again naturally in the course of the therapy*. I trust that if there *is* any significance and relevance to my persistent feeling or thought, then I can anticipate that it will occur again as a matter of course. I need be in no hurry! And I am open to the feeling or thought not recurring, too.

As I think about it, I realise that what I have just said is not entirely truthful. I think, if I'm honest, that there have been times when a feeling or thought has been persistent within a single therapy session, and that I have voiced such feelings or thoughts on occasions. As I said that, I wondered if there might be a further link with intuitive moments, or with empathic guessing or inference. If there is persistence within one therapy session, I think that I tend to perform an 'internal check' in order to satisfy a few questions, such as 'Is this my stuff or the client's?' and 'Am I able to access unconditional positive regard and empathy to the full?', and so on. If my feeling or thought seems to be acting as a barrier to empathic understanding, then I need to find a way to remove it. This *may* mean voicing it. It might also mean that I 'make it recede' and empathy comes to the fore. I like the visualisation of a sound recording mixing desk, with several sliding volume controls – I can lower the volume of my own thought or feeling and pump up the empathy!

We could also look at how the therapist's persistent feeling or thought is introduced to the counselling relationship. Rogers said that he would usually accompany any such communication *with his own feelings or thoughts about his disclosure*. I wonder if I can think of an example of this. I know that at times I might say something like 'I'm not at all sure where this is coming from, yet . . .' or 'I'm a little unsure whether I've understood you correctly – there would seem to be something around . . .'. Maybe they are not such good examples. I feel a little reluctant to share this, because I think that it is a real exception rather than some kind of 'rule' that I follow. I remember one client I found myself disliking pretty intensely. This feeling persisted. In the end (and after a great deal of agonising and exploring), I voiced something of how I did not look forward to meeting with him. In some respects, he seemed quite relieved to have this out in the open. Clients ain't daft! They sense things. It transpired that on this occasion my disclosure that I was really struggling to experience unconditional positive regard was quite helpful both to the client and to the therapeutic relationship. My disclosure was not said in a blaming way, and I was very careful to own my words (while at the same time being aware that no matter how much I own my words, that isn't necessarily how they will be heard). Carl Rogers also felt that the owning of words was important. There is a difference between saying 'I am experiencing boredom' and 'You are a boring person', although I still think that the former is often nevertheless heard as the latter – one excellent reason for caution around self-disclosing, in my view. Anyway, it turned out that this particular client had no friends at all, and was 'disliked by everybody' – that was his problem as he saw it. So having someone voicing their experiencing of dislike, yet in a compassionate and respectful way (doesn't that sound strange?) actually gave him an opportunity to explore how he was experienced by others – the real 'nitty-gritty' of why he was there in the first place.

I feel uncomfortable now. I do want to stress that this example is exceptional, in my experience. I do not regularly experience dislike for clients and then voice it!

I have just thought of another example, but it feels pretty exceptional again. Some years ago I recall a client who I experienced as powerfully exuding two things from every single pore in her body – sex and vulnerability. It was not just sex, but 'raw sex' – a real sexy sexiness. And not just vulnerability, but a really raw fragility. Now bearing in mind that I was at that time being investigated under a professional complaints procedure, I understandably felt that I had to be pretty careful about anything I said. As a professional sanction, I was required to change to an accredited supervisor, who was psychodynamic. 'What are you worried about?', he said, 'I frequently say things like "I'm having sexual thoughts about you and I'm wondering where they're coming from" – it's all part of transference and counter-transference'. Thanks a lot – very helpful! Fortunately, I always tape recorded sessions with this client, and I think that this was a significant factor in my eventually 'going for it' and voicing something of how I experienced her. Let me take a step back a moment. One outcome for me was that I realised that a simultaneous mixture of vulnerability and sex was a pretty powerful cocktail for me to experience. I reflected on this a great deal, including exploring my issues in personal therapy. I learned a great deal about myself and grew through this process. So to an extent, I can say that my experiencing of this client represented 'my stuff', and I don't think that stuff that is purely mine has any place in therapy (at least, not when I am the therapist!). On the other hand, there was something about this client. It *was she* who was exuding the sex and vulnerability, not me. It was not 'my fantasy' – I was absorbing signals *from her*. So clearly there was also an element that was not simply 'my stuff'. I felt some kind of 'charge' or 'energy' in the space between us. I was pretty confident that she experienced this, too, although I was less confident as to how conscious this experiencing was to her. Anyway, in the end I 'went for it'. It was very scary for me and it felt as if my only security was the tape recording. As it happened, it turned out to be good. It was therapeutic. This client had repeatedly experienced dissatisfying liaisons and constantly asked herself the question 'Why me?'. I want to be absolutely clear here that we didn't get into some kind of 'she was asking for it' scenario. However, what did unfold was a real opportunity for this client to explore how the 'signals' that she emitted might have consequences for her. In other words, she became more self-aware and as a result became more empowered, more able to make choices. I'm not sure that I've done justice here, yet I hope that you catch something of my meaning.

You know, given that to me both of the above examples feel exceptional, you can probably conclude from this that only very rarely do I personally self-disclose. Like Rogers, I believe that being congruent or 'transparent' or 'genuine' does not mean 'impulsively blurting out every passing feeling' or that 'anything goes'. I also believe that there is an absolute, direct link between 'transparency' and the degree to which unconditional positive regard and empathic understanding are deeply embedded as natural aspects of my way of being. By this I mean that impulsively blurting out every passing feeling does not necessarily have to mean that I am incongruent in that moment. However, to me it *would* mean that I have yet to deeply incorporate unconditional positive regard and empathy in my therapeutic way of being. It might mean that perhaps I have more self-exploring to undertake, too.

Can I think of a less extreme example? Let us say that a client 'has a go at me', saying that I don't really understand and, after all, that I am being paid to just sit and listen. One of my inner reactions is of righteous indignation and defensive-ness (this is an affront to my self-concept, so I feel a need to protect it), so I gently let the client know that truly I *am* trying really hard to enter their world of feelings and personal meanings. Am I being congruent? No, because I am denying myself access to those respectful and empathic capacities that I hold so dear. Okay, so I respect this client and empathise with the frustration, or anger, or feelings of hopelessness that are being expressed. Am I being congruent now? No, because I am denying myself access to my feelings of indignation, and they are real enough, too. Congruence means allowing myself access to *all* significant parts of me in the moment, and making choices based on this accessing. Thus, for me, *any self-disclosure involves choice on my part* – what I will choose to disclose and what I will not choose to disclose, why and when and what . . .

I'm reminded of reading a transcript of an interview that Carl Rogers gave a month or two before he died. He was asked if he had any regrets. He replied that as he himself was a *disciplined* person, he had mistakenly assumed that other client-centred therapists were self-disciplined practitioners, too. He realised that this was far from always being the case, and he wished that he had placed greater emphasis on client-centred therapy being a disciplined approach. For me, the therapist who regularly self-discloses lacks the discipline to put the commitment to stay within the client's frame of reference fully into practice.

Having stated that Rogers believed that any *persistent* feeling or thought that arose within a therapeutic relationship probably belonged in that relationship, this reminds me a little of applying the 'Two Circles' diagram to frames of reference as we did earlier. There is 'material' that is clearly the client's, material that is clearly mine (these two would be in the outer segments), and material that has arisen between us (in the overlap segment, or in the hyphen of the I–thou relationship). I feel okay about voicing anything that is the client's or that has arisen in the space between us, yet I am more cautious about anything that is clearly in *my* segment. However, if it is persistent and I have checked it out, then I may well share it.

I guess that if I were to dredge up my most important agenda here, it would be about not 'letting it all hang out' and 'in your face' self-disclosing – this was never what was intended by 'transparency' or congruence or authenticity or being genuine. So I come back to the degree to which unconditional positive regard and empathic understanding are deeply embedded, and a self-aware way of being that does not deny or distort significant experiencing.

22 How does your process of empathising link with what Dave Mearns has written about configurations of self?

Hmm, well, I don't want to be disrespectful to Dave Mearns, but just to set the record straight, Carl Rogers used the term 'configurations of self' as early as 1959, so I don't think it is accurate to consider this term wholly as a creation of Dave's! Incidentally, Carl also said that he didn't *particularly* care for configurations of self theory, but that he was fine with those who did (he stated that he was more interested in the process of becoming than in somewhat fixed self-structures). Nevertheless, I personally find the idea of *configurations of self* pretty helpful, although whether or not the use I make of this theory is quite what was intended I do not know.

There are a couple of things that occur to me, and one of these is an issue that I am still in the process of working through, so it feels like unfinished processing to me.

The first issue (one that I feel is pretty sorted for me) concerns how helpful I find looking at configurations of self in supervision, specifically with a view to exploring the nature of my unconditional positive regard and empathic understanding. An example that I have used before is that of the client who self-injures. In this instance, there is a 'client self' who is doing the cutting, a 'client self' who is being cut, and a 'client self' who is 'overseeing' the action between the 'hurter' and the 'hurt'. Once a self (or a configuration of self) has been symbolised, it becomes a self-as-object – something identified, something named – something that can be broken down into its component parts and examined and explored. The self-as-process is doing the exploring. I find it helpful in supervision to explore the degree to which I am engaging with each of these aspects of the client's whole. For instance, I might discover that I feel somehow 'seduced' into empathising with the victim and the hurt, and as a consequence I may be neglecting the hurter or the self-as-process. Alternatively, I might believe that the hurter, the perpetrator, is more in need of the core conditions if change is to be generated, so the hurt and the processor become neglected. Or I might be so in awe of the client's motivation to change that I engage with the processor and neglect the hurter and the hurt. Whatever, what I appreciate is the opportunity – a framework or strategy, if you like – to explore the depth and extent of my engagement with clients. And if I discover that I am neglecting certain 'configurations of self' in a client's self-structure, I can also explore whether or not there might be links with configurations within my own self-concept. So I do find the notion of configurations of self very helpful as a strategy or process that I can use to try to maximise the extent to which the whole of me is open to the whole of the client, including maximising the unconditional nature of my positive regard and striving to ensure empathy with and understanding of all significant aspects of the client.

The second issue I'm not so sure about. I'm not so sure about it because it feels maybe directive, or it seems as if the therapist is taking the lead, or something, yet I don't want these thoughts to get too much in the way of exploring the issues. Let me stray into an anecdote. There is a large international gathering of person-centred people. Some are known to all (the 'big names', the 'A list'), others are known to many (the 'B list'), some are known to some (the 'C list', the 'minor celebrities'), and yet others are not known at all, or certainly did not feel known (the 'D list'). Some people were very evident – visible and speaking out. Others were somewhat hidden in the outer rings of the circles, saying nothing. I asked if Venya was present. Venya had come from Russia and we had exchanged emails but had never met 'in the flesh', so to speak. We connected. I later approached several people and engaged with them, and as a consequence they engaged with me and moved on to engage with others. There was another person from Poland, and I recall saying something like 'I realise that this is my need and I don't want to pressure you. I want to say that I feel I'm really missing out through not hearing you speak'. This person then spoke for some time – and in part expressed gratitude for feeling welcomed for the first time.

I thought about these experiences, and wondered whether such *community* processes might in any way be relevant to experiencing a group of configurations of self – a *community of selves* that make up a whole person. If I think about clients, there are certain self-aspects that are equivalent to 'A list celebrities' – being very visible and outspoken. There are also hidden, coy or silent aspects of self, and often a good many in between. Might these (or some of these) hidden or silent selves appreciate some kind of recognition or welcome? Might these selves be wondering whether or not it is safe to come out and play, and yearn for an invitation? I'm reminded of another international person-centred gathering at which the first few hours were spent negotiating seating arrangements. Many people engaged meaningfully with this (it triggered all kinds of feelings and thoughts about inclusion and exclusion, power and equality, for instance). Others tolerated the process, and two social workers who were interested in the person-centred approach left within the first 24 hours of what was a five-day meeting. Why? Because *they didn't know the rules*. They didn't know what to expect. Most of all, they didn't feel either acknowledged or welcomed. Could this apply to configurations of self within an individual's self-concept? If an aspect of self feels neglected, ignored or unwelcome, might it just go away?

I'm really not sure about this one. To welcome a self that is barely visible feels somewhat directive – no, it doesn't feel somewhat directive, it feels *directive*. Welcoming a barely visible self could be received by the client as ignoring, sabotaging or hijacking the self in full flow at that moment, for instance, and it is my choice to do so. Yet just before he died, Rogers reiterated (here I'm thinking of the chapter he wrote with Nat Raskin for *Current Psychotherapies*) that from time to time he would tentatively venture his sensing of feelings and personal meanings of which the client seemed only dimly aware, or even complely unaware – 'edge of consciousness' material. And it seems to me that such 'hidden' selves could be regarded as 'lurking at the edge of awareness'. I don't want to come across as too patronising or arrogant here, but I can honestly say that over the years my trust in venturing such tentative 'feelers' (tentative tentacles!) has grown somewhat. I have followed my client's lead when I have done so, and, whatever it sounds like, it has usually proved to be both relevant and helpful.

Perhaps I should also mention that, fairly often in fact, clients identify their own 'configurations of self' and want to explore them, sometimes also wanting to explore the nature of the inter-relationships between their selves. That is okay – they sing, I accompany.

I want to return to a more specific focus on my process of empathising, and say something about the links between empathic understanding, congruence and unconditional positive regard.

23 What, then, do you see as being the relationship between empathic understanding and congruence and unconditional positive regard?

I have a dim recollection that I said I would answer the question on therapist self-disclosure in two ways, and only answered it in the glib one. Maybe here is my opportunity to give the other version.

Let us assume that we are right at the start of a training course in client-centred therapy. Let us also assume for the moment that there is not a stunningly high degree of unconditional positive regard or empathic understanding being experienced or communicated – otherwise why the need to embark on a training programme? Now for me this does not have to mean that course members are necessarily being *incongruent*. However, it may well be that unconditional positive regard and empathic understanding are not as deeply embedded as fundamental and integral aspects of the self as they could be. For instance, if I respond to your communication with a 'This is how it is for me'-type response, this does not have to mean that I am denying or distorting anything.

My hope would be that as person-centred learning progresses, so unconditional positive regard and empathic understanding *do* become more deeply embedded as integral, fundamental and authentic aspects of the therapist. As this happens, so disclosure to clients from the internal frame of reference of the therapist becomes less and less likely. In other words, I am more likely to respond to your disclosure with an authentic attempt at letting you know that I am respectfully striving to understand your world of feelings and meanings.

Another way of saying this is that unless empathic understanding is authentic, it is less likely to be of value. At the same time, Rogers stated that the best way to communicate unconditional positive regard is via empathy.

24 Do you see each of the core conditions as having different relevance or importance to different phases of the therapeutic process?

Gosh – yes and no, I suppose! Now why did I say that?

I guess that the 'Yes' feels kind of heady – something about clients entering therapy being vulnerable, anxious and defensive. The theory – and my experience as both therapist and client – is that the client's receiving of unconditional positive regard serves to dissolve conditions of worth and serves to dissipate defensiveness. Very often it seems to me that unconditional positive regard features very highly in the early stages of therapy. Yet very often therapy moves into a real searching for meaning, and here maybe empathy and understanding feel somehow more significant. Neither respect nor empathy are likely to be of much value if they lack authenticity, so I guess I'd see congruence as vital throughout.

Yet I have a strong 'No' response, too. This is based on the artificiality of breaking down the core conditions into component parts. I don't want to lose sight of the whole person. Although we break things down into manageable portions in order to digest them, let us not forget that a spark plug sure ain't the whole engine, and the car won't run if we don't put the motor back together again! So if we do take things apart in order to examine them, let's not forget to put them back together.

25 How do you empathise with silence, or the client who says very little? Is it true that the client-centred therapist should never be the one to break a silence?

How do I empathise with silence? Well, as I have said, there's far more to the universe of communication than just the solar system of the spoken word. Thus although in silence there is no *verbal* communicating, there might still be some paraverbal communication (such as sighing, or specific fast or slow or deep or shallow breathing) and lots of non-verbal communication. So I guess that mostly I use my eyes. For instance, facial expressions can tell us a great deal. I know that sometimes I shift my body shape to approximate to that of the client's in the hope that this will give me some insight into their possible sensations. Sometimes I might actually shift, and sometimes I just *imagine* sitting that way and what it would feel like. Sometimes I also stare at my hands or imagine my face being set in the same way. I don't have any rules or set techniques other than to strive to get into the client's world of feelings and meanings in whatever way feels appropriate at the time.

I recall someone asking Carl Rogers about the client who seemingly doesn't say very much, and he replied that *most* clients don't say very much! He reckoned that most clients say only a few things of central or deep significance in a therapy session. Some surround these significant disclosures with many other words, and some surround them with silence.

Where does 'The client-centred therapist should never be the one to break a silence' come from? Was it something you were told on your skills course? My guess is that this was justified on the grounds of not intruding into the client's space or something. Yes? Okay. 'Empathic guessing' and 'empathic inference' have been referred to a few times already. Rogers wrote about these in (among other places) a chapter called 'The Silent Young Man', and in the transcripts you will see that Carl Rogers broke just about every silence. So wherever the idea came from that client-centred therapists don't break silences, it didn't come from Rogers! I guess that not being intrusive would be based on a notion of respect for a client's space, which on the face of it sounds okay. Yet venturing an empathic guess could be respectful, too. Perhaps there could be something about main-taining and enhancing the 'two persons in contact' element through empathic inference as well. I would estimate that I probably speak first after half or more of the silences that I meet with in therapy, and almost always with some attempt at empathy – either empathic understanding if the non-verbal and paraverbal signals are fairly clear (for instance, I might say something like 'You're feeling really fidgety right now'), or with an empathic guess if the signals are less clear. Very occasionally I might even self-disclose! I have sometimes said something like 'I really can't sense with any accuracy what's going on for you right now, but I am interested if or when you're ready to speak . . .'.

There's something about this *how* to empathise that I feel uncomfortable with, at least sometimes. I think that this is about feeling okay with the core conditions being stipulated – it being necessary (and sufficient) to be congruent, respectful and empathic – yet feeling ill at ease when it comes to *stipulating* how. So the fact that Rogers broke nearly all silences doesn't have to mean that you should do so, too. And the fact that I break half or more of the silences doesn't mean that this has to be an objective for you. So by all means be authentically and respectfully empathic, but be so in your own way.

26 Do you ever not experience empathy, or experience a loss of empathy, and if so, what do you do about it?

Who, me? Of course not, never! Yes, of course I do. Maybe I can reframe the question to ask *why* I might not experience empathy, or *why* I might lose it. The first thought that occurs to me is that I might lose empathy if one of my own thoughts, feelings or sensations is getting in the way, in which case I need to deal with it. I have already talked about this – maybe whatever it is recedes as I pump up the empathy volume, or maybe I voice something. I find it more difficult to sustain empathy if I'm extremely tired or distracted, so I need to take care of myself if I'm striving to be fully present for clients.

It seems to me that a crucial factor here is therapist *self-awareness*. Sometimes loss of empathy feels quite sudden, a kind of surprised realisation – 'where did that go?'. In such instances it feels quite easy to become empathic once again. I simply refocus on striving to sense what the client is experiencing in this moment. At other times there is a kind of creeping, growing awareness of feeling less connected with either the client or my empathy, or with both, although again internal monitoring and accessing feels like the crucial factor in reconnecting.

Then there is the question of how to reconnect. Very often it really is as simple as turning my attention to what the client is feeling and what meanings might be present in the moment. There has been something in the background here that feels really embarrassing – 'aw shucks' and all that. I have a framed picture of Carl Rogers on my wall, and sometimes when I realise that I need to reconnect, I glance at it. This feels embarrassing because maybe it sounds like hero worship or something 'sad'. Well, I clearly have great admiration for Carl Rogers, but that's not it. The picture acts as a kind of instant trigger for me. To me, the picture symbolises what Carl stood for, like the necessity and sufficiency of the core conditions. So one quick glance and bam – it all comes flooding back! I believe that I can undertake this instant reconnecting with my capacity to strive to be the core conditions with or without a picture – it just happens to be there and it helps. I guess that almost anything could be used to symbolise the core conditions – a plant, the clock, the box of tissues, my own awareness, my mind . . .

So maybe something common to all of us is an awareness that empathy has been lost or is fading. However, *how* we reconnect with our empathic selves and our clients might be more personal.

27 Isn't there something about setting aside your own feelings and thoughts or, better still, shouldn't your baggage be left outside the room?

Not for me, no. To me, that doesn't make much sense. You look taken aback. Okay, let me try and explain. First of all I'm not entirely happy with the word 'baggage', as it doesn't exactly feel like unconditional acceptance – rather, it feels somewhat derogatory and dismissive. I'm sure we wouldn't dismiss a client's baggage so readily, so maybe we can afford ourselves the same respect that we offer to clients. Secondly, how is it possible to speak of therapists being *whole persons* in the relationship if they leave chunks of themselves outside? I'm reminded of the earlier theme about client-centred discipline, too – the discipline to live out my stated commitment to striving to be the therapist conditions, that includes staying within the client's frame of reference as much as I am able to. Also I wrote a piece relatively recently in which I came up with the idea of an empathy *muscle*. Don't use your empathy muscle and it will atrophy and wither. Use and exercise your empathy muscle regularly and it will easily be strong enough to carry baggage with no problem.

I don't mean to sound too abrupt, but it's late and I'm feeling a bit worn out. As we only have a few minutes left, I'd like to make copies of the article on 'Setting Aside Or Erupting?' available to those who would like to explore this theme further, which might make it more acceptable to move quickly on. Would that be okay?

28 I'm really interested in the use of imagery and metaphor. Do you use them?

Yes. I realise that I responded with a spontaneous and definite 'yes', yet it occurs to me that my use of imagery, analogy and metaphor is something that has very much grown over the years. I guess there might even have been a time when I didn't use them at all. Maybe I've grown in confidence, or something. Maybe I've just grown! Or I am growing . . . One thing I love is the diversity of communication available to us. Why not exploit all of the rich ways of communicating at our disposal?

29 Can you give any examples?

Okay, I can think of two examples from the last couple of days that I feel quite pleased with! Now that's interesting – why do I say *'pleased'*? I guess it might sound as if I feel smug about having 'done my job well' or something, but that's not it. It might be *part* of it, but I think it's much, much more to do with feeling pleased with the reaction from the clients – like their response to me let me know that I really was with them in their world, and that feels really enriching, meaningful (to both of us) and worthwhile. Anyway, one client is a shop worker, and she had described (several sessions previously) how she would be at work and her mother would appear peering pleadingly through the front window, and just how difficult the client found this. This week she was reflecting on how her relationship *with herself* in relationship to her mother had changed over the course of her therapy. I recall saying something like:

> It doesn't feel like there are shutters or walls between you any more. It feels more like the glass shop window. Sometimes you just wave at her because you are very busy, and that feels okay now. Sometimes you are busy serving someone else, and it feels okay for mom to wait in the queue. Hell, sometimes you even shut up shop – and even that feels okay now, too.

The client became quite excited – it seemed as if I had captured her feelings and meanings in a way that really made sense to her. It felt *right* to her. Now if I try to access my own process of empathising, then it seems clear to me that although my mind wasn't lost in words and experiences my client had talked about several weeks before, my concentrated focus on her world as she experienced it clearly involved the specific area of my memory *as it pertained to that client*, and I could readily access this part of me. It feels a bit like being primed again – when I am with a particular client, I am primarily modulated to access memories *that relate to my knowledge of that client*. Of course I can access other memories if need be, yet they are maybe further removed from my immediate experiencing. So it feels as if when the client communicates to me, what I absorb 'floats around' in my knowing of that client, and sometimes connects with something that seems to have meaning *from within her own experience* that I have absorbed and retained. I realise that I am speaking off the cuff, but it feels as if *something* like that was going on within me.

As I was speaking, I was thinking ahead, too, and it occurs to me that my second example is different, and that feels good and interesting. It feels good because it illustrates diversity, and it interests me because this time the metaphor came from within me, not from my recall of the client's experience. This second client had been talking about how in some ways she resented her husband and felt disappointed because he didn't really understand her. She also spoke of how she didn't want to expose herself fully to him because he might feel hurt, so she was protecting his feelings. At one point (and I'm aware that this is only a small part of our total communication) I said something like:

> It feels a bit like a question in a maths paper – on the one hand you give him the numbers and the equals sign, but on the other hand you omit the plus, minus, multiply and divide signs because he might feel hurt by them. Then you feel disappointed when he doesn't come up with the right answer . . .

You know, as I said those words just then they sounded very judgemental! It really wasn't like that at all at the time, with the client. I'm feeling defensive! In context, my communication was fine, and the client responded with something along the lines of 'Wow! What a brilliant way of putting it! Yes, that's it *exactly*!'. And you know, I feel pretty confident that this enhanced awareness will make a real difference, if only at times, to the client's experiencing of herself and the way that she relates to her partner. Now if I think about what my processing was *at the time*, it feels as if I was again trying my hardest to absorb all that she was communicating, and again *what* I absorbed was 'floating around' my mind seeking some kind of referent. What emerged might have been from me, but it was very much related to trying to sense and feel how the client experienced herself and her relationship with her partner.

I've just remembered another example. I said to a client (I'll call her Jane for now) something like:

> It feels like you're in a tug of war with yourself. On one end of the rope there's all these big, hefty Janes and at the other end one tiny Jane. Yet somehow Baby Jane seems able to hold her ground.

This had real meaning for the client, and it seemed to really capture 'where she was at'. Anyway, yes, I think that when it seems to be an accurate way of communicating empathic understanding, I make use of metaphor and imagery quite a lot.

It occurs to me that the examples I have given don't quite do justice to the full process, as more often than not there is a *follow-up* to the use of metaphor. Even if, from the client's reaction, I am confident that I have accurately sensed the feeling, I need to determine whether ('perception check') I have also caught the meaning accurately, and vice versa. As with most if not all empathic communications, clients seem to take my words and then self-reference, and if we have established a therapeutic working alliance, they will feel free to come back with 'No, that's not quite it. . . . It's more like . . .'. Indeed, they may even choose to ignore my offering if it is not relevant or appropriate to them, in which case I believe that they nevertheless very much experience me as striving my hardest to sense how they experience themselves and their world, which is therapeutic in itself. Incidentally, I also believe that they feel safe and secure enough in our relationship to know that I will not take umbrage if they do ignore my metaphor.

In summary, I guess that the congruent and respectful use of metaphor (by which I mean that I am not simply using a technique, and am humble and tentative enough not to be adamant about appropriateness) is one of many effective ways of letting a client know that I really am committed to trying to empathically sense their feelings and meanings.

30 One last question? On the subject of being worn out, you spoke a while back about empathy being hard work. How do you avoid burnout?

Funnily enough, although I hear quite a few people talking about it, the prospect of 'burnout' is not something that bothers me personally. And you know, as I think about it, the people who talk about burnout tend to be teachers, lecturers, social workers, and so on – I don't recall too many *counsellors* talking about it. Maybe right now could be a good example. I've been talking for quite a while and I feel pretty tired, yet I also feel a sense of energy because I have enjoyed engaging with you, I have enjoyed and appreciated your engaging with me, and I also hope that I'll get some useful material out of this session! So I guess I'd say I'm something like 80% tired, fatigued or exhausted (because I've talked a lot and given out a lot) and 20% energised, invigorated or excited. When I have been with clients it usually feels different – perhaps nearer to fifty-fifty. There's something about being empathic (and congruent, and respectful) that is simultaneously draining but also edifying, or nourishing. It is draining because I put a great deal of effort into it. It is energising because it really is a privilege and honour to be allowed or welcomed into the private world of feelings and meanings of another person – I grow through this, too. So I experience both fatigue and exhilaration at one and the same time. It reminds me a little of what Rogers said about entropy and syntropy. Entropy is the ongoing decay of the universe, and there has been a long-held belief that ever since the Big Bang our universe has been dying. Yet ongoing trends towards growth, development and ever-increasing complexity have been acknowledged, too, and this is syntropy. So maybe the *effort* I put into empathising is like *entropy*, and the *privilege* and *personal growth* is like *syntropy*.

I thank you all.

Empathic understanding: in the necessary and sufficient conditions

For what we are about to receive

In Chapters 1, 2 and 3 we have to some degree looked at what empathic understanding is, in greater length have looked at what empathy is not, and have shared an interview about the process of empathising. Perhaps we can now place empathy back in the context of the necessary and sufficient conditions for therapeutic personality change as a whole.

Over the years, there has been much debate over just how many conditions there actually are – four, five, six, seven or eight. For instance, there would be only four conditions if conditions one and two (contact and client incongruence) were to be seen as *pre*-conditions, and only five conditions if condition one was regarded as a *pre*-condition. If either 'spirituality' or 'intuition' were to be seen as an additional condition, there would be seven in all, and if both were to be added, there would be eight. As Carl Rogers spoke of 'spirituality' (a term he viewed with some caution) as falling within conditions four and five and of 'intuition' as falling (mostly) within the realms of empathic understanding, these are not treated as separate or additional conditions in this book. Although there is truth in the statement that Carl Rogers sometimes wrote of condition one and sometimes of conditions one and two being *pre*-conditions in some senses, he nevertheless retained them within the entire version – likewise here.

Although there is, as we shall see, a logical rationale for the conditions appearing in the order that they do, it is worth remembering the rich interface between them. Just as Natalie Rogers was fearful of the essence of the I–thou relationship being lost through a nit-picking focus on minutiae, let us retain the essence of a client-centred relationship as a complex whole as we strive to place empathic understanding back within the context of this therapeutic interweave.

Empathy and contact

I was trying to think about myself as a therapist in the therapeutic hour – as I seem to myself, and as I seem to be seen by others. I think one of the outstanding things is that I'm very much present to the client. I'm not sure entirely what I mean by that, but the main thing that's going on in me is my concern with, and attention to, and listening to the client. I'm very much present; nothing else matters much. That's why I can do a demonstration interview – it doesn't matter how many other people are around; for me there's just this one person that exists. What develops is a feeling of connectedness, and this is often very strongly felt by the client as well as by me, as though

there is some kind of a real bond between us. That grows out of the fact
that I do enter so fully into the client's personal world.

> (Rogers and Russell 2002: p. 280)

We shall now take a brief sidetrack of sorts. In his 1957 paper 'The necessary and
sufficient conditions of psychotherapeutic personality change' (Rogers 1957a),
Carl used the term *psychological contact*, and in his 1959 chapter 'A theory of
therapy, personality, and interpersonal relationships, as developed in the client-
centered framework' (Rogers 1959) he used the term *contact*. There has since been
much academic debate as to why the word 'psychological' was dropped. Well, it
wasn't. For the sake of historical accuracy if for no other reason, let it be known
that Carl Rogers wrote the 1959 chapter:

> much earlier than when it was published. I think it was written about
> 1953.
>
> (Frick 1989: p. 90)

So the 1959 version was penned *before* the 1957 paper. It was the slow process
between authorship and publication of a chapter in a *book* (to which other
psychologists were also contributing chapters) that meant that the 1959 version
appeared *after* the 1957 paper was presented and published in a *journal*. Thus the
debate about Carl 'dropping' the word 'psychological' makes no sense at all – the
more accurate detail is that he *added* the word! 'In which case', you might ask,
'why did Carl Rogers *add* the word "psychological" to his definition of two
persons being in contact?'. The answer is that no one really seems to know for
sure, although some speculate that it was because the 1957 paper was prepared
for the *Journal of Consulting Psychology*! The main reference used in this book is the
1959 version, which Carl himself said:

> was a very strenuous effort on my part to make some sense out of all
> my professional experience and personal experience up to that point.
>
> (Frick 1989: p. 90)

Although in the definitions that follow the use of the term 'psychological' would
seem, in context, to make sense, sometimes for the sake of brevity and to avoid
needless repetition only the word 'contact' is used. The meaning in each instance
remains the same.

Look – no couch!

The counselling process begins *before therapist and client actually meet*. In my own
practice, for instance, all would-be clients receive a leaflet that explains something
about me together with a basic overview of client-centred therapy. Rogers said:

> In the kind of therapeutic approach that makes sense to me, the more
> the individual knows of the climate I want to create, and of the
> principles that I believe to be operative, the more intelligently he
> can participate in the experience of therapy . . . For me, that is one
> reason for feeling that it is a more potent or widely useable theory: that
> it is not destroyed by people becoming aware of what is happening.
>
> (Rogers 1989: pp. 124–5)

Now imagine having a time machine and travelling back to the era when client-centred therapy was first being formulated. What might have been the norms at that time? Imagine that you arrive for your first appointment to be instructed by your analyst to lie:

> I've always found the notion of sitting behind a person who's lying on a couch and can't see [me] somewhat repugnant . . . I guess I felt this way because of many of the psychological connotations: I'm the expert; I'll observe what's going on in you, but I don't want to be present as a person; you adopt an unusual posture because this is now a very unusual situation, and it should produce all kinds of material you don't understand. All of that seemed a bit repugnant to me . . .
>
> (Rogers and Russell 2002: p. 246)

In client-centred therapy, however, even before the counsellor and the client meet, the therapist believes that:

> the opportunities for new learning are maximized when we approach the individual without a preconceived set of categories which we expect him to fit.
>
> (Rogers 1951: p. 497)

We probably take it pretty much for granted that we both sit, at even heights, face to face, yet this in itself is a statement about power sharing even as contact begins. Therapist values, attitudes and beliefs have great significance both *prior to* and at the *beginning* of a therapeutic relationship, for it:

> can be initiated by one party, without waiting for the other to be ready.
>
> (Rogers 1961: p. 336)

The therapist needs to be ready.

Look, listen and learn!

Sitting upright, face to face, at an even height communicates the therapist's respect for equality – the counsellor is not in sole control or in a superior position. This 'simple' or basic first impression gathers substance as the client-centred therapist refrains from taking a client case history, engaging in assessment or diagnostic procedures, or otherwise testing the client in some way. Nor is the therapist firing questions at the client. Rather, the client-centred therapist focuses their attention and their respectfully empathic capacities on striving to sense the client's world of feelings and meanings:

> Both client and therapist are implicitly directed to fix their attention on the client. This means that the therapist characteristically looks, listens, and otherwise attends to the client; the client characteristically looks into space or at some object in the room so as to focus on himself. The client, then, is in the spotlight, directed by the total situation and the power of the therapist's steady attentiveness to be conscious of himself. This kind of attention to the client and waiting for his initiative . . . energizes the situation, is an enormously powerful living out of the

belief in the organismic self, and casts the client steadily back on himself.

(Barton 1974: p. 186)

Thus statements such as 'This reminds me of' or 'It seems to me' or 'Have you considered . . .?' or 'I think that . . .' or 'This is clearly . . .' are not characteristic of client-centred therapy, in part because they shape the relationship around the therapist. Rather, the *client*-centred therapist *guards against* external intrusions into the client's world of feelings and meanings – the underlying message is that *you and only you matter to me*.

Relationship

In actual fact, Carl Rogers and his colleagues initially defined this first necessary and sufficient condition as neither 'contact' nor 'psychological contact' – the term used was *'relationship'*. However, this soon changed to 'two persons are in contact' due to the term *relationship* being experienced as too open to confusion and misunderstanding. For instance, some people seemed to interpret 'relationship' as meaning *deep* relationship, or *good* relationship, or *therapeutic* relationship, whereas:

> The present term has been chosen to signify more clearly that this is the *least* or minimum experience which could be called a relationship.
>
> (Rogers 1959: p. 207)

Carl Rogers never entirely abandoned the term *relationship*, though. In his later years, he often referred to Martin Buber and the notion of an 'I-thou' relationship:

> When there is a real trendless immediacy in the relationship, when you're aware of nothing but this person and he's aware of nothing but you, and there is a deep sense of communication and unity between the two of you, that's the thing I refer to as I–Thou relationship.
>
> (Rogers and Evans 1975: p. 25)

What, then, of such issues as age, gender, race, and social status? Rogers wrote that:

> . . . it would appear that such elements as the sex, appearance, or mannerisms of the counselor play a lesser role than might be supposed. When the counselor is favorably perceived, it is as someone with warmth and interest for the client, someone with understanding.
>
> (Rogers 1951: p. 69)

In effect, the therapist striving, and achieving to a significant degree, to be the 'core' client-centred conditions is more significant than social factors. When looking at empathy we saw that our own frames of reference contain within them counterparts to the perceptions of others, which enabled Carl to say:

> As human beings trying to cope with life, to understand it and learn from it, we have vast pools of commonality to draw on. It makes no difference that I am an older white middle-class American male, and you may be yellow or black or communist or Israeli or Arab or Russian

or young or female. If we are openly willing to share, then there is a large area in which understanding is possible.

(Rogers 1978: p. 122)

Bearing in mind that the sixth necessary and sufficient condition is that the client perceives, *at least to a minimal degree*, the unconditional positive regard and empathic understanding of the therapist, it made sense to Rogers to define 'relationship' in *minimal* terms, thereby avoiding assumptions and value judgements about *good, deep,* and so on (and note, incidentally, the interconnection between conditions one and six). Returning to, clarifying and expanding on this, Carl wrote:

> This difference may be quite minimal and in fact not immediately apparent to an observer. Thus it might be difficult to know whether a catatonic patient perceives the therapist's presence as making a difference to him. But it is almost certain that at some physiological level he does sense or subceive this difference.
>
> (Rogers and Truax 1967: p. 99)

Two fundamental characteristics of relationship or (psychological) contact were deemed to be *safety* and (especially as the encounter progresses) *free communication*. How, then, might a safe and free relationship come about?

> And this basic security is not something the client believes because he is told, not something about which he convinces himself logically, it is something he experiences, with his own sensory and visceral equipment.
>
> (Rogers: 1951: p. 209)

Rogers wrote:

> We know that if the therapist holds within himself attitudes of deep respect and full acceptance for this client as he is, and similar attitudes toward the client's potentialities for dealing with himself and his situations; if these attitudes are suffused with a sufficient warmth, which transforms them into the most profound type of liking or affection for the core of the person; and if a level of communication is reached so that the client can begin to perceive that the therapist understands the feelings he is experiencing and accepts them at the full depth of that understanding, then we may be sure that the process is already initiated.
>
> (Rogers 1961: pp. 74–5)

Carl gave us several models of client-centred therapeutic process (*see* Chapter 5). In the 'seven-stage' version he stated that when in the first stage a client is unlikely to enter therapy, as there is a fixity and remoteness to the client's experiencing and an unwillingness to communicate the self (communicating instead only externalised feelings and meanings). In such a condition, clients do not embrace, recognise or own feelings and meanings due to their rigid self-concepts (rigid 'personal constructs' or 'configurations of self') and, as a defence, 'close and communicative relationships are construed as dangerous' and internal awareness is blocked. The client therefore does not experience in awareness any desire to change:

> In a deeply defended person . . . change is difficult to bring about. Change tends to occur more in the person who is already somewhat open-minded and searching in his approach to life.
>
> (Rogers and Evans 1975: p. 73)

(Again, then, we can see the potential in *open-minded seekers* prior to them even entering therapy.) For the deeply defended client, life tends to be perceived through the past rather than the present (so there is little or no 'immediacy'), and thinking is 'black and white' rather than gradational. Everything is everyone else's problem. It is unlikely that persons will either enter or benefit from therapy at this stage:

> We seem to know very little about how to provide the experience of being received for the person in the first stage, but it is occasionally achieved in play or group therapy where the person can be exposed to a receiving climate, without himself having to take any initiative, for a long enough time for him to experience himself *as received*.
>
> (Rogers 1961: p. 133)

Carl wondered, very tentatively, whether it might be more helpful to try to establish contact with very young clients, or 'extremely immature or regressed' people with conditional rather than unconditional positive regard:

> It seems clear that some immature or regressed clients may perceive a conditional caring as constituting more acceptance than an unconditional caring.
>
> (Rogers 1989: pp. 14–15)

Readers interested in stage-one characteristics might find some of the more recent work on pre-therapy helpful, such as *Pre-Therapy* by Garry Prouty, Dion van Werde and Marlis Portner (Prouty *et al.* 2002).

On several occasions Rogers pointed out that the condition that two persons are in contact could in fact be considered a *pre*-condition, for without this engagement in a relationship the predictions about establishing a therapeutic relationship that in turn would have predictable consequences in terms of personality change would be unlikely to follow. In terms of establishing whether or not contact exists:

> The simplest method of determination involves simply the awareness of both client and therapist. If each is aware of being in personal or psychological contact with the other, then this condition is met . . . The first condition of therapeutic change is such a simple one that perhaps it should be labeled an assumption or a precondition in order to set it apart from those that follow. Without it, however, the remaining items would have no meaning, and that is the reason for including it.
>
> (Rogers 1957a: pp. 95–103)

This could be restated using the '*If–then*' formula of client-centred theory, wherein necessary and sufficient condition one is the first 'if'. *If* two people make a perceived difference to one another, *then* the actualising of a potential becomes possible. (This formulaic version would continue 'and *if* one person is incongruent and *if* the other is congruently respectful and empathic, and *if* conditions four and five are perceived by the client in this relationship, *then* a client-centred therapeutic relationship ensues, and *if* this relationship exists, *then* predictable outcomes in client personality change occur'.) Carl Rogers wrote that only when clients experience themselves as fully *received* can they move on to stage two, wherein expression (albeit about non-self topics) begins to flow:

> I believe I can state this assumed condition in one word. . . . I shall assume that the client experiences himself as being fully *received*. By this I mean that whatever his feelings – fear, despair, insecurity, anger; whatever his mode of expression – silence, gestures, tears or words; whatever he finds himself being in this moment, he senses that he is psychologically *received*, just as he is, by the therapist. There is implied in this term the concept of being understood, empathically, and the concept of acceptance. It is also well to point out that it is the client's experience of this condition which makes it optimal, not merely the fact of its existence in the therapist.
>
> (Rogers 1961: pp. 130–31)

In stage two the client still externalises and feels little or no personal responsibility. Feelings may be exhibited but are disowned or are seen as past objects, rigid constructs are seen as facts, differentiation of feelings and meanings is limited or over-generalised; and, although contradictions may be expressed, they are not seen by the client as contradictions. If communication is not blocked – that is, if the client again feels fully *received* – then there is movement towards stage three. Having stated that one aspect of a client being received is free communication, Carl Rogers believed that a major factor in *impeding* interpersonal communication 'is our very natural tendency to judge, to evaluate, to approve or disapprove . . .'. This tendency becomes more sharp or 'heightened' when feelings and emotions are deeply involved:

> So the stronger our feelings the more likely it is that there will be no mutual element in the communication. There will just be two ideas, two feelings, two judgments, missing each other in psychological space . . . This tendency to react to any emotionally meaningful statement by forming an evaluation of it from our own point of view is, I repeat, the major barrier to interpersonal communication.
>
> (Rogers 1961: pp. 331–4)

Here, then, is a direct link between *contact* (or *relationship*, or *receiving*) and *empathic understanding*, for Rogers gives us his solution to this major communication barrier:

> Real communication occurs, and this evaluative tendency is avoided, when we listen with understanding.
>
> (Rogers 1961: pp. 331–4)

Although writing about group tensions at the time, there might nevertheless be a link with the client-centred therapist striving to receive clients fully and the place of supervision, personal therapy and peer networking, for Rogers stated that a *third party* can be of great help. If the third party is able to be non-evaluative and listen with understanding, like the supervisor clarifying the therapist's views and attitudes with an empathic acceptance, then the therapist will experience this acceptant understanding and be less prone to exaggeration, less defensive, less critical and more self-accepting. As a result the therapist is less inhibited about openly exploring how best to get as close as possible to being the therapist conditions.

The situational context is that clients arrive for therapy with an expectation that this wise healer will be of help. They feel distressed at what they perceive as their inability to cope with life. What have *clients* reported as being helpful in terms of the establishment of contact (or the beginning of a therapeutic relationship)? How might *clients* feel received? Rogers found that two words summarised this, namely *impersonal* and *secure*. How could a client-centred therapist be experienced as *impersonal*? Carl clarified as follows:

> This is obviously not intended to mean that the relationship was cold or disinterested. It appears to be the client's attempt to describe this unique experience in which the *person* of the counselor – the counselor as an evaluating, reacting person with needs of his own – is so clearly absent . . . the relationship is experienced as a one-way affair in a very unique sense. The whole relationship is composed of the self of the client, the counselor being depersonalized for purposes of therapy into being 'the client's other self'. It is this warm willingness on the part of the counselor to lay his own self temporarily aside, in order to enter into the experience of the client, which makes the relationship a completely unique one, unlike anything in the client's previous experience.
>
> (Rogers 1951: p. 208)

Clearly, in 1951 Carl took the view that therapist focus was entirely towards the client. While the construct of congruence developed over the years to include greater emphasis on therapist self-awareness too, Rogers never lost (as we shall see) his 'disciplined' attention to the client (usually to the extent of elimination of awareness of even an audience of hundreds).

Turning now to clients feeling safe and secure, Carl Rogers highlighted the importance of all three therapist conditions. With reference to *unconditional positive regard*, he maintained that a 'thoroughly consistent acceptance' was not at all linked with approval. Rather, it is an *absolute assurance* that the therapist will not evaluate, interpret, probe or foist personal opinion upon clients. As a consequence, clients feel less need to be defensive or to pretend anything in order to please their counsellor. Acceptance may be experienced by the client as supportive, yet it is not the therapist's intent to offer support. The therapist is alongside the client, not pushing from behind or towing – he or she is 'permitting' clients to choose any direction that they wish to take.

> The client does not feel that the therapist 'likes' him, in the usual sense of a biased and favourable judgment, and he is often not sure . . .

whether he likes the therapist . . . There is simply no evidence upon which such a judgment could be based. But that this is a secure experience, in which the self is deeply respected, that this is an experience in which there need be no fear of threat or attack – not even of the subtlest sort – of this the client gradually becomes sure.

(Rogers 1951: pp. 208–9)

However, Carl linked client feelings of security in the therapeutic relationship with *empathy*, too:

It is this which provides the ultimate in psychological safety, when added to the other two. If I say that I 'accept' you, but know nothing of you, this is a shallow acceptance indeed, and you realise that it may change if I actually come to know you. But if I understand you empathically, see you and what you are feeling and doing from your point of view, enter your private world and see it as it appears to you – and still accept you – then this is safety indeed. In this climate you can permit your real self to emerge, and to express itself in varied and novel formings as it relates to the world. This is a basic fostering of creativity.

(Rogers 1951: p. 358)

Opening channels

Most clients *want* to enter therapy, although they would of course prefer not to be in a position where seeking therapy becomes necessary. When clients do feel in need of help, they go for it. In the context of education and student-centred learning, Rogers wrote:

I think that you pick up a book or you go to a person because you want to learn something, which is quite different from teaching in the ordinary sense . . . teaching usually does mean the imparting of information, which is quite different from seeking what you want to learn . . . it is different from being visited by a seeker. . . . I can tell you it's a very different experience to be asked by a class to lecture on some given topic that they think I'm expert on, than it is to be pouring that same lecture on to them when they haven't asked for it. . . . I favor much more being visited by seekers than seeking to put something across.

(Rogers 1989: p. 195)

The phrase 'you can read me like a book' comes to mind. In client-centred therapy, the client picks up their own book, and the book is them. The therapist is alongside clients as they compose, decompose and recompose themselves. Clients may well have been 'squashed by the system' and expect their counsellor to be the equivalent of a teacher, delivering didactic lectures and filling them full of knowledge. However, the therapist knows that the natural curiosity driven by their actualising tendency can be reawakened.

Perhaps the beginning of therapy could be characterised by the counsellor inviting the client to begin reading the book that is the client:

> *I have found it enriching to open channels whereby others can communicate their feelings, their private perceptual worlds, to me*. Because understanding is rewarding, I would like to reduce the barriers between others and me, so that they can, if they wish, reveal themselves more fully.
>
> ([Rogers 1961: p. 19)

Carl was keen to meet people as human beings *in a process of becoming*, rather than seeing people as bound up in the past (his or the client's) or through having his own aims to impose upon or guide the client's future. Perhaps the most extensive guidance with regard to the client-centred therapist establishing *contact* in a *relationship* in which the client feels *received* is to be found in the section entitled 'How can I create a helping relationship?' in *On Becoming A Person* (Rogers 1961: pp. 50–57). In these extracts, Carl Rogers offers ten questions that he had asked of himself when reflecting upon how he could be of most therapeutic help:

> a. Can I *be* in some way which will be perceived by the other person as trustworthy, as dependable or consistent in some deep sense?
>
> (Rogers 1961: p. 50)

In part this could be to do with aspects such as keeping appointments, being on time and respecting confidentiality. However, to *act* as consistent when inner experience is not in harmony with outer expression would, in the longer term, be perceived as inconsistent and untrustworthy. Thus Carl preferred to think in terms of being 'dependably real' rather than being 'rigidly consistent' (although once again it can be stressed that if inner experiencing is to match outer expression, the client-centred therapist will have conditions four and five – unconditional positive regard and empathic understanding – deeply integrated within this congruence). In summary, Carl said that when he was a unified and integrated person:

> I can *be* whatever I deeply *am*.
>
> (Rogers 1961: p. 51)

> b. Can I be expressive enough as a person that what I am will be communicated unambiguously?
>
> (Rogers 1961: p. 51)

In other words, *if* the therapist is in denial of (or is distorting to awareness) inner experiencing – that is, if the therapist is incongruent – *then* the client will perceive mixed or contradictory messages. This could be linked with the notion of reciprocal empathy. If the therapist's words are somehow out of synchronisation with paraverbal or non-verbal communicating, the client will sense this at some level.

> One way of putting this which may seem strange to you is that if I can form a helping relationship to myself – if I can be sensitively aware of and acceptant towards my own feelings – then the likelihood is great that I can form a helping relationship toward another.
>
> (Rogers 1961: p. 51)

Rogers believed that it is tremendously difficult to transparently be that which one is, and one consequence is that the facilitation of the growth of others through the therapist means that the therapist must grow, too. While this may be enriching, it is often also painful. Thus one characteristic that could be deemed necessary for the client-centred therapist is *courage*, for the therapist might fear being changed or influenced, or becoming lost in the world of the other. Carl believed the experience of client-centred therapy to be so wonderful, and yet quoted the possibility of personal change *for the therapist* as being a reason why it is not more widespread.

> c. Can I let myself experience positive attitudes toward this other person – attitudes of warmth, caring, liking, interest, respect?
>
> ([Rogers 1961: p. 52)

If I allow myself to freely experience such positive feelings toward another, might they somehow trap me? Might demands be made of me or might I be disappointed as an outcome of trusting? Carl Rogers believed that fears such as these are one explanation for a tendency to be 'professionally aloof' and to work in diagnostic, 'objective' ways. Yet:

> It is a real achievement when we can learn, even in certain relation-ships or at certain times in those relationships, that it is safe to care, that it is safe to relate to the other as a person for whom we have positive feelings.
>
> (Rogers 1961: p. 52)

> d. Can I be strong enough as a person to be separate from the other?
>
> (Rogers 1961: p. 52)

Carl stated that the therapist needs to be able to embrace *self*-respect as well as respect for clients, owning (and if needs be, communicating) personal feelings and meanings yet being clear that they belong to the therapist and not to the client. Is the therapist personally 'together' enough to be able to enter the world of client feelings and meanings without identifying with them, without drowning in or becoming depressed or engulfed by them? The therapist needs to be strong enough to really receive a client's anger or rage without being 'destroyed' by it, or to accept a client's feeling of love without becoming entrapped by that love, or somehow enslaved by it. When therapists experience and own their *personal power*, then they can engage with clients at deeper levels due to not being afraid of losing themselves.

> e. Am I secure enough within myself to permit him his separateness?
>
> (Rogers 1961: pp. 52–3)

Clearly linking with the last question, Rogers really needed to know whether he could permit people the *freedom to be* – and if therapists believe that clients should follow their advice or be dependent upon them, then clearly this is not so. He quoted a small, unpublished study conducted by Richard Farson, in which it was discovered that:

> The less well adjusted and less competent counselor tends to induce conformity to himself, to have clients who model themselves after him.
>
> (Farson 1955)

However, better-adjusted (that is, more congruent) counsellors do not interfere with the freedom of clients to be themselves.

> f. Can I let myself enter fully into the world of his feelings and personal meanings and see these as he does?
>
> (Rogers 1961: p. 53)

Carl Rogers experienced an immersion in the world of clients so deep that all desire to evaluate or judge them dissipated. Sensitive and accurate empathic understanding is not just of the obvious, but also of implicit, dimly perceived and confused meanings, and on an ongoing basis.

> g. Can I receive him as he is? Can I communicate this attitude? Or can I only receive . . . conditionally?
>
> (Rogers 1961: p. 54)

Disapproval of some feelings, whether openly or silently, is not a hallmark of client-centred therapy. Ideally, the therapist is open to and accepting of each and every facet or aspect that the client chooses to be in therapy. Carl Rogers, if he experienced evaluative feelings, would explore where they might have come from:

> And when – afterward and sometimes too late – I try to discover why I have been unable to accept him in every respect, I usually discover that it is because I have been frightened or threatened in myself by some aspect of his feelings. If I am to be more helpful, then I must myself grow and accept myself in these respects.
>
> (Rogers 1961: p. 54)

We see once again, then, the value of personal growth activities for the therapist (or trainee therapist), the value of good, person-centred supervision, and the value of networking with like-minded peers.

> h. Can I act with sufficient sensitivity in the relationship that my behavior will not be perceived as a threat?
>
> (Rogers 1961, p. 54)

Rogers explained that this was not about being hypersensitive to clients, although physiological research evidence indicates that even communicating a word only slightly stronger in meaning than that which the client used does produce a threat reaction. Rather:

> It is simply due to the conviction based on experience that if I can free him as completely as possible from external threat, then he can begin to experience and to deal with the internal feelings and conflicts which he finds threatening within himself.
>
> (Rogers 1961: p. 54)

> i. Can I free him from the threat of external evaluation
>
> (Rogers 1961: p. 54)

Related to the previous question, this one is specifically to do with the punishment and reward systems that nearly all (if not all) people meet with from birth to death – *good* or *bad*, *pass* or *fail*, *worthy* or *shameful*, and so on. Rogers believed that such evaluations are not at all facilitative of personal growth and therefore play no part in a client-centred therapeutic relationship – and this includes positive judgements, for as we have already seen, the right to praise (to cite just one example) implies the right to chastise or withhold praise.

> So I have come to feel that the more I can keep a relationship free of judgment and evaluation, the more this will permit the other person to reach the point where he recognizes that the locus of evaluation, the center of responsibility, lies within himself.
>
> (Rogers 1961: p. 55)

The client-centred therapist strives to reach a state of congruence wherein there is no inner evaluation of clients, let alone communicated judgements. Ideally, there is not even any *desire* to be evaluative – the optimum desire is to authentically and respectfully understand.

> j. Can I meet this other individual as a person who is in a process of *becoming*, or will I be bound by his past and my past?
>
> (Rogers 1961: p. 55)

Any preconceived ideas or notions that the therapist holds about a client serve to limit what that client can be in the therapeutic relationship. Rogers quoted Martin Buber, who wrote about *confirming the other:*

> Confirming means . . . accepting the whole potentiality of the other . . . I can recognize in him, know in him, the person he has been . . . *created* to become . . . I can confirm him in myself, and then in him, in relation to this potentiality that . . . can now be developed, can evolve.
>
> (Rogers and Buber 1957)

As Carl stated, if the client is perceived as someone already bound by the past, already categorised and fixed in time, this is very, very different from receiving a person who is in a *process of becoming*.

In summary, Carl Rogers wrote of his exploration of personal characteristics relevant to entering or initiating therapeutic contact:

> . . . the optimal helping relationship is the kind of relationship created by a person who is psychologically mature. Or to put it another way, the degree to which I can create relationships which facilitate the growth of others as separate persons is a measure of the growth I have achieved in myself. In some respects this is a disturbing thought, but it is also a promising or challenging one. It would indicate that if I am interested in creating helping relationships, I have a fascinating lifetime job ahead of me, stretching and developing my potentialities in the direction of growth.
>
> (Rogers 1961: p. 56)

Contact – chocks away!

It is perhaps to be regretted that almost inevitably the six necessary and sufficient conditions for therapeutic personality change are represented as a vertical yet linear hierarchy – a *top-to-bottom* list with each aspect following the one above it.

Although there is a rationale for certain conditions preceding others (for instance, without relationship there is no process or outcomes, and unconditional positive regard and empathic understanding need to be perceived as authentic), when placing empathy back in the context of all six necessary and sufficient conditions, it will become evident that all six conditions interweave, very much relating to and being integrated with one another. A slightly less linear diagrammatic representation might be that within the context of two individuals who are in psychological contact.

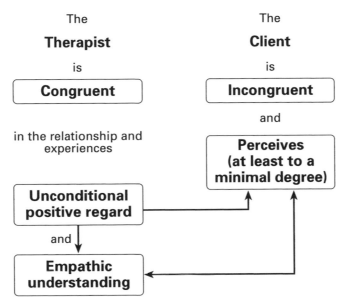

Figure 4.1 An illustration of the interface between counsellor and client in the process of client-centred therapy.

Thus the counsellor is congruent in this therapeutic relationship. Integral to this congruence is the experiencing of unconditional positive regard towards and empathic understanding of an incongruent client, who perceives this empathic understanding (at least to a minimal degree, and in part through a process of reciprocal empathy) and, via that understanding, the unconditional positive regard of the therapist.

Experience will be defined (in different ways) shortly. As Rogers wrote of *contact* meaning that the therapist and client 'each make a perceived or subceived difference in the experiential field of the other', it might be helpful to explore briefly the meaning of some of the terms used. We can begin with '*each makes a perceived . . .*'.

Carl Rogers (and others, such as Hebb and Riesen) believed that the 'impingement of stimuli' and the 'meaning given to the stimuli' are 'inseparable parts of a single experience' (that is, the instant we *experience* something we attach a *meaning* to it):

> perception is a hypothesis or prognosis for action which comes into
> being in awareness when stimuli impinge on the organism. When we
> perceive 'this is a triangle', 'that is a tree', 'this person is my mother', it

means that we are making a prediction that the objects from which the stimuli are received would, if checked in other ways, exhibit properties we have come to regard, from our past experience, as being characteristic of triangles, trees, mother.

(Rogers 1959: p. 199)

Thus Carl reckoned that the terms *perception* and *awareness* were more or less synonymous, although he tended to use the term *perception* in the more detailed sense that emphasised the stimulus (that which is being perceived), whereas *awareness* included the symbolisations and meanings associated with that which is being perceived, as well as memory traces and physiological changes connected to the perception.

With regard to the meaning of *'each makes a perceived or subceived . . .'*, Rogers drew on the work of McCleary and Lazarus, who stated that:

even when a subject is unable to report a visual discrimination he is still able to make a stimulus discrimination at some level below that required for conscious recognition.

(McCleary and Lazarus 1949: pp. 171–9)

Carl incorporated the process of discrimination without awareness, known as *subception*, into client-centred personality theory, stating that:

It is this capacity which, in our theory, permits the individual to discriminate an experience as threatening, without symbolization in awareness of this threat.

(Rogers 1959: p. 200)

One of the capacities of human beings that distinguishes us from many other animals is that we do have a significant capacity for awareness:

We have a rich capacity for symbolization, which I believe is synonymous with awareness. We have the capacity for a rich subjective living. We have the capacity for an organismic valuing process. That is, we are able to test objects and events in terms of our own experiencing and can gauge the values they have for enhancing us or for being destructive in our experience.

(Rogers 1989: p. 120)

Carl Rogers saw *symbolisation* and *awareness* (and indeed *consciousness*) as meaning essentially the same thing:

consciousness (or awareness) is the symbolization of some of our experience. Awareness is thus seen as the symbolic representation (not necessarily in verbal symbols) of some portion of our experience. This representation may have varying degrees of sharpness or vividness, from a dim awareness of something existing as ground, to a sharp awareness of something which is in focus as figure.

(Rogers 1959: p. 198)

When asked by Richard Evans whether his own view of consciousness paralleled the theory promulgated by Sigmund Freud, Carl Rogers replied:

> I feel that whether or not I totally agree or disagree (with Freud's concept of the unconscious) is, in itself, not very helpful. I'd rather point out the way I conceptualize . . .
>
> (Rogers and Evans 1975: p. 7)

And Carl did point out the way he thought of awareness when he said:

> I'd prefer to think of a range of phenomena: first, those in sharp focus in awareness right now – the height of consciousness; secondly, a range of material which could be called into consciousness, that you really know and can call into consciousness but you don't have in 'figure' right now – it is in the 'ground' or background; then, finally, some phenomena which are more and more dimly connected with awareness, to material that is really prevented from coming into even vague awareness because its coming into awareness would damage the person's concept of himself.
>
> (Rogers and Evans 1975: p. 6)

When there is an absence of psychological defensive processes, experience can be symbolised freely and is hence deemed to be *available to awareness*. In defining *accurate symbolisation*, Rogers wrote:

> The symbols which constitute our awareness do not necessarily match, or correspond to, the 'real' experience, or to 'reality'. Thus the psychotic is aware of (symbolizes) electrical impulses in his body which do not seem in actuality to exist. I glance up quickly and perceive a plane in the distance, but it turns out to be a gnat close to my eye. It seems important to distinguish between those awarenesses which, in common-sense terms, are real or accurate and those which are not.
>
> (Rogers 1959: p. 198)

Stating that the examples given above are, effectively, *perceptual hypotheses* that can be checked (for instance, by brushing at the speck it might become more clear that it was a gnat and not an aeroplane), and while acknowledging this link between symbolisation and perception, Rogers nevertheless was able to state that:

> when we speak of accurate symbolization in awareness, we mean that the hypotheses implicit in the awareness will be borne out if tested by acting on them.
>
> (Rogers 1959: p. 199)

The notion of two people being in contact will be examined again when we look at the sixth necessary and sufficient condition (the client *perceiving*, at least to a minimal degree, the fourth and fifth conditions, namely the *unconditional positive regard* of the therapist, and the *empathic understanding* of the therapist). For the moment, let us note a subtle yet significant difference. Person A (who is relatively incongruent) and person B (who is relatively congruent) are in contact. However, person A is unable or unwilling (for whatever reason) to *perceive* (even to a minimal degree) the authentic unconditional positive regard and empathic understanding of person B. In this scenario, it is far less likely that a therapeutic relationship will be a consequence of their contact.

As we relocate empathic understanding within all six necessary and sufficient conditions, let us also note, in summary, that integral to the nature of this *contact*, this *psychological contact*, this *relationship* between two people, is the commitment of the therapist to experience genuine respect and empathy for and of the client's world of feelings and personal meanings.

Empathy and client incongruence

One meaning of congruence and incongruence relates to geometric shapes. If two shapes can be exactly superimposed they are said to be *congruent*; if they do not exactly fit over one another and match, they are said to be *incongruent*.

Carl Rogers believed that the human organism has one, and only one, basic motivation – to maintain and enhance itself. This 'function' is 'undertaken' by the *actualising tendency*:

> This is the inherent tendency of the organism to develop all its capacities in ways which serve to maintain or enhance the organism.
> . . . It should also be noted that it is the organism as a whole, and only the organism as a whole, which exhibits this tendency . . . The self, for example, is an important construct in our theory, but the self does not 'do' anything. It is only one expression of the general tendency of the organism to behave in those ways which maintain and enhance itself.
> . . . It might also be mentioned that such concepts of motivation as are termed need-reduction, tension-reduction, drive-reduction, are included in this concept. It also includes, however, the growth motivations which appear to go beyond these terms: the seeking of pleasurable tensions, the tendency to be creative, the tendency to learn painfully to walk when crawling would meet the same needs more comfortably.
>
> (Rogers 1959: p. 196)

Carl Rogers and human nature

What it means to be a human being is of central significance in client-centred theory. When asked about where he would place himself in a continuum stretching from humanistic to psychoanalytic and behavioural, Rogers said of psychoanalytic thinking and practice:

> I think that it really is a damaging cultural influence, as it gets spread more widely in an oversimplified form. I think Freud's view that man is by nature evil and that the id is a terrible force which, if released, would play havoc, could lead to damage.
>
> (Rogers and Evans 1975: p. 71)

In response to where Rogers would locate himself, his view was that:

> there is a flavor of existentialism developing which is definitely positive in its view of man, and I certainly would have to place myself there.
>
> (Rogers and Evans 1975: p. 71)

Carl Rogers would not say, though, that he was a 'pure' existentialist because he thought that:

> some of the existentialists take the point of view that man really has no nature, but it seems to me that he has – I have taken the point of view that man belongs to a particular species. He has species characteristics. One of those, I think, being the fact that he is incurably social; I think he has a deep need for relationships. Then I think that simply because man is an organism he tends to be directional. He's moving in the direction of actualizing himself. So, for myself, I really feel man does have a describable nature.
>
> (Rogers 1989: p. 66)

Returning to the theme of psychoanalysis, Carl Rogers despised the view that at the 'heart' of human beings lie instincts and impulses that must be controlled, because his own experience led him to believe that:

> if we can release what is most basic in the individual, it will be constructive . . . when you get to what is deepest in the individual, that is the very aspect that can be most trusted to be constructive or to tend toward socialization or toward the development of better interpersonal relationships.
>
> (Rogers 1989: p. 58)

He stated many times that of course 'negative' feelings will be disclosed in psychotherapy, and that each and every one of us has the capacity for 'evil' behaviour. However:

> It is well to bear in mind that I also have a capacity to vomit, for example. Whether I, or anyone, will translate the impulses into behavior depends, it seems to me, on two elements: social conditioning and voluntary choice.
>
> (Rogers 1989: p. 254)

So the fact that we all have the *potential* to behave in antisocial ways does not have to mean that such ways are our base nature. On the contrary:

> I find in my experience no . . . innate tendency toward destructiveness, toward evil . . . if the elements making for growth are present, the actualizing tendency develops in positive ways. In the human these elements for growth are not only proper nutrition etc., but a climate of psychological attitudes.
>
> (Rogers 1989: pp. 253–4)

In terms of the 'nature versus nurture' debate, Rogers said that:

> I see members of the human species . . . as *essentially* constructive in their fundamental nature, but damaged by their experience.
>
> (Rogers 1981: p. 238)

Some years later:

> Viewed from the external, scientific, objective perspective, man is determined by genetic and cultural influences. I have also said that

in an entirely different dimension, such things as freedom and choice are extremely real.

(Rogers 1989: p. 132)

Growth is not a balancing act!

Very often, mental health is seen as trying to achieve *balance*. Yet the actualising tendency is *not* seeking balance – far from it. Balance in this sense would stem, thwart or stifle evolution and growth:

> I think the organism is definitely not trying to achieve *stasis*. I think that even organisms as low as the flatworm are always trying to look for more enriching stimuli, more complicated stimuli . . .
>
> (Rogers and Evans 1975: p. 3)

Richard Evans queried whether Carl meant that actually at some deep level human beings seek rather than strive to avoid tensions. Rogers replied:

> Always. I think there is plenty of evidence now that this is true. In man, we call it curiosity.
>
> (Rogers and Evans 1975: p. 3)

It was fundamental to Carl's belief system that the central motivating aspect of every person is a tendency toward *maintaining* the entire organism and, whenever possible, *enhancing* the organism, too – and given the right conditions, growth will be in a constructive not a destructive direction:

> I've said sometimes that my generally optimistic point of view could be glandular; I don't know. One facet of this is certainly true – there's no doubt that my interest in growth is positive.
>
> (Rogers and Evans 1975: p. 71)

The self

As a human being grows, a *portion* of the whole organism (and both Carl Rogers and Sigmund Freud were in agreement that this was a relatively small fraction of the whole) becomes *differentiated* as the *self*. The actualising tendency now 'gets to work' on the self, or self-concept – so the *self* now strives to maintain and enhance itself, often at the expense of the organism as a whole.

> Following the development of the self-structure, this general tendency toward actualization expresses itself also in the actualization of that portion of the experience of the organism which is symbolized in the self. If the self and the total experience of the organism are relatively congruent, then the actualizing tendency remains relatively unified. If self and experience are incongruent, then the general tendency to actualize the organism may work at cross purposes with the subsystem of that motive, the tendency to actualize the self.
>
> (Rogers 1959: pp. 196–7)

I confess to 'having an agenda' here, for it alarms me that so many practitioners who undertook client-centred training seem unaware of the fundamental distinction between the actualising tendency and self-actualising. Imagine, if you will, the self being akin to a small yolk in a large egg. The egg is the entire organism, and the yolk is but a differentiated part. The actualising tendency is a motivational force for the whole egg, self-actualising for the yolk only (examples to illustrate this follow soon).

Going back to geometric shapes, incongruence means that there is a mismatch between one shape, the entire human organism, and the other shape, that portion of the whole organism that has become differentiated as the self. Another way of illustrating this might be to see the two shapes as *experience* and our *perception* of that experience. If our perception of our experiencing is at odds with organismic experiencing (that is, experiencing by the total organism and not just the differentiated part of it known as the self), we are in a state of incongruence.

When exploring and defining *experience*, Carl Rogers differentiated between *experience as a noun*, *experience as a verb,* and *experiencing a feeling. Experience* as a *noun* means:

> all that is going on within the envelope of the organism at any given moment which is potentially available to awareness. It includes events of which the individual is unaware, as well as all the phenomena which are in consciousness. Thus it includes the psychological aspects of hunger, even though the individual may be so fascinated by his work or play that he is completely unaware of the hunger; it includes the impact of sights and sounds and smells on the organism, even though these are not in the focus of attention. It includes the influence of memory and past experience, as these are active in the moment, in restricting or broadening the meaning given to various stimuli. It also includes all that is present in immediate awareness or consciousness. It does not include such events as neuron discharges or changes in blood sugar, because these are not directly available to awareness. It is thus a psychological, not a physiological definition.
>
> (Rogers 1959: p. 197)

Stating that in the past he had used phrases for experience (as a *noun*) like *organismic experiencing* and *sensory and visceral experiences* (the viscera are the large cavities and organs in the body), Rogers stated that other synonyms included *experiential field* or *phenomenal field*. It was the *'totality'* of the concept of experiencing that he wished to convey – that experience means more than simply *conscious* experiencing. He also clarified that experience in this sense refers to *here and now* experience rather than an accumulation of historic experiencing. Thus as a *noun*, experience refers to *an* experience.

Experience as a *verb* means simply that sensory or physiological events happening in the moment are received (or impact upon) the organism. Again Rogers clarified:

> Often this process term is used in the phrase 'to experience in awareness', which means to symbolize in some accurate form at the conscious level the above sensory or visceral events. Since there are varying degrees of completeness in symbolization, the phrase is often 'to experience more fully in awareness', thus indicating that it is the

extension of this process toward more complete and accurate symbolization to which reference is being made.

(Rogers 1959: pp. 197–8)

In other words, experiencing may be denied symbolisation in awareness altogether. However, if experiencing is granted conscious recognition yet only in a distorted fashion, there are clearly *degrees* of distortion. Another way of putting this is to consider that rather than an individual either being totally aware or totally unaware, there is a continuum between these two extremes, different degrees of conscious awareness (again, examples follow shortly). Finally, there is Carl's definition of *experiencing a feeling*:

> This is a term which has been heavily used in writings on client-centered therapy and theory. It denotes an emotionally tinged experience, together with its personal meaning. Thus it includes the emotion but also the cognitive content of the meaning of that emotion in its experiential context. It thus refers to the unity of emotion and cognition as they are experienced inseparably in the moment. It is perhaps best thought of as a brief theme of experience, carrying with it the emotional coloring and the perceived meaning to the individual. Examples would include 'I feel angry at myself', 'I feel ashamed of my desires when I am with her', 'For the first time, right now, I feel that you like me'. This last is an example of another phenomenon which is relevant to our theory, and which has been called *experiencing a feeling fully*, in the immediate present.
>
> (Rogers 1959: p. 198)

Where does all of this *experiencing* take place, especially experiencing that the individual is aware or conscious of (experience that is symbolised)? Originally, Carl Rogers was not particularly interested in a concept as 'woolly' or meaningless or ambiguous as the *self*. However, as client after client after client kept referring to being aware of experience and meanings in this thing they called a 'self', he eventually thought that to honour them he had best acknowledge it in some way:

> I think I'll talk a little bit about the concept of *self*, because that's a very central aspect of my theory. It seemed to me that self could be defined in many, many different ways, and many of them acceptable, but what definition would be most profitable for consideration in research and in thinking? It seemed to me that the most useful definition was that the self was composed of the perceptions that the individual had of the 'I' or the 'me', the perceptions of the self in relationship to others and to the environment, and the values that were attached to those perceptions. So it was a definition which was all based on perceptions by the individual, and those could be determined – could conceivably be measured, though for a long time I didn't see how they could be measured.
>
> (Rogers and Russell 2002: pp. 247–8)

Rogers also said that although social conditioning was undoubtedly a factor in the building of a self, the self was more fluid than a fixed, built construction. He came to this conclusion in part through his experiencing of clients and in part through being a client himself. On being a client, Rogers said:

> I could go away from an interview feeling, 'I really am a worthwhile person; I do like myself; I'm okay'; but then some little incident could flip that feeling over while it was still tentative and newborn, and I would feel, 'I'm really no good; I'm a fraud; I'm not much good'. It was quite fascinating to me – not very comfortable, but fascinating – to look back a little bit and see how, in a moment of time, that whole gestalt could shift. And that happened in my clients, too. Very definitely.
>
> (Rogers and Russell 2002: pp. 247–8)

Unlike many other psychologists, however, Rogers opted not to tread the path of trying to predict or determine physical brain structures (for instance, Freud believed that it was only a matter of time before the physical locations of the id, ego and superego would be found in the human brain), or to relate human psychological development to specific phases or ages (such as Freud's psychosexual theories or Piaget's 'life-span' stages). Nor was Rogers overly engaged with colleagues and students working on the notion of different selves, and configurations of self:

> It should be mentioned that some students of mine stressed too much, perhaps, the notion of different selves, although I don't quarrel with that idea particularly. This idea hasn't had as much meaning for me as the notion of a more complex unified picture of self which keeps expanding to include new experiences, even though some experiences may be shut out as being too threatening. It seems that each of us endeavors to preserve the concept or picture that he has of himself and that a sharp change in that picture is quite threatening. Any change destroys some of the security that we feel we need.
>
> (Rogers and Evans 1975: p. 17)

It might be that evidence from neuroscience comes close to corroborating Carl's notion of a shifting or fluid self. Professor Susan Greenfield wrote:

> There is no 'centre' for consciousness – after all, that would mean we have a complete mini-brain in a brain.
>
> (Greenfield 2000: p. 176)

Likewise, there is no single centre (or 'mini-brain') for unconsciousness, or for being non-conscious. With regard to a sense of self, Greenfield thinks in terms of what she calls a *transient self* – in other words, the kind of fluid self that Carl Rogers wrote and spoke of. According to Greenfield:

> At some global level of organization, then, beyond a single cell and beyond any rigidly demarcated edifice in the brain, we might imagine an assembly of neurons so large that it does not respect any particular anatomical boundary, and at the same time one that is fast and evanescent enough to permit the formation of another, different and dominant assembly a moment later.
>
> (Greenfield 2000: p. 176)

It could just be, then, that a personal construct (George Kelly) or a configuration of self (Rogers, then Mearns) is analogous to an assembly of neurons that at any given time can become dominant. At the same time, the sense of self might be transient, a constant and complex flow within the brain. Perhaps the notion of rigidity of constructs could be linked with the notion that:

> Connectivity between and among assemblies of neurons creates in advance a dominant assembly that prevents competing assemblies from forming.
>
> (Greenfield 2000: p. 181)

Incidentally, such neuroscientific thinking might also be linked with Carl's notion of the client-centred therapist being disciplined and the idea of the therapist being modulated for receiving clients empathically. Of modulation:

> We know that neurotransmitters can say more than just yes or no – they can 'modulate' a target cell, not giving a signal directly but, more subtly, putting the cell on red alert.
>
> (Greenfield 2000: p. 100)

Furthermore, it may even be possible that unless cells have a modulated status, they may not react at all, and:

> when we speak of modulation, it is always with reference to two events occurring one after the other within a limited timeframe.
>
> (Greenfield 2000: p. 100)

Might this relate to a client disclosing, followed by the therapist's perception of that disclosure, followed by the therapist's attempt to understand that disclosure, followed by the therapist communicating the outcome of this process?

Anyway, Carl Rogers thought very much in terms of an *ongoing process of becoming a person*, yet this did not prevent the research and development of a client-centred theory of personality.

Having explored the notion of a self, we can now turn to definitions of *self-experience, self, concept of self* and *self-structure*. *Self-experience* was considered to be the raw data out of which the concept of self is formed and organised. In defining *self, concept of self* and *self-structure*, Rogers wrote:

> These terms refer to the organized, consistent, conceptual gestalt composed of perceptions of the characteristics of the 'I' or 'me' and the perceptions of the relationships of the 'I' or 'me' to others and to various aspects of life, together with the values attached to these perceptions. It is a gestalt which is available to awareness though not necessarily in awareness. It is a fluid and changing gestalt, a process, but at any given moment it is a specific entity which is at least partially definable in operational terms . . . The term self or self-concept is more likely to be used when we are talking of the person's view of himself, self-structure when we are looking at this gestalt from an external frame of reference.
>
> (Rogers 1959: p. 200)

Carl Rogers also wrote of the *ideal self*, which denotes the self-concept that is valued most highly and that a person would most like to have.

So the human infant is born with an *actualising tendency*, a motivational force akin to the inherent potential of an acorn to become a flourishing oak tree (and an acorn *does* strive to become a flourishing oak tree – and *given the right conditions*, it *will* become all that it can be). This actualising tendency is constructive and tends towards growth and complexity. As Carl wrote, we are less likely to actualise our potential for migraine than our potential for tasty food. Thus the actualising tendency, striving to maintain and enhance the whole human organism, is positive and trustworthy in nature. However, a portion of the whole organism becomes differentiated, *through interaction with the environment*, as the self. This led Rogers and his colleagues to think about how this self might develop and be constructed, and a key emergent idea was the notion of a *regard complex*:

> The regard complex is a construct defined by Standal as all those self-experiences, together with their interrelationships, which the individual discriminates as being related to the positive regard of a particular social other.
>
> This construct is intended to emphasize the gestalt nature of transactions involving positive or negative regard, and their potency. Thus, for example, if a parent shows positive regard to a child in relationship to a specific behavior, this tends to strengthen the whole pattern of positive regard which has previously been experienced as coming from that parent. Likewise specific negative regard from this parent tends to weaken the whole configuration of positive regard.
>
> (Rogers 1959: pp. 208–9)

As the self is becoming differentiated from the organism as a whole, certain human needs become evident. The *need for positive regard* is a need to feel liked, loved, valued, appreciated, and considered as being of worth, especially by 'significant others' (such as close family or carers initially, and later by friends, teachers, and so on). First we shall consider the meaning of *positive regard*, then the *need for positive regard*:

> In general, positive regard is defined as including such attitudes as warmth, liking, respect, sympathy, acceptance. To perceive oneself as receiving positive regard is to experience oneself as making a positive difference in the experiential field of another.
>
> It is postulated by Standal that a basic need for positive regard, as defined above, is a secondary or learned need, commonly developed in early infancy. Some writers have looked upon the infant's need for love and affection as an inherent or instinctive need. Standal is probably on safer ground in regarding it as a learned need. By terming it the need for positive regard, he has, it is believed, selected out the significant psychological variable from the broader terms usually used.
>
> (Rogers 1959: p. 208)

Client-centred theorists believe that most human relationships are *conditional*. Conditional relationships involve values and judgements, as in *'I like you when . . .'*,

'I love you when . . .', *'I value you most when . . .'*, *'I really appreciate it when you . . .'*, and so on. The reverse, of course, is also true: *'I don't like it when you . . .'*, *'I hate it when you . . .'*, and so on. Now due to the desire (or need) for positive regard from significant others, the young human being quickly learns that it is usually best to adapt to the values and judgements of others, to behave in ways that are more likely to result in positive, not negative, regard – better to be loved than to be hated . . .

Another development is the *need for positive self-regard*:

> This term is used to denote a positive regard satisfaction which has become associated with a particular self-experience or a group of self-experiences, in which this satisfaction is independent of positive regard transactions with social others. Though it appears that positive regard must first be experienced from others, this results in a positive attitude toward self which is no longer directly dependent on the attitudes of others. The individual, in effect, becomes his own significant social other.
>
> (Rogers 1959: p. 209)

In other words, we value *ourselves* via the values that we have taken on (or 'introjected') from *others*. Thus the young girl might learn that *caring* for dolls and *pleasing others* is very acceptable to a parent, as is *being seen but not heard*. Due to this child's need to feel liked and loved by and acceptable to that parent, she mostly behaves in a way that elicits responses from the parent that feel internally pleasurable, rather than risking the likelihood of unpleasant internal feelings that are a consequence of parental disapproval. However, if 'being seen but not heard' is internalised (or introjected) by this young girl as a 'good' value – albeit based on the evaluative attitudes of a parent and not on her own, direct, organismic valuing – then she may come to value herself for being seen and not heard even when the relevant parent is not around. Consider another example. How often do we (in the UK, at least) come across the perception that men are less adept than women at expressing their feelings? The answer is very often. Does the aphorism 'big boys don't cry' also 'ring a bell'? Suppose that a young boy injures himself while playing, and hears a significant adult remark 'big boys don't cry'. Needing to be valued, the boy therefore doesn't weep. This value may then be introjected *as if* it were direct organismic valuing, so that the next time he suffers an injury while playing, the boy feels good about himself for holding back his tears even though there are no significant adults around. He thus satisfies his need for positive *self*-regard. The *organismic* experience (based on an *internal locus of evaluation*) of the young girl is that sometimes it feels good to smash up a doll and scream. The *organismic* valuing (again based on an *internal locus of evaluation*) of the boy is that sometimes it feels good to sob and get cuddled. In both instances, however, their *perceptions* of their experiencing come to include *introjected values*, and organismic valuing is therefore denied or distorted. The individual still experiences a positive feeling about *self* through feeling that they have lived up to their introjected values – values that have effectively become *independent* of significant others (in effect, the individual has become his or her own significant other).

Two further dimensions complete the gestalt of an individual's regard complex:

Need for self-regard. It is postulated that a need for positive self-regard is a secondary or learned need, related to the satisfaction of the need for positive regard by others.

Unconditional self-regard. When the individual perceives himself in such a way that no self-experience can be discriminated as more or less worthy of positive regard than any other, then he is experiencing unconditional positive self-regard.

(Rogers 1959: p. 209)

Other definitions relevant to the preceding scenario are of *locus of evaluation* and *organismic valuing*. We shall begin with *locus of evaluation*:

This term is used to indicate the source of evidence as to values. Thus an internal locus of evaluation, within the individual himself, means that he is the center of the valuing process, the evidence being supplied by his own senses. When the locus of evaluation resides in others, their judgment as to the value of an object or experience becomes the criterion of value for the individual.

(Rogers 1959: p. 210)

Organismic valuing refers to:

an ongoing process in which values are never fixed or rigid, but experiences are being accurately symbolized and continually and freshly valued in terms of the satisfactions organismically experienced; the organism experiences satisfaction in those stimuli or behaviors which maintain and enhance the organism and the self, both in the immediate present and in the long range. The actualizing tendency is thus the criterion. The simplest example is the infant who at one moment values food, and when satiated, is disgusted with it; at one moment values stimulation, and soon after, values only rest; who finds satisfying that diet which in the long run most enhances his development.

(Rogers 1959: p. 210)

We have learned, though, that *self-actualising* (maintaining and enhancing the *self-concept*, formulated around our *conditions of worth*) – despite feeling good within the self – can nevertheless serve to *thwart* the *actualising tendency* (based on *organismic valuing processes*). In such instances, the individual is in a state of *incongruence*:

It seems to me that the infant is not estranged from himself. To me it seems that the infant is a whole and integrated organism, gradually individual, and that the estrangement that occurs is one that he learns – that in order to preserve the love of others, parents usually, he takes into himself as something he has experienced for himself, the judgments of his parents: just like the small boy who has been rebuked for pulling his sister's hair goes around saying, 'bad boy, bad boy'. Meanwhile, he is pulling her hair again. In other words, he has introjected the notion that he is bad, where actually he is enjoying the experience, and it is this estrangement between what he is experiencing and the concepts he links up with what he is experiencing that seems to me to constitute the basic estrangement.

(Rogers 1989: p. 70)

All of our introjected values, derived from our regard complex, combine to form what Carl Rogers and his colleagues termed our *conditions of worth*. Conditions of worth are the values and attitudes that have become part of the way in which we see ourselves – our self-concepts. If self-actualising thwarts the actualising tendency, this may generate tension or anxiety that, if it is admitted to conscious awareness or comes close to being admitted to conscious awareness, may result in the individual seeking counselling.

Rogers credited Stanley Standal with developing the idea of *conditions of worth*. Just as Standal coined the term *unconditional positive regard* as an alternative to acceptance, so he replaced 'introjected values' with conditions of worth:

> The self-structure is characterized by a condition of worth when a self-experience or set of related self-experiences is either avoided or sought solely because the individual discriminates it as being less or more worthy of self-regard . . . A condition of worth arises when the positive regard of a significant other is conditional, when the individual feels that in some respects he is prized and in others not. Gradually this same attitude is assimilated into his own self-regard complex, and he values an experience positively or negatively solely because of these conditions of worth which he has taken over from others, not because the experience enhances or fails to enhance his organism.
>
> (Rogers 1959: p. 209)

If *unconditional* positive regard is experienced, then valuing may be based on an internal locus of evaluation (that is, on whether or not the experiencing maintains or enhances the organism). However, if valuing is based on values 'adopted' from others, then there is no reference to organismic valuing. This in turn links with symbolisation, for if I value an experience (as positive or negative) as if it were referenced to my own actualising tendency when in fact my referencing is 'second hand' (that is, to values introjected from others), then my symbolisation is inaccurate. In other words:

> An experience may be perceived as organismically satisfying, when in fact this is not true. Thus a condition of worth, because it disturbs the valuing process, prevents the individual from functioning freely and with maximum effectiveness.
>
> (Rogers 1959: p. 210)

When there is a discrepancy between *organismic experiencing* and the *self as perceived*, the individual is in a state of *incongruence* – that is, the physiological organism and conscious awareness are not coexisting in harmony:

> When such a discrepancy exists, the state is one of incongruence between self and experience. This state is one of tension and internal confusion, since in some respects the individual's behavior will be regulated by the actualizing tendency, and in other respects by the

self-actualizing tendency, thus producing discordant or incomprehensible behaviors.

(Rogers 1959: p. 203)

Note again the crucial distinction between the actualising tendency and self-actualising.

When the individual is incongruent, the term *vulnerability* is used to describe the potentialities of this state, and when individuals are not consciously aware of this incongruence, they are:

> potentially vulnerable to anxiety, threat, and disorganization. If a significant new experience demonstrates the discrepancy so clearly that it must be consciously perceived, then the individual will be threatened, and his concept of self disorganized by this contradictory and unassimilable experience.
>
> (Rogers 1959: p. 204)

A consequence of experiencing anxiety because the concept of self and actual experience are incongruent is the *defensive* psychological processes of *denial* or *distortion*. *Self*-actualising demands that we maintain and enhance the way that we see ourselves – the tendency is to both maintain and enhance the *self-concept*. So the concept of self must be *defended* against anything that is potentially threatening to it.

Carl Rogers defined *anxiety* as:

> a state of uneasiness or tension whose cause is unknown. From an external frame of reference, anxiety is a state in which the incongruence between the concept of self and the total experience of the individual is approaching symbolization in awareness. When experience is *obviously* discrepant from the self-concept, a defensive response to threat becomes increasingly difficult. Anxiety is the response of the organism to the 'subception' that such discrepancy may enter awareness, thus forcing a change in the self-concept.
>
> (Rogers 1959: p. 204)

Threat, on the other hand, occurs when there is an experiencing (a perception or anticipation) of incongruence:

> If we think of the structure of the self as being a symbolic elaboration of a portion of the private experiential world of the organism, we may realize that when much of this private world is denied symbolization, certain basic tensions result. We find, then, that there is a very real discrepancy between the experiencing organism as it exists, and the concept of self which exerts such a governing influence on behavior. This self is now very inadequately representative of the experience of the organism. Conscious control becomes more difficult as the organism strives to satisfy needs which are not consciously admitted, and to react to experiences which are denied by the conscious self. Tension then exists, and if the individual becomes to any degree aware of this tension or discrepancy, he feels anxious, feels that he is not united or integrated, that he is unsure of his direction.
>
> (Rogers 1951: pp. 510–11)

Thus *self-actualising* demands that the concept of self be *defended* against any perceived threats to it:

> Defense is the behavioral response of the organism to threat, the goal of which is the maintenance of the current structure of the self. This goal is achieved by the perceptual distortion of the experience in awareness, in such a way as to reduce the incongruity between the experience and the structure of the self, or by the denial to awareness of an experience, thus denying any threat to the self. Defensiveness is the term denoting a state in which the behaviors are of the sort described.
>
> (Rogers 1959: pp. 204–5)

The two main processes of defence in client-centred personality theory are those in which experiences may be *denied* accurate symbolisation in our awareness, and those in which they may be granted only a *distorted* symbolisation. (Incidentally, another way of understanding what is meant by 'symbolisation' might be to consider the notion of *pre*-symbolisation! You know that feeling you get when there is something niggling away yet you 'can't quite put your finger on it'? Well, when it feels evasive or illusive it has yet to be accurately symbolised in awareness – and when you *have* put your finger on it, you have symbolised it in your awareness.)

If an experience is threatening to the concept of self, it may be *denied* entry to conscious awareness as a way of defending the self, and the more important our self-beliefs are to us, the more rigid and tough will be our defences against any threats to them. Another defensive option that the self has is to permit experiencing into awareness, yet only in a *distorted* fashion. In other words, the organism cannot deny crying because the tears felt real, wet and salty enough. However, they were shed because someone nearby must have been peeling onions or some grit had found its way into the eye socket, or the person has a cold coming on – they were not due to any distress:

> I gradually came to realize that one denies awareness – which seemed to me a better term than repression – to those feelings which do not fit into the current concept of self. If I see myself in such and such a fashion – as poor at mathematics, for example – and I find myself solving a mathematical problem, well, that was luck, or chance. That wasn't me. I have to deny or distort in my awareness things that don't fit my concept of self . . . The self-concept becomes a sort of screening device, determining what is admitted to awareness.
>
> (Rogers and Russell 2002: p. 248)

There is also a notion of psychological defence known as *pre-perception* or *'subception'*, which is the non-conscious denial of experience to awareness. Experimental support for this notion arose from the flashing of subliminal images in front of volunteers who were wired up to various physiological measuring devices. If, for instance, an individual was scared of spiders, there was a measurable physiological reaction to the image of a spider even though the picture was flashed on screen more quickly than the eye could see. In other words, an experience that is *subceived* (or pre-perceived) as threatening to the self may be denied to awareness without the individual even being consciously aware of threats to their self-concept or their defensive psychological processes.

> We feel internally threatened when we recognize that something we're experiencing – anger or hatred or positive feelings, something that we're denying to awareness – is getting close to awareness. And there, the concept of *subception*, which was to my mind pretty well confirmed by our research, was a very valuable concept. Without any conscious awareness, one can *subceive* certain things in the environment. That's true physically; it's also true psychologically: one could subceive the fact that 'I am unhappy in my marriage' or that 'I don't like my mother', or whatever. And it's then that a person feels anxious because the experience one is having is almost breaking through into the concept of self.
>
> (Rogers and Russell 2002: p. 252)

Carl Rogers summarised much of this when he wrote:

> It may help to clarify this basic concept of incongruence if we recognize that several of the terms we are defining are simply different vantage points for viewing this phenomenon. If an individual is in a state of incongruence between self and experience and we are looking at him from an external point of view, we see him as vulnerable (if he is unaware of the discrepancy) or threatened (if he has some awareness of it). If we are viewing him from a social point of view, then this incongruence is psychological maladjustment. If the individual is viewing himself, he may even see himself as adjusted (if he has no awareness of the discrepancy) or anxious (if he dimly subceives it) or threatened or disorganized (if the discrepancy has forced itself upon his awareness).
>
> (Rogers 1959: p. 204)

Another fundamental aspect of client-centred personality theory is that *experiencing remains available to conscious awareness given the right conditions*, and especially given the relative absence of threat to the self-concept (there therefore being less need to defend the self). When writing of this, Carl Rogers mentioned that he was aware of the pressure of the seat on his buttocks. Say this in a group and suddenly *everyone* will become aware of the pressure on their own buttocks (and indeed you may well have become aware of the pressure on your own buttocks as you read this last sentence). Thus the organismic experience of sitting is *being experienced* and although not necessarily in awareness, it is nevertheless *available to awareness*.

There are those who argue that because Carl was so passionate about client-centred therapy (and the person-centred approach) as *a way of being*, learning and understanding theory is therefore not really necessary at all – that if the individual can *be* (or strive with some success to be) the therapist (or core) conditions, client-centred *theory* is almost an irrelevance. Such arguments are sometimes supported by comments written by Rogers early in his career, such as the following:

> I might at [this] point say a word or two about theory. I remember very clearly telling my students at Ohio State, 'I have no use for psycho-

logical theory'. Later I came to change that point of view, but I realized eventually that I had no use for psychological theory which was not my own . . . So I really thought that I had no use for psychological theory. Later I came to realize its significance.

(Rogers and Russell 2002: pp. 246–7)

Please note the phrase 'Later I came to change that point of view . . . Later I came to realize its significance'! These very same people who dismiss theory often also claim that they engage as *whole* persons with other *whole* persons, yet somehow see no contradiction whatsoever when omitting a significant part (that is, *theory*) from the *whole* that is client-centred therapy. Yet how do we *understand* or *empathise* with another person? It is very clear indeed that to Rogers both unconditional positive regard and empathic understanding need to be integrated and deeply held beliefs and values residing within the client-centred therapist, and that congruence means the absence of denial or distortion with regard to any significant aspects of self-experience. To perceive another person means to filter the 'stimuli' from them through our own feelings and meanings – and if a client-centred way of being is deeply integrated, those sensations and meanings will include cognitive as well as emotional and physiological dimensions. So many students of Rogers have made comments along the lines of 'I was searching for something, and when I picked up and read a book by Carl Rogers, it *spoke* to me. Here, at last, was someone who seemed to understand what it *feels* like to be me, what it *means* to be me'. Perhaps this kind of comment is only to be expected. After all, *we taught Carl Rogers all that he knew*. Client-centred theory grew out of listening to and striving to understand hundreds upon hundreds of clients, and in turn that theory can now inform how we continue to strive to understand one another.

Lest there be any misunderstanding, it is not the contention here that client-centred therapists are fully aware of theory when meeting with clients. The client-centred therapist experiencing 'Aha! – that sounded like an introjected value' or 'Hmm, this subceived threat to the concept of self is clearly distorted in symbolisation' would hardly be paying extraordinary attention to the client's world of feelings and personal meanings! Rather:

> There needs to be an intellectual context into which what he or she is doing fits.

(Rogers and Russell 2002: p. 272)

This is not to say, though, that such issues as introjection, subception, threat, self-concept, distortion, symbolisation and so on are altogether irrelevant, for it might be useful and appropriate to make links with theory in supervision or in discussion with colleagues. However, our knowledge (hopefully or ideally derived from experiential, person-centred learning) is unarguably a significant part of our whole being, guiding and informing us even if in non-conscious ways, or in ways that are less prominent than or secondary to our integrated embodiment of the therapist conditions.

In illustration of the relevance of client-centred *theory* to client-centred *practice*, person-centred *supervision* has as a primary focus the extent to which the therapist is able to *be* the core conditions in the counselling relationship. It might, for example, be helpful for the therapist to reflect upon the nature of a particular client's defensive processes, to explore whether significant client experiencing might be denied to or distorted in awareness. After all, clients are unlikely even to seek therapy unless they are incongruent. Thus person-centred supervision is an opportunity to give more prominence to client-centred theory than would be the case in therapy itself, and through doing so the therapist might just become more open to experiencing that client more fully. The therapist is usually (and ideally) so closely attuned to *client* experiencing *in the moment*, and *moment by moment*, that self-reflecting activities and supervision are a wonderful opportunity to explore more deeply the therapist's own experiencing and underpinning values, attitudes, beliefs and practices, with a view to maintaining and enhancing the quality of presence of the therapist conditions. An understanding of incongruence is invaluable and indispensable in terms of maximising therapist congruence, more deeply experiencing unconditional positive regard, and empathically understanding the nature of the whole person that is the client – and the therapist.

Empathy and therapist congruence

> The order in which these therapeutic conditions are described has some significance because they are logically intertwined. In the first place, the therapist must achieve a strong, accurate empathy. But such deep sensitivity to the moment-to-moment 'being' of another person requires that the therapist first accept, and to some degree prize, the other person. That is to say, a sufficiently strong empathy can hardly exist without a considerable degree of unconditional positive regard. However, since neither of these conditions can possibly be meaningful in the relationship unless they are real, the therapist must be, both in these respects and in others, integrated and genuine within the therapeutic encounter. Therefore, it seems to me that genuineness or congruence is the most basic of the three conditions.
>
> (Rogers 1989: p. 11)

The third, fourth and fifth necessary and sufficient conditions (congruence, unconditional positive regard and empathic understanding) became known as the *core conditions* of client-centred therapy; others have called these the *therapist conditions*. At various times in his career Carl Rogers was asked which of these core conditions was the most important. Here is one version he gave about *congruence* in 1975:

> In the ordinary interactions of life – between marital and sex partners, between teacher and student, employer and employee, or between colleagues, it is probable that congruence is the most important element. Such genuineness involves letting the other person know 'where you are' emotionally. It may involve confrontation, and the personally owned and straightforward expression of both negative and positive feelings. Thus congruence is a basis for living together in a climate of realness.
>
> (Rogers 1975: p. 9)

One definition of congruence is that the two shapes of the *whole organism* and that differentiated part of it which is known as the *self* correspond (the actualising tendency and self-actualising being as one) – that is, when our experiencing and our perceptions of that experiencing are in harmony. Congruence, then, is described as a *state* – the 'state' is that either the two shapes do overlap exactly, or they do not.

Some writers have defined this 'accurate overlap' of organism and self, or experience and the perceiving of that experience, as meaning that the congruent therapist's 'outward' communications are consistent with their inner sensations and meanings. Although this makes sense and is to some extent true, it is also only part of the whole picture. For instance, if therapist A's inner feeling is a fatigued disinterest linked with personal meanings (including values) about boredom, and the outward expression of this is to say 'I am bored', then therapist A could be deemed to be being congruent. However, it would also be true to say that the congruence of therapist A does not evidence a great deal of unconditional positive regard or empathic understanding as deeply embedded and integrated personal characteristics. For therapist A to voice boredom without having accessed (that is, in *denial* of) client-centred attitudes, beliefs and values could be deemed to be more representative of a state of *incongruence*. Now therapist B steps in and does not communicate any feeling of boredom because to do so would hardly be client-centred. Is therapist B congruent? Not necessarily! If the feelings and meanings of fatigued disinterest through boredom are denied access to awareness (and hence cannot be voiced) because they would be too threatening to therapist B's self-concept (therapist B's self-concept is of being a *good* client-centred practitioner), then therapist B is also incongruent. Therapist C has an awareness of feeling tired and losing interest, yet also accesses a deeply held client-centred way of being, and will make a choice about whether or not to voice boredom depending upon the outcome of processing *both* significant experiences and their meanings. Again, there are those who seem to see congruence as meaning 'harmony' because the two shapes of organism and self correspond. In fact, *being congruent might mean embracing conflict* – conflict that is usually avoided through psychological processes of denial and distortion. Gradational rather than 'black-and-white' thinking (pertinent to *extensionality*, which as we shall see shortly was considered by Rogers to be one of the five ingredients of congruence) might mean *being many things simultaneously*:

> It is not simple to achieve such reality. Being real involves the difficult task of being acquainted with the flow of experiencing going on within oneself, a flow marked especially by complexity and continuous change.
>
> (Rogers 1989: pp. 11–13)

Returning to the theme of the therapist voicing to a client a feeling of boredom in that client's presence, Carl stated that during such moments he would strive to remain in ongoing touch with his own *changing* feelings and meanings. For instance, he would *own* his feeling of boredom ('own' in the sense that he would view his feeling as his own feeling, and not as some putative 'fact' about the client). Nor is this feeling of boredom fixed, rigid or permanent, nor does it exist in isolation. (This might be an example of what is meant by *extensionality* as a feature of congruence – boredom is an abstraction from the whole. This is not a boring client and nor is the boredom continuous.) Rogers would include in his verbalisation his discomfort in expressing his feeling of boredom and his distress at even feeling it and, as he does so, he realises that actually 'behind' this *apparent* boredom is a feeling of distance between himself and the client, and that he would wish for greater proximity. This, too, is voiced, and he would most definitely not feel bored as he eagerly (and with some apprehension) awaited the client's reactions to such disclosures:

> I also feel a new sensitivity to him now that I have shared this feeling which has been a barrier between us. I am far more able to hear the surprise, or perhaps the hurt, in his voice as he now finds himself speaking more genuinely because I have dared to be real with him. I have let myself be a person – real, imperfect – in my relationship with him.
>
> (Rogers 1989: p. 12)

While ideally it might be considered best if the therapist experienced no (apparently) negative feelings at all towards clients (although Rogers was honest enough to say that he thought all therapists did experience negative feelings or thoughts about their clients from time to time), Carl Rogers believed that when they are present it can be more harmful to keep them hidden than to voice them. Although such negative feelings might at first glance seem potentially harmful or damaging to clients, congruence is a better policy than dishonesty, pretence and false posturing.

> It is not easy for a client, or for any human being, to entrust his most deeply shrouded feelings to another person. It is even more difficult for a disturbed person to share his deepest and most troubling feelings with a therapist. The genuineness of the therapist is one of the elements in the relationship that make the risk of sharing easier and less fraught with dangers.
>
> (Rogers 1989: p. 13)

We shall see when we explore therapeutic process in Chapter 5 that one phase is the discovery that the core of the personality is positive – likeable. If we can trust that at heart clients are inherently positive, perhaps congruent therapists can have a similar faith in themselves. If I know myself and am aware of what lies within me, I can come to know that fuelling my apparently negative or hostile feelings are more understandable and acceptable features of my being. Thus the more I have engaged in my own self-development activities, the less likely it is that I will have to discover my potential through and with clients.

Being many things simultaneously might be illustrated by looking at the notions of *self-as-object* and *self-as-process*. George Kelly (Kelly 1955) wrote about *personal constructs*, and more recently Dave Mearns (Mearns 2002) has written about *configurations of self* – a phrase coined by Carl Rogers in his 1959 paper. Just as Rogers differentiated between *self-concept* (how individuals see themselves) and *self-structure* (how individuals see themselves viewed by another person), so *individuals* can view aspects of *themselves*. Suppose that person A is a mother of two children who is training to become a client-centred therapist. One 'personal construct' that she may have is composed around parenting. Another configuration of self is constructed around being a student. Person A can look at each of these constructs, and in doing so they become somewhat objectified – hence *self-as-object*. Indeed, she could draw diagrams of each of these constructs, mapping and charting their relevant components. Now suppose that on a training day one of her children feels a little ill – not dramatically ill, just slightly poorly. The 'good mother' personal construct tells her to miss out on a day's training in order to care for her child, yet her 'good student' configuration tells her to attend her course. Thus there is a conflict *between two of the self-structures* of person A. Every individual is made up of far more than just two personal constructs or configurations – and let us not forget organismic valuing either! However, there is also the *self-as-process* to consider. If it is possible to identify and examine aspects of the self-structure, what about *the self that is doing the exploring*? In the previous example of therapists A, B and C, only therapist C really utilised self-as-process. Therapist A was influenced only by a tired, bored aspect of self, and therapist B was influenced only by client-centred conditions of worth. It is the *self-as-process* that enables access to *all* significant configurations of self *and* organismic valuing, and the self-as-process that is free to make informed choices based on the sensory and visceral evidence available to it.

> I have learned that my total organismic sensing of a situation is more trustworthy than my intellect . . . I have found that when I have trusted some inner non-intellectual sensing, I have discovered wisdom in the move . . . As I gradually come to trust my total reactions more deeply, I find that I can use them to guide my thinking . . . I think of it as trusting the totality of my experience, which I have learned to suspect is wiser than my intellect. It is fallible I am sure, but I believe it to be less fallible than my conscious mind alone.
>
> (Rogers 1961: pp. 22–3)

Two other common misunderstandings occur, both concerning the 'communication' of congruence. All too often it seems to be 'forgotten' that therapist communication of congruence (or client perception of therapist congruence, even to a minimal degree) was *not* one of the original (1957 or 1959) necessary and sufficient conditions. It is necessary (and sufficient) only that the client perceives the unconditional positive regard and empathic understanding of the therapist as being genuine. The crucial importance of congruence is that the unconditional positive regard and empathic understanding of the therapist are real – that is, authentic and genuine. There can be nothing more inaccurate or misleading than the notion of 'communicating congruence' for, as we have seen, congruence means a state of being wherein there is no denial to awareness or distorted symbolisation in awareness of significant sensory or visceral experiencing. Imagine therapist D 'communicating congruence' (and think about how therapeutic it might be for the client) by saying to that client 'Right now I am neither denying symbolisation of significant sensory or visceral experience to my awareness, nor granting only a distorted symbolisation in my awareness of significant sensory or visceral experiencing'! Personally, I would rather see a different counsellor! Maybe therapist C.

Yet none of this is intended to detract from the following:

> Being genuine also involves the willingness to be and to express, in my words and my behavior, the various feelings and attitudes which exist

in me. It is only in this way that the relationship can have *reality*, and reality seems deeply important as a first condition. It is only by providing the genuine reality which is in me, that the other person can successfully seek for the reality in him. I have found this to be true even when the attitudes I feel are not attitudes with which I am pleased, or attitudes which seem conducive to a good relationship. It seems extremely important to be *real*.

(Rogers 1961: p. 33)

It may be worth remembering that a congruent communication of an attitude with which the therapist does not feel pleased could well include the feeling of displeasure that accompanies the attitude.

Carl Rogers defined congruence as a *cluster of factors* that included *congruence of self and experience, openness to experience, psychological adjustment, extensionality* and *maturity*.

The first aspect of congruence is *congruence of self and experience*, typified by individuals who constantly revise and reorganise their concept of self to include characteristics that would previously have been inconsistent with self:

Thus when self-experiences are accurately symbolized, and are included in the self-concept in this accurately symbolized form, then the state is one of congruence of self and experience. If this were completely true of all self-experiences, the individual would be a fully functioning person . . . If it is true of some specific aspect of experience, such as the individual's experience in a given relationship or in a given moment of time, then we can say that the individual is to this degree in a state of congruence. Other terms which are in a general way synonymous are these: integrated, whole, genuine.

(Rogers 1959: p. 206)

Openness to experience, the second factor, occurs when the self-structure experiences no threat to itself. In other words, openness to experience is the exact opposite of defensiveness:

whether the stimulus is the impact of a configuration of form, color, or sound in the environment on the sensory nerves, or a memory trace from the past, or a visceral sensation of fear, pleasure, or disgust, it is completely available to the individual's awareness. In the hypothetical person who is completely open to his experience, his concept of self would be a symbolization in awareness which would be completely congruent with his experience. There would, therefore, be no possibility of threat.

(Rogers 1959: p. 206)

In other words, complete and total openness to experience would only be possible if there was also a complete absence of denial and distortion.

Thirdly, *psychological adjustment* means that:

the concept of the self is such that all experiences are or may be assimilated on a symbolic level into the gestalt of the self-structure. Optimal psychological adjustment is thus synonymous with complete congruence of self and experience, or complete openness to experience. On the practical level, improvement in psychological adjustment is equivalent to progress toward this end point.

(Rogers 1959: p. 206)

If there is a complete absence of denial and distortion, the individual experiences no threat and there is therefore no need to feel either vulnerable or anxious.

Extensionality is the fourth ingredient of congruence, and it is a term derived from general semantics, meaning that the individual is aware that any experience is an *abstraction* from the whole and not the whole itself (the map is not the territory, for instance), and thus abstractions are by their very nature limited and differentiated (a fairly comprehensive and novel view of general semantics can be found in the 'Null-A' science fiction books written by AE van Vogt). The extensional individual will give 'pride of place' to facts rather than concepts, and will be aware that such facts are 'anchored' in space and time (as in 'what is true today might not be so true tomorrow'), evaluation is multiple and varied (gradational, not simply 'left or right' or 'black or white' or 'right or wrong'), and inferences and abstractions will be constantly tested against reality.

The fifth and final component in this congruence cluster is *maturity*, defined by Carl Rogers as meaning that the person:

> perceives realistically and in an extensional manner, is not defensive, accepts the responsibility of being different from others, accepts responsibility for his own behavior, evaluates experience in terms of the evidence coming from his own senses, changes his evaluation of experience only on the basis of new evidence, accepts others as unique individuals different from himself, prizes himself, and prizes others. If his behavior has these characteristics, then there will automatically follow all the types of behavior which are more popularly thought of as constituting psychological maturity.

(Rogers 1959: p. 207)

In summary, Rogers wrote:

> Congruence is the term which defines the state. Openness to experience is the way an internally congruent individual meets new experience. Psychological adjustment is congruence as viewed from a social point of view. Extensional is the term which describes the specific types of behavior of a congruent individual. Maturity is a broader term describing the personality characteristics and behavior of a person who is, in general, congruent.

(Rogers 1959: p. 207)

So congruence is a state of being comprised of several factors. If the *direct communication of congruence* is utterly irrelevant, might there nevertheless be issues about *congruent communications*? The answer is yes, because in terms of the 'core' conditions for client-centred therapy, congruence was originally placed first of the three *because of the critical importance of both unconditional positive regard and empathic understanding being authentic* and being *perceived by the client as genuine*. We have seen that when writing of psychological adjustment, Carl Rogers stated that 'On the practical level, improvement in psychological adjustment is equivalent to progress toward this end point'. The would-be client-centred therapist, then, is likely to engage in processes that aim to optimise the likelihood of therapist congruence in counselling relationships.

The therapist who passes judgements, evaluates, diagnoses, praises, gives opinions, reassures and problem solves (and so on) may well be being congruent. However, unconditional positive regard (in particular) and empathic under-standing do not feature as prominent or primary components of this particular congruence. Therapists who, for instance, tell their clients what they would do if faced with a similar situation may well be being 'real' and genuine. However, 'transparency' in client-centred therapy means that the client will see uncondi-tional positive regard and empathic understanding residing at the heart of the therapist's presence. In client-centred therapy there is no professional counselling façade or mask. The 'naked' or 'uncovered truth' reveals unconditional positive regard and empathy as deeply embedded and integrated aspects of the therapist's way of being.

Within this client-centred way of being, congruence is described as a *state* (whereas unconditional positive regard is seen as an *attitude*, and empathic understanding as a *process*). Congruence can be viewed as a *state of being* because at any given moment the person either is or is not congruent, or more accurately, could be located somewhere along the continuum between absolute congruence and total incongruence. However, this does not have to mean that congruence is a *fixed* or *static* state. Indeed Carl Rogers said that if this were so there would be no client-centred therapists at all!

> Another point worth noting is that the stress is upon the experience *in the relationship*. It is not to be expected that the therapist is a completely congruent person at all times. Indeed if this were a necessary condition there would be no therapy. But it is enough if in this particular moment of this immediate relationship with this specific person he is completely and fully himself, with his experience of the moment being accurately symbolized and integrated into the picture he holds of himself. Thus it is that imperfect human beings can be of therapeutic assistance to other imperfect human beings.
>
> (Rogers 1959: p. 215)

It is a condition, then, that *during the therapy session* client-centred counsellors strive to be as congruent as it is possible for them to be, to come as close to the ideal of congruence as is possible. Concluding that in terms of any given moment congruence is a *state*, and that client-centred therapists strive to maximise and *be* in that state during counselling sessions, does not have to mean that there are no *process* dimensions to congruence. We might ask, for example, what factors and activities might be conducive to maximising congruence during the therapy session. Perhaps a few possibilities might be worth mentioning here.

If therapist A in the recent example could be seen as congruent but not especially client-centred, and if therapist A would like to work towards becoming more client-centred, then looking for ways of developing authentic (that is, congruent) attitudes and beliefs that value the dignity, capacity and potential resources of the client (as explored further when looking at unconditional positive regard) might make sense. Carl Rogers wrote that we need to *permit* ourselves to be empathic with another, adding that the word 'permit' may seem strange to many, for:

> Our first reaction to most of the statements which we hear from other people is an immediate evaluation, cause or judgment, rather than an understanding of it. When someone expresses some feeling or attitude or belief, our tendency is, almost immediately, to feel 'That's right', or 'That's stupid', 'That's abnormal', 'That's unreasonable', 'That's incorrect', 'That's not nice'. Very rarely do we permit ourselves to *understand* precisely what the meaning of his statement is to him. I believe this is because understanding is risky. If I let myself really understand another person, I might be changed by that understanding. And we all fear change. So as I say, it is not an easy thing to permit oneself to understand an individual, to enter thoroughly and completely and empathically into his frame of reference. It is also a rare thing.
>
> (Rogers 1961: p. 18)

Thus a link between congruence and empathic understanding is *danger*, for personal (that is, *therapist*) change will in all likelihood be a consequence of accurate empathic understanding of clients. Unless therapists are open to the possibility of personal change *to themselves* through engaging in counselling *as the counsellor*, empathic understanding would be threatening to the concept of self (and hence the self would strive to defend itself against this threat). Thus we can see the vital importance of *therapist self-development*, for Rogers wrote of being empathic:

> In some sense it means that you lay aside your self; this can only be done by persons who are secure enough in themselves that they know they will not get lost in what may turn out to be the strange or bizarre world of the other, and that they can comfortably return to their own world when they wish.
>
> (Rogers 1980: p. 143)

Rogers saying that *in some sense* therapists 'set aside their self' may well be significant. Perhaps it is symptomatic of the shift from a 'total' laying aside of the self (1951), to greater therapist transparency (during the 'intervening years'), to eventual modification into the self being disciplined around respect and empathy (1987).

In other words, the congruence of a client-centred therapist includes both an openness to the possibility of being changed oneself through offering therapy *and* the personal strength and inner security to boldly (yet sensitively) go where no one has gone before – often not even the client – safe in the knowledge that there is a way back.

> I know that I have some knowledge of who I am, so I can let myself go into the world of this other person – even though it's a frightening, crazy, bizarre world – because I know I can come back to my world and be myself. When one is not very secure in oneself, one can get tangled into the world of the other and not know who's who. That's a very painful situation.
>
> (Rogers and Russell 2002: p. 281)

There are many ways in which Carl Rogers linked therapist congruence and empathic understanding with a journey of personal growth for the therapist. For instance, empathic understanding is a quality that the therapist *offers*, rather than something that is drawn from the therapist by specific types of clients or concerns. If empathy comes more easily with certain clients or problems, the therapist is *evaluating* and *judgemental* (and certainly not unconditional), so *becoming non-evaluative* is a major aspect of the journey towards becoming a congruent client-centred therapist. How might someone learn to become less judgemental? It is perhaps the case that each and every (trainee) client-centred therapist needs to find ways of achieving both broad (social, political and cultural, for instance) and personal (through receiving therapy or engaging in equivalent activities themselves and through counselling practice) understanding of human nature – and although Carl Rogers shunned the mantle of 'guru', there is nevertheless much to learn from studying his work.

It is also helpful if the therapist is authentically interested in the other person, demonstrating 'an extraordinary kind of concentrated attention' (Rogers 1951: p. 349). So often in training groups, for instance, student A speaks and student B says something about their own experience or thoughts, with the consequence that student A feels completely unheard, and wonders whether it was worth saying anything at all in the first place. Students C and D then pitch in with *their* own thoughts, feelings and experiences, so actually *no one* feels heard or that there is even an attempt to value the speakers by striving to understand them. As Rogers said, 'Without attention there can be no understanding and hence no communication' (Rogers 1951: p. 349) – and this means *paying attention to the client's world of feelings and meanings*. Yes, congruence *does* include therapists being aware of their inner flow of sensations and meanings. However, if the therapist's attention is either focused on one specific point that has significance *to the therapist* or is devoted to how the therapist will *respond*, then he or she is not attending as fully as possible to the client, and 'this is not communication in any sense of the word' (Rogers 1951: p. 349). Linked with therapist unconditional positive regard, the client-centred therapist conveys an *authentic* valuing of clients in part through this precious attention to the *client's* world of feelings and personal meanings.

> The accuracy of the therapist's empathic understanding has often been emphasised, but more important is the therapist's *interest* in appreciating the world of the client and offering such understanding with the willingness to be corrected. This creates a process in which the therapist gets closer and closer to the client's meanings and feelings, developing an ever-deepening relationship based on respect for and understanding of the other person.
>
> (Rogers and Raskin 1989: p. 171)

Thus the journey towards becoming a client-centred therapist involves the development of a *genuine desire* to respect others through paying attention to them – not 'laying aside your own self' because *Carl Rogers* or a *facilitator* says that this is how it should be. A genuine desire will be integrated and fully present and will be based upon the internal locus of evaluation of the therapist rather than being a value that is introjected from some significant other – not even Rogers. Another way of putting this is that *therapist* needs for positive regard and positive self-regard based on introjected values are not the sole determinants of *client-centred* therapist behaviour.

However, respect goes two ways, for Carl Rogers believed that *therapist self-respect* was an important factor, too. Having stated the importance of being able to enter the private world of another without becoming 'lost' in it, he furthered this thinking when he wrote:

> Can I be strong enough as a person to be separate from the other? Can I be a sturdy respecter of my own feelings, my own needs, as well as his? Can I own and, if need be, express my own feelings as something belonging to me and separate from his feelings? Am I strong enough in my own separateness that I will not be downcast by his depression, frightened by his fear, nor engulfed by his dependency? Is my inner self hardy enough to realise that I am not destroyed by his anger, taken over by his need for dependence, nor enslaved by his love, but that I exist separate from him with feelings and rights of my own? When I can freely feel this strength of being a separate person, then I find that I can let myself go much more deeply in understanding and accepting him because I am not fearful of losing myself.
>
> (Rogers 1961: p. 52)

Client-centred therapist congruence therefore includes personal strength and robust self-respect – a capacity to absorb the strong feelings of another without either being threatened by or drowning in them, or being 'seduced' into meeting the needs of the other. Carl continued:

> The next question is closely related. Am I secure enough within myself to permit him his separateness? Can I permit him to be what he is – honest or deceitful, infantile or adult, despairing or over-confident? Can I give him the freedom to be? Or do I feel that he should follow my advice, or remain somewhat dependent on me, or mould himself after me? In this connection think of the interesting small study by Parson which found that the less well-adjusted and less competent counsellor tends to induce conformity to himself, to have clients who model themselves after him. On the other hand, the better-adjusted and more competent counsellor can interact with a client through many interviews without interfering with the freedom of the client to develop a personality quite separate from that of his therapist. I should prefer to be in this latter class, whether as parent or supervisor or counsellor.
>
> (Rogers 1961: pp. 52–3)

If an element of a therapist's self-concept is 'knowing the way' for clients, then that therapist is likely to lead clients, advise them, foster dependency or even see him- or herself as a role model to be copied, followed and admired. Therefore a crucial factor in client-centred therapist congruence is not only the strength and self-respect that enables 'uncluttered' entry to the world of the other, but also the strength to permit the other person the dignity of being whoever it is that they are.

To the therapist, it is a new venture in relating. He feels, 'Here is this other person, my client. I'm a little afraid of him, afraid of the depths in him as I am a little afraid of the depths in myself. Yet as he speaks, I begin to feel a respect for him, to feel my kinship to him. I sense how frightening his world is for him, how tightly he tries to hold it in place. I would like to sense his feelings, and I would like him to know that I understand his feelings. I would like him to know that I stand with him in his tight, constricted little world, and that I can look upon it relatively unafraid. Perhaps I can make it a safer world for him. I would like my feelings in this relationship with him to be as clear and transparent as possible, so that they are a discernible reality for him, to which he can return again and again. I would like to go with him on the fearful journey into himself, into the buried fear, and hate, and love which he has never been able to let flow in him. I recognise that this is a very human and unpredictable journey for me, as well as for him, and that I may, without even knowing my fear, shrink away within myself from some of the feelings he discovers. To this extent I know I will be limited in my ability to help him. I realise that at times his own fears may make him perceive me as uncaring, as rejecting, as an intruder, as one who does not understand. I want fully to accept these feelings in him, and yet I hope also that my own real feelings will show through so clearly that in time he cannot fail to perceive them. Most of all I want him to encounter in me a real person. I do not need to be uneasy as to whether my own feelings are 'therapeutic'. What I am and what I feel are good enough to be a basis for therapy, if I can transparently *be* what I am and what I feel in relationship to him. Then perhaps he can be what he is, openly and without fear.'

(Rogers 1961: pp. 66–7)

As well as self-respect as a constituent of therapist congruence, *self-understanding* is absolutely crucial, too. To be authentic means that:

no inner barriers keep him from sensing what it feels like to be the client at each moment of the relationship; and that he can convey something of his empathic understanding to the client. It means that the therapist has been comfortable in entering this relationship fully, without knowing cognitively where it will lead, satisfied with providing a climate which will permit the client the utmost freedom to become himself.

(Rogers 1961: p. 185)

Carl Rogers wrote of intercultural work, stating how 'I understand your feelings but I've never personally oppressed you' or 'I can understand because I've been oppressed, too' may be natural responses to disadvantaged people, but such expressions are of no help whatsoever. Effectiveness for Rogers needed to include:

the realisation and ownership of 'I think white'. For men trying to deal with women's rage, it may be helpful for the man to recognise 'I think male'. In spite of all our efforts to seem unprejudiced, we actually carry within us many prejudiced attitudes.

Rage needs to be *heard*. This does not mean that it simply needs to be listened to. It needs to be accepted, taken within, and understood empathically. While the diatribes and accusations appear to be deliberate attempts to hurt the whites – an act of catharsis to dissolve centuries of abuse, oppression, and injustice – the truth about rage is that it only dissolves when it is really heard and understood, without reservations. Afterwards, the blacks or other minority members change in what seems a miraculous way, as though a weight has been lifted from their shoulders.

To achieve this kind of empathic listening the white needs to listen to his own feelings, too, his feelings of anger and resentment at 'unjust' accusations. At some point he too will need to express these, but the primary task is to enter empathically the minority world of hate and bitterness and resentment and to know that world as an understandable, acceptable part of reality.

(Rogers 1978: pp. 133–4)

There is also a *'practice makes perfect'* element to therapist personal growth, as Rogers believed that:

> Experienced therapists offer a higher degree of empathy to their clients than those less experienced . . . Evidently, therapists do learn, as the years go by, to come closer to their ideal of a therapist and to be more sensitively understanding.
>
> (Rogers 1980, p. 148)

Stating that this journey of personal growth 'puts a heavy demand on the therapist as a person', Rogers, albeit with some discomfort, reluctantly concluded (as had Bergin and Jasper, and Bergin and Solomon) that those therapists who felt relaxed and confident in interpersonal relationships offer greater empathic understanding than therapists who are less well integrated. Rogers wrote of counsellor training (Bergin and Jasper 1969; Bergin and Solomon 1970):

> If the student is to become a therapist, the more he has been able to achieve of empathic experiencing with other individuals, the better will his preparations be. There are undoubtedly countless avenues to this end.
>
> (Rogers 1951: p. 437)

Thus the process of becoming more congruent can be a broad-ranging activity, including the study of anthropology, sociology, cultural studies, gender studies, politics, literature, role taking in dramatic productions ('method acting'), and humanistic psychology courses. However, Carl Rogers also wrote that only rarely does *reading* something change behaviour to a significant degree, and that study was therefore not the only way of learning to become more integrated:

> It can come simply through the process of living, when a sensitive person desires to understand the viewpoint and attitudes of another.
>
> (Rogers 1951: p. 437)

'Have an open mind', 'try out something client-centred (or person-centred) experientially, if only in a small way, or even to try and prove it wrong – and judge the consequences for yourself . . .'.

But if you try them at all, you open yourself to visceral learnings that may change your behavior and change you.

(Rogers 1978: p. 103)

There has in some instances been a regrettable tendency only to define congruent as meaning transparent, real, genuine, authentic, and being without front, professional mask or façade, without mentioning the absolute necessity (and sufficiency) of the primary place of unconditional positive regard and empathic understanding within client-centred therapist congruence. Yet just about *any* practice could be justified in the name of congruence if the fundamental necessity of the other two core conditions is omitted: 'I was just being congruent . . .'.

Empathy and therapist unconditional positive regard

In the last section, we saw that Carl Rogers believed congruence to be the most important element of the 'core' conditions 'in the ordinary interactions of life'. At the same time, he said of unconditional positive regard:

But in certain other special situations, caring or prizing may turn out to be the most significant. Such situations include non-verbal relationships – parent and infant, therapist and mute psychotic, physician and very ill patient. Caring is an attitude which is known to foster creativity – a nurturing climate in which delicate, tentative new thoughts and productive processes can emerge.

(Rogers 1975: p. 9)

Unconditional positive regard is placed second in the sequence of the three therapist conditions because a 'prerequisite' is that the respect of the counsellor for a client is authentic (that is, congruent), and one of the most effective ways of communicating (or indeed of the client perceiving) unconditional positive regard is through the process of the therapist striving to empathise with and understand the client from within that client's internal frame of reference.

Carl Rogers did not commence his career in psychology with a particularly strong belief in the capacity of clients:

The more I have worked with people, both in individual therapy and in encounter groups, the more respect I've come to have for man as a person worthy of respect and a person of dignity. This value that I've come to place on the human being is something that really grows out of my experience. I didn't start with that high a regard for the individual person.

(Rogers and Evans 1975: p. 74)

In terms of *becoming* client-centred, there would appear to be a *process* along the lines of the more client-centred attitudes and beliefs are put into practice, the more meaning and strength those beliefs come to have. What might begin as a *'leap of faith'* moves toward being a more solid *belief*, which moves toward *trusting* in that belief, which moves toward a *knowing* – and client-centred therapists implement their faith–belief–trust–knowing not by *doing*, but through a way of *being*. It may well also be that the client needs to feel fully received in each phase of development in order to move on to the next one. As they enter counselling courses, trainees tend to be rather passive. As experience is gained they gradually become more inclined to listen to their clients and display some willingness to allow their clients autonomy (self-direction). However, in the early stages this passivity may be characterised by the trainee counsellor attempting to 'keep out of the way' of clients by trying to avoid any of their own feelings or thoughts straying into the session. Sometimes clients perceive this passivity as indifference rather than prizing – they might even feel rejected as they perceive their counsellor as not caring much about them, hardly seeing themselves as individuals worthy of their therapist's respect. Sadly, some counsellor training seems to be built around passivity and counsellors 'keeping out of the way', and Rogers wrote:

> This misconception of the approach has led to considerable failure in counseling – and for good reasons . . . Hence the counselor who plays a merely passive role, a listening role, may be of assistance to some clients . . . but by and large his results will be minimal, and many clients will leave, both disappointed in their failure to receive help and disgusted with the counselor for having nothing to offer.
>
> (Rogers 1951: p. 27)

We have learned that clients entering therapy are likely to be psychologically defensive because they are vulnerable and anxious. They are vulnerable and anxious because their ability to deny or distort experiencing that is threatening to their self-concept is no longer entirely successful, so there is a perceived or subceived state of tension – the client is *incongruent*. Denied or distorted experience is most likely to be permitted access to conscious awareness (and therefore be open to expression, exploration and ownership) if there is an absence or relative absence of perceived threat. Yet empathic understanding by a therapist might in itself (that is, *without* authenticity and respect) feel *very* threatening to clients. If the therapist might be thought of as an 'alter ego mirror' for a moment, then no matter how 'magical' the mirror, the client might not *want* to see or meet with their 'reflection' or 'doppelganger' (as stated in Chapter 2, in some sense there is nothing more 'confrontational' than accurate empathic understanding). If, for instance, a therapist senses that underlying a client's anger is deep pain, the client might at that moment be willing to engage with their rage yet feel really fearful of connecting with their hurt. Although it may be true that profound, sensitive and delicate empathic understanding would in this instance *include* a sensing of the client's fear of the pain that lies behind their anger, there is another crucial factor, too, namely that the therapist *respects* the client's defensive needs (prizing not prising again).

When Rogers defined congruence as a *state*, he defined unconditional positive regard as an *attitude*. This respectful attitude means a variety of things at a variety of levels. The importance of congruence is that unconditional positive regard is genuine. The importance of empathic understanding is that it is the best way to communicate unconditional positive regard and be respectful. What different meanings, then, can be ascribed to this precious and rare (because most relationships are conditional) attitude of *unconditional* positive regard?

First, the client-centred therapist holds a deep belief in the *actualising tendency* as the motivating force in human beings. Furthermore, the actualising tendency is viewed as trustworthy and positive, for given the right conditions people are *by nature* co-operative, social, constructive, trustworthy, creative, responsible, complex and fluid forward-moving entities. This fundamental acceptance of the actualising tendency as the primary human growth factor enables the client-centred therapist to see the *positive potential* in anyone and everyone. For sure, not everyone does meet with the 'right' conditions (that is, the conditions most conducive for growth), yet everyone has within them the *potential* to become all that they could be – and given the right conditions that 'becoming' will be in a positive direction.

Carl Rogers stated that his belief in the positive nature at the heart of all human beings was the single most *revolutionary* and *radical* aspect of client-centred therapy. For so long, *feelings* and *thoughts* (or at least, feelings and thoughts deemed to be antisocial, harmful, negative or just plain wrong) were considered as things to be managed, controlled, suppressed, changed, repressed, altered or influenced. From most religions through to Freudian psychoanalysis, human beings were viewed with notions of 'original sin', and unless the 'raging id' was to be controlled, the consequence would be brutal mayhem. Likewise *behaviour* was also something to be evaluated (by those who knew best, of course) and either encouraged (or reinforced), controlled (or managed) or eliminated altogether, depending upon the social, political and psychological mores and norms of the time.

The revolutionary view of psychologist Carl Rogers that in fact the underlying motivational tendency in human beings was positive and trustworthy met – and still meets – with a great deal of scepticism, disbelief and scorn. On many, many occasions Rogers was asked how *anyone* could be so *naive* as to believe that human beings are inherently 'good' when there is clearly so much 'bad' happening in the world around us. He tended to respond to such scepticism, disbelief and scorn in two ways. He acknowledged that *of course* there was the possibility that the human race would eventually destroy itself and even the planet earth in one way or another – for instance, the proliferation of weaponry and environmental damage hardly bodes well. However, he pointed out that there was also the possibility that we would not destroy either ourselves or the planet, and in the meantime he chose to put all of his energy into the potential for positive growth rather than destruction. Secondly, he stated that *of course* therapists encountered many of the more unpleasant and unpalatable aspects of living. Indeed, this very difficulty with human nature is why so many people seek counselling in the first place. However, his experience over and over and over again was that if he could be genuinely respectful and empathically understanding of the apparently negative, aggressive and hostile aspects of human nature and behaviour – if he could be a sensitive companion to such feelings and thoughts without judging, condemning, and trying to control or change them – then *inevitably* he discovered that 'behind' or 'beneath' them are the positive attributes of the actualising tendency. This exactly matches my own experience. In some 30 years as a counsellor I have met with a high number (thousands) of clients who have disclosed anger, hostility, hate, rage and aggression, yet without fail it has transpired that behind their anger lies hurt and pain – and behind their hurt and pain lies a desperate need to simply love and be loved, value and be valued, like and be liked, and so on.

Having said that *given the right conditions* the human organism is essentially trustworthy, co-operative, responsible, creative and forward moving, it is all too apparent that very often people do not encounter conditions that are fully conducive to healthy growth. As was stated earlier on when looking at the notion of incongruence, all too often interpersonal relationships are conditional – there are strings attached. *Unconditional* positive regard ideally has no strings attached, and it is this very unconditional acceptance that, if experienced, tends to reduce feelings of threat and therefore the need for defensiveness (as will be more fully explored when we look at therapeutic process in Chapter 5).

Another link between unconditional positive regard and empathic understanding is that the client-centred therapist *understands* (and accepts) that were *any* human being to have the same history and experience that this particular client has, they would in all likelihood exhibit similar feelings, thoughts and behaviours. There is a kind of 'there but for the grace of fate go I' feel to it. In other words, get the good breaks and chances in life and certain consequences are likely, but get only bad breaks or be deprived of opportunities, and outcomes might be equally predictable – and very *understandable*. Linking congruence *and* unconditional positive regard *and* empathic understanding, let us not forget the importance of therapists offering and being the core conditions *to and for themselves*. If therapists can understand themselves they are likely to be self-accepting, too.

> One way of putting this is that I feel I have become more adequate in letting myself *be* what I *am* . . . It seems to me to have value because the curious paradox is that when I accept myself as I am, I change. I believe that I have learned this from my clients as well as within my own experience – that we cannot change, we cannot move away from what we are, until we thoroughly accept what we are. Then change seems to come about almost unnoticed.
>
> (Rogers 1961: p. 17)

We saw when looking at incongruence how all too often our conditions of worth can be formulated around the values and judgements of significant others, and that these conditions of worth may thwart the positive nature of the actualising tendency. To be congruent, client-centred therapists will have explored and reappraised their conditions of worth, retaining and reintegrating those introjected values that have personal meaning and discarding those introjected values that do not. To be a *client-centred* therapist, unconditional positive regard will have become deeply embedded within the concept of self – not because Carl Rogers said so, but because both the *giving* and the *receiving* of unconditional positive regard have been personally experienced as meaningfully helpful. Unconditional positive regard will be valued from an internal locus of experiential evaluation and be a primary constituent of the therapist's self-concept as an attitudinal way of being that ideally is in complete harmony with the actualising tendency.

If the client-centred practitioner holds an unswerving faith in the primacy and positive nature of the actualising tendency, other consequences follow. Many of these repercussions could be deemed 'political' in nature (and it was only late in his career that Carl Rogers acknowledged the political dimensions of client-centred therapy), such as a belief in the *right* of the individual to *self-determination* or *autonomy*. A belief in the sanctity of the *dignity* of human beings follows, too. If the therapist believes that it is the *therapist* who knows best what is good for the client, then that therapist is blatantly not respectful of the capacity, resources and potential within human beings for autonomy and dignity.

As already stated, Carl Rogers believed one of the most effective ways to communicate unconditional positive regard is to be empathic, as:

> it would appear that for me, as counsellor, to focus my whole attention and effort upon understanding and perceiving as the client perceives and understands, is a striking operational demonstration of the belief I have in the worth and the significance of this individual client. Clearly the most important value which I hold is, as indicated by my attitudes and my verbal behaviour, the client himself. Also the fact that I permit the outcome to rest upon this deep understanding is probably the most vital operational evidence which could be given that I have confidence in the potentiality of the individual for constructive change and development in the direction of a more full and satisfying life. As a seriously disturbed client wrestles with his utter inability to make any choice, or another client struggles with his strong urges to commit suicide, the fact that I enter with deep understanding into the desperate feelings that exist, but do not attempt to take over respons-ibility, is a most meaningful expression of basic confidence in the forward-moving tendencies in the human organism.
>
> We might say, then, that for many therapists functioning from a client-centred orientation, the sincere aim of getting 'within' the attitudes of the client, of entering the client's internal frame of reference, is the most complete implementation which has thus far been formulated, for the central hypothesis of respect for and reliance upon the capacity of the person.
>
> (Rogers 1951: pp. 35–6)

Clients really are worth the therapist's efforts in striving to understand them, and the fact that the 'non-directive' counsellor respects clients' self-direction and self-responsibility is meaningful evidence of the therapist's trust in the inner capacity and resources of clients, who thus perceive (to whatever degree) a very precious and rare respect and trust in themselves.

> If the counsellor maintains this consistently client-centred attitude, and if he occasionally conveys to the client something of his under-standing, then he is doing what he can to give the client the experience of being deeply respected. Here the confused, tentative, almost incoherent thinking of an individual who knows he has been eval-uated as abnormal is really respected by being deemed well worth understanding.
>
> (Rogers 1951: p. 44)

So the client perceives that the therapist holds and lives by an attitude which indicates that the client is worth the effort of striving to understand him or her. Again, this could be reformulated using an '*If–then*' approach. *If* person A strives to the utmost to understand person B, *then* person B is likely to perceive how respectful person A is of person B.

> Acceptance does not mean much until it involves understanding. It is only as I *understand* the feelings and thoughts which seem so horrible to you, or so weak, or so sentimental, or so bizarre – it is only as I see them as you see them, and accept them and you, that you feel really free to explore all the hidden nooks and frightening crannies of your inner and often buried experience. This *freedom* is an important condition of the relationship. There is implied here a freedom to explore oneself at both conscious and unconscious levels, as rapidly as one can dare to embark on this dangerous quest. There is also a complete freedom from any type of moral or diagnostic evaluation, since all such evaluations are, I believe, always threatening.
>
> (Rogers 1961: p. 34)

The intertwined nature of unconditional positive regard and empathic understanding comes across very clearly in the following passage written by Carl Rogers in 1980. He states that a consequence of *empathic understanding* is that:

> the recipient feels valued, cared for, accepted as the person that he or she is. It might seem that we have here stepped into another area, and that we are no longer speaking of empathy. But this is not so. It is impossible to accurately sense the perceptual world of another person unless you value that person and his or her world – unless you, in some sense, care. Hence, the message comes through to the recipient that 'this other individual trusts me, thinks I'm worthwhile. Perhaps I *am* worth something. Perhaps *I* could value *myself*. Perhaps I could care for myself'.
>
> (Rogers 1980: pp. 152–3)

Stating that 'The highest expression of empathy is accepting and non-judgmental . . . because it is impossible to be accurately perceptive of another's inner world if you have formed an evaluative opinion of that person', Carl Rogers suggests an experiential way of testing this:

> If you doubt this statement, choose someone you know with whom you deeply disagree and who is, in your judgment, definitely wrong or mistaken. Now try to state that individual's views, beliefs, and feelings so accurately that he or she will agree that you have sensitively and correctly described his or her stance. I predict that nine times out of ten you will fail, because your judgment of the person's views creeps into your description of them.
>
> Consequently, true empathy is always free of any evaluative or diagnostic quality. The recipient perceives this with some surprise: 'If I am not being judged, perhaps I am not so evil or abnormal as I have

thought. Perhaps I don't have to judge myself so harshly.' Thus the possibility of self-acceptance is gradually increased.

(Rogers 1980: p. 154)

Quoting RD Laing (and likening Laing to Buber), Rogers agreed with the notion that 'the sense of identity requires the existence of another by whom one is known' (Laing 1965), and that our very existence is affirmed or confirmed by interaction ('contact') with or being *fully received* by another person. Authentic, respectful, empathic understanding serves to give this 'needed confirmation that one does exist as a separate, valued person with an identity' (Rogers 1980: p. 155).

Having spent so long exploring a necessary and sufficient condition, the comments made by Carl Rogers when he was interviewed for RTE Dublin television in 1985 might seem a little surprising. The interviewer, John Masterson, said to him:

> One of the requirements you place upon yourself is to unconditionally accept and have unconditional positive regard for the person who you are trying to help. That presumably must be very difficult in some cases . . .
>
> (Rogers 1985)

To which Carl Rogers replied:

> No, that's a slight misunderstanding. What I have said is that I am very fortunate in the relationship if that feeling and attitude exists in me. I am very fortunate and the client is very fortunate if I can feel I really accept you fully just as you are.
>
> (Rogers 1985)

This comment was and is extremely liberating for me. For many years my introjected value, my condition of worth, was that experiencing and offering the conditions was both necessary and sufficient, and if I *failed* I would give myself (or be given – maybe by a supervisor or by peers) a hard time. The client-centred conditions had become, at times at least, a stick with which to beat myself, or to be beaten with, or, worse still, as a trainer a stick with which I would beat trainees! Now, some 28 years after first postulating the *necessary and sufficient client-centred conditions* in 1957, here was Carl saying *'No, that's a slight misunderstanding'*. Yet how qualitatively different it is – how uplifting – to feel *fortunate* if I could succeed to some degree as compared with feeling a *failure* if I did not. Rogers continued:

> Sometimes that's very difficult. I held an interview with a young man in South Africa. I didn't know anything about him. It turned out he was an officer in the South African army. Now for me, that meant a real . . . It meant that I stretched my empathic abilities to their very limit, to try to be with him, to try to understand him, to try to be caring toward him. I didn't feel I did too well – and yet I've learned since that interview really changed the course of his life.
>
> (Rogers 1985)

Once again, then, we see how it is that imperfect human beings can be of help to other imperfect human beings – and actually what Rogers said in 1985 was not especially new, as he wrote eight years earlier (in 1978):

> It is not, of course, possible to feel such an unconditional caring all the time. A therapist who is real will often have very different feelings, negative feelings toward the client. Hence it is not to be regarded as a 'should' that the therapist *should* have an unconditional positive regard for the client. It is simply a fact that unless this is a reasonably frequent ingredient in the relationship, constructive client change is less likely.
>
> (Rogers 1978: p. 10)

To think (and experience) in terms of it being *fortunate* if unconditional positive regard frequently exists in the therapeutic relationship feels far more conducive to developing such a respectful attitude than to experience a sense of *failure* if a *condition* is not met.

Now let us turn to empathic understanding, the most effective medium through which clients are able to perceive the deep respect of the therapist for them, this ongoing desire of the therapist to understand, this caring and valuing of the client, and this freedom to explore in a non-judgemental relationship.

Empathy and empathic understanding

> We know from our research that such empathic understanding – understanding with a person, not about him – is such an effective approach that it can bring about major changes in personality.
>
> (Rogers 1961: p. 332)

We have learned that congruence is a *state* and that unconditional positive regard is an *attitude* that is authentic if congruent. The third of the therapist conditions, namely empathic understanding, is a *process* through which congruent unconditional positive regard is most effectively conveyed. Returning again to what Carl Rogers had to say about the relative importance of each of the therapist conditions:

> Then, in my experience, there are other situations in which the empathic way of being has the highest priority. When the other person is hurting, confused, troubled, anxious, alienated, and terrified, or when he or she is doubtful of self-worth, uncertain as to identity, then understanding is called for. The gentle and sensitive companionship of an empathic stance – accompanied, of course, by the other two attitudes – provides illumination and healing. In such situations deep understanding is, I believe, the most precious gift one can give to another.
>
> (Rogers 1975: p. 9)

If the therapist is client-centred, *then* unconditional positive regard and empathic understanding will be deeply held and integrated personal characteristics, especially evident in the therapy hour. *If* the therapist experiences an authentic (that is, congruent) attitude of unconditional positive regard towards a client, *then* the most effective expression of this will be through the process of empathic understanding. This deep understanding is not an understanding *about*, but a delicate sensing of the personal feelings and meanings that reside in another person:

> My belief is that, by extending sensitive empathy to another person, it enables him or her to come forth and gain a better understanding of himself and what direction he is going.
>
> (Rogers and Teich 1992: p. 55)

Empathy and client perception of therapist unconditional positive regard and empathic understanding

Although a counselling relationship can be initiated by the therapist providing the core conditions:

> Unless the attitudes I have been describing have been to some degree communicated to the client, and perceived by him, they do not exist in his perceptual world and thus cannot be effective. Consequently it is necessary to add one more condition to the equation . . . It is that when the client perceives, to a minimal degree, the genuineness of the counselor and the acceptance and empathy which the counselor experiences for him, then development in personality and change in behavior are predicted.
>
> (Rogers and Stevens 1967: p. 96)

Two people are in contact (or psychological contact, if preferred), each making a perceived difference in the experiential (or perceptual, or phenomenological) world of the other. We have already seen that:

> In a deeply defended person . . . change is difficult to bring about. Change tends to occur more in the person who is already somewhat open-minded and searching in his approach to life.
>
> (Rogers and Evans 1975: p. 73)

Thus clients entering therapy are likely to be *seekers* – they will be in stage two of the process. The therapist is congruent in the relationship and the client is incongruent, being vulnerable or anxious. The therapist experiences unconditional positive regard towards and engages in a process of empathic understanding of the client. The notion of reciprocal empathy links with client perception of the therapist's empathy. Rogers said:

> It seems to me that the moment where persons are most likely to change, or I even think of it as the moments in which people *do* change, are the moments in which perhaps the relationship is experienced the same on both sides.
>
> (Rogers 1989: p. 53)

The therapist's perception of the client is motivated by an authentic desire to continually respect and understand that client's world from within the client's internal frame of reference. While therapists 'perception check' their sensing of clients' feelings and personal meanings, the notion of reciprocal empathy indicates that, to some degree, therapists rely on the capacity of their client to empathise with and understand their therapist's attempts to empathically understand that client. As Rogers wrote when reviewing various studies:

> it is the attitudes and feelings of the therapist, rather than his theoretical orientation, which are important. His procedures and techniques are less important than his attitudes. It is also worth noting that it is the way in which his attitudes and procedures are *perceived* which makes a crucial difference to the client, and that it is this perception which is crucial.
>
> (Rogers 1961: p. 44)

Whether through the therapist's ongoing desire to understand or via reciprocated empathy, for a therapeutic relationship and engagement in a therapeutic process to occur, the authentic unconditional positive regard and empathic understanding of the therapist must be perceived by the client *at least to a minimal degree* (or in terms of the therapist striving to communicate respect and empathy, then *at least to a minimal degree* such communication needs to be achieved):

> We know that if the therapist holds within himself attitudes of deep respect and full acceptance for this client as he is, and similar attitudes toward the client's potentialities for dealing with himself and his situations; if these attitudes are suffused with a sufficient warmth, which transforms them into the most profound type of liking or affection for the core of the person; and if a level of communication is reached so that the client can begin to perceive that the therapist understands the feelings he is experiencing and accepts them at the full depth of that understanding, then we may be sure that the process is already initiated.
>
> (Rogers 1961: pp. 74–5)

What does '*at least to a minimal degree*' actually *mean*? Does it mean that if during a therapy session a therapist communicates, say, 20 times, and three or four of those communications could be deemed authentically respectful and empathic, then the condition of *at least to a minimal degree* has been met (despite 16 or 17 communications *not* being respectful or empathic)? Does it mean that if the therapist is authentically respectful and empathic for 5 minutes out of 50 this condition has been met? Does it mean that if every therapist communication is characterised by *low* levels of authentic unconditional positive regard and empathic understanding, then at least to a minimal degree the sixth condition has been achieved? Carl Rogers wrote that:

> This struggle to achieve the client's internal frame of reference, to gain the center of his own perceptual field and see with him as perceiver, is rather closely analogous to some of the Gestalt phenomena. Just as, by active concentration, one can suddenly see the diagram in the psychology text as representing a descending rather than an ascending stairway or can perceive two faces instead of a candlestick, so by active effort the counselor can put himself into the client's frame of reference. But just as in the case of the visual perception, the figure occasionally changes, so the counselor may at times find himself standing outside the client's frame of reference and looking as an

external perceiver at the client. This almost invariably happens, for example, during a long pause or silence on the client's part. The counselor may gain a few clues which permit an accurate empathy, but to some extent he is forced to view the client from an observer's point of view, and can only actively assume the client's perceptual field when some type of expression again begins.

(Rogers 1951: p. 32)

There are nearly always 'some clues' that the therapist can absorb or sense. Equally, while the therapist might *constantly strive* to empathically understand, it is extremely unlikely that deep empathic understanding will be experienced as *constant*. There may be some clues, too, as to how training in client-centred therapy helps to bring about increased empathy. When comparing types of counsellor communications in written papers before and after training, even back in 1951 Carl Rogers found that there was increased 'reflection' (from around 50% to around 85%), and decreased interpretation (from approximately 19% to around 11%). Support, and moralising were reduced to zero (from around 8% and 5% respectively), and diagnosing fell from about 18% to 3%. Studying counselling practice produced similar results. 'Reflection' increased (from around 11% to 59%), as did information giving (from 2.5% to 3.9%) and 'innocuous' communications such as acceptance or silence (from around 5% to 11%). There was reduced interpretation (from 22% to 15%), support (from 15% to 5%), 'oughtness' (from 20% to 3%), information seeking (from 16% to 2%) and personal opinion (from 9% to 1%) (Rogers 1951: p. 455).

Degrees of empathy

The notion of low, intermediate and high levels of empathy was briefly introduced in Chapter 1. In summary:

At a relatively low level of empathic sensitivity the therapist responds with clarity only to the patient's most obvious feelings. At an inter- mediate level, the therapist usually responds accurately to the client's more obvious feelings and occasionally recognizes some that are less apparent, but in the process of tentative probing, he may anticipate feelings which are not current or may misinterpret the present feelings. At a higher level, the therapist is aware of many feelings and experiences which are not so evident, but his lack of complete understanding is shown by the slightly inaccurate nature of his deeper responses . . . At a very high level of empathic understanding, the therapist's responses move, with sensitivity and accuracy, into feelings and experiences that are only hinted at by the client.

It is this sensitive and accurate grasp and communication of the patient's inner world that facilitates the patient's self-exploration and consequent personality growth.

(Rogers and Truax 1967: p. 106)

Yet further clues about degrees of empathic understanding might be gained from the work of Barrett-Lennard (quoted by Carl Rogers) who developed a Relation- ship Inventory (Barrett-Lennard 1959) that in part tried to measure the degree to which clients felt empathically understood. Thus a scale from high to low empathy looked like this:

- He appreciates what my experience feels like to *me*.
- He tries to see things through my eyes.
- Sometimes he thinks that I feel a certain way because he feels that way.
- He understands what I say from a detached, objective point of view.
- He understands my words but not the way I feel.

Over the years, a variety of writers from a range of different orientations have come up with a variety of empathy 'rating scales' or other strategies for evaluating the degree of empathy (or empathic understanding) deemed to be present in a (therapeutic) relationship. While in person-centred counsellor training there may well be an across-the-board agreement that developing empathy is a fundamental aspect of the learning, there would appear to be little or no agreement on measuring or evaluating the *degree* of empathy necessary as evidence of practice good enough to merit the award of a diploma, despite the UK Government's focus on 'benchmarking'. It is also true to say that, over the years, there has been some resistance to empathy 'rating scales' in person-centred circles. For instance, some evaluation mechanisms have been perceived as overly mechanistic – 'add 0.5 to stage three if a later element is present', 'subtract 0.25 if an earlier element is missing', and so on.

However, formulating *some* kind of structure for exploring the degree of empathy that might (or might not) be present in counselling relationships might still be helpful for two main reasons. First, therapists in the UK (and maybe this is especially true of 'person-centred' counsellors or those in existential and 'humanistic' schools) have tended to be rather 'anti-evaluation' because such monitoring is reminiscent of 'Big Brother'-type meddling. Yet as therapists perhaps we *should* be accountable – to our clients, to ourselves, and to our profession (and to our employer, if we have one). Secondly, any activity that has the potential to develop a client-centred therapeutic way of being is worth engaging in. Ongoing monitoring of the degree to which therapists are able both to experience empathy and to communicate their sensing of client feelings and meanings could be valuable in terms of personal and professional development. Here, then, is an attempt to define some of the significant characteristics of empathic understanding (this time from low to high) (Vincent 2002c):

- Level A: The therapist communicates some *understanding* of what the client is experiencing in the moment, if only in a somewhat diluted way.
- Level B: The therapist communicates some sensing of what the client is *feeling* in the here and now, if only in a somewhat diluted way.
- Level C: The therapist communicates some sensing of what the client's present experiencing *means* to the client, if only in a somewhat diluted way.
- Level D: The therapist communicates some *understanding* of what the client is experiencing in the moment, at approximately the same intensity as the client's disclosure.
- Level E: The therapist communicates some sensing of what the client is *feeling* in the here and now, at approximately the same intensity as the client's disclosure.

- Level F: The therapist communicates some sensing of what the client's present experiencing *means* to the client, at approximately the same intensity as the client's disclosure.
- Level G: The therapist communicates some sensing of what the client is *feeling* in the moment, and a degree of *understanding* why the client feels as she does, if only in a somewhat diluted way.
- Level H: The therapist communicates some sensing of what the client is *feeling* in the here and now, some *understanding* of why the client feels as she does, and a degree of sensing what this experiencing *means* to the client, if only in a somewhat diluted way.
- Level I: The therapist communicates some sensing of what the client is *feeling* in the moment, and a degree of *understanding* why the client feels as she does, at approximately the same intensity as the client's disclosure.
- Level J: The therapist communicates some sensing of what the client is *feeling* in the here and now, some *understanding* of why the client feels as she does, and a degree of sensing what this experiencing *means* to the client, at approximately the same intensity as the client's disclosure.
- Level K: The therapist communicates some sensing of here-and-now client *feelings* at an intensity similar to the client's disclosure, and feelings of which the client seems only dimly aware.
- Level L: The therapist communicates some sensing of what the client is *feeling* in the here and now, some *understanding* of why the client feels as she does, and a degree of sensing what this experiencing *means* to the client, at approximately the same intensity as the client's disclosure, together with sensing feelings and meanings of which the client seems only dimly aware.

Anything presented as a hierarchy carries with it the danger that, to give an example, *empathy level L* is far more desirable than *empathy level A* – and the whole picture isn't as clear as that. For instance, there can be a tendency (especially within person-centred circles) to value feelings ('gut stuff') above thoughts ('head stuff'). Yet any valuing of one aspect of empathic understanding above another hardly sits comfortably with *unconditional* positive regard of whatever a client is experiencing from moment to moment. Suppose that in a first session with a client I notice upon reflection that many of my communications are characteristic of empathy level A. Does this mean that I have to give myself a hard time and beat myself up? No, it doesn't have to mean that I must come down really heavily on myself. It could be, for instance, that if I replay the tape inserting alternative communications at *empathy level L*, I realise that to have done so at the time would have been inappropriate and insensitive in the extreme to what the client was experiencing in that moment. Had I communicated at empathy level L throughout, maybe my main intent would have been to demonstrate my own cleverness or something – certainly not client-centred!

Something akin to this formulation of degrees of empathic understanding might be of use to (trainee) therapists when analysing a tape-recorded client session. The tape can be paused at any point and the therapist's communication can then be explored, looking at the degree to which empathy was experienced, the extent to which empathic understanding was communicated, the links between the levels of experienced and communicated empathy and where the client and therapist are in terms of therapeutic process, and the level, appropriateness and sensitivity of each and every spoken word.

In the previous example, it was stated that a constant striving for *level L* empathic understanding in a first session might be more to do with a therapist needing to prove their own cleverness than with demonstrating respect for and sensitivity to the client. On the other hand, if a therapist realises upon reflection that only understanding (*level A*) was communicated when feelings and meanings were evident, then there is an opportunity to explore why this might have been, and why at the time the counsellor was unable to experience and offer, say, *level C* empathy. Again, let this format come with a health warning. If this kind of reflecting on practice becomes too mechanistic, the essence of the I–thou relationship could well be lost.

Carl Rogers seemed keen to take a step back from rating each and every therapist communication on empathy scales. Perhaps like his daughter Natalie he feared that a microscopic obsession with detail ran the risk of losing the essence of the I–thou relationship – and Rogers definitely feared (as we have already seen) that a possible consequence of focusing too minutely on counsellor responses could lead to a proliferation of cold, wooden techniques and skills training. Perhaps it might make sense, then, to transpose something that he wrote when asked how long counsellor training should be. He said:

> some training is better than none, more training is better than some.
>
> (Rogers 1951: p. 442)

Our answer, then, to the question 'What does *at least to a minimal degree* actually mean?' could be:

- *Some unconditional positive regard is better than none.*
- *More unconditional positive regard is better than some.*

and

- *Some empathic understanding is better than none.*
- *More empathic understanding is better than some.*

Finally, Rogers wrote:

> And with clients in therapy, I am often impressed with the fact that even a minimal amount of empathic understanding – a bumbling and faulty attempt to catch the confused complexity of the client's meaning – is helpful, though there is no doubt that it is most helpful when I can see and formulate clearly the meanings in his experiencing which for him have been unclear and tangled.
>
> (Rogers 1961: pp. 53–4)

In Chapter 5 we shall explore the impact of some empathy being better than none and more being better than some in relation to client-centred therapeutic process. The journey of untangling begins . . .

Summary

Contact

Carl Rogers and his colleagues learned that a describable, sequential and orderly process, similar for all clients, was a predictable consequence of incongruent clients coming into contact (that is, being in a relationship within which each person is aware of the other) with a therapist for whom authentic respect and empathy were key personal traits. When the therapist is client-centred and the client is incongruent and contact has been established:

> then we may be sure that the process is already initiated.
>
> (Rogers 1961: pp. 74–5)

Client incongruence

We have seen that Carl Rogers believed it was essential for therapists to understand the intellectual context of therapy, and this includes an in-depth understanding of the concepts behind all of the necessary and sufficient conditions for personality change, including the nature of human beings and how they experience, and a concept of the self – if only because it is so relevant to clients (and this is about *client*-centred therapy). Incongruence, wrote Carl Rogers:

> results when the individual's experience is quite discrepant from the way he has organized himself.
>
> (Rogers and Evans 1975: pp. 19–20)

The conditions of worth embedded in a client's self-concept may well have served to thwart his or her actualising tendency.

Therapist congruence

Congruence is a construct composed of five main elements, namely congruence of self and experience, openness to experience, psychological adjustment, extensionality and maturity.

> It means that he is *being* himself, not denying himself. No one fully achieves this condition, yet the more the therapist is able to listen acceptantly to what is going on within himself, and the more he is able to be the complexity of his feelings without fear, the higher the degree of his congruence.
>
> (Rogers and Stevens 1967: pp. 90–1)

However, it is crucial to understand that:

> Being real does not involve us doing anything we want to do; it means a disciplined approach.
>
> (Rogers and Russell 2002: p. 284)

Therapy cannot be considered to be client-centred if therapist congruence does not as the primary feature include deeply integrated and experienced respect and empathy. As this too is so crucial, let us again read that:

> For therapy to occur the wholeness of the therapist in the relationship is primary, but a part of the congruence of the therapist must be the experience of unconditional positive regard and the experience of empathic understanding.
>
> (Rogers 1959: p. 215)

Therapist unconditional positive regard

Unconditional positive regard is about valuing a person without judging that person as good or bad, or as more or less worthy of respect:

> Unconditional positive regard, when communicated by the therapist, serves to provide the nonthreatening context in which the client can explore and experience the most deeply shrouded elements of his inner self. The therapist is neither paternalistic, nor sentimental, nor superficially social and agreeable. But his deep caring is a necessary ingredient of the 'safe' context in which the client can come to explore himself and share deeply with another human being.
>
> (Rogers 1989: p. 14)

Therapist empathic understanding

Carl wrote that:

> to enter deeply with this man into his confused struggle for selfhood is perhaps the best implementation we now know for indicating the meaning of our basic hypothesis that the individual represents a process which is deeply worthy of respect, both as he is and with regard to his potentialities.
>
> (Rogers 1951: p. 45)

Entering deeply, sensitively and delicately into a client's world means sensing their feelings and personal meanings as if you were that client, yet having a strong enough sense of your own selfhood to ensure that you remain a companion and do not get lost in the other's world.

The client perceives . . .

The client-centred therapist has *presence* in the relationship – so much so that nothing other than the client's world of feelings and meanings matters very much to the therapist during the client's time. There is a very real sense of two people being deeply connected (or at least the client senses that the therapist is striving towards a profound engagement). In striving to enter fully the private world of another individual, the therapist to some degree relies upon the client to empathise with and understand this effort – hence the notion of reciprocal empathy.

So the client-centred therapist strives to be in a *state* of *congruence* when with clients, a state that incorporates an authentic (because it is congruent), deeply held *attitude* of *unconditional positive regard* that is perceived by clients through a *process* of *empathic understanding*.

If
a person is incongruent,
then
they might seek counselling.

If
the therapist is congruent,
then
unconditional positive regard will, in harmony with the actualising
tendency, be deeply embedded and integrated within the therapist's
self-concept.

If
the therapist is congruently respectful,
then
the client will perceive this through the process of being empathically
understood by the therapist.

If
the client receives congruent unconditional positive regard and empathic
understanding,
then
a therapeutic process will unfold.

In other words,

If
the therapist can
be
the core conditions,
then
the client will
do
the rest.

Chapter 5

Empathic understanding and therapeutic process

Therapist be, client do

Carl Rogers wrote of empathy:

> We know from our research that such empathic understanding –
> understanding *with* a person, not *about* him – is such an effective
> approach that it can bring about major changes in personality.
>
> (Rogers 1961: p. 332)

Having looked earlier at the notion of two people coming into (at least minimal)
contact, and how a therapeutic *relationship* might be established in which the client
feels *received*, we can now begin to fathom what these 'major changes in
personality' might be and how they might come about. Rogers wrote that
client-centred therapeutic process is:

> [a] unique and dynamic experience, different for each individual, yet
> exhibiting a lawfulness and order which is astonishing in its generality.
>
> (Rogers 1961: p. 74)

Carl became increasingly disinterested in questions such as 'Does client-centred
therapy *apply* to this, that or the other, and will it *cure* these ailments?'. While
acknowledging that, for some individuals, hospitalisation, drugs or other medical
aid might well be appropriate, he wrote:

> client-centered therapy is very widely applicable – indeed in one sense
> it is applicable to all people. An atmosphere of acceptance and respect,
> of deep understanding, is a good climate for personal growth . . . This
> does *not* mean that it will *cure* every psychological condition, and
> indeed the concept of cure is quite foreign to the approach . . . Yet a
> psychological climate which the individual can use for deeper self-
> understanding, for a reorganization of self in the direction of more
> realistic integration, for the development of more comfortable and
> mature ways of behaving – this is not an opportunity which is of use
> for some groups and not for others. It would appear rather to be a point
> of view which might in basic ways be applicable to all individuals, even
> though it might not resolve all the problems or provide all the help
> which a particular individual needs.
>
> (Rogers 1951: p. 230)

While Rogers was not in any way opposed to legitimate and reasonable
questioning of client-centred therapy, he also proposed that the asking of such
questions was akin to enquiring whether gamma rays would be an appropriate

application or cure for chilblains! It is perhaps interesting to note that the 'Ah yes, but does this type of therapy apply to or cure A, B and C through to X, Y and Z?' type enquiry is only rarely encountered as a critique of approaches *other* than client-centred therapy, when in fact there is little or no evidence whatsoever to suggest that client-centred therapy (or person-centred counselling) is any less effective than any other approach. Indeed, there is a growing body of national and international evidence (e.g. Friedli *et al.* 1977: pp. 1662–5; King *et al.* 2000: pp. 1383–8) that in fact client-centred therapy is the most effective approach.

In Southampton Primary Care Trust, from 2003 to 2004, client-centred therapy proved to be the most effective therapeutic approach, with over 90% of patients showing improvement (as measured by CORE, Clinical Outcomes in Routine Evaluation), less than 8% showing no change, and 1% showing deterioration. Gestalt and transactional analysis faired similarly. Psychodynamic, structured/ brief and integrative approaches all showed between 70% and 80% improvement, with 10–20% of patients (approximately) showing no change and with 0–10% showing deterioration. Cognitive–behavioural therapy showed less than 70% improvement, with 30% of patients showing no change, and no patient showing deterioration (Vincent 2004).

Nevertheless, Carl Rogers consistently remained disinterested in comparing client-centred with other approaches, preferring instead to write and speak of his own view: 'I'll just speak about what you call my end of it' (Rogers and Evans 1975: p. 116) and 'rather than try to comment on such approaches . . . I'd rather say something about my own view' (Rogers and Evans 1975: p. 5). Rather than engage in such wrangling, he instead approached those issues by stating that client-centred therapy is a describable *process*, with a *core of sequential orderly events* that tend to be present *for all* clients.

> The practice of person-centred therapy dramatizes its differences from most other orientations. Therapy begins immediately; with the therapist trying to understand the client's world in whatever way the client wishes to share it.
>
> (Rogers and Raskin 1989: p. 172)

While some counsellors may begin by greeting clients and then launching into taking case histories, or meeting some other administrative, bureaucratic, 'professional' or 'therapeutic' needs, the client-centred therapist *begins immediately* by authentically respecting and striving to understand. Thus if clients look anxious and seem uncertain where to begin, client-centred therapists are likely to voice something of their sensing of where clients are, in the moment. So *therapy begins immediately* - and provided that the necessary and sufficient conditions exist, a process will unfold that has predictable outcomes. *If* the necessary and sufficient conditions (A) are met, *then* a therapeutic process (B) ensues, and *if* a therapeutic relationship (B) exists over time, *then* there are predictable outcomes (C) in terms of personality change. What, then, is the process, and what outcomes can be predicted?

Process models

> It has been the experience of many, counselors and clients alike, that when the counselor has adopted in a genuine way the function which

he understands to be characteristic of a client-centered counselor, the client tends to have a vital and releasing experience which has many similarities from one client to another.

(Rogers 1951: p. 49)

How can we explore and understand the process of client-centred therapy? Carl Rogers wrote that:

> Therapy has always been interesting to me as a process of change . . . I realize I'm a person who is greatly interested in process, and I don't know quite how to put this, but I know part of my concern about philosophy of science grew out of the fact that there is no objective way of testing process. You have to slice the process and take a cross-section of it at point *A* and point *B* and see the difference. That's the process. But there is no such thing as an objective study of a long-itudinal changing process; and that's just an intriguing aspect about science – that . . . you can prove that change takes place; you can prove the nature of that change, but you don't study the process as it occurs. You study it by taking samples at different points and investigating those.

(Rogers and Russell 2002: p. 270)

Several 'models' of therapeutic process can be found within the writings of Carl Rogers, mostly consisting of or constructed around these 'slices' or cross-sections into the process, and hence there is a marked tendency to present process as a sequence of stages or phases. The earliest model of non-directive process was composed of just three phases, namely *release, insight*, and *positive action* based on insight:

> In the earliest description of client-centered therapy, I pictured the process as composed primarily of three steps . . . first, in the release of expression, the release of personal feelings . . . Following this emotional catharsis, insight tended to develop into the origin and nature of the difficulties being experienced by the client. Such insight was followed by the making of positive choices and decisions in regard to the problem elements of the client's life – an emotional re-education involved in the practice of applying the newly gained insights in reality.

(Rogers 1989: p. 23)

However, Carl soon learned from the data gained through research that greater emphasis needed to be placed on looking into the notion of the client's self and how clients experienced processes of disorganisation and reintegration within the self.

> I think those concepts were the first rather simple attempt to understand what was going on. I suppose, in looking back on them, I feel they were oversimplified but accurate so far as they went. That release of pent-up feelings is important – there's much more to it than that:

the development of insight is important, and I don't feel therapy is complete unless it evolves into some sort of positive action.

(Rogers and Russell 2002: p. 246)

In 1951, Rogers published his findings on how *clients* experienced the process. It seems in keeping with client-centred philosophy that given the primacy of the belief and trust in the dignity, resources and rights of clients (as compared with the authority and knowledge of therapies or therapists), *clients* are placed at the heart of describing and understanding how therapy is experienced. Remember that it was through listening to and striving to understand clients that Rogers learned much of what became formulated as client-centred therapy and the person-centred approach – it could be said that *his clients taught Carl Rogers all that he knew* (or at least much of what he came to know). A five-stage model emerged from researching what clients had to report about their therapy. They tended to experience:

1 responsibility
2 the discovery of denied attitudes
3 a reorganising of the self
4 progress
5 ending.

Later, in 1961, a seven-stage model was developed that is included here because it is still used by many programmes in the UK as the 'main' training resource. People in *stage 1* are unlikely to refer themselves for therapy, as they are remote from their experience and in a 'fixed' state of being. They are unwilling to communicate in personal depth (instead referring only to external factors), and their inner communications are also significantly blocked, so both feelings and meanings are neither recognised nor owned. Relational depth is viewed as dangerous and there is no desire to grow, as basically people at this stage do not see themselves as having a problem.

It is noteworthy that throughout this model of process (other than between stages 6 and 7) Rogers saw it as essential that *the client must experience being fully received in one stage prior to moving into the next*. Again we see the *'If–then'* nature of client-centred theory. *If* clients experience themselves as fully received in one stage, *then* they can move into the next. If clients do feel fully received in stage 1, *stage 2* involves a slight increase in disclosure, although issues are still viewed as external to the self. Feelings are disowned and are sometimes seen as belonging to the past, and there is little sense of responsibility for them. The organisation of the self is still rigid or fixed and seen as a static fact. Even if contradictory feelings and meanings are verbalised, there is no awareness of their contradictory nature. *If the client feels fully received* in stage 2, then there is a movement in *stage 3* towards talking about the self, although the self is still viewed as an object and often only seen through the way in which others view that self. Feelings and meanings may be described, but they are not felt as present in the here and now and they are mostly unacceptable, being seen as somehow 'bad', shameful or embarrassing. There is still a sense of remoteness from experiencing, although some feelings are spoken of and there may be some recognition of personal constructs within the self, albeit still as objects. There is slightly less generalising and some recognition of contradictions, although any choices that the client has made are usually seen as inadequate. If the client feels fully received in stage 3, then they can move into stages 4 and 5, which according to Carl Rogers:

> constitute much of psychotherapy as we know it.
>
> (Rogers 1961: p. 139)

When clients feel that they have been welcomed, received, and understood and accepted just as they are, then in *stage 4* personal constructs tend to become slightly less rigid and feelings begin to flow more readily:

> The client describes more intense feelings of the 'not-now-present' variety. Feelings are described as objects in the present. Occasionally feelings are expressed as in the present, sometimes breaking through almost against the client's wishes. There is a tendency toward experiencing feelings in the immediate present, and there is distrust and fear of this possibility. There is little open acceptance of feelings, though some acceptance is exhibited. Experiencing is less bound by the structure of the past, is less remote, and may occasionally occur with little postponement. There is a loosening of the way experience is construed. There are some discoveries of personal constructs; there is the definite recognition of these as constructs; and there is a beginning questioning of their validity. There is an increased differentiation of feelings, constructs, personal meanings, with some tendency toward seeking exactness of symbolization. There is a realization of concern about contradictions and incongruences between experience and self. There are feelings of self-responsibility in problems, though such feelings vacillate. Though a close relationship still feels dangerous, the client risks himself, relating to some small extent on a feeling basis.
>
> (Rogers 1961: pp. 137–8)

Once clients feel fully received in stage 4, yet further processing is set in motion, characterised in *stage 5* by freely expressed and experienced current feelings, despite these often being accompanied by feelings of distrust and fear, and hence often surprise. However, there is an increasing desire to both own and be such feelings, as personal constructs move toward reduced fixity and so can be questioned and explored. Clients exhibit fewer generalisations and strive to be more precise when differentiating between feelings and personal meanings, and begin to face their incongruence and contradictions. As greater acceptance of self-responsibility accrues, so there are more meaningful internal dialogues.

At this point, Rogers stepped aside from his narrative to clarify that this is not necessarily a smooth, even process – clients might 'take two steps forward, one step back', and can journey back to previous stages as new disclosures occur. The idea is more of a general trend and movement towards greater self-actualisation which, if the client feels fully received in stage 5, enters *stage 6*. Carl stated that this stage was frequently dramatic and very distinctive:

> A feeling which has previously been 'stuck', has been inhibited in its process quality, is experienced with immediacy now. A feeling flows to its full result. A present feeling is directly experienced with immediacy and richness. This immediacy of experiencing, and the feeling which constitutes its content, are accepted. This is something which is not

something to be denied, feared, struggled against. There is a quality of living subjectively in the experience, not feeling about it. Self as an object tends to disappear. Experiencing, at this stage, takes on a real process quality. Another characteristic of this stage of the process is the physiological loosening which accompanies it. The incongruence between experience and awareness is vividly experienced as it disappears into congruence. The relevant personal construct is dissolved in this experiencing moment, and the client feels cut loose from his previously stabilized framework. The moment of full experiencing becomes a clear and definite referent. Differentiation of experiencing is sharp and basic. In this stage, there are no longer 'problems', external or internal. The client is living, subjectively, a phase of his problem. It is not an object.

(Rogers 1961: pp. 145–50)

Carl also believed that there was almost an *irreversibility* about such moments of immediacy of experience that are fully accepted. As there is a sense of the outcomes of this process by now being irreversible, it is no longer so necessary (although it is still very helpful) for the client to feel fully received prior to entering *stage* 7, in which experiencing of new feelings is both rich and immediate, both inside and outside of counselling. Clients access, own and directly trust their inner experiencing, and are more *process* than *object* beings, becoming more here-and-now oriented. As there is less need for defensive denial or distortion, self-configurations may be reformulated and loosely held, in that they will be constantly checked against ongoing experiencing:

> He has changed, but what seems most significant, he has become an integrated process of changingness.

(Rogers 1961: p. 158)

In summary, then, within this model, clients were deemed to experience a 'loosening' of feeling, changes in their manner of experiencing, shifts from incongruence to congruence, changes in the manner and extent to which they communicated the self in a receptive climate, a 'loosening' of the 'cognitive maps' of experience, changes in their relationship to 'their problems', and changes in their manner of relating.

A ten-point overview (that included much of the personality theory covered in Chapter 4 of this book) of what *clients* bring to *client*-centred therapeutic process was included in Carl's 1959 paper entitled 'A theory of therapy, personality, and interpersonal relationships, as developed in the client-centered framework'. This charted the characteristics of the human infant, the development of self, the development of need for positive regard, the development of need for self-regard, the development of conditions of worth, the development of incongruence between self and experience, the development of discrepancies in behaviour, experiences of threat and the process of defence, the process of breakdown and disorganisation, and the process of reintegration. It is the last two, namely the *process of breakdown and disorganisation* and the *process of reintegration*, that are especially relevant once a client has entered therapy. In the same paper, Carl posited the characteristics of a *deteriorating* relationship (with the self and others) and the process whereby relationships (again, with the self and others) *improve*.

With both *reintegration* and the process of an *improving* relationship, client experiencing of the client-centred *therapist* conditions (authentic respect and empathy) are at the core. Having quoted the necessary and sufficient conditions for therapeutic process, Rogers stated that when they were in place a process with 12 characteristic steps would unfold.

In *step 1*, clients become increasingly expressive (verbally, paraverbally and non-verbally). In *step 2*, such expressions move towards having more reference to the self (rather than to others or to situations). In *step 3* there is increasingly accurate symbolisation of experience, characterised by greater awareness of the differences between feelings and perceptions, and between self, others and contextual factors – and there is greater awareness of the interplay between all of the preceding steps. In *step 4* clients become more aware of incongruence between their self-concepts and their experience. In *step 5*, the threat of such incongruence is experienced.

(At this point in the process, Carl Rogers highlights therapist unconditional positive regard for clients as a key factor, stating that clients are only able to experience the threat that awareness of their incongruence brings because the therapist is as accepting and respectful of incongruence as of congruence, and as accepting of anxiety as of being anxiety free.)

In *step 6*, clients are now able to be fully aware of and experience feelings that have up until now been either denied or distorted. In *step 7* there follows a process of reorganising the self-concept in order to include previously denied or distorted experiences. In *step 8* this ongoing reorganising of the self-concept produces more and more congruence between self and experience, so experience that had up until now been too threatening to symbolise accurately in awareness can now be incorporated.

(A further consequence of *step 8* is that fewer new or ongoing experiences will be perceived as threatening, and hence there will be less denial or distortion of perceptions, and therefore less need to be defensive.)

In *step 9*, as a result, the client is more open to experiencing the unconditional positive regard of the therapist without attendant feelings of threat. In *step 10*, clients begin to feel a sense of unconditional positive *self*-regard. In *step 11*, clients move towards experiencing themselves as the primary locus of evaluation. In *step 12*, organismic valuing becomes more prominent as valuing based on a reaction to conditions of worth recedes.

Having offered this 12-step model of therapeutic process, Carl Rogers then stated that he was in no doubt that students would be puzzled by the absence of any statements about the 'mechanisms' (skills, techniques or procedures) that bring such processes about. He again pointed us to the '*If–then*' nature of his theory – *if* these conditions are present, *then* these processes are a natural and understandable (and predictable) consequence. In experiential terms, *if* you drop a pencil, *then* it will fall to the floor. You might not fully understand the theory of gravity or other relevant factors, yet you can say with some certainty that *if* the condition of standing on the planet earth is met, *then* when you let go of the pencil it will drop – not stay where it is, float off sideways or shoot upwards. In the early years of developing client-centred therapy, Rogers tended to write in terms of *propositions*, *hypotheses* and *speculation*. However, subsequent research has tended to strongly support, corroborate and validate client-centred beliefs, theory and practices.

In addition to stating that client-centred therapy *begins immediately*, Carl Rogers also wrote that to begin with the process must be *initiated*, and that *therapist attitude* is of primary importance – that client-centred therapists hold a deep respect *for* and full acceptance *of* clients *as they are in the here and now*. The client-centred therapist also believes fundamentally in clients having the innate potentialities for 'dealing' or coping with themselves and their situations. Rogers wrote that:

> If these attitudes are suffused with a sufficient warmth, which trans-forms them into the most profound type of liking or affection for the core of the person; and if a level of communication is reached so that the client can begin to perceive that the therapist understands the feelings [the client] is experiencing and accepts [the client] at the full depth of that understanding, then we may be sure that the process is already initiated.
>
> (Rogers 1961: pp. 74–5)

Rather than enquire as to whether or not client-centred therapeutic process served specific ends, Rogers was more inclined to research into the *nature* of this process, looking for its inherent characteristics and what direction or directions it tends to take, and what (if any) the natural end-points of the process might be. He wrote:

> *I enjoy the discovering of order in experience*. It seems inevitable that I seek for the meaning or the orderliness . . . Thus I have come to see both scientific research and the process of theory construction as being aimed toward the inward ordering of significant experience . . . So I have come to recognize that the reason I devote myself to research, and to the building of theory, is to satisfy a need for perceiving order and meaning, a subjective need which exists in me.
>
> (Rogers 1961: pp. 74–5)

In his book entitled *Client-Centred Therapy* (Rogers 1951), Carl had already attempted to gain understanding of the fundamental processes and to honestly and openly describe them. He was clear that he was keen to avoid 'warping' his ideas about process in order that they might somehow fit with or 'prove' any preconceived theories or clinical needs. Carl noted that there *were* certain characteristics of client movement, such as an increase in insightful statements, an increase in maturity of reported behaviour, an increase in positive attitude as therapy progresses, a change in perception of and acceptance of the self, the incorporation of previously denied experience into the self-structure, and a shift in the locus of evaluation from outside to inside the self. In other words, as therapy proceeds *the therapeutic relationship changes, personality structure changes*, as does *behaviour* and the *physiological condition* of the client. Carl wrote that even though such reflections upon therapeutic process might prove to be faulty descriptions, they were nevertheless part of an authentic attempt to understand what happens when clients and counsellors engage in a client-centred therapeu-tic experience.

Pride of place here is given to the five-stage model that Carl presented in 1961, first as it, too, is based on how *clients* reported their experience of therapy, and second, because it is my own favourite, mainly because it feels more *personal*. The five stages are as follows: *Stage 1*, experiencing of the potential self; *Stage 2*, the full experiencing of an affectional relationship; *Stage 3*, the liking of one's self; *Stage 4*, the discovery that the core of personality is positive; and *Stage 5* being one's organism, one's experience.

Experiencing of the potential self

Experiencing of the potential self could more accurately be described as the *awareness of experiencing* or even the *'experiencing of experience'*. In client-centred therapy, given the relative absence of implied or actual threat from the counsellor, clients could explore experiencing *as they actually experienced it*, and without any strong need to *deny* or *distort* that experiencing in an attempt to 'fit' it into the existing concept of self. Thus if as a client I do not feel too threatened by my therapist, *I can be me*, in the here and now – I can experience at different levels *who I am* and *how I am* being me. As much of such experiencing may well represent an extreme contradiction to the client's existing concept of self, there is often a theme along the lines of 'I cannot deny that this is how I am, yet it feels very at odds with who I am', and very often therapists hear phrases from clients such as 'It was out of character' or 'I was beside myself with . . .' or 'It's not like me at all', 'I don't know what came over me', and so on. Such expressions may be part of a process that could be formulated along the lines of 'I am experiencing a me that feels quite different from any part of my experience to date', and this moves tentatively towards 'Well, maybe there are quite a few different selves within me – or at least I have within me rather more contradictions or differences than I had realised', and this in turn moves towards 'I felt certain that it was not possible for me to *be* all of my experiences or all of my selves, because some of those experiences and some of those selves just seemed too at odds with one another – yet I am beginning to think that, just maybe, I can be all of my selves and all of my experiencing'. In other words, the client moves *away from* being a conglomeration of disparate, separate and often conflicting selves and *towards* being a community within which each and every self can be valued and embraced, celebrating difference where appropriate, rather than demanding absolute conformity and balance – several selves *being as one* rather than *being one self*, if you will.

Carl Rogers wrote of one client, *Mrs Oak*, who described her own process as:

> No longer thinking about my problems or working on them – but enjoying the feeling *process* - and certainly learning something.
> (Rogers 1961: p. 78)

Carl described how, at first, client experiences may be examined by the client without him or her particularly associating them with the self, while later in the counselling process they become assimilated into the client's self structure. Mrs Oak described her therapy as a kind of 'song without words', a kind of poem, only to become dismissive of this, discounting or 'writing off' her experience as an unacceptable form of self-indulgence. All too often the therapist hears clients say things like there being so many other people 'out there' far more deserving of help than the client him- or herself, other people with far more serious problems, concerns and situations. Very often clients report that they feel 'bad' about 'wasting the therapist's time' and feel guilt and shame about this focus on their selves. Yet such feelings and thoughts meet with authentic therapist respect (which does not mean that the therapist agrees with the client) and empathic understanding (not reassurance), and as a consequence clients tend to move into a new phase – one that Rogers summarised as follows:

> *I came here to solve problems, and now I find myself just experiencing myself.*
> (Rogers 1961: p. 80)

Mrs Oak said that it was somewhat akin to blind people learning Braille – that is, it was like *learning a new way of experiencing*. There follows a shift *away from* mechanical or *practical problem solving* and *doing* and *towards being* a *process* of *experiencing*. At a cognitive or intellectual level, clients often report that the experiencing of experiencing somehow seems 'wrong' (hence the feelings of guilt or shame), and yet *at an experiential level the experiencing of experiencing feels good*. It may well be that early in the therapeutic process this experiencing of experience is a kind of free-flowing, 'go for it', 'no holds barred – just be it' way of being, because it feels so good, so new, so unusual, so different, and so liberating that clients really revel in it. However, as therapy progresses, more attention is paid by the client to exploring and organising this experiencing, and then assimilating or incorporating their self-discoveries into the self-concept.

The full experiencing of an affectional relationship

A second characteristic of client-centred therapeutic process is:

> the learning, on the part of the client, to accept fully and freely and without fear *the positive feelings of another*.
> (Rogers 1961: pp. 80–1)

Research findings indicated that this learning does not occur in every single client-centred therapeutic experience, although it is more frequent in long-term counselling – yet even then the pattern is not uniform. However, this learning *is* more likely to be a feature of *successful* therapy. The full experiencing of an affectional relationship is *deep* and *significant*, and yet it is often *not verbalised*. For Mrs Oak, for instance:

> This letting the counsellor and his warm interest into her life was *undoubtedly one of the deepest features of therapy*.
> (Rogers 1961: p. 81)

Carl Rogers wrote that the phenomenon of clients fully receiving therapist unconditional positive regard is definitely *not* one of transference and counter-transference. Rather, it is something *mutual* and *appropriate* (whereas so-called transference and counter-transference are characteristically one-way and inappropriate to the 'realities' of the situation). Rogers believed that there was a discernible trend towards client-centred therapists becoming less afraid of their positive (and negative) feelings towards clients (which in turn may be linked with more activities aimed at increasing therapist *congruence*). For example, at the beginning of therapy the client-centred counsellor *receives* clients with a deep respect for their innate *potential*:

> As a result [the therapist] feels toward the client a warm, positive, affectional reaction. This poses a problem for the client who often . . . finds it difficult to accept the positive feeling of another. Yet once accepted the inevitable reaction on the part of the client is to relax, to let the warmth of liking by another person reduce the tension and fear involved in facing life.
>
> (Rogers 1961: p. 82)

As tension and fear dissipate, so the therapeutic process unfolds, and the attitude of the therapist becomes more one of *awe* as the counsellor is witness to a process that is experienced like watching a beautiful butterfly emerging from a chrysalis, or seeing a foal determinedly taking its first few stumbling steps, or a newly hatched baby turtle struggling to reach the sea. Indeed, Carl Rogers wrote that being with the awesome actualising of potential can engender far deeper feelings of connection – not just with or to the client but with the whole brotherhood and sisterhood of human beings, or even a kind of universal, spiritual connection and sense of belonging.

Yet being in awe of clients is, according to Carl Rogers, not quite the same as feeling *love*, for he stated that profound unconditional positive regard is even more basic or fundamental than love, sexual desires or parental commitments to children.

> It is the simple outgoing human feeling of one individual for another . . . caring enough about the person that you do not wish to interfere with [another's] development, nor to use [others] for any self-aggrandizing goals of your own. Your satisfaction comes in having set others free to grow in [their] own fashion.
>
> (Rogers 1961: p. 84)

We have seen that on entering therapy clients are likely to be vulnerable, anxious, incongruent, defensive, and lacking in self-esteem. Thus clients are likely to believe that they do not deserve or merit being respected or prized by others. They especially do not believe that they merit *unconditional* prizing. Most if not all of their relationships up until now will have been *conditional*, not unconditional, so this experiencing with their therapist may well be something clients feel that they have never received, or at least a quality they have no recollection of ever having experienced. Clients may also try to minimise the significance of their therapist to them, for prizing from a significant other is threatening. If a client's self-concept (largely constructed around introjected conditions of worth) dictates to them that they are not worthy of being openly and freely valued, then there will understandably follow a defence against the full receiving of therapist unconditional positive regard. However:

> To discover that it is *not* devastating to accept the positive feeling from another, that it does not necessarily end in hurt, that it actually 'feels good' to have another person with you in your struggles to meet life – this may be one of the most profound learnings encountered by the individual, whether in therapy or not.
>
> (Rogers 1961: p. 85)

This aspect of the process does not end here – for if a client becomes openly able to experience therapist unconditional positive regard (that is, if there is no longer a defensive need to protect the self-concept by denying or distorting perceptions of being unconditionally valued) then, as a consequence, the client becomes more able to value others. Rogers summarised this by writing that:

> It appears possible that one of the characteristics of deep or significant therapy is that the client discovers that it is not devastating to admit fully into his own experience the positive feeling which another, the therapist, holds towards him. Perhaps one of the reasons why this is so difficult is that essentially it involves the feeling that 'I am worthy of being liked' . . . This aspect of therapy is a free and full experiencing of an affectional relationship which may be put in generalized terms as follows: 'I can permit someone to care about me, and can fully accept that caring within myself. This permits me to recognise that I care, and care deeply, for and about others'.
>
> (Rogers 1961: p. 86)

Thus in early process stages the client might distort the experiencing of unconditional positive regard from their therapist by being cynical (for instance, their therapist is being paid to like them, or is simply being professional and doing his or her job). To admit that I am a person worthy of the effort of a significant other trying to respectfully understand me might be very threatening to the concept of self. Yet once I can *allow* myself to be unconditionally prized, I discover that I, too, can allow myself to care deeply and unconditionally about others.

The liking of one's self

So in successful client-centred therapy, *negative self-attitudes decrease* and *positive self-attitudes increase* – there is a gradual *increase in self-acceptance* and a correlated *increase in acceptance of others*. However, Rogers wrote that such statements do not entirely or accurately portray the complete picture, as they could be interpreted as implying a *grudging* or *reluctant* acceptance, whereas in successful client-centred therapy the client not only becomes more accepting of self, but actually comes to *like* that self. Carl qualified this statement, too, when he wrote that:

> This is not a bragging or self-assertive liking; it is rather a quiet pleasure in *being* one's self.
>
> (Rogers 1961: p. 87)

This process *may* be a consequence of the full experiencing of an affectional relationship, the experiencing in full awareness of therapist unconditional positive regard and empathic understanding. There may be a sense of spontaneity or of relaxation, of enjoyment or a 'primitive joie de vivre' – a basic appreciation of *being oneself*. It is possible that this is a way of being that clients *rediscover* rather than something new that is discovered (that is, if self-actualising has served to thwart the actualising tendency, reconnecting with the actualising tendency may well feel like a 'new' experience). Carl Rogers made references to analogies such as the frisky, gamboling lamb, and the playful dolphin leaping and splashing in the water, in an attempt to capture this joyful awakening.

Very often when clients enter therapy they soon disclose that they feel 'dysfunctional' or that they 'don't even know who the real me is' any more, or that they are 'in pieces' and broken – a kind of 'I'm not working properly any more' theme. However, this can move towards experiencing themselves as more healthy, whole, 'functional' people who, rather than rejecting, denying, suppressing or distorting their feelings can actually *be* and *live* them. Carl Rogers stated that this is:

> an important and often overlooked truth about the therapeutic process. It works in the direction of permitting the person to experience fully, and in awareness, all of his reactions including his feelings and emotions. As this occurs, the individual feels a positive liking for himself, a genuine appreciation of himself as a total functioning unit, which is one of the important *end points* of therapy.
>
> (Rogers 1961: p. 90)

The discovery that the core of personality is positive

Carl Rogers wrote that:

> One of the most revolutionary concepts to grow out of our clinical experience is the growing recognition that the innermost core of man's nature, the deepest layers of his personality, the base of his 'animal nature', is positive in nature – is basically *socialised, forward-moving, rational* and *realistic*.
>
> (Rogers 1961: pp. 90–1)

Rogers did not expect his findings to be accepted by our modern culture. Most major religions, for instance, permeate societies with the concept that we are all fundamentally sinful and in need of either punishment, forgiveness or repentance – and that only by something akin to a miracle (or death) can we become inherently good. We have to be saved from our inner natures. Likewise in the so-called 'caring' professions, where the influence of Sigmund Freud and his psychoanalytic followers, for example, indicated that if the 'id' enshrined within us were to be uncontrolled or unrestrained, the resultant mayhem would be catastrophic. Behaviourists, too, seemed to see people as things to be restrained, controlled and managed, and more cognitive approaches set about correcting our faulty thinking. Overwhelmingly, whether it be religion or psychology, *people simply could not be trusted to direct their own lives*. Carl Rogers therefore instigated thorough research activities in an attempt to validate his deeply held views based on his own experiences as a client-centred psychotherapist. Of religions and psychoanalysis he wrote:

> The whole problem of therapy, as seen by this group, is how to hold these untamed forces in check in a wholesome and constructive manner, rather than in the costly fashion of the neurotic.
>
> (Rogers 1961: p. 91)

Furthermore, such views that at heart human beings are irrational, unsocialised, self-destructive and destructive of others tended to be (and perhaps still are, by many) accepted without question. Although his own beliefs could be seen as radical and revolutionary, blasphemous or just plain wrong, Rogers was not entirely alone. He was able to quote Abraham Maslow, for instance, who posited that 'anti-social' emotions (such as hostility and jealousy) result from the *frustration* of *more basic impulses* for love, security and belonging, which are in themselves desirable. Rogers also referred to Montagu, who postulated that *co-operation*, rather than struggle, is the basic law of human life. Yet the alleged 'caring' resources of so-called 'advanced' societies seemed to serve to 'keep us under control' or 'keep us under cover', or both. Rogers stated that *of course* hostile and anti-social feelings are continually uncovered in counselling, yet must one therefore assume that these indicate that our deeper and therefore basic natures are somehow 'bad' or even 'evil'? Rogers believed that feelings associated with hostility (or feelings that are deemed to be 'anti-social') were not represent-ative of either our deepest or even our strongest essence. Rather:

> the inner core of man's personality is the organism itself, which is essentially both *self-preserving* and *social*.
>
> (Rogers 1961: p. 92)

In the case of Mrs Oak, beneath her first layer of defence lay *bitterness* and a desire for *revenge*. A 'murderous feeling of hatred and a desire to get even' was behind the 'socially controlled surfaces of her behaviour'. Considerably later on (after 23 sessions), Mrs Oak returned to this theme, this time feeling somewhat emotion-ally blocked, and when the 'barriers burst', the feeling has become *less* anti-social and more like a 'desire to get back at the world which has cheated her, a deep experience of having been hurt'. At this deeper level, it is very clear indeed that she had no desire to put her murderous feelings into action. In fact, she *disliked* those feelings and *wished to be rid of them*. Three sessions later, 'reaching far down into herself', the following consecutive themes were expressed:

> I'm going to talk about myself as *self*-ish
> but with a new connotation to the word.
>
> I've acquired an acquaintance with the structure of myself,
> *know myself* deeply.
>
> As I descend into myself, I discover something exciting,
> a *core* that is *totally without hate*.
>
> It can't be a part of everyday life – it may even be *abnormal*.
>
> I thought first it was just a *sublimated sex drive*.
>
> But no, this is more *inclusive, deeper* than sex.
>
> One would expect this to be the kind of thing one would discover
> by going up into the thin realms of *ideals*.
>
> It seems to be something that is the *essence* that lasts.
>
> (Rogers 1961: pp. 99–100)

This almost 'mystical' client summary, by the next session, has become clearer, more concise, and the *difficulty of the experiencing is acknowledged* by the client. Rogers wrote:

> Her feeling goes against the grain of her culture, [yet] she feels bound to say that the *core* of herself is *not bad, nor terribly wrong,* but something *positive. Underneath* the layer of *controlled surface behaviour,* underneath the *bitterness,* underneath the *hurt,* [underneath the murderous hate and desire for revenge] is a *self that is positive,* and that is *without hate.* This I believe is the lesson which our clients have been facing us with for a long time, and which we have been slow to learn.
>
> (Rogers 1961: p. 101)

The deeper one digs into oneself, the less one has to fear. Instead of discovering 'bad' or 'terribly wrong', there is a core self which is deeply socialised, without hate, and wanting to neither punish nor reward others.

In looking at 'negative', 'aggressive' or 'hostile' aspects of human behaviour, at this point I turn from Carl's narrative about Mrs Oak towards my own experience. After thousands upon thousands of client hours, I can state that *not once* have I ever experienced anger, rage, hostility or aggression as being at the core of, or being the essential essence of, a human being. *Inevitably* there were other factors underlying and generating such apparently 'negative' emotions and thoughts. Take anger, for instance. In terms of the belief system that forms a significant part of a self-concept, there was often the thought that something was unfair or unjust. Outrage at injustice may not only be an understandable reaction, it might even be considered honourable (and hence acceptable or even highly desirable) in certain contexts. Who decides which contexts make anger at injustice acceptable or unacceptable? Let us say, however, that in one particular instance the perceived injustice seems 'irrational' or 'unreasonable' – maybe exhibited by a 'selfish' individual who is seemingly unable to see past or beyond their own needs and desires, and to hell with anyone else. Is this person inherently bad? Or might this person feel deprived of the love and affection that they had been led to believe was their birthright?

Quite often when engaging in a debate about the basic nature of human beings, someone will say 'Ah . . . but what about Hitler? How can anyone conceivably say that he was innately good?'. Well, personally I cannot conceive that Adolf Hitler was inherently bad at the time he was conceived! I wonder, for instance, how things might have been if this young Adolf, with his creative flair and artistic temperament, had not been so cruelly rejected (as he perceived it) for art college.

Being one's organism, one's experience

The client-centred approach to therapy, Carl Rogers wrote, is:

> a process whereby man *becomes* his organism – *without self-deception, without distortion.*
>
> (Rogers 1961: p. 103)

It is because this process and its outcomes are *experiential* (a way of being) that it does not readily lend itself to accurate symbolisation in words alone. Indeed, Carl stated that if this experience were to be understood in words alone, this would be but an abstract from or a distortion of the totality of this way of being (in the way that a map of a piece of land is not the territory itself). Yet we try. There is a sense in which the process of client-centred therapy is towards a rediscovery of what once was – getting back in touch with the organismic valuing process that reigned prior to the onset of a concept of self that was formulated through introjected values and the sometimes tyrannical rule of consequent conditions of worth. Historically, *prior to* the advent of client-centred therapy, clients might have asked themselves things like 'What do others think I should do?', 'What would my parents suggest that I should do?', 'What would my culture tell me to do?' and 'What do I think I should or ought to do?'. Yet clients may not necessarily *act* in or *be in accord with* the 'shoulds' and 'oughts' of others (individually or as groups). Even when in rebellion against external loci of evaluation, the person is nevertheless *acting in terms of those external expectations*, and often those expectations have been *introjected into the self-concept*. (I am reminded of the woman who, having felt liberated from a lousy relationship for some months, suddenly froze in the middle of a clothes-shopping expedition. She suddenly realised that she was buying exactly the opposite of what she used to buy for, or on behalf of, her ex-partner. It struck her like a thunderbolt that actually it was the ex-partner's values that were still the governing factor, only this time it was in opposition to them. She was dumbstruck as she realised that she did not know what she herself liked.)

During client-centred therapy, clients are more likely to ask 'How do I *experience* this?' and 'What does it *mean* to me?' and 'If I behave in a certain way, *how do I symbolise the meaning* which it will have for me?'. Rogers termed this openness to experiencing and the accurate perception and symbolisation of that experiencing '*realism*', in the sense that the person is able to access all significant valuing and achieve a realistic 'balance' based on this absence of denial or distortion. It does seem that a great deal of human energy is expended on trying to be, feel and do in a consistent manner. For instance, the recently bereaved should only feel sorrow, even if a part of them (no matter how small) perhaps feels a sense of relief at being unburdened from an exhausting and inconvenient duty of care. If this feeling of relief is admitted to awareness (that is, if it is symbolised in consciousness), it is accompanied by deep guilt for being so shameful, callous and selfish as to have an emotion that one is not expected to feel. Rogers gave several of his own examples, including the following:

> I have thought I must feel only *love* for my parents, but I find that I experience both love and *bitter resentment*. Perhaps I can be that person who *freely experiences both love and resentment*.

And:

> I have thought I was only *bad* and *worthless*. Now I experience myself at times as one of *much worth* – at other times as one of *little worth* or usefulness. Perhaps I can be a person who *experiences varying degrees of worth*.
>
> (Rogers 1961: p. 104)

For another client:

> I have held the conception that *no-one could really love me for myself*. Now I experience the *affective warmth of another* for me. Perhaps I can be a person who is *loveable by others* – perhaps I *am* such a person.
>
> (Rogers 1961: p. 104)

And for yet another:

> I have been brought up to feel that *I must not appreciate myself – but do*. I can *cry* for myself, but I can *enjoy* myself, too. Perhaps *I am a richly varied person* whom I can enjoy and for whom I can feel sorry.
>
> (Rogers 1961: p. 104)

Mrs Oak said:

> I have thought that in some deep way I was *bad*, that the most basic elements in me must be *dire* and *awful*. I don't *experience* that *badness* but rather a *positive desire to live and let live. Perhaps I can be that person who is, at heart, positive.*
>
> (Rogers 1961: p. 104)

Rogers believed that a fundamental ingredient that makes much of this process of becoming possible is complete and free *awareness of experiencing*. There is an *elimination* (or at least minimising) of *denial and distortion to awareness*. There is an awareness of *what is actually experienced* through all of the body and mind ('sensory and visceral' experiencing) – not simply the crumbs that are left over after our needs to defend our self-concepts (and conditions of worth) have screened out or filtered our experience in an attempt to either 'make it fit' or deny it altogether. This, then, is about *realising full potential*, with:

> the enriching element of awareness freely added to the basic aspect of sensory and visceral reaction. The person comes to *be* what he *is*, as clients so frequently say in therapy . . . the individual comes to *be* – in *awareness* – what he *is* - in experience. He is, in other words, a *complete and fully functioning human organism.*
>
> (Rogers 1961: pp. 104–5)

Carl Rogers was very passionate about what he saw as the appalling way in which people had hitherto been treated in the name of health, based on socialised beliefs that we are all inherently 'bad' and untrustworthy. He wrote that:

> When a man is *less* than fully man – when he *denies to awareness* various aspects of his experience – then indeed we have all too often reason to fear him and his behaviour, as the present world situation testifies.
>
> (Rogers 1961: p. 105)

In other words, if we believe that we are by our very innermost natures evil or hostile or murderous or licentious, then we are likely to treat ourselves and each other in ways designed to prevent our animalistic ways holding sway. Yet Rogers believed that:

> There is no beast in man. There is only man in man, and this we have been able to release.
>
> (Rogers 1961: p. 105)

If we can dare to believe that our innermost natures, based on a constructive actualising tendency, are actually *positive* in nature, then rather than control, manage, manipulate, educate, change or suppress the human psyche, our innate potential becomes open to being *released* (through the provision of appropriate conditions). Here then is a real message of hope:

> If we can *add to* the sensory and visceral experiencing which is characteristic of the whole animal kingdom, the gift of a *free and undistorted awareness* of which only the human animal seems fully capable, we have an organism which is *beautifully and constructively realistic*. We have then an organism which is as *aware* of the demands of the culture as it is of its own physiological demands for food or sex – which is just as *aware* of its desire for friendly relationships as it is of its desire to aggrandize itself – which is just as *aware* of its delicate and sensitive tenderness toward others, as it is of its hostilities toward others. When man's *unique capacity of awareness* is thus functioning freely and fully, we find that we have not an animal whom we must fear, not a beast who must be controlled, but an organism *able to achieve*, through the remarkable integrative capacity of its central nervous system, a *balanced, realistic, self-enhancing, other-enhancing* behaviour as a resultant of all these elements of *awareness*.
>
> (Rogers 1961: p. 105)

Carl believed that 'fully functioning' people are to be *trusted* and that their behaviour is *constructive*. In today's world, fully functioning people might not be perceived as entirely conventional and they might not always blindly conform to accepted norms and mores, yet Rogers believed that by our very natures we are both individuals (and hence individualised) and social beings (and hence socialised). He also acknowledged the difference between *deep conviction* and *truth*. He wrote of his own deep truths, and asked readers to look deep inside themselves and see whether or not the deep convictions of Carl Rogers resonate within.

Therapist be

If the therapist initiates certain conducive conditions, *then* a process can unfold in which the innate potential of human beings may be released. Carl Rogers stated that:

> In client-centered therapy we are deeply engaged in the prediction and influencing of behavior, or even the control of behavior. As therapists, we institute certain attitudinal conditions, and the client has relatively little voice in the establishment of these conditions. We predict that if these conditions are instituted, certain behavioral consequences will ensue in the client.
>
> (Rogers and Evans 1975, p. lxxvi)

It is *the therapist* who takes the core conditions to the relationship, and *the therapist* who establishes a contact based on those conditions. So, having placed the therapist conditions back together again in Chapter 4, now let us to some degree take them apart again! As we move through these conditions, they do in fact interface very much throughout.

Congruence and process

The greater the degree of counsellor congruence (that is, congruence between experience, awareness, and communication), the more likely it is that the client will experience *clear* and therapeutic communication. If all verbal, paraverbal and non-verbal therapist expressions spring from congruence and unity, then there is much less likelihood that these cues will have an ambiguous or unclear meaning to the client. In order to achieve relatively high degrees of congruence, therapists need to listen to themselves, to get in touch with their own inner, complex, diverse and ever-changing attitudes and feelings, and to hear themselves with *acceptance* – as Rogers said, to become more adequate at *being* the person you *are*, and that just as clients need to feel received in order to grow, so therapist self-acceptance is necessary for personal growth.

The more clearly the counsellor communicates, the more *the client* responds to the counsellor with clarity. Thus even though clients are likely to be relatively incongruent, nevertheless their responses will have *more* clarity and congruence than would be the case if the counsellor's communications were themselves incongruent, confusing or ambiguous. The greater the degree of counsellor congruence (for instance, the less the counsellor is presenting a professional façade or a role that has to be defended), the less need there is for that counsellor to be defensive, and hence the greater the counsellor's ability to listen accurately and hear the client. This links *congruence* to *empathic understanding*, because at least to this degree clients feel empathically understood – that whether client expressions have been defensive or congruent, the counsellor has understood clients more or less as they see themselves. This also links congruence and empathic understanding to *unconditional positive regard*, as the client who feels really heard and to some degree understood (which is a consequence of the counsellor respecting the client) is likely to experience positive feelings (regard) *towards their counsellor* (the mutual respect and understanding that Rogers called 'the experiencing of an affectional relationship'). As a consequence of this there are fewer barriers to communication – the need for defensiveness gradually decreases. In other words, it is therapeutic if clients experience their therapists as real:

> I would like my feelings in this relationship with him to be as clear and transparent as possible, so that they are a discernible reality for him, to which he can return again and again . . . I recognize that this is a very human and unpredictable journey for me, as well as for him, and that I may, without even knowing my fear, shrink away within myself, from some of the feelings he discovers. To this extent I know I will be limited in my ability to help him. I realize that at times his own fears may make him perceive me as uncaring, as rejecting, as an intruder, as one who does not understand. I want fully to accept these feelings in him, and yet I hope also that my own real feelings will show through so clearly that in time he cannot fail to perceive them. Most of all I want him to encounter in me a real person. I do not need to be uneasy as to whether my own feelings are 'therapeutic'. What I am and what I feel are good enough to be a basis for therapy, if I can transparently *be* what I am and what I feel in relationship to him. Then perhaps he can be what he is, openly and without fear.

And the client, for his part, goes through far more complex sequences which can only be suggested. Perhaps schematically his feelings change in some of these ways. 'I'm afraid of him. I want help, but I don't know whether to trust him. He might see things which I don't know in myself – frightening and bad elements. He seems not to be judging me, but I'm sure he is. I can't tell him what really concerns me, but I can tell him about some past experiences which are related to my concern. He seems to understand those, so I can reveal a bit more of myself. But now that I've shared with him some of this bad side of me, he despises me. I'm sure of it, but it's strange I can find little evidence of it. Do you suppose that what I've told him isn't so bad? Is it possible that I need not be ashamed of it as a part of me? I no longer feel that he despises me. It makes me feel that I want to go further, exploring *me*, perhaps expressing more of myself. I find him a sort of companion as I do this – he seems really to understand.

But now I'm getting frightened again, and this time deeply frightened. I didn't realize that exploring the unknown recesses of myself would make me feel feelings I've never experienced before. It's very strange because in one way these aren't new feelings. I sense that they've always been there. But they seem so bad and disturbing I've never dared to let them flow in me. And now as I live these feelings in the hours with him, I feel terribly shaky, as though my world is falling apart. It used to be sure and firm. Now it is loose, permeable and vulnerable. It isn't pleasant to feel things I've always been frightened of before. It's his fault. Yet curiously I'm eager to see him and I feel more safe when I'm with him.

I don't know who I am any more, but sometimes when I *feel* things I seem solid and real for a moment. I'm troubled by the contradictions I find in myself – I act one way and feel another – I think one thing and feel another. It's very disconcerting. It's also sometimes adventurous and exhilarating to be trying to discover who I am. Sometimes I catch myself feeling that perhaps the person I am is worth being, whatever that means.

I'm beginning to find it very satisfying, though often painful, to share just what it is I'm feeling at this moment. You know, it is really helpful to try to listen to myself, to hear what is going on in me. I'm not so frightened any more of what *is* going on in me. It seems pretty trustworthy. I use some of my hours with him to dig deep into myself to know what I *am* feeling. It's scary work, but I want to *know*. And I do trust him most of the time, and that helps. I feel pretty vulnerable and raw, but I know he doesn't want to hurt me, and I even believe he cares. It occurs to me as I try to let myself down and down, deep into myself, that maybe if I could sense what is going on in me, and could realize its meaning, I would know who I am, and I would also know what to do. At least I feel this knowing sometimes with him.

I can even tell him just how I'm feeling toward him at any given moment and instead of this killing the relationship, as I used to fear, it seems to deepen it. Do you suppose I could be my feelings with other people also? Perhaps that wouldn't be too dangerous either.

You know, I feel as if I'm floating along on the current of life, very adventurously, being me. I get defeated sometimes, I get hurt sometimes, but I'm learning that those experiences are not fatal. I don't *know* exactly *who* I am, but I can feel my reactions at any given moment, and they seem to work out pretty well as a basis for my behavior from moment to moment. Maybe this is what it *means* to be *me*. But of course I can only do this because I feel safe in the relationship with my therapist. Or could I be myself this way outside of this relationship? I wonder. I wonder. Perhaps I could . . .'.

(Rogers 1961: pp. 67–9)

Thus clients are now able to hear themselves *and* receive the communications from their counsellor more clearly, and hence the process of reciprocal empathy has been established.

This means that to some degree the process of therapy occurs in each and that the outcomes of therapy will to that same degree occur in each; change in personality in the direction of greater unity and integration; less conflict and more energy utilizable for effective living; change in behaviour in the direction of greater maturity.

(Rogers 1961: p. 344)

Relating congruence to interpersonal communication, Carl Rogers proposed a 'tentative statement of a general law' stating that, assuming an ongoing minimal willingness to be in contact and receive communication from each other:

The greater the congruence of experience, awareness and communication on the part of one individual, the more the ensuing relationship will involve: a tendency toward reciprocal communication with a quality of increasing congruence; a tendency toward more mutually accurate understanding of the communications; improved psychological adjustment and functioning in both parties; mutual satisfaction in the relationship.

(Rogers 1961: p. 344)

It follows that the opposite is true, too – that the greater the *incongruence* of the therapist, the greater the 'disintegration' of accurate empathic understanding, resulting in psychological 'maladjustment' and hence a dissatisfying therapeutic relationship for both counsellor and client.

Congruence: Fay, an illustration

Fay was referred for counselling by her GP, having 'fallen apart' as a consequence of the termination of a relationship. She was in torment, with one part (her self-structure) trying its best to convince another part (a part that was desperately yearning to be with her partner again) that actually she was better off without him – battle was raging within.

Let me be honest here: I abhor violence and aggression in whatever way, shape or form – I am a confirmed pacifist and coward, and proud of being both. Therefore one judgement would be that *of course* Fay would be better off without her former partner, for he sounded like a 'nasty piece of work'! However, to voice such an opinion within the therapy session would not have been congruent, as my client-centred *discipline* means that while such an opinion might very fleetingly enter my mind (mostly due to a physiological reaction of disgust at some of his reported behaviour), my primary commitment is to strive to enter *Fay's* world of feelings and meanings. My anti-violence values are neither denied nor distorted in the moment – they are simply irrelevant, or at least relatively irrelevant in relation to Fay's world of feelings and meanings. Perhaps the notion of taking precedence might illustrate client-centred discipline. In counselling, the client's frame of reference *takes precedence over* the counsellor's frame of reference. If there is no psychological denial or distortion on the part of the therapist, unconditional positive regard, in large part communicated through empathic understanding, remains authentic and is largely communicated through being empathic.

Unconditional positive regard and process

To begin with, Carl Rogers wrote that:

> When the therapist is experiencing a positive, acceptant attitude toward whatever the client is at that moment, therapeutic movement or change is more likely to occur.
>
> (Rogers 1980: p. 116)

This view that the client experiencing respect from the counsellor was therapeutically helpful moved from being a *belief* in the early days into being a therapeutic value based on solid experience and data, for:

> Research indicates that the more this attitude is experienced by the therapist, the greater the probability that therapy will be successful.
>
> (Rogers 1978: p. 10)

So what is this change that is more likely to occur if clients experience unconditional positive regard, and how might this change occur?

> In therapy, with its climate of acceptance and safety and its freedom to explore one's feelings whatever they may be, it becomes possible for the client to experience the feelings that have not been admitted into his concept of self. Once experienced in an accepting climate, they can gradually be incorporated into his self-picture, and he thereby achieves more unity and integration between the person he organismically is and the self he perceives himself as being.
>
> (Rogers 1989: p. 24)

Thus we can see that client experiencing of unconditional positive regard from the therapist enables the client to inwardly go where they may not have been before. There is a hitherto unexperienced freedom to explore the self:

> When the counselor perceives and accepts the client as he is, when he lays aside all evaluation and enters into the perceptual frame of reference of the client, he frees the client to explore his life and

experience anew, frees him to perceive in that experience new meanings and new goals.

(Rogers 1951: p. 48)

As a consequence of this freedom to self-explore in an acceptant climate, self-perceptions may change and be incorporated into a new and ever-changing 'picture' of the self. However, Rogers said:

> I believe that the person can only accept the unacceptable in himself when he is in a close relationship in which he experiences acceptance. This, I think, is a large share of what constitutes psychotherapy – that the individual finds that the feelings he has been ashamed of or that he has been unable to admit into his awareness, that those can be accepted by another person, so then he becomes able to accept them as a part of himself.

(Rogers 1951: p. 71)

As we have seen, the client-centred therapist does not have to be *perfect* in this experiencing of respect that clients perceive – if this were so, there would be no client-centred therapists. Yet we have also already encountered the conviction that *imperfect* human beings can be of help (especially if they are aware and accepting of their imperfections). Although it may be true that more empathy is better than some, and some is better than none, at the heart of the empathic process is a therapist who is *striving* to understand:

> The accuracy of the therapist's empathic understanding has often been emphasised, but more important is the therapist's *interest* in appreciating the world of the client and offering such understanding with the willingness to be corrected. This creates a process in which the therapist gets closer and closer to the client's meanings and feelings, developing an ever-deepening relationship based on respect for and understanding of the other person.

(Rogers and Raskin 1989: p. 171)

Thus the therapist attitude of unconditional positive regard precedes empathic understanding in the necessary and sufficient conditions because any understanding is born out of the genuine desire to understand, based on a belief that people are worth the effort. This desire and genuine interest, as we have seen in Chapter 4, is ongoing through each moment.

> Acceptance does not mean much until it involves understanding. It is only as I *understand* the feelings and thoughts which seem so horrible to you, or so weak, or so sentimental, or so bizarre – it is only as I see them as you see them, and accept them and you, that you feel really free to explore all the hidden nooks and frightening crannies of your inner and often buried experience. This *freedom* is an important condition of the relationship. There is implied here a freedom to explore oneself at both conscious and unconscious levels, as rapidly as one can dare to embark on this dangerous quest.

(Rogers 1961: p. 34)

Here, then, we see an interweave between respect and empathy. Respect is communicated through a belief that the other person is worth the effort of striving to understand, and without understanding that respect means little. The therapist appreciates the client enough to want to understand, and, as the client reveals more, and as each disclosure meets with non-evaluative unconditional positive regard and a degree of empathic understanding, further respect is evident as the therapist strives to respectfully engage with and understand this untangling client.

While it has been stressed that the client-centred therapist does not intend to offer support, clients may well nevertheless experience the counselling relationship as supportive, because they will feel *prized* through perceiving the therapist's attitude of unconditional positive regard and consequent striving to empathise – an attitude that is based at least in part upon a belief in the primacy of the actualising tendency, and the right and capacity of clients to self-direction and dignity. Thus one consequence of sensitive and respectful empathy is that it helps to *dissolve feelings of alienation*, because the client feels deeply accepted as the person he or she is from moment to moment:

> A second consequence of empathic understanding is that the recipient feels valued, cared for, accepted as the person that he or she is . . . It is impossible to accurately sense the perceptual world of another person unless you value that person and his or her world – unless you, in some sense, care. Hence, the message comes through to the recipient that 'this other individual trusts me, thinks I'm worthwhile. Perhaps I *am* worth something. Perhaps *I* could value *myself*. Perhaps I could care for myself'. It is, I believe, the therapist's caring understanding . . . which has permitted this client to experience a high regard, even a love, for himself.
>
> (Rogers 1980: p. 152)

Remember that Carl Rogers believed that it was *impossible* for a judgemental therapist to perceive a client's world of feelings and meanings with accuracy.

Thus we begin to see some of the process consequences of respectful empathic understanding. Through the client perceiving the attitude of unconditional positive regard that underpins the therapist's striving to understand, there is a tendency for feelings of alienation to dissolve, an increase in client positive self-regard, and an increase in client self-acceptance.

The following quote demonstrates why Carl Rogers believed unconditional positive regard to be a necessary precursor to empathic understanding:

> In the emotional warmth of the relationship with the therapist, the client begins to experience a feeling of safety as he finds that whatever attitude he expresses is understood in almost the same way that he perceives it, and is accepted. He then is able to explore, for example, a vague feeling of guiltiness which he has experienced. In this safe relationship he can perceive for the first time the hostile meaning and purpose of certain aspects of his behavior, and can understand why he has felt guilty about it, and why it has been necessary to deny to awareness the meaning of this behavior. But this clearer perception is in itself disrupting and anxiety-creating, not therapeutic. It is evidence

to the client that there are disturbing inconsistencies in himself, that he is not what he thinks he is. But as he voices his new perceptions and their attendant anxieties, he finds that this acceptant alter ego, the therapist, this other person who is only partly another person, perceives these experiences too, but with a new quality. The therapist perceives the client's self as the client has known it, and accepts t; he perceives the contradictory aspects which have been denied to aware-ness and accepts those too as being a part of the client; and both of these acceptances have in them the same warmth and respect. Thus it is that the client, experiencing in another an acceptance of both these aspects of himself, can take toward himself the same attitude. He finds that he too can accept himself even with the additions and alterations that are necessitated by these new perceptions of himself as whole. He can experience himself as a person having hostile as well as other types of feelings, and can experience himself in this way without guilt. He has been enabled to do this (if our theory is correct) because another person has been able to adopt his frame of reference, to perceive with him, yet to perceive with acceptance and respect.

(Rogers 19 p. 41)

Unconditional positive regard: Fay, an illustration

It felt very important that Fay's yearning to be with her ex-per be heard, accepted and understood if she were to feel fully received and therefore free to move on, so the therapist conditions easily took precedence over extraneous thoughts or opinions that, albeit momentarily, I may have had. was not at all accepting of the desperately yearning part of herself. Indeed, every time she made reference to this aspect of herself, words like 'silly, stupid, and mad' were mentioned – and she herself felt that this part of her needed to be ated, crushed, exiled or even killed. Yet if I had also judged this part of her to worthy (or if I had somehow colluded with or condoned her value judgem about herself), then in effect her *whole person* would not have been welcome therapy.

Metaphorically speaking, a barrier burst during the ninth n, and Fay felt able to use the entire 50 minutes to disclose how she had ed in 'perverse, often brutal, sexual activities' with her ex-partner. Although part of her 'ex-orcism' Fay needed to be sexually explicit, I shall not be. e it to say that physically she would lose handfuls of hair and end up y battered and bruised as a consequence of these episodes. Worse still f very often she would 'willingly' go along with or even instigate these hid ex sessions. Fay felt shame and embarrassment, and she felt violated and yet how could she be angry or blame her ex-partner when she had tly so willingly engaged in this way?

It is my deeply held conviction that had Fay not fel received during previous sessions – that is, had her *longing-to-be-with-him*elf as well as her *better-off-without-him* self not met with compassion, ace, respect and understanding, she would not have felt safe and sec ugh as a whole person to make the disclosures she did (to a male coun t that).

Both at the time and in retrospect, it was clear that th session was a real turning point for Fay, a truly cathartic experience. She later described how hard it had been for her to disclose such shameful acts and her feeling of being so dirty and unclean, especially as she had co-operated so seemingly fully with her own defilement. Again it felt really important to hear, accept and understand her *I-blame-myself* self, as unless she felt fully received once again she would have been less likely to move on. (I stress that my intent is never to move anyone on – my intent is to strive to be the therapist conditions, and I know that, if I am successful, clients are likely to move on as a consequence.)

In the ninth session, Fay was realising the degree to which she had been manipulated, coerced and exploited by her ex-partner (she had clearly spent a great deal of time in between these two sessions processing her feelings and thoughts) From the moment she arrived (she seemed to breeze into the room rather than trudge) Fay seemed lighter in mood – genuinely smiling at times, for instance, and making more open eye contact – and derogatory riders about her own 'sillins' had all but disappeared. She had remembered how her ex-partner had a video of sexual performances with a previous partner (a 'benchmark' that she was expected to live up to), and she recalled how he would express feeling let down, and would mope or become withdrawn if Fay was not prepared to engage in certain sex activities (as previous partners had), so that she felt as if she really was the huge disappointment her partner made her out to be. Fay was able to recall many, many examples of the 'emotional blackmail' that had been used on her. This time though she approached her self (or selves) with compassionate understanding rather than self-blame, also being able to locate her way of being in time (that is, she was developing a kind of extensional congruence).

We can now turn to empathic understanding and process.

Empathic understanding and process

> The focus condition for therapy is that the therapist is experiencing an accurate empathic understanding of the client's world as seen from the inside. To sense the client's private world as if it were your own, but without losing the 'as if' quality – this is empathy, and this seems essential to therapy. To sense the client's anger, fear, or confusion as if it were your own, yet without your own anger, fear, or confusion getting bound up in it, is the condition we are endeavoring to describe. When the client's world is this clear to the therapist, and he moves about in it freely, then he can both communicate his understanding of what is clearly known to the client and can also voice meanings in the client's experience of which the client is scarcely aware.
>
> (Rogers 1961: p. 284)

It's just you, you all the time!

In a sense, *we're* 'all over the place' at the beginning of therapy, while the client-centred counsellor is primarily intent on entering clients' worlds as they see them – hence *you, you . . .* However:

> The therapist is not interested in everything <u>equally</u>. He is interested in that which is real, important, significant, or fundamental.
>
> (Barton 1974: p. 191)

And what is most real, important, significant and fundamental to client-centred therapists is the internal frame of reference of clients. How, then, does the process of *beginning* to enter the frame of reference of clients commence? A fundamental striving to hear, understand and voice something of that empathy is a good start, for the experience of Carl Rogers indicated that:

> simply to listen understandingly to a client and to attempt to convey that understanding were potent forces for individual therapeutic change.
>
> (Rogers and Evans 1975: p. 124)

However, the therapist might well face some difficulties in striving to be this way, just one being that:

> much of the client's expression is confused and expressed in such private symbolism that it is difficult to enter into his perceptual field and see experience in his terms.
>
> (Rogers 1951: p. 44)

Rogers pointed out that entering the internal frame of reference of clients is not the 'royal road to learning', and that there are *many* hindrances, another being that the therapist can *mostly* only perceive what is in the conscious awareness of clients, so it follows that the greater the degree of client incongruence, the more incomplete the picture that is available to awareness. Nevertheless:

> If some of that [understanding] can be communicated to the client, that I do really see how you feel and understand the way you feel, that can be a most releasing kind of experience. To find that here is a real person who really accepts and understands sensitively, and accurately perceives just the way the world seems to me – that just seems to pull people forward. The effect that can have is really fascinating. It is this aspect that enables the process of therapy to go on.
>
> (Rogers and Evans 1975: pp. 29–30)

Although Carl wrote of 'empathic inference' and 'empathic guessing', he also cautioned us that:

> The more we try to infer what is present in the phenomenal field but not conscious (as in interpreting projective techniques), the more complex grow the inferences until the interpretation of the client's projections may become merely an illustration of the clinician's projections.
>
> (Rogers 1951: p. 495)

The imperfection of communication is another drawback. We know from our own experience that communicating (verbally, paraverbally and non-verbally) precisely what we think, what we feel, and our personal meanings, is nigh-on impossible – yet in striving to empathically understand clients, client-centred therapists have no choice but to rely on their accurate perceiving of client (verbal, non-verbal and paraverbal) communications. However, the situation is not hopeless. For instance, we can, albeit with some caution, let our own frames of reference be a tentative guide to the world of feelings and meanings of others, for as Rogers wrote:

> It is possible to achieve, to some extent, the other person's frame of reference, because many of the perceptual objects – self, parents, teachers, employers, and so on – have counterparts in our own perceptual field, and practically all the attitudes toward these perceptual objects – such as fear, anger, annoyance, love, jealousy, satisfaction – have been present in our own world of experience . . . Hence we can infer, quite directly, from the communication of the individual, or less accurately from observation of his behaviour, a portion of his perceptual and experiential field.
>
> (Rogers 1951: pp. 495–6)

In process terms, we have seen that the more therapist unconditional positive regard is perceived by clients, the less defensive clients feel that they need to be. The less defensive they feel they need to be, the more experiences become available to conscious awareness and the more free the expression of feelings and meanings becomes. A similar process emerges when therapists offer their empathic perceptions to clients:

> Empathic understanding, when it is accurately and sensitively communicated, seems crucially important in enabling the client more freely to experience his inward feelings, perceptions, and personal meanings. When he is thus in contact with his inward experiencing, he can recognize the points at which his experience is at variance with his concept of himself and, consequently, where he is endeavoring to live by a false conception. Such recognition of incongruence is the first step towards its resolution and the revision of the concept of self to include the hitherto denied experiences. This is one of the major ways in which change becomes possible and a more complete integration of self and behavior is inaugurated.
>
> (Rogers 1989: p. 16)

Given that clients (and especially self-referred clients) are usually motivated towards change in the first place (that is why they sought counselling), it makes a great deal of sense for the therapist to behave in ways that are conducive to clients feeling as free as possible to express and explore themselves:

> It should also be added that the dynamic results – for the client and for the learning of the therapist – which are achieved in client-centred therapy when even a portion of the perceptual field is communicated have led us to feel that here is a way of viewing experience which is much closer to the basic laws of personality process and behaviour. Not only does there result a more vivid understanding of the meaning of behaviour, but the opportunities for new learning are maximized when we approach the individual without a preconceived set of categories which we expect him to fit.
>
> (Rogers 1951: pp. 496–7)

In the process of therapy, a therapist striving to empathise with and understand clients clearly has consequences, one of which is:

> By pointing to the possible meanings in the flow of another person's experiencing, you help the other to focus on this useful type of referent, to experience the meanings more fully, and to move forward in the experiencing.
>
> (Rogers 1980: p. 142)

We have also seen how Carl Rogers believed that empathy, at its best, is mutual in the counselling relationship. Furthemore, he linked this to growth when he said:

> perhaps in the moment where real change takes place, then I wonder if it isn't reciprocal in the sense that I am able to see this individual as he is in that moment and he really sees my understanding and acceptance of him. And that I think is what is reciprocal and is perhaps what produces change.
>
> (Rogers 1989: p. 55)

Carl believed that:

> The whole task of psychotherapy is the task of dealing with a failure in communication.
>
> (Rogers 1961: p. 330)

Thus the process is that incongruent clients have lost their *inner* communication abilities, and this in turn hinders their self-expression (that is, the quality of their relationships with others). Incongruent clients are, due to their defensive psychological denial and distortion mechanisms, no longer consciously aware of certain aspects of their experiencing, and such denial and distortion will of course be reflected in their outer as well as their inner communications. Thus:

> The task of psychotherapy is to help the person achieve, through a special relationship with a therapist, good communication within himself. Once this is achieved he can communicate more freely and more effectively with others. We may say then that psychotherapy is good communication, within and between men. We may also turn that statement around and it will still be true. Good communication, free communication, within or between men, is always therapeutic.
>
> (Rogers 1961: p. 330)

In 1961, when considering group tensions, Carl wrote about how empathic understanding can overcome:

> the insincerities, the defensive exaggerations, the lies, the 'false fronts' which characterize almost every failure in communication.
>
> (Rogers 1961: p. 336)

When people both realise and accept that the sole – the only – authentic intent of the therapist is to respectfully understand, then the need for defensive processes rapidly dissipates. As barriers to both inner and outer communicating dissolve, so clients are enabled to reach their inner truths and engage in more fulsome relationships with others. Rogers likened this movement toward inner truth with accessing data – that is, the more accurate the information one has, the more one is able to make informed, self-directed, responsible choices. For sure, our anxieties, doubts and fears might serve to trigger our defensive processes that are designed to deny us access to the 'facts' yet nevertheless Carl Rogers believed that *the facts in themselves are always friendly*:

> *The facts are friendly* . . . as I look back, it seems to me that I regarded the facts as potential enemies, as possible bearers of disaster. I have perhaps been slow in coming to realize that the facts are *always* friendly. Every bit of evidence one can acquire, in any area, leads one that much closer to what is true. And being closer to the truth can never be a harmful or dangerous or unsatisfying thing. So while I still hate to readjust my thinking, still hate to give up old ways of perceiving and conceptualizing, yet at some deeper level I have, to a considerable degree, come to realize that these painful reorganizations are what is known as *learning*, and that though painful they always lead to a more satisfying because somewhat more accurate way of seeing life.
>
> (Rogers 1961: p. 25)

Having stated that client perception of a therapist's genuine and acceptant empathy helps serve to reduce blocks to communication, it follows that the opposite is true:

> When we don't feel understood, we're defensive.
>
> (Rogers and Teich 1992: p. 59)

An example of this might arise through therapeutic work with couples. All too often, neither partner really hears the other because each is indignantly defending their own stance, and as a consequence neither feels heard or understood. Very often, in the first few sessions with a couple, each client only experiences respectful empathy from the therapist. Very often, couples have later reported how, when they departed from sessions at the beginning of therapy, their conversation was typified by 'See, the counsellor agreed with *me*', 'Oh no! The counsellor was clearly on *my* side!', and so on. In fact the therapist did not *agree* with anyone, but clients reacted in this way to having felt a contact – to having felt received, heard and understood.

As the therapist strives to experience and offer empathic understanding, client change and growth may be fostered in many different ways. Reviewing a transcript of a half-hour therapy session conducted by Carl Rogers, Nat Raskin and Rogers identified several characteristics of Carl's communications:

> The interview exemplifies the therapeutic mechanism of empathy backed by genuineness and unconditional positive regard. It helps the client to (1) examine her problems in a way that shifts responsibility from others to herself, (2) experience emotions in the immediacy of the therapy encounter, (3) accept aspects of self formerly denied to awareness, and (4) raise her general level of self-regard.
>
> (Rogers and Raskin 1989: p. 178)

Carl's initial verbalisations were open. He did not try to reassure the client, ask questions or delve into their history, or take on the responsibility of providing a solution or focusing therapy (thereby living by his belief in the inner resources of clients). As a consequence of perceiving that she has been understood, the client broadens her view. Rather than talking *about* her emotions, she experiences them in the here and now. She moves towards talking about herself and away from narrative concerning others, and as a consequence of experiencing unconditional positive regard from Carl, she moves towards greater self-acceptance. Their relationship becomes more reciprocal.

What psychological purpose is served by attempting to duplicate, as it were, the perceptual field of the client in the mind of the counsellor? Clients experience therapists in a variety of ways, but some common threads do emerge. For one client:

> Notice how the significant theme of the relationship is, 'we were mostly *me* working together on my situation as I found it.' The two selves have somehow become one while remaining two – 'we were *me.'* This idea is repeated several times. . . . The impression is that the client was in one sense 'talking to herself', and yet that this was a very different process when she talked to herself through the medium of another person.
>
> (Rogers 1951: p. 39)

For another client:

> Perhaps it would be accurate to say that the attitudes which she could express but could not accept as a part of herself became acceptable when an alternate self, the counselor, looked upon them with acceptance . . .
>
> (Rogers 1951: p. 39)

We have already seen that *if* conducive therapeutic conditions prevail, *then* clients are more readily able to access their ('self-referent') psychological and physiological inner flow of experiencing. Indeed, you may even have reacted to the idea of a momentary pause in order to try it and experience it for yourself. Very often therapists are confident in their sensing of something difficult or even fearful going on within clients, yet feel less secure about the content and personal meanings embraced by the client's here-and-now experiencing. In order to perception check, the therapist might volunteer a *tentative* voicing of their empathic grasping, while being constantly alert and willing to move with the client's feelings and personal meanings that emerge as a consequence of the therapist *striving* to understand. Rogers wrote:

> Now here's what has interested me about this . . . it's always puzzled me. Some words fit what is being experienced and others do not. I've noticed that, in trying to check my understanding, I might say: 'Is it that you are so angry at your father?', 'No, that's not quite it.' 'Is it that you hate him?', 'No.' 'Is it that you feel contempt?', 'Yes! Contempt. That's what I feel.' That's the kind of response, as though at some level there is a verbal form which really fits what's being experienced. It's quite striking sometimes. I don't pretend to understand it, but it's

something that has intrigued me. How is it that, if he hadn't been able to describe his feelings, one word fits exactly and another word quite similar to it doesn't fit at all? Puzzling . . .

(Rogers and Teich 1992: pp. 62–3)

Especially when reflecting in retrospect upon their developmental movements over the course of a diploma in client-centred therapy, trainees have often related how at the beginning it felt safer to rely on implementing skills that they had learned or been taught than to venture into communicating empathic under-standing. The *fear* of owning that a perception might be inaccurate seemed to far outweigh the *embarrassment* of not quite getting a technique right. We have seen, too, that Rogers believed a further risk of the therapist experiencing and communicating empathy is that the *therapist* might change through the process as well as the client. Yet to understand another empathically is edifying in at least two fundamental respects:

> I find when I am working with clients in distress, that to understand the bizarre world of a psychotic individual, or to understand and sense the attitudes of a person who feels that life is too tragic to bear, or to understand a man who feels that he is a worthless and inferior individual – each of these understandings somehow enriches me. I learn from these experiences in ways that change me, that make me a different and, I think, a more responsive person.
>
> (Rogers 1961: pp. 18–19)

Thus counsellors themselves may be enriched through being this therapist core condition. However, even greater significance is the humbling privilege of being allowed to enter the private world of feelings and meanings – both 'courage and kindness and love and sensitivity' and terror, oddity, disillusionment and tragedy – and to be in awe of oneself and the other as this provision of understanding generates greater client self-understanding and self-acceptance. Being an active party in the change and growth of another is indeed an awesome, humbling privilege.

Empathy: Fay, an illustration

My genuine and respectful striving to empathically sense and understand Fay's world of feelings and meanings was central to the whole of our journey together. From moment to moment I would not only hang on her every word and expression, but also really try to sense what lay beneath or behind her commun-ications, and would tentatively venture what I believed I had gleaned. My own process was an internal seeking. What does this *shame*, for instance, actually *feel* like? What meanings are attached to feeling this way? As a consequence of seeking to sense such feelings and meanings, some empathic understanding was ventured, accompanied by a very close attendance to how Fay received my communications (verbally, paraverbally and non-verbally). I tried to follow her ongoing leads in every way.

Sometimes my sensing was very accurate, and that felt *good* to Fay – to feel honestly accepted and understood. At other times, I was a little off-beam – and that felt good to Fay, too, for in clarifying for me (which she willingly did), things crystallised for her as well. As we moved towards ending, she was able to say:

> I was a person who allowed myself to be exploited, and now I can see why. I feel warmth toward that me now, rather than despising and loathing myself. I do regret what happened, yet I am different now. I still feel very lonely, but I do not miss him: I miss some aspects of togetherness that being in a relationship can bring. I realise, in fact, that I do not miss him at all – rather, I miss the hopes and dreams, the potential that I initially saw in that relationship. I do not hate him anymore; I feel a little pitying of him (and pity doesn't feel too good, maybe I still need to work on that). I realise that at times I sensed his vulnerable inner self, and it was *that* part of him that I held out hope for. I have moved on now . . . It feels like I am a new, more real, me.

Client do

When asked what he was working towards as a therapist, Carl Rogers said:

> For myself, I have a rather simple definition, yet one which I think has a good many implications. I feel that I'm quite pleased in my work as a therapist if I find that my client and I, too, are – if we are both moving toward what I think of as greater openness to experience.
>
> (Rogers 1989: p. 75)

Client movement towards greater openness to experience is enabled through clients becoming better able to listen to and participate fully in their inner communications, and hence becoming more perceptive about themselves, others and the worlds that they inhabit. This *process of becoming* could hardly be described as static, and Rogers said that he felt rewarded when this process gets under way. The client also becomes more realistic and socialised:

> simply because one of the elements which he can't help but actualize in himself is the need and desire for closer human relationships; so for me, this concept of openness to experience describes a good deal of what I would hope to see in the more optimal person . . .
>
> (Rogers 1989: p. 75)

We have seen that client-centred therapy can be formulated as *if* (A) *then* (B), *if* (B) *then* (C), where (A) is a relationship founded on contact between an incongruent (vulnerable, anxious and defensive) client and an authentic, respectful and understanding therapist, and in which the client at least to some degree perceives that respectful empathy. (B) can represent the capacity and potential of clients, which is released as a consequence of being in (A). What is the outcome of (A) plus (B), (C)? Put simply, if the therapist can *be* the core conditions, what do clients *do*? In essence, client potential for self-actualising that is in harmony with the organism as a whole can be released and realised:

> The individual will reorganize himself at both the conscious and deeper levels of his personality in such a manner as to cope with life

more constructively, more intelligently, and in a more socialized as well as a more satisfying way.

(Rogers 1961: p. 36)

Carl Rogers was able to base his comments on solid research findings, and could state that even *limited* client experience of client-centred therapy would produce changes in personality (attitudes and behaviour) that were both significant and profound. The outcome is a more integrated and effective individual. So it had been discovered that even relatively minimal engagement with client-centred therapy results in fewer symptoms of what might, in other approaches to helping, be labelled as 'neurosis', 'psychosis' or other 'personality problems' of some description. Client-centred therapy results in clients moving towards having a more realistic perception of the self. They become more like their ideal selves, self-valuing increases, they exhibit higher levels of confidence and greater autonomy, there is improved self-understanding, they become more open to experiencing (that is, there is a marked decrease in denial and distortion), and they become more accepting of the sovereignty and humanity of others. Furthermore, there are behavioural changes as a result of psychological growth. Clients become less frustrated by and recover more quickly from stress, they are more mature and less defensive, and they are more flexible and creative. As Rogers stated:

> there can no longer be any doubt as to the effectiveness of such a relationship in producing personality change.
>
> (Rogers 1961: p. 36)

One client said 'Now that I see all that, I guess it's really up to me', and Carl Rogers said:

> I think as the person becomes aware of these various factors in his background, he can make realistic and sensible choices as to how he's going to both live with and transcend the circumstances of the past . . . I don't believe in free will in the sense that a person's free to do anything, but to deny the reality of the significance of choice as the strict behaviorists do is totally unrealistic . . . the person can and does choose significantly. I've always been rather pleased that our researchers in psychotherapy have shown that the element that probably changes most in individual therapy is the person's concept of himself. He moves toward being a more confident self, more acceptant of himself; self-hatred decreases. As a person feels genuinely confident of who he is and what he's capable of doing, he goes ahead to utilize his past, his conditioning, and the biological basis of his being in more constructive ways.
>
> (Rogers and Evans 1975: pp. 76–7)

Both at the time but even more so in retrospect, Rogers realised the importance of research into 'delinquents' undertaken by Bill Kell, through which it was discovered (much to the astonishment of the research team) that in terms of predicting future behaviour, self-understanding *far* outweighed such factors as family environment and other social and economic concerns:

Actually, it was so shocking to us at the time of the original study – this was many years ago – that we frankly didn't believe the findings.

(Rogers and Evans 1975: pp. 77–9)

As a result of this scepticism, Helen McNeil replicated the research, and again self-understanding proved to be the most significant factor. Yet again the research team could hardly believe these findings, and so embarked on a sub-analysis, about which Rogers wrote:

Again, self-understanding . . . correlated more highly with their be-havior. So the essence of that seemed to be that the person who is realistic about himself, who has a pretty good comprehension of what he's up against, and what factors have influenced his behavior, has a much better chance of controlling his behavior. He can make choices. At the time I didn't appreciate the full depth of that study. It's taken a good many years for me to realize that our findings were more significant than I thought at the time.

(Rogers and Evans 1975: pp. 77–9)

So if a fundamental aspect of client-centred therapy is increasing awareness, acceptance and reorganising of the self, here was evidence from 'beyond' therapy that corroborated the social significance of this.

In expressing a real trust in the potential (the actualising tendency) that resides in all of us, Rogers said:

it seems to me that you could speak of the goal toward which therapy moves, and I guess the goal toward which maturity moves in an individual, as being *becoming*, or being knowingly and acceptingly that which one most deeply is.

(Rogers 1989: p. 62)

The client-centred therapist is committed to an ongoing, moment-by-moment striving to authentically and respectfully empathise with clients. How might clients experience this?

the concentrated attentiveness to his person, concerns, and his feelings is an extraordinary event in the client's life, naturally increasing his sense of substantiality and reality and lending dignity to his self and his concerns.

(Barton 1974: p. 192)

From first contact onwards, clients ideally experience both a meaningful and an acceptant connection and feel fully received for whoever and however they are. All disclosures are met with respect and an attempt at understanding, rather than being contemptuously dismissed as 'silly', as is so often the case with both their own inner and external communications. As self-expression meets with an authentic, acceptant empathy, so clients feel increasingly safe in further verbalising what is going on for them, and they gradually begin to incorporate previously denied or distorted experiencing into their self concepts. This in turn influences their behaviour (actions or inaction):

> As I listened to therapeutic interviews . . . I realized that when a person is functioning well and has profited a lot from therapy, you find they're open to their experience. They're able to express what they're feeling – what they're experiencing – and to assimilate that into their concept of themselves and guide their behavior in terms of it.
>
> (Rogers and Russell 2002: pp. 256–7)

As therapy progresses, clients gradually adopt a client-centred way of being *towards themselves*. In a sense, the *process* of client-centred therapy could almost be represented by the therapist 'core' conditions in reverse – that is, a congruent therapist experiences unconditional positive regard towards and strives to empathically understand an incongruent client. As a consequence of perceiving these conditions from the therapist, clients move towards experiencing greater self-acceptance (especially as a consequence of feeling respected by a significant other) and deeper self-understanding (especially as a consequence of being reciprocally involved in a process of genuine empathic striving). As a consequence of greater self-acceptance and self-understanding, more and more experience becomes available to awareness, and hence may become integrated aspects of the client's concept of self:

> The self is now more congruent with the experiencing. Thus, the persons have become, in their attitudes toward themselves, more caring and acceptant, more empathic and understanding, more real and congruent. But these three elements are the very ones that both experience and research indicate are the attitudes of an effective therapist. So we are perhaps not overstating the total picture if we say that an empathic understanding by another enables a person to become a more effective growth enhancer, a more effective therapist, for himself or herself.
>
> (Rogers 1980: p. 159)

No surrender!

We have seen that upon entering therapy clients are likely to be vulnerable, anxious and tense as a consequence of their defensive psychological denial and distortion processes no longer being able to entirely prevent *some* awareness of their incongruence. Conversely:

> We may say that freedom from inner tension, or psychological adjustment, exists when the concept of self is at least roughly congruent with all the experiences of the organism.
>
> (Rogers 1951: p. 513)

Rogers gave the example of a woman who denied her sexual impulses (this was in the 1950s) because she believed them to be taboo, or improper, or immoral. Therefore they *had* to be denied because they simply could not coexist with her view of herself as she *should* be, namely 'proper' and moral. Many of her *shoulds* and *oughts* were values introjected both from significant others and from her society. However, as therapy progresses the client is able to accept and understand both of these facets of herself. She can accept her sexual desires as 'normal' *and* respect and understand cultural moral values as being significant in her world.

She can more fully *be* herself rather than being in conflict, strenuously striving to achieve victory for one aspect of herself over another – even though neither facet of herself is overly willing to surrender or accept defeat:

> The feeling of reduction of inner tension is something that clients experience as they make progress in 'being the real me' or in developing 'a new feeling about myself' . . . The cost of maintaining an alertness of defense to prevent various experiences from being symbolized in consciousness is obviously great.
>
> (Rogers 1951: p. 513)

In a medical practice, many patients are referred to counselling because either they have told their doctor, or their doctor has told them, that they are *depressed*. These clients often arrive at their first therapy session already having been prescribed antidepressant medication. Thus there is a very powerful underlying value that the patient *should not* be depressed. Usually the patients themselves feel that there is something wrong with them (which is why they went to see their GP in the first place), and this is confirmed by their having been prescribed medication that is intended to 'help them get better' or 'recover' from their depression.

Yet, more often than not, there really would be something odd or weird about these clients if they did *not* feel depressed. As human beings who have experienced or are experiencing difficult or tragic circumstances, *of course* they feel low. When these clients experience authentic unconditional positive regard and empathic understanding from their counsellor, they begin to realise that they are not 'abnormal' after all – being received as human beings who are worthy of respect *as and how they are* means that they can begin to accept themselves as and how they are. Furthermore, the feelings, thoughts and behaviours that they believe they should not have or ought not to have begin to seem understandable and therefore normal, acceptable and real consequences of living – they begin to reconnect with humanity. 'Depressed' clients very often describe their feeling of exhaustion – even getting up in the morning just feels like too much trouble, and the effort involved in getting through another day just doesn't feel worthwhile. Yet very often it seems that a huge portion of the inner energy available to them has been spent in fighting and trying to vanquish their sinister, bizarre, unacceptable feelings and thoughts. Their inner experience is of wanting to stay curled up in bed feeling miserable and alone, but there is an internal voice saying 'pull yourself together, stop being a self-indulgent wimp, and get on with life – there are many people worse off than you!'. Other internal dialogues might be characterised by, on the one hand, feeling insignificant and worthless while, on the other hand, there is a wailing, plaintive 'What about me?'. Many thoughts or wishes might be dismissed as silly, irrational or needless. All in all, a massive amount of energy is expended in striving to resolve this exhausting inner battle that rages on – no wonder that there is a reluctance to emerge from the cocoon of sleep, or a yearning to sleep for ever. Yet when clients begin to *be* their depression rather than fighting it, there is a huge metaphorical 'inner sigh' as the client relaxes into a greater acceptance of their wholeness, and gradually the vast stores of energy used up in a conflict that *inevitably* tends towards there being winners and losers, victors and vanquished, can gradually be recouped or recharged and redirected towards the more constructive tendencies in all of us. As more experiencing becomes available to consciousness, the client can move towards celebrating the diversity of all of their selves, rather than striving to impose a harsh regime in which some selves have to surrender while others bask in their tyrannical glory:

> The term 'available to consciousness' . . . is deliberately chosen. It is the fact that all experiences, impulses, sensations are *available* that is important, and not necessarily the fact that they are present in consciousness. It is the organization of the concept of self *against* the symbolization of certain experiences contradictory to itself, which is the significant negative fact. Actually, when all experiences are assimilated in relationship to the self and made a part of the structure of the self, there tends to be *less* of what is called 'self-consciousness' on the part of the individual. Behavior becomes more spontaneous, expression of attitudes is less guarded, because the self can accept such attitudes and such behavior as a part of itself. Frequently a client at the beginning of therapy expresses real fear that others might discover his real self. 'As soon as I start thinking about what *I* am, I have such a terrible conflict at what I am that it makes me feel awful. It's such a self-depreciation that I hope nobody ever knows it . . . I'm afraid to act natural, I guess, because I just don't feel as though I like myself.' In this frame of mind, behavior must always be guarded, cautious, self-conscious. But when this same client has come to accept deeply the fact that 'I am what I am', then she can be spontaneous and can lose her self-consciousness.
>
> (Rogers 1951: p. 515)

From four to one against to two to one on!

As a consequence of experiencing the therapist conditions and of feeling received, there is less potential for clients to experience tension or anxiety, and so they become less vulnerable. Due to there being a reduced possibility of threat to the self-concept, the self-concepts of clients change, as they are able to assimilate previously denied or distorted experiencing. The self-concept has become more flexible and inclusive, and the client has moved away from feeling unworthy of respect (a feeling that was based largely on standards introjected from others) and towards greater self-acceptance. There is less and less need for defensiveness as each new disclosure meets with therapist respect and understanding and, more importantly, self-respect and self-understanding:

> One study showed that at the beginning of therapy, current attitudes toward self were four to one negative, but in the final fifth of therapy self-attitudes were twice as often positive as negative.
>
> (Rogers 1961: p. 65)

Clients become more self-directing as both they and their therapists trust in the wisdom of the whole organism – the client's internal locus of evaluation. Reduced defensiveness (due to the absence of threat from both the client's own self-concept and an acceptant, understanding therapist) leads to clients becoming more adaptable, as their behaviour becomes increasingly based on more accurate and complete data (as less and less is denied or distorted) – the facts are always friendly! Clients become more open to their full experiencing, and the *ideal* self changes, too, becoming more realistic and therefore the person's aims are more achievable. The chasm between clients as they want to be and clients as they are is greatly diminished, which means that they are less and less likely to have 'out-of-character' experiences – and this feeling of enhanced self-control in turn leads to yet greater self-acceptance:

> More of the total experience of the organism is directly incorporated into the self; or more accurately, the self tends to be discovered in the total experience of the organism. The client feels he is his 'real' self, his organic self.
>
> (Rogers 1951: pp. 531–2)

As a consequence, tension of *all* types (physiological as well as psychological) continues to decrease as each new step on this journey of personal development continues to meet with unconditional prizing and empathy – from the therapist and, increasingly, internal acceptance and understanding. As a consequence of feeling yet further self-acceptance, clients become more accepting of others (and the ideals that they project on to others also become more realistic and therefore achievable). In large part due to the reduced need to protect their own concepts of self, others are now perceived as unique and separate individuals, worthy of their own dignity:

> The best definition of what constitutes integration appears to be this statement that all the sensory and visceral experiences are admissible to awareness through accurate symbolization, and organizable into one system which is internally consistent and which is, or is related to, the structure of self. Once this type of integration occurs, then the tendency toward growth can become fully operative, and the individual moves in the directions normal to all organic life. When the self-structure is able to accept and take account in consciousness of the organic experiences, when the organizational system is expansive enough to contain them, then clear integration and a sense of direction are achieved, and the individual feels that his strength can be and is directed toward the clear purpose of actualization and enhancement of a unified organism.
>
> (Rogers 1951: pp. 513–14)

The client rejects the desire to control or manipulate others, preferring instead to *be* the greater valuing and enrichment that increased awareness of self and others brings. The 'fully functioning person' is able both to accurately perceive and to assimilate and integrate all relevant experiencing:

> If I feel hatred, I can assimilate the fact that I'm a hating person; I hate this individual. If I feel love, I'm able to recognize that that's what I'm feeling. If I'm frightened, the same, so that when one is really open to all of the experiences going on in the organism, one is a fully functioning person.
>
> (Rogers and Russell 2002: p. 252)

Being fully functioning – open to a multiplicity of experiences – involves a readiness to engage with both here-and-now factors (such as social and cultural expectations, satisfactory and unsatisfactory relationships with others, physical requirements, legal demands, and so on) and the personal feelings and meanings that have grown out of the past. Every new stimulus is evaluated in terms of the whole person, which means that the actualising tendency and self-actualising are now harnessed together rather than being in conflict.

Rogers wrote that the valuing process which resides in the 'fully functioning' or 'psychologically mature' person is the most sensitive such process that exists in the entire universe as we know it. He was also keen to point out that although this valuing process does operate on whether or not experiencing maintains and enhances (self-)actualisation, it is nevertheless *not selfish*, because the fundamental nature of human beings is highly *social* (see, for instance, Rogers 1989: pp. 137–8). Nor are fully functioning individuals entirely 'perfect' – we have already learned that Rogers saw the human being not as seeking stasis only, and that openness to experience and self-acceptance includes being non-defensive about failings and negative feelings as well as positive qualities:

> This emerging person would not bring utopia. He would make mistakes, be partially corrupted, go overboard in certain directions.
> (Rogers and Evans 1975: pp. 173–4)

However, the 'person of tomorrow' – the fully functioning, psychologically mature person – would tend to live in (or create) a culture (and subcultures) within which certain values, beliefs and attitudes have significant emphasis. Non-defensive openness to self and experience would apply in all relationships (with partners, family and at work, for instance), and the 'style' of interpersonal relationships would be more deeply rooted in the principles of the person-centred approach. There would be greater respect for the dignity of all people, irrespective of such issues as social status, gender, ethnicity, or perceived and real disadvantages. Groupings of people (in communities, at work, in education, and so on) would be less gargantuan and impersonal, and greater respect towards and reciprocity with nature would feature. Resources of all kinds would be shared with greater equity, and there would be less materialism for its own sake, and less greed. There would be reduced authoritarianism and bureaucracy, with sensitivity to human needs outweighing administrative and control needs, and leadership would become less significant. Concern for those who would benefit from assistance would be genuine and caring, and scientific creativity would become more humane, too. Indeed, there would be greater creativity of all kinds in all areas. The human being has moved:

> toward the exploration of self and the development of the richness of the total, individual, responsible human soma – mind and body.
> (Rogers and Evans 1975: pp. 173–4)

Thus in becoming more open to self and others, in becoming more aware of self and others, in becoming more acceptant of self and others, and in becoming more understanding of self and others, people become more able to live their lives meaningfully, more:

> able to love; able to receive love.
> (Rogers and Russell 2002: p. 258)

Chapter 6

Afterwords

> Our civilization does not yet have enough faith in the social sciences to utilize their findings.
>
> (Rogers 1961: p. 335)

We have found that if a therapist can initiate a helping relationship with an incongruent person, founded on the therapist experiencing authentic unconditional positive regard and empathic understanding and the client perceiving these therapist qualities, then this is all that is necessary and sufficient for people to change and grow. Have faith! Trust in the resources and capacities of clients! As Rogers said:

> I'm pretty good at living and letting live, so the fact that people have held different points of view from mine has never troubled me much. Criticisms that come from honest viewpoints which differ quite sharply from mine are OK with me. I suppose the kind of criticism that I've liked least is the notion, which used to be expressed in many ways over the course of years, that the client-centered approach was a pretty shallow approach. It might do for dealing with superficial problems, but it really had not much to do with the deeper problems of personality. That is just not true. This criticism has probably troubled me more than any other I can think of, because I don't think I am superficial. You can't do anything but laugh off the fact that a lot of criticism is based on a complete lack of understanding of what I've been doing and what my associates have been doing.
>
> (Rogers and Evans 1975: p. 114)

The would-be therapist is faced with a choice – either we can utilise what we know to control, manipulate, teach and, effectively, depersonalise and enslave people (often without their knowing) into becoming what we or society wish them to be, or we can opt for freedom, to provide conditions that are conducive to growth, creativity and autonomy (whether it be for individuals or groups of people):

> There are really only a few psychologists who have contributed ideas that help to set people free, making them psychologically free and self-responsible, encouraging them in decision making and problem solving.
>
> (Rogers and Evans 1975: pp. 65–6)

Rogers stated that he was particularly disappointed in academic psychologists working within the 'great' learning establishments, and conceded that he:

> had very little influence on academic psychology, in the lecture hall, the textbook, or the laboratory. There is some passing mention of my thinking, my theories, or my approach to therapy, but, by and large,

I think I have been a painfully embarrassing phenomenon to the academic psychologist. *I do not fit.* Increasingly I have come to agree with that assessment.

(Rogers and Evans 1975: p. 126)

Rogers nevertheless maintained that *of course* we need to remain sceptical, to continually engage in research and question our findings, theories and practices – yet in doing so we can remain committed to the sovereignty of the person. As Carl said (in response to Richard Evans):

And I like your word, *commitment*, because I think that's probably better than my use of *belief* or *conviction*.

(Rogers and Evans 1975: p. 69)

Indeed, when asked to state who had most shaped his work, Carl had no hesitation in saying that it was his clients, and that he had paid attention to their 'raw data' that emerged from the tape recording of sessions and the subsequent reflection upon the experiencing:

I feel that almost everything I've learned about personality, about interpersonal relationships, about personality dynamics came not from books or from some charismatic teacher, but from direct experience with the people I've been working with . . . I guess something I take some pride in is that I always wanted to get close to the raw data.

(Rogers and Evans 1975: p. 109 and pp. 112–13)

So what of the future? Of his own developmental trends, Rogersl said:

I will introduce quite a personal note here. I find I'm not much good at planning my future. I *ooze* toward my future. That is one term I like to use because it reminds me of an amoeba that sends out these protoplasmic tentacles and then if it runs into something aversive, it draws back, but it flows in the direction that seems rewarding. And I feel that's the nature of my professional life.

(Rogers and Evans 1975: p. 117)

Of the world, Rogers acknowledged that evidence of negativity is all around us, yet he remained optimistic:

To me, the person who offers the most hope in our crazy world today, which could be wiping itself out, is the individual who is most fully aware – most fully aware of what is going on within himself: physiologically, feeling-wise, his thoughts; also aware of the external world that is impinging on him. The more fully he is aware of the whole system, or perhaps context – the more hope there is that he would live a balanced human life without the violence, the craziness, the deceit, the horrible things that we tend to do to each other in the modern world.

(Rogers 1989: pp. 188–9)

And what of the future of client-centred therapy? We face threats from without – those who ignore the evidence of research that sometimes they themselves have sponsored, evidence which time and time again demonstrates that what is most helpful is a therapeutic relationship founded on authentic respect and under-standing. We face threats from within – those who lack the discipline of Rogers and view congruence as meaning that anything goes, agencies responsible for the validation of training (governmental and academic) who reduce an exquisite intimacy to the mechanical implementation of skills, and those who lack the courage or support to hold on to the therapist conditions as both necessary and sufficient. Yet just as the client-centred therapist trusts in the capacity of clients, maybe we need to trust in our own resources, too:

> I hope you would find it possible to trust your deepest self. I think that the follies and tragedies of the world result from people being fearful of themselves, not knowing themselves, and certainly being unwilling to trust the deepest level of selfhood.
>
> (Rogers 1989: pp. 197–8)

There *is* hope.

I simply say with all my heart:

> *Power to the emerging person and the revolution he carries within.*
> (Rogers and Evans 1975: p. 175)

Emerge!

> *Why be the magician when you can be the magic?*

Bibliography

Barrett-Lennard G (1959) *Dimensions of the client's experience of his therapist associated with personality change.* Unpublished doctoral dissertation, University of Chicago, Chicago.

Barton A (1974) *Three Worlds of Therapy.* Mayfield Publishing Company, Palo Alto, CA.

Bergin A and Jasper L (1969) Correlates of empathy in psychotherapy: a replication. *J Abnorm Psychol.* 74: 477–81.

Bergin A and Solomon S (1970) Personality and performance correlates of empathic understanding in psychotherapy. In: J Hart and T Tomlinson (eds) *New Directions in Client-Centered Therapy.* Houghton Mifflin, Boston, MA.

Bergman DV (1951) Counseling method and client responses. *J Consult Psychol.* 15: 216–44.

Berwick K and Rogers C (1983) *Carl Rogers Counsels Keith Berwick.* CSP/CCTPCAS video.*

Budman SH and Gurman AS (1988) *Theory and Practice of Brief Therapy.* Guilford Press, New York.

Farber B, Brink D and Raskin P (eds) (1996) *The Psychotherapy of Carl Rogers.* Guilford Press, New York.

Farson R (1955) *Introjection in the psychotherapeutic relationship.* Unpublished doctoral dissertation, University of Chicago, Chicago.

Fiedler FE (1953) Quantitative studies on the role of therapists' feelings toward their patients. In OH Mowrer (ed.) *Psychotherapy Theory and Research.* Ronald Press, New York.

Frick W (1989) *Humanistic Psychology: conversations with Maslow, Murphy, Rogers.* Wyndham Hall Press, Bristol. This includes a conversation with Carl Rogers (1969).

Friedli K, King M, Lloyd M and Horder J (1997) Randomised controlled assessment of non-directive psychotherapy versus routine general practitioner care. *Lancet.* 350: 1662–5.

Greenfield S (2000) *Brain Story.* BBC books, London.

Hampshire Association for Counselling (2001) *Counsellor Resources Directory.* Hampshire Association for Counselling.

Kelly G (1955) *The Psychology of Personal Constructs.* Norton, New York.

King M, Sibbald B, Ward E *et al.* (2000). Randomised controlled trial of non-directive counselling, cognitive behaviour therapy and usual general practitioner care in the management of depression as well as mixed anxiety and depression in primary care. *BMJ.* 321: 1389–92.

Kirschenbaum H (1979) *On Becoming Carl Rogers.* Delacorte Press, New York.

Laing RD (1965) *The Divided Self.* Pelican, Middlesex, UK.

McLeary RA and Lazarus RS (1949) Autonomic discrimination without awareness. *J Pers.* 18: 171–9.

Mearns D (2002) *Developing Person-Centred Counselling.* Sage Publications, London.

Mellor-Clark J, Connell J, Barkham M and Cummins P (2001) Counselling outcomes in primary health care: a CORE system data profile. *Eur J Psychother Counsell Health.* 4: 65–86.

Nelson-Jones R and Patterson CH (1975) Measuring client-centred attitudes. *Br J Guidance Counsell.* 3: 228–36.

Patterson CH (2000) *Understanding Psychotherapy: fifty years of client-centred theory and practice.* PCCS Books.

Patterson CH and Watkins CE Jnr (1996) *Theories of Psychotherapy.* HarperCollins College Publishers, New York.

Prouty G, van Werde G and Portner M (2002) *Pre-Therapy.* PCCS Books, Llangarron, Ross-on-Wye.

Rogers C (1937) The clinical psychologist's approach to personality problems. *Family.* 18: 233–43.

Rogers C (1939) *Clinical Treatment of the Problem Child.* Houghton Mifflin, Riverside Press, Cambridge, MA.

Rogers C (1942) *Counseling and Psychotherapy.* Houghton Mifflin, Boston, New York.

Rogers C (1946) Significant aspects of client-centered therapy. *Am Psychol.* 1: 415–22.

Rogers C (1951) *Client-Centered Therapy.* Constable, London.

Rogers C (1957a) The necessary and sufficient conditions of therapeutic personality change. *J Consult Psychol.* 21: 95–103.

Rogers C (1957b) A therapist's view of the good life. *Humanist.* 17: 299–300.

Rogers C (1959) A theory of therapy, personality, and interpersonal relationships, as developed in the client-centered framework. In: S Koch (ed.) *Psychology: a study of a science. Study 1. Volume 3. Formulations of the person and the social context.* McGraw-Hill, New York, Toronto, London.*

Rogers C (1961) *On Becoming a Person.* Constable, London.

Rogers C (ed.) (1967) *The Therapeutic Relationship and its Impact: a study of psychotherapy with schizophrenics.* The University of Wisconsin Press, Madison, Milwaukee and London.

Rogers C (1975) Empathic: an unappreciated way of being. *Counsel Psychol.* 5: 2–10.

Rogers C (1978) *Carl Rogers on Personal Power.* Constable, London.

Rogers C (1980) *A Way of Being.* Houghton Mifflin, Boston, New York.

Rogers C (1981) Notes on Rollo May. In: H Kirschenbaum and V Land Henderson (1989) *Carl Rogers: Dialogues.* Hougton-Mifflin, Boston.

Rogers C (1985) *Personally Speaking: Carl Rogers.* John Masterson interview, RTE Dublin television programme, recorded on 23 November 1985.*

Rogers C (1986) *Politics and Innocence: a humanistic debate.* Saybrook, Dallas, Texas.

Rogers C (1987a) Client-centered? Person-centered? *Person-Centered Rev.* 2: 11–13.

Rogers C (1987b) What areas of person-centred theory or practice require further research? 2: 252.

Rogers C (1989) *Carl Rogers: dialogues* (H Kirschenbaum and V Land Henderson, eds). Houghton Mifflin, Boston.

Rogers C and Buber M (1957) In: H Kirschenbaum and V Land Henderson (1989) *Carl Rogers: Dialogues.* Houghton Mifflin, Boston.

Rogers C and Frick W (1967) Conversation with Carl. In: *Humanistic Psychology: conversations with Maslow, Murphy, Rogers.* Wyndham Hall Press, Bristol, Indiana.

Rogers C and Stevens B (1967) *Person to Person: the problem of being human.* Souvenir Press, Guernsey, Channel Islands.

Rogers C and Truax C (1967) The therapeutic conditions antecedent to change: a theoretical view. In: C Rogers, E Gendlin, D Kiesler and C Truax (eds) *The Therapeutic Relationship with Schizophrenics.* Wisconsin Press, Madison, WI.

Rogers C and Evans R (1975) *Carl Rogers: the man and his ideas.* EP Dutton, New York.

Rogers C and Evans R (1981) *In Dialogue with Carl Rogers.* Praeger, New York.

Rogers C and Ryback D (1984) One alternative to nuclear planetary suicide. *Counsel Psychol.* 12: 3–12.

Rogers C and Barfield GS (1985) *Empathy: an exploration.* Center for Studies of the Person video.*

Rogers C and Sanford R (1985) *In Conversation.* Center for Studies of the Person video.*

Rogers C and Raskin N (1989) Person-centred therapy. In: R Corsini and D Wedding (eds) *Current Psychotherapies.* Peacock, Itasca, Illinois.

Rogers C and Teich N (1992) Conversation with Carl Rogers (1985). In: N Teich (ed.) *Rogerian Perspectives: collaborative rhetoric for oral and written communication.* Ablex Publishing Corporation, Norwood, New Jersey.

Rogers C and Russell D (2002) *Carl Rogers: the quiet revolutionary.* Penmarin Books, Roseville, CA.

Rogers C, Kohut H and Erickson M (1986) A personal perspective on some similarities and differences. *Person-Centered Rev.* **1**: 125–40.

Rogers N (1997) Counselling and creativity: an interview with Natalie Rogers. *Br J Guidance Counsell.* **25**: 263–73.

Seeman J (1949) A study of the process of nondirective therapy. *J Consult Psychol.* **13**: 157–67.

Shlien J (1986) In: *Reflection of Feelings* by Carl Rogers. *Person-Centered Rev.* **1**: 375–7.

Slack S (1985) Reflections on a workshop with Carl Rogers. *J Human Psychol.* **25**: 35–42.

Snyder WU (1943) An investigation of the nature of non-directive psychotherapy, PhD thesis, Ohio State University, Ohio.

Standal S (1954) *The need for positive regard: a contribution to client-centered theory.* Unpublished doctoral dissertation. University of Chicago, Chicago.*

Tolman M (1990) *Single Session Therapy: maximizing the effect of the first (and often only) therapeutic encounter.* Jossey-Bass, San Francisco.

Vincent S (1998) *PC or Not PC . . . that is the question.* BAPCA AGM address, 14 March 1998.*

Vincent S (1999a) *On becoming person-centred?* Masters dissertation, University of East Anglia, Norwich.*

Vincent S (1999b) On becoming more whole? Necking in Glasgow. *Person-Centred Pract.* **7**: 95–101.

Vincent S (1999c) *A Training Trip Through Empathy.* CCTPCAS publication.*

Vincent S (2000a) From Whole to Hole to Whole. In: *Person-To-Person*, BAPCA Newsletter.* pp. 8–10.

Vincent S (2000b) *Failing or Fortunate? Striving for revising.* CCTPCAS publication.*

Vincent S (2001) Setting Aside or Erupting? *Person-Centred Pract.* **9**: 97–104.

Vincent S (2002a) Client-centred therapy: conditions, process and theory. *Person-Centred Pract.* **10**: 35–44.

Vincent S (2002b) *Empathy: an inter-view.* CCTPCAS publication.*

Vincent S (2002c) *Evaluating Empathy?* CCTPCAS publication.*

Vincent S (2004) Approach and outcomes. In: *Southampton Primary Care Counselling Services Summary Report 2004.* * Southampton Primary Care Counselling Service.

Index